HYDRODYNAMICS AND HEAT TRANSFER IN FLUIDIZED BEDS

S. S. ZABRODSKY

HYDRODYNAMICS AND HEAT TRANSFER IN FLUIDIZED BEDS

Translation Editor *Frederick A. Zenz*

THE M.I.T. PRESS
Massachusetts Institute of Technology
Cambridge, Massachusetts, and London, England

Originally published as
Gidrodinamika i Teploperenos v Kipyaschem Sloye
by Fizmatgiz, Moscow-Leningrad, 1963

Translated by *Scripta Technica, Inc.*

Library of Congress Catalog Card Number: 66-22353
Printed in the United States of America

Preface to the American Edition

Professor Zabrodsky's book on Hydrodynamics and Heat Transfer in Fluidized Beds presents the most complete and comprehensive review of this field existing to date in the Russian presses. The author has contributed to this field through independent research and through a continuing interest over the better part of his professional career and is well qualified to authoritatively represent the Russian work.

The translation of Zabrodsky's book is a significant effort in surmounting the obstacles of language and of availability of journals. The author has presented the key results, conclusions and data from a tremendous volume of references in a chronological, correlated and organized fashion. Professor Zabrodsky, who kindly supplied a considerable amount of additional material for the American edition, is quite familiar with most of the U. S. work in this field and references to the principal books and publications in this country are abundant. He has drawn from these American sources and also criticizes them. Thus, Zabrodsky presents a blend of Russian and Western work, and now this translation enables English-speaking researchers to afford themselves the benefits of this blend. The American publishers are to be congratulated for their choice and their interest in this work.

Roslyn Harbor, N. Y. *F. A. Zenz*
February, 1966

Foreword to the Translated Edition

In order to make the presentation as complete as possible and to include the most recent work published in the periodical literature, the author found it necessary to make some small changes in and additions to the original text of the monograph.

The author expresses his thanks to the publishers for their courtesy in agreeing to make these changes in the English translation.

S. S. Zabrodsky

Foreword

The use of fluidized bed equipment has opened wide possibilities for improving various industrial technologies. When brought to a fluidized state, granular materials possess the property of fluidity, the hydraulic resistance of such a bed is essentially independent of the fractional size analysis of the granular material and the capabilities of the bed as a heat transfer medium are exceptional.

The application of fluidized beds represents one approach to the problem of increasing the rate of heat transfer between visual walls and process streams. In many cases the properties of fluidized beds permit conversion of less desirable batch processes to continuous operations. Automation is also facilitated by carrying out industrial processes in fluidized beds.

However, the development and wide introduction of fluidized bed equipment has been hindered by insufficient detailed study of the hydrodynamics and heat transfer in such beds and also by the delay in systematizing and generalizing the available material that has been published in a wide variety of periodicals.

The purpose of the present book is to acquaint readers with features of the hydrodynamics and heat transfer in fluidized beds, to provide information required for a better understanding of the complicated processes that occur in fluidized bed equipment, and thus to provide workers in various fields with the background required for the creative application of fluidized beds in their branches of industry.

Formulas and graphs suitable for practical engineering calculations are provided, as well as other necessary reference data.

Certain processes which are special cases of fluidization have long been known [90, 191], but only in the last 15 years or so has wider use been made of fluidization as a result of more careful study of the properties of fluidized beds. The stimulus for such studies was the success achieved in the practical application of fluidization to the cracking of petroleum, which led to a technical revolution in this field.

At present there is great interest in fluidization techniques. Dozens of papers are published each year on fluid bed processes, and the number is accelerating. The total number of publications, not counting the patent literature, long ago passed a thousand. More than 70 organizations in the Soviet Union are carrying out investigations in fluidization.

Naturally, a great deal of theoretical and experimental material has been accumulated in these investigations. This makes the need for monographs based on this material particularly pressing. However, up to now only one monograph has been published in the USSR—"Protsessy v kipyashchem sloye" ("Processes in Fluidized Beds"), by

Syromyatnikov and Volkov [942]. This valuable book, which appeared in 1959, contains very little material published after 1957. A large part of the book is taken up with descriptions of special applications of fluidization: calcination, gasification, combustion of fuels, drying, and ore beneficiation, which made it impossible to deal in detail with the fundamentals of fluidization.

Special applications of fluidization are dealt with in the recently published books by Leyzerovich et al. ("Obzhig tsinkovykh kontsentratov" - "Roasting of Zinc Concentrates," Metallurgizdat, 1959) and by Zeynalov and Bagirov ("Kataliticheskiy kreking s pylevidnym katalizatorom" - "Catalytic Cracking with Powdered Catalysts," Azerneftneshr., 1958) and also in the handbook "Primeneniye protsessov obzhiga v kipyashchem sloye v SSSR" - "The Use of Fluidized Bed Calcining Processes in the USSR" (Izd. Ts IIN Tsvetmet, Moscow, 1960).

Of the foreign monographs on processes in fluidized beds, the first was the book "Fluidization" by Reboux, which appeared in French in 1954; a Russian translation entitled "Kipyashchiy sloy" ("Fluidized Beds") was published by the Central Institute of Information on Nonferrous Metallurgy in 1959.

In 1956, a collection of papers edited by Othmer was published in the USA; the Russian translation "Protsessy v kipyashchem sloye" ("Processes in Fluidized Beds") was put out by Gostoptekhizdat in 1958.

"Fluidization," a monograph by the Polish scientist Ciborowski, appeared in Warsaw in 1957.

A book entitled "Fluidization" by Esayyan and Esayyan was published in Romania in 1959, and a monograph of the same name by Leva appeared in the USA (Russian translation: Gostoptekhizdat, 1961). Finally, in 1960 the monograph by Zenz on processes in fluidized beds appeared in the USA; this in many respects consists of a collection of papers by this author published in 1957. Of the foreign books listed above, the monograph by Leva deserves the most attention, not so much for the coverage of recent material, which was limited to only a few 1958 publications, but because of its attempt to explain fully the physical fundamentals of the processes in fluidized beds, and to provide guidance to specialists in different fields in applying this new technical medium. As mentioned above, this is a very real problem which still exists for the author of the present book, but an attempt has been made to include the more recent material from Soviet and foreign investigations. In the same way as Leva and Zenz, the author thinks that it is useful to give in a book on fluidization a brief account of the other states of granular materials as well, since they are also encountered in systems with fluidized beds, and permit establishing certain comparisons and generalizations.

The main deficiency of Leva's book is the complete absence of an account of the numerous investigations by the scientists of the Soviet Union or of the countries of the peoples' democracies. In a number of cases this leads to incorrect, out-of-date interpretations of phenomena or important problems. Thus, for example, Leva throws little light on spouting beds; they are mentioned as a recent innovation hailing from Canada, whereas we have known about air-fountain equipment for more than 20 years (the VNIGI gas generator, then the VTI dryer, etc.). By omitting the work of Soviet investigators, Leva has presented an outdated discussion of the mechanism of heat transfer in fluidized beds with internal heating surfaces. The data on the stability limits for polydisperse materials are incorrect, no information is given on the convenient interpolation formulas for the hydrodynamics of fluidization proposed by Todes and co-workers, out-of-date information is given on the entrainment of material from beds, on the effective thermal conductivity of fluidized beds, etc., etc. Other omissions and errors in Leva's original monograph are indicated by Gel'perin's footnotes in the Russian translation of the book.

The most important problems in the investigation of fluidization are: hydrodynamics, heat and mass transfer, and the carrying out of various industrial processes in fluidized beds.

Soviet scientists have made great contributions to the study of the hydrodynamics of fluidized beds. The work of Syromyatnikov and Volkov, Todes and co-workers, Fedorov, Smirnov and Yerkova, Romankov, Martyushin, Mukhlenov, and others must be specially mentioned.

The Soviet Union leads other countries in the study of heat and mass transfer (the work of Lykov and his school). Soviet scientists have also done a great deal in the more restricted field of heat and mass transfer in fluidized beds. Well-known work has been carried out by Fedorov, Chechetkin, Syromyatnikov and Baskakov, Todes and co-workers, Rychkov and Shakhova, Sharlovskiy, Gel'perin and Aynshteyn, and others.

Among the investigations dealing with the use of fluidized beds for carrying out various industrial processes, we must mention the work of Syromyatnikov and Volkov, Leyzerovich, Romankov and Mukhlenov and co-workers, Orochko, Khodorov, Pis'men, and others.

Finally, some work on hydrodynamics and heat transfer in fluidized beds was carried out with the participation of the author in the Staff Laboratory for Heat Transfer in Disperse Media in the Institute of Heat and Mass Transfer of the Academy of Sciences of the Belorussian SSR.

The author will be grateful to readers for pointing out inaccuracies and omissions. Critical comments should be sent to Gosenergoizdat, Shlyuzovaya nab., d. 10, Moscow Zh-114.

S. S. Zabrodsky

Translator's Note

The transliteration of Russian names has been carried out following the "British" system for Cyrillic-English Transliteration, outlined in Table A of British Standard 2979 (British Standards Institution, London, 1958), except that accents are not used.

Throughout the text, "degrees" or "°" always refers to degrees Celsius (centigrade).

Most Frequently Used Symbols

a thermal diffusivity, $m^2 \cdot hr^{-1}$;
A effective thermal diffusivity, $m^2 \cdot hr^{-1}$;
c specific heat, $kcal \cdot kg^{-1} \cdot deg^{-1}$;
C constant, proportionality coefficient;
C_D drag coefficient for flow around a sphere;
d_g hydraulic diameter, m;
d particle diameter, m;
d_e equivalent particle diameter, m;
D_T, D_{bed} diameter of tube, of fluidized bed, m;
D diffusion coefficient (diffusivity), $m^2 \cdot hr^{-1}$;
D_0 coefficient of molecular diffusion, $m^2 \cdot hr^{-1}$;
D_c coefficient of convective diffusion, $m^2 \cdot hr^{-1}$;
f resistance coefficient (friction factor);
F particle surface area, heat transfer surface, m^2;
F_{sec} cross-sectional area of bed, m^2;
g acceleration of gravity, $m \cdot sec^{-2}$;
G_M weight of material in bed, kg;
G fluid mass flow rate, $kg \cdot m^{-2} \cdot sec^{-1}$ or $kg \cdot m^{-2} \cdot hr^{-1}$;
$G_{1.s}$ minimum mass flow rate of fluidization, $kg \cdot m^{-2} \cdot sec^{-1}$;
h height above gas-distributing grid, m;
H_0 initial bed height, m;
H bed height, m;
l path length, m;
L distance between two rows of particles, m;
m bed porosity;
$m_{1.s}$ porosity of a monofractional bed at minimum fluidization;
M weight flow rate of material, $kg \cdot m^{-2} \cdot hr^{-1}$ or $kg \cdot m^{-2} \cdot sec^{-1}$; molecular weight;
n number of particles in sample; exponent in equation (1-3);
N fluidization number;
p, P pressure, $kg \cdot m^{-2}$;
P weight of sample of material, kg;
Q quantity of heat transferred per unit time, $kcal \cdot hr^{-1}$;
Q_M flow rate of material, $kg \cdot hr^{-1}$ or $kg \cdot sec^{-1}$;
Q_m flow rate of medium, $kg \cdot hr^{-1}$ or $kg \cdot sec^{-1}$;
q_τ quantity of heat transferred per m^2 of wall surface in time τ, $kcal \cdot m^{-2}$;
R bed radius, m;
S surface area of particles per m^3 of bed, m^{-1};
t temperature of medium, °C;
T temperature of medium, °K;
t_{sh} bed temperature measured with a thermocouple with a hot junction shielded from the particles, °C;
t_{meas} bed temperature measured with a thermocouple with an exposed hot junction, °C;
U pulsation velocity, $m \cdot sec^{-1}$;
V volume of particle, m^3;
V_{bed} volume of bed, m^3;
V_M volume of material in bed, m^3;
w_M overall velocity of motion of the material, $m \cdot sec^{-1}$;
$w_{1.s}$ superficial fluid velocity at minimum fluidization, $m \cdot sec^{-1}$;
w_{ff} velocity of free fall, $m \cdot sec^{-1}$;
w_f superficial fluid velocity, $m \cdot sec^{-1}$;
w_e mean-stream velocity between the particles of the bed, $m \cdot sec^{-1}$;
w_n component of the particle velocity normal to the heat transfer surface, $m \cdot sec^{-1}$;
W_m water equivalent of the medium, $kcal \cdot hr^{-1} \cdot deg^{-1}$;
W_M water equivalent of the material, $kcal \cdot hr^{-1} \cdot deg^{-1}$;
W_g water equivalent of gas, $kcal \cdot hr^{-1} \cdot deg^{-1}$;
α particle heat transfer coefficient, $kcal \cdot m^{-2} \cdot hr^{-1} \cdot deg^{-1}$;
α_{eff} effective particle heat transfer coefficient, $kcal \cdot m^{-2} \cdot hr^{-1} \cdot deg^{-1}$;
α_{wall} wall (surface) heat transfer coefficient, $kcal \cdot m^{-2} \cdot hr^{-1} \cdot deg^{-1}$;
α_{max} maximum wall heat transfer coefficient, $kcal \cdot m^{-2} \cdot hr^{-1} \cdot deg^{-1}$;
α_{conv} convective component of the heat transfer coefficient, $kcal \cdot m^{-2} \cdot hr^{-1} \cdot deg^{-1}$;
α_{cond} conduction component of heat transfer coefficient, $kcal \cdot m^{-2} \cdot hr^{-1} \cdot deg^{-1}$;
α_r radiant component of the heat transfer coefficient, $kcal \cdot m^{-2} \cdot hr^{-1} \cdot deg^{-1}$;
γ_M particle density, specific weight of nonporous material, $kg \cdot m^{-3}$;
γ_m density of medium, $kg \cdot m^{-3}$;
γ_g density of gas, $kg \cdot m^{-3}$;
γ_{bulk}, γ_{bed} bulk density (weight "as poured") of material, $kg \cdot m^{-3}$;

Δ	finite difference;
δ_w	mean distance from wall to surface of particles in the layer nearest to it, m;
δ	thickness of layer of medium, m;
ε	degree of blackness;
λ	resistance coefficient in formula (1-3);
λ_g	molecular thermal conductivity of gas, $kcal \cdot m^{-1} \cdot hr^{-1} \cdot deg^{-1}$;
λ_r	radiant component of thermal conductivity, $kcal \cdot m^{-1} \cdot hr^{-1} \cdot deg^{-1}$;
λ_m	molecular thermal conductivity of medium, $kcal \cdot m^{-1} \cdot hr^{-1} \cdot deg^{-1}$;
λ_{eff}	effective thermal conductivity, $kcal \cdot m^{-1} \cdot hr^{-1} \cdot deg^{-1}$;
λ_f	filtrational component of thermal conductivity, $kcal \cdot m^{-1} \cdot hr^{-1} \cdot deg^{-1}$;
μ	dynamic viscosity, $kg \cdot sec \cdot m^{-2}$;
ν	kinematic viscosity, $m^2 \cdot hr^{-1}$, $m^2 \cdot sec^{-1}$;
ϑ	particle temperature, °C;
Θ	the same, °K;
$\vartheta_{f.b.}$	temperature of particles at core of fluidized bed, °C;
$\Theta_{f.b.}$	the same, °K;
φ_1	particle shape factor according to Zhavoronkov;
$\varphi=\varphi_1^2$	shape factor according to Leva.

Subscripts

g	gas;
M	material;
fl	fluid;
f.b.	fluidized bed;
l.s	refers to the limit of stability (the minimum velocity for fluidization);
m	medium;
bed	bed;
wall	wall;
t	tube;
e	equivalent;
e.h.	equivalent hydraulic.

Contents

Introduction

If a granular material is fed through a side tube 3 (Fig. I-1) onto a grid 2 in a vertical tube 1 it will lie in the form of a so-called fixed bed. The surface of the bed will be inclined at the poured angle of repose of the material.

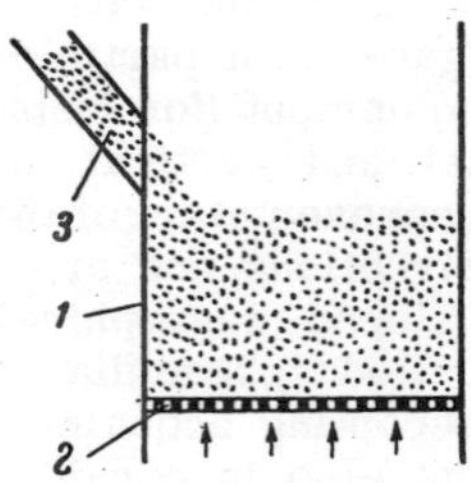

FIG. I-1. Test equipment with a bed of granular material.
1) body of equipment; 2) screen; 3) side tube for supply of material.

Now let us pass an ascending gas stream through the bed, gradually increasing its velocity. At first the charge of material remains stationary and the bed resistance increases with increase in gas velocity. When the force resisting the flow of gas is equal to the bed weight there is no further increase in the hydraulic resistance of the bed, and further increase in gas velocity leads to expansion of the bed, corresponding to the so-called limit of stability of a fixed bed and marking its transition to the fluidized state. The bed becomes suspended and expands, and the contact and close adjacency of the particles to one another ceases as they become mobile. The surface of the bed levels itself just like a liquid, and if the walls confining the bed contain openings, jets of material will flow through them. For this reason the bed is said to be fluidized, since it possesses fluidity. As the gas velocity through the fluidized bed is increased still further, gas bubbles are formed and intensive mixing of the bed begins, with a churning action reminiscent of a boiling liquid, giving rise to the frequently used term "boiling bed," which is justified for this stage of fluidization. On increasing the gas velocity still more, carry-over of particles from the tube finally occurs (pneumatic transport of material): the fluidized bed undergoes transition to a bed entrained by the fluidizing stream, which is usually termed a suspensoid, although suspension (floating, buoying) of particles, in the strict sense, has already occurred in the transition from the fixed to the fluidized bed. Some authors [163, 916] have taken into account this suspension of particles at the limit of stability, and imply by the term "suspended bed" a fluidized bed. However, they are inconsistent, for they do not include in the concept of a "suspended bed" either the regime of pneumatic transport or the fall of a multiplicity of particles in an ascending or descending gas stream (a "falling bed"). Against the term "fluidized bed" Mukhlenov [916] has raised the objection that this name is not suitable for gas-liquid and liquid-liquid systems, "since liquids cannot be fluidized." However, this objection is groundless. For a gas-liquid system ("foam bed") it is obviously quite appropriate to speak of a fluidized bed of gas bubbles that is moving and flowing like a liquid. It is also acceptable to apply this term to liquid-liquid systems. Thus, for any two-phase system (and primarily for the "classic" solid-gas system), a bed as a whole acquires in the fluidized state the properties of fluidity and mobility whether the bed consists of solid particles, liquid drops, or gas bubbles; the material of which the elements of the bed are comprised (for instance, solids) does not itself acquire fluidity.

It seems that we must either use the concept "suspended bed" in the wider sense indicated above or, better yet, use the term "suspended bed" for a bed in the pneumatic transport regime, in agreement with long-established tradition. In any case, "fluidized bed" is the best of the terms that have been used to describe the state intermediate between the fixed bed and the regime of pneumatic transport.

It is also possible for various other states of two-phase systems (gas—solid particles) to exist in the auxiliary parts of fluidized bed equipment (for instance, dense beds moving under gravity and horizontal and vertical pneumatic transport). In order to achieve a clear representation of the behavior of two-phase systems moving in various directions and at different flow intensities, Zenz [490, 625-626] developed a graphical method for describing the systems using a so-called "phase diagram." As Zenz has noted, this is useful not only in becoming acquainted initially with operations in two-phase systems, but also in considering apparent contradictions in experimental and operating data. As applied to equipment being developed, this graphical method can be used for comparing the pressure balances of the streams. The details can be found in Zenz's papers [490, 625-626]. A brief description of the phase diagram is presented here as taken mainly from [490].

The gas velocity is plotted along the axis of the phase diagram (ascending streams to the right and descending streams to the left of the origin). The pressure drop per unit length of tube, $\Delta P/H$, is plotted on the ordinate. If the pressure in the upper parts of the tube is lower than that below, $\Delta P/H$ is regarded as positive and is plotted upward. In the opposite case, the pressure gradient is plotted downward. Each line on the diagram is drawn for a constant weight flow rate of the total stream of material in the tube M (kg/m^2-sec); i.e., this flow rate is used as the parameter. Since the pressure drop over a section of the tube depends not only on the velocities of the streams of material and gas and the method in which they are introduced, but also partly on the particle dimensions and the densities of the gas and material, at least some of the lines will shift on the phase diagram when any of these quantities varies. The mass flow rate of solid material is indicated by the letter M, with numerical subscripts, placed near the line. In Fig. I-2 the larger numbers correspond to the larger mass flow rates. The direction of the motion (upward or downward) is indicated by the vertical arrows on the lines.

For wide ranges of $\Delta P/H$ and gas velocity w_f it is most convenient to draw the lines on a logarithmic scale, but in such a case it is impossible to pass through the coordinate origin without breaking the continuity of the lines. The graphs are therefore drawn using two scales on each of the axes: logarithmic far from the axis and linear in the immediate neighborhood of the axis, i.e., at small gas velocities

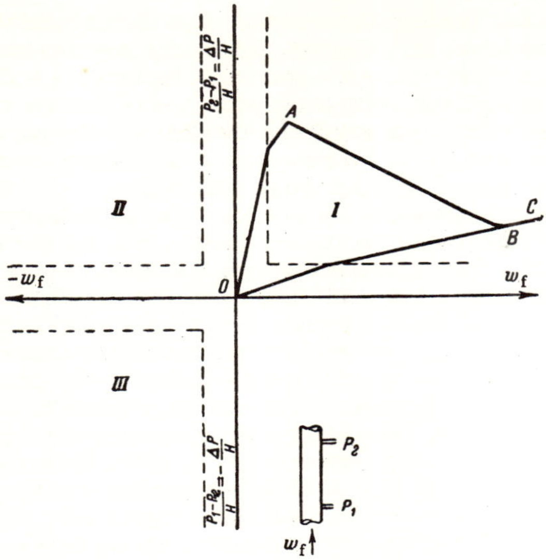

FIG. I-2, Zenz phase diagram.

and pressure drops. The cross-shaped region of the graph for which the linear scales are used is separated from the rest of the graph by the dotted lines in Fig. I-2. Some distortion results from the use of two scales on the phase diagram: at the boundaries between the regions with the linear and logarithmic scales the smooth lines show inflection points which would not occur if the graph were drawn on a single scale.

The whole phase diagram fills three quarters of the coordinate plane (quadrants *I*, *II* and *III*). It will be shown later (Chapters 1, 2 and 3) that quadrants *II* and *III* are used for plotting the downward cocurrent flow of a material and a gas. The greatest interest lies in quadrant *I*, where the gas stream is directed upward and the pressure in the upper part of the tube is lower than in the lower parts. Quadrant *I* is therefore used for plotting a number of processes: cocurrent upward motion of a material and a gas (vertical pneumatic transport), "falling beds" (the material moves downward, while the gas stream passing through the "bed" moves upward), the flow of an ascending gas stream through a fixed bed of material, the upward flow of a pure gas along a tube, and finally all stages of fluidized beds.

By way of an example, let us see how the process that we have considered for a gas permeating through a fixed bed and the idealized process of fluidization of the material up to the point of entrainment appear on the phase diagram (Fig. I-2). This whole process is shown in Fig. I-2 by the line *OAB*. Here the line *OA* (with an inflection point at the transition from the linear to the logarithmic scale) shows the program of interstitial flow, during which the pressure drop increases continuously as the superficial fluid velocity increases. The point *A* is the limit of stability. The section *AB* corresponds to the region of fluidization for the particular bed, where the pressure drop over the whole bed becomes independent of the superficial fluid velocity; the pressure drop per unit height of the bed decreases as this velocity increases. The line OC (which also has an inflection point at the change of scale) gives the relationship between the hydraulic resistance of the same tube and the fluid velocity in the absence of solid particles in the tube. Thus the point B, corresponding to the intersection of the line OC and the line for the fluidized bed, represents the limiting state of the latter, or its greatest degree of expansion; i.e., it corresponds to such a low volumetric concentration of solid particles that the hydraulic resistance per unit height of the bed practically does not depend on the presence of these few particles.

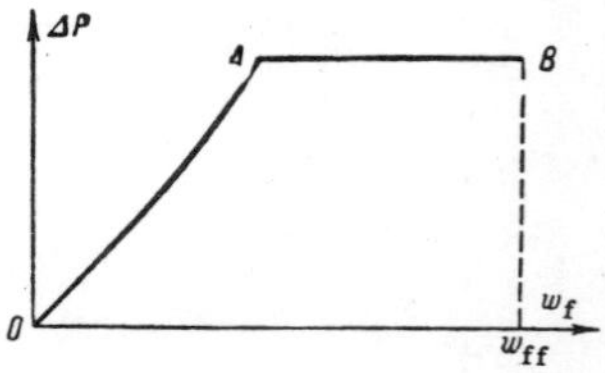

FIG. I-3. Diagram showing relationship between bed pressure drop ΔP and fluid velocity w_f.

Figure I-3 shows a second type of graph which is more frequently used in the study of the hydrodynamics of dense and fluidized beds; this presents the relationship between the rate of fluid throughput and the pressure drop over the whole bed of material; the quantity of material above the screen remains constant. The form of the lines OA for flow through a stationary bed is similar on the two types of graphs (Figs. I-2 and I-3), since in this stage the bed is not expanded (H = const), and ΔP differs from $\Delta P/H$ only by a constant factor. However, the line for the fluidized bed AB is horizontal in Fig. I-3, since in the ideal case the pressure drop over the whole fluidized bed remains constant after passing the limit of stability at A and is equal to the "hydrostatic" pressure of the column of fluidized material. The point B (Fig. I-3) corresponds to the limit for the existence of the fluidized bed, or the point at which it is entrained. The abscissa of this point, w_{ff}, is the free fall velocity of the particles (in a practically unconfined space). In the ideal case of uniform expansion of the fluidized bed, the bed will be destroyed (i.e., entrained) only after the gas velocity reaches the value w_{ff}.

In the Foreword some of the possible advantages of carrying out processes in fluidized beds were mentioned. Such possibilities are not without important disadvantages also (longitudinal mixing of the material, erosion of the tube). We can only conveniently go into a detailed consideration of the advantages of fluidized beds, their drawbacks, and methods of overcoming the drawbacks after a careful consideration of the hydrodynamics and heat transfer in real fluidized beds.

It should be noted that the application of fluidization is economically advantageous for many processes.

It has been stated [320] that despite the present far from complete progress in applying fluidization it is probably still the chemical industry's main achievement in the last decade.

Chapter 1

Fixed Beds and the Limit of Stability

1.1 Flow Through a Stationary Fixed Bed

The flow of liquids and gases through stationary fixed beds of granular materials has long been studied, and a huge number of papers have been published on this subject. Many authors have treated fluid permeation as an internal problem (flow in channels between the particles [36, 37, 54, 92]), while others have regarded it as an external problem (external flow around the particles [14, 282, 393]). In considering permeation as flow in channels between the particles we must remember that all these channels of complicated shapes intercommunicate, combining to form a single overall channel of small hydraulic diameter with special transverse agitation of the flowing medium which is not identical with simple turbulent mixing. Here the flow does not become hydrodynamically stable even at the end of the "channel." There are the downstream regions of poorly streamlined bodies everywhere, and the points of contact of the particles are sources of artificial turbulance in the stream. The shapes and cross sections of the channels depend on the type of packing of the particles. It is interesting that even in the simplest case of spherical particles of uniform diameters, the porosity of the bed (the volumetric fraction of the empty spaces) is not characterized by a single type of packing or channel shape, or, consequently, by a single value of the hydraulic resistance of the bed. Borishanskiy [729] has shown experimentally that the change from an octahedral to a tetraoctahedral packing of spheres at the same porosity leads to an increase in the hydraulic resistance under otherwise equal conditions.

The treatment of fluid permeation as an external flow around the particles is more easily justified, since it is not unusual to have unstabilized flows in external situations. However, the problem is greatly complicated, when we have to deal with other than individual, widely separated particles, by the conditions of stream expansion downstream of the particles [50], by the presence of large numbers of centers promoting turbulence in the stream at low Re, and finally by the different (larger) resistance coefficient for the particles, which in any case depends not only on the bed porosity but also on the type of packing.

Recently, Bernshteyn, Pomerantsev and Shagalova [727] introduced a so-called generalized method for calculating the aerodynamic resistance of packed cross sections. These authors started with a bundle of "jetlike" streams and a bed of granular material, and assumed that the main phenomenon determining the resistance to the motion of the stream in the bed (the packed cross section) is the periodic contraction of the jets in passing through the parts of the cross section where the particles are closest together and their subsequent expansion.

The viewpoints taken in the internal, external, and "jet" flow approaches are useful for a qualitative analysis of permeating phenomena. However, in order to derive analytical design formulas on the basis of any of these concepts with our present knowledge of the subject, it is necessary to make such simplifying assumptions that the results are not of great value. For practical calculations, therefore, it is necessary to use empirical or semiempirical expressions (with experimentally determined coefficients, exponents, etc.); i.e., it is necessary to use, in essence, interpolation formulas. The discrepancies in the values of the hydraulic resistance of a fixed bed calculated by the various formulas are not large and can serve as a subject which is quite simple and which is at the same time described by a large mass of experimental material (see below).

For calculating the hydraulic resistance of a stationary fixed bed, Zhavoronkov [36, 37, 39] has proposed an expression of the type

$$\Delta P = 2f_e \frac{H}{d_{e.h.}} \cdot \frac{\gamma_m w^2}{g}, \tag{1-1}$$

where H = bed height, m;
γ_m = density of the flowing medium, kg/m³;
g = acceleration of gravity, m/sec²;
$w = w_f/m_0$, the mean-stream velocity in the interstices between the particles, m/sec;
w_f = superficial velocity of the permeating fluid (the stream velocity before the entrance to the bed), m/sec;
m_0 = porosity of the fixed bed;
$d_{e.h.} = \frac{2}{3} \cdot \frac{m_0}{1-m_0} \varphi_1 d_e$, the effective hydraulic diameter of the channels formed by the particles;
d_e = diameter of the sphere equivalent to the particle in volume;
$\varphi_1 = \sqrt{0.205F/V^{2/3}}$, the particle shape factor;
F_1 and V = surface area (m²) and volume (m³) of the particle, respectively;
f_e = resistance coefficient.

For $Re_e < 4$, $f_e = 34/Re_e$ [39]; and for $4 < Re_e < 80$,

$$f_e = 27.8/Re_e + 0.8, \tag{1-2}$$

where $Re_e = wd_{e.h}/\nu$;
ν = fluid kinematic viscosity.

It is, however, more accurate to calculate ΔP from the data of Leva [988], and others, who have generalized a wide range of experimental material and taken into account the relationship between the resistance coefficient of a fixed bed in the transitional and turbulent zones of the particle surfaces (Fig. 1-1).

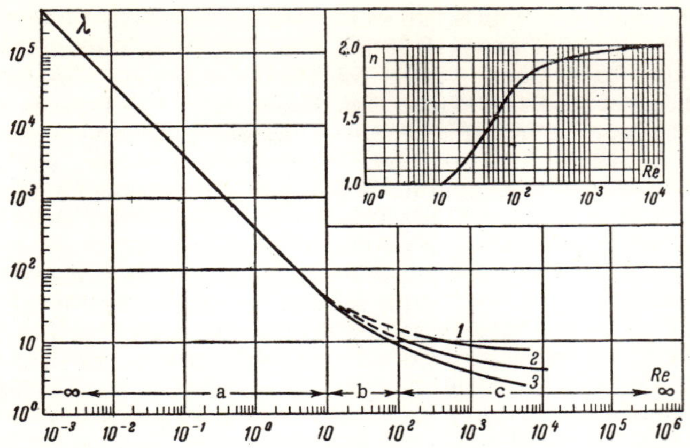

FIG. 1-1. Relationship between the resistance coefficient λ and the Reynolds number Re.
a) laminar regime; b) transition regime; c) turbulent regime; 1) aloxite, granules of molten MgO, etc.; 2) alundum, alumina, etc.; 3) cement, porcelain, glass, etc.; n—exponent in Eq. (1-3) in transition region calculations [788].

For a bed of particles of arbitrary shape in any regime of flow:

$$\Delta P = \lambda \frac{H}{d_e} \cdot \frac{w_f^2 \gamma m}{2g} \left[\frac{(1 - m_0)^{3-n}}{m_0^3} \varphi^{3-n} \right], \tag{1-3}$$

where $\varphi = 0.205 F/V^{2/3}$, the shape factor of the particles (dimensionless);
$\lambda = f_1(\mathrm{Re})$, the resistance coefficient (dimensionless);
$n = f_2(\mathrm{Re})$, an exponent.

The other symbols are the same as in Eq. (1-1); λ and n can be read from the graph in Fig. 1-1.

The derivation of Eq. (1-3) can be found in the work of Ciborowski [788], among others.

Although this is of no consequence in calculations using the graphs in Fig. 1-1, it should be noted that it is only by convention that we apply the term "turbulent" to the regime of flow at Re $>$ 100; this is fundamentally incorrect. However, many other authors [662, 788, 932] use this terminology. Usually the limited applicability of the concept of "turbulence" to flow through fixed beds is not emphasized.

In the experimental determination of the relationship between the hydraulic resistance of a stationary fixed bed and the superficial fluid velocity, we observe initially a direct proportionality (up to about Re = 10), corresponding to so-called laminar flow. Then the transitional regime begins, and finally at Reynolds numbers of the order of 7000 the resistance law becomes quadratic ($\Delta P \sim w^2{}_f$). It must be assumed that then the viscous forces are negligible compared with the inertia effects; this is the basis for calling this the turbulent regime. However, we can only speak of a turbulent stream of the fluid and its corresponding transfer coefficients for the spaces between neighboring granules, and not for the whole complex "channel" consisting of the fixed bed taken as a whole.

On the other hand, in the so-called laminar regime, flow is laminar only in the general sense that here the effects of the viscous forces predominate over those of the inertia forces, thus determining the linear nature of the resistance law. However, it must not be forgotten that there is substantial transverse or radial mixing of the flowing medium in the laminar regime. As Baum has shown, such "radial mixing" is almost the same in the laminar, transitional and even turbulent regimes of flow [10]. This is to be expected, since for two-phase systems composed of solid particles and a flowing medium, at such high concentrations of solid particles as occurs in fixed beds, the transverse mixing of the flowing medium is determined largely by the dimensions of the particles rather than by the regime of flow between them, since within each space between the particles transfer is sufficiently intensive as a result of separation of flow in the passage of the medium around the particles.

It is impossible to justify the use of the term turbulent flow for the whole range of values of $Re > 10$. At the beginning of this zone there cannot be total disruption of the flow stability or the appearance of sustained turbulence; also here there are at least two exponents in the resistance law. Obviously it is more valid to assume that there is a greatly extended transitional regime with local but not general turbulence in the flow between the particles as a result of the progressive development of flow separation.

1.2 Some Characteristics of Particles and Beds

Let us discuss the characteristics of particles and beds that appear in the design formulas. There are certain difficulties in determining these characteristics. These include the calculated dimensions and shape factors of the particles, their degree of surface roughness, and the porosity of the bed (the volumetric fraction of unoccupied space).

1. Calculated particle dimensions

The calculated equivalent particle diameter $d_e = \sqrt[3]{6V/\pi}$ is easily determined for a bed consisting of material of a narrow size range after finding the volume of a single particle V by counting the number n of particles in an arbitrary sample of weight G: $V = G/n\gamma_M$, where γ_M is the apparent particle density (the bulk weight for porous materials).

In most cases γ_M can be found from tables, but in the absence of suitable data, it can be determined by standard methods using a pycnometer. However, with the pycnometer it is possible to find the volume of the particles (n particles) directly, so that it is no longer necessary to determine the bulk weight.

An additional difficulty arises in determining V, γ_M or the porosity m of a bed of porous material. In this case a part of the "measuring" fluid enters into the pores, and the pycnometer method underestimates the volume of the particles and overestimates the bed porosity. Preliminary sealing of the pores with paraffin is only useful for coarse particles, whose porosity is larger than that of fine particles, whose dimensions are of the same order as the pore dimensions. An indirect method is therefore usually used to determine the values of V, γ_M or the bed porosity for fine porous particles. The idea behind this method is as follows: under simple laboratory conditions several experiments are carried out to determine the hydraulic resistance ΔP of a particular bed, giving it various degrees of compaction (porosity) from one experiment to another by artificial means (for instance, by shaking). As can be seen from Eq. (1-3), d_e, γ_m and φ are constants in these experiments. Only the bed height H and the porosity m vary, as a result of which the resistance coefficient λ, the fluid velocity w_f, and the exponent n vary. The last three quantities are functions only of the Reynolds criterion, while m is a function of the bulk weights of the bed and the particles ($m = 1 - \gamma_{bed}/\gamma_M$). As a result, in the present case

$$\Delta P = f(H, \mathrm{Re}, \gamma_{bed}, \gamma_M). \tag{1-4}$$

From this we see that it is possible in principle to obtain γ_M, regardless of the porosity of the material, from experiments in which it is easy to determine the numerical values of ΔP, H, γ_{bed}, and Re. The specific details of the method for finding γ_M in this way, as developed by Ergun [274, 275], can be found in Leva's monograph [988].

The experimental determination of V and γ_M in laboratory equipment at room temperature and atmospheric pressure is of practical interest when the process in the projected equipment occurs under complicated conditions (for instance, at elevated temperatures) and it is difficult or impossible to duplicate these conditions in the laboratory for measuring the operating ΔP.

For rounded particles approaching a spherical shape, it is possible to determine the calculated particle diameter of a narrow fraction (retained between screens of adjacent numbers) not from the formula $d_i = (6V_i/\pi)^{1/3}$, but as the arithmetic mean $d_i = (d + d)/2$ or the geometric mean $d_i = (d_1 d_2)^{1/2}$ of the dimensions of the openings of these screens. The hole sizes of various screens are given in Table 1-1.

For polydisperse beds with a wide size range of particles, it is necessary to calculate a representative mean diameter which would make it possible to use the equations obtained for narrow fractions. For wide fractions the method of averaging is of importance. Reboux [511] showed that the best agreement with the concept of the hydraulic channel diameter which was used in deriving Eqs. (1-1) and (1-3) was obtained by calculating the mean diameter d from the expression

$$\frac{1}{d} = \Sigma \frac{X_i}{d_i} = \frac{X_1}{d_1} + \ldots + \frac{X_n}{d_n}, \tag{1-5}$$

where X_i is the weight fraction of a narrow fraction of diameter d_i.

This relationship has been used by several authors [210, 332, 422]. Leva [988] mentions that another widely used method of calculating the mean diameter

$$d = \Sigma X_i d_i \tag{1-5'}$$

Table 1-1

Systems of Screens

Cost 2584-50[1] wire screens		Taylor screens	US Bureau of Standards[3]		British Standard[4]		IMM[5]		German Standard[6]	
No.	Side of opening in screen, mm	mesh[2]	No.	Side of opening in screen, mm	No.	Side of opening in screen, mm	No.	Side of opening in screen, mm	No.	Side of opening in screen, mm
									1	6.00
		3½	3¼	5.66						
5	5.00									
		4	4	4.76						
4	4.00	5	5	4.00						
		6	6	3.36						
					5	3.353				
3.3	3.30									
		7	7	2.83					2	3.00
					6	2.812				
2.8	2.80						5	2.540		
					7	2.411				
									2½	2.40
		8	8	2.38						
2.3	2.30									
					8	2.057				
2	2.00	9	10	2.00					3	2.00
1.7	1.70									
		10	12	1.68	10	1.676				
							8	1.600		
									4	1.50
		12	14	1.41	12	1.405				
1.4	1.40									
							10	1.270		
1.2	1.20	14	16	1.19	14	1.204			5	1.20
							12	1.059		
									6	1.02
1	1.00	16	18	1.00	16	1.003				
085	0.85				18	0.853				
		20	20	0.84						
							16	0.795		
									8	0.75
		24	25	0.71						
07	0.70				22	0.699				
							20	0.635		
0.6	0.60				25	0.599			10	0.60
		28	30	0.59						
									11	0.54
05	0.50	32	35	0.50	30	0.500				
									12	0.49
									14	0.43
042	0.42	35	40	0.42	36	0.422	30	0.424		
									16	0.385
0355	0.355									
		42	45	0.35	44	0.353				
							40	0.317		
03	0.30	48	50	0.297	52	0.295			20	0.300
025	0.25	60	60	0.25	60	0.251	50	0.254	24	0.25
021	0.21	65	70	0.21	72	0.211	60	0.211		
									30	0.20
018	0.18	80	80	0.177	85	0.178	70	0.180		
							80	0.160		
015	0.15	100	100	0.149	100	0.152			40	0.150
							90	0.139		
0125	0.125	115	120	0.125	120	0.124	100	0.127		

Table 1-1 (continued)

Cost 2584-50[1] wire screens		Taylor screens	US Bureau of Standards[3]		British Standard[4]		IMM[5]		German Standard[6]	
No.	Side of opening in screen, mm	mesh[2]	No.	Side of opening in screen, mm	No.	Side of opening in screen, mm	No.	Side of opening in screen, mm	No.	Side of opening in screen, mm
0105	0.105	150	140	0.105	150	0.104	120	0.104	50	0.120
		170	170	0.088	170	0.089			60	0.102
									70	0.088
0085	0.085						150	0.084		
0075	0.075	200	200	0.074	200	0.076			80	0.075
					240	0.066				
0063	0.063	250	230	0.062			200	0.063		
									100	0.060
0053	0.053	270	270	0.053	300	0.053				
		325	325	0.044						
0042	0.042									
		400	400	0.037						

[1]Wire grid screens for the abrasives industries (Cost = All-Union State Standard).
[2]Mesh = number of openings per linear inch.
[3]US National Bureau of Standards, LS-584 and ASME-11.
[4]British standard BS 410:1943.
[5]Institute of Mining and Metallurgy, London. This series of screens has largely been replaced by the British standard series.
[6]German standard DIN 1171.

In Soviet powder-analysis practice it is usual to characterize screens by the grid opening in microns (μ) without indicating the number of the screen (for instance, an 88 screen, a 200 screen, etc.).

gives rather larger values of the representation mean diameter, i.e., smaller values of the calculated bed resistance. Thus it is better to use Eq. (1-5) from the point of view of allowing a safety factor in hydrodynamic calculations (but not in heat transfer calculations).

2. Shape factors

In order to calculate the shape factors (coefficients) φ and φ_1 appearing in Eqs. (1-1) and (1-3) one must know the surface area of the particles. The surface area can be calculated easily only for particles of regular geometric shape (cubes, cylinders, etc.). The shape factor for a cube is $\varphi = 1.23$. It is given in Fig. 1-2 for cylinders and rings. For particles of irregular shape it is practically impossible to make a direct calculation of the surface area, and the shape factors for these are not usually obtained from the formula given earlier, but from laboratory experiments in which the hydraulic resistance of a bed of the particles is measured in the laminar region of flow ($Re < 10$), where roughness has no effect. The shape factor is then the only unknown in Eqs. (1-1) or (1-3). Its calculated value is then used for calculating the bed resistance under other more complicated operating conditions.

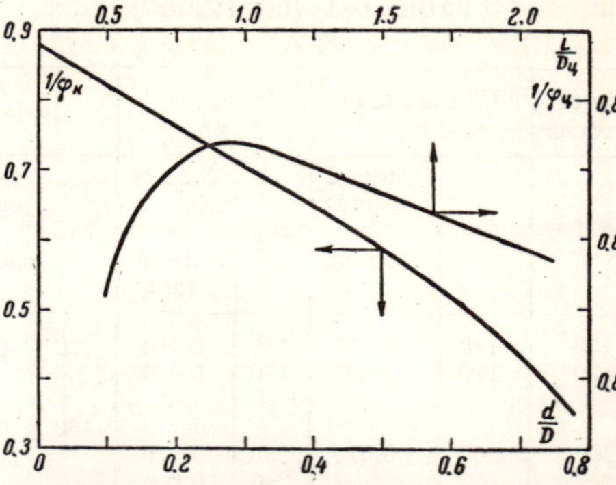

FIG. 1-2. Shape factors for cylinders (φ_{cyl}) and (φ_{ring}) as functions of their dimensions [988]. L—height of cylinder; D_{cyl}—diameter of cylinder; D and d—external and internal diameters of ring.

3. Particle roughness

Roughness of the particles has two effects. First, it has an effect on the bed porosity in all regimes of flow. On packing (charging) a bed, rough particles have a tendency to come to rest more quickly than smooth particles of the same shape, and a less dense packing results, since friction between the particles impedes settling of the bed (Fig. 1-3). Thus particle roughness leads indirectly to a decrease in the

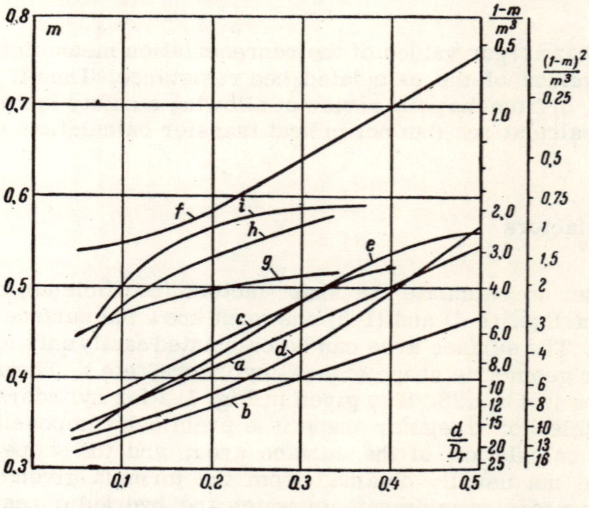

FIG. 1-3. Porosity (m) of dense beds of various granular materials as a function of the ratio of the particle diameter to tube diameter (D_T) [988].
Spherical particles: *a*) smooth particles of uniform dimensions; *b*) smooth, mixture; *c*) clay particles.
Cylindrical particles: *d*) smooth particles of uniform dimensions; *e*) alundum, uniform dimensions; *f*) ceramic Raschig rings; *g*) fused magnetite (catalyst); *h*) fused alundum; *i*) aloxite.

hydraulic resistance of the bed. Second, however, the roughness of the particles, just like roughness of channel walls, leads to a direct increase in the resistance coefficient in the transitional and turbulent regimes of flow (starting at Re $\approx$ 10). Thus, for instance, according to the data of Leva, the resistance coefficients for beds of clay and cement particles reached values 1.5 times larger than for beds of smooth glass spheres, while for beds of fused MgO, the values were 2.3 times larger.

For porous materials the heights of the protuberances on the particle surfaces are determined by the pore dimensions. For approximate calculations it is possible in practice to estimate the effect of the particle roughness of a given material on the resistance coefficient by comparing the material with "reference" samples corresponding to the three branches of the curve in Fig. 1-1. For more exact calculations the position of the new branch of the curve $\Delta P = f(\mathrm{Re})$ corresponding to the particular material can be determined from laboratory experiments in which ΔP is found in the zone of Re close to the working conditions after determining the shape factor from experiments in the laminar zone.

4. Porosity of the bed

The porosity of the bed depends on many factors: the method of loading the material, the roughness and shape of the particles, the ratio of the particle diameter to the bed diameter, and the size analysis of the material.

If the material is loaded into the bed from a considerable height the bed becomes denser. The bed is also made more dense by vibrating it during loading. The least dense packing of the particles is obtained by "wet charging," i.e., when the column (apparatus) is initially filled with water to some depth, and the material gradually poured into it. After some time such a freely deposited bed undergoes a contraction [988].

The porosity of mixed (polydisperse) beds (e.g., of spheres of different sizes) is smaller than that of a bed of uniform elements; the small spheres fill the spaces between the larger ones [788]. A high accuracy in the calculated values of the hydraulic resistance cannot be expected for beds with wide size fractions of particles.

It is known that the bed porosity depends to a considerable extent on the particle shape. Spheres give the least free volume on random loading. The effect of the roughness of the particle surface on the bed porosity has been mentioned in the preceding section.

The effect on the ratio of the particle diameter to the bed (column) diameter d/D_T is connected with the directing action of the walls, as a result of which the packing of the particles in the rows near the walls differs from that in the middle of the bed. The packing near the walls is usually less dense. Naturally the effect of the walls on the mean-bed porosity is particularly important at large values of d/D_T, since there are then only a small number of rows of particles far from the walls which are unaffected by the walls (Table 1-2). Nonuniform porosities over the bed also occur far from the walls due to random packing of the particles during charging of the material. The types of packing are distributed in the bed according to the law of probability (on a Maxwellian curve). Packing nonuniformity in a dense bed is observed as a nonuniformity in the distribution of the permeating fluid velocity. The nonuniformity of the bed porosity and of the velocity distribution in it have been studied by a number of Soviet scientists [6, 13, 101]. This work has been partly dealt with in Chudnovskiy's monograph [175].

Table 1-2

Bed Porosities

$\frac{d}{D_T}$	Ceramic spheres	Smooth spheres	Smooth cylinders	Raschig rings
0.1	0.40	0.33	0.34	0.55
0.2	0.44	0.37	0.38	0.58
0.3	0.49	0.40	0.44	0.64
0.4	0.53	0.43	0.50	0.70
0.5	—	0.46	0.56	0.75

Aerov and Umnik [6] have measured the velocity distribution over the whole bulk of a fixed bed of 6-mm rough spheres by determining the advance of a front of colored surfaces on the bed particles as they absorbed a special ingredient from the flowing gas. In these experiments the velocity at the walls appeared to be 30–70% larger than the average for the bed at $Re > 40$. When the central part of the bed was considered separately, the velocity profile there also appeared to be nonuniform, and the maximum deviation of the local values from the mean reached ±18%. Figure 1-4 shows a plan of the velocity distribution in the reactor model tested. From Fig. 1-4 it can be concluded that the velocity profiles in fixed beds measured by various investigators (see, for instance, [963]) scarcely differ in accuracy and reproducibility.

For beds consisting of spheres of different dimensions and beds of particles of irregular shapes we can expect still less uniform distributions of the porosity and fluid velocity. The entry conditions of the stream into the bed (type of grid, etc.) affect the velocity distribution in the lower part of the bed. The velocity distribution is further complicated when the bed does not operate isothermally [175].

From what has been said about the complicated and still almost quantitatively unknown relationships between the porosity of a fixed bed and a whole series of other factors, it follows that any accurate values for the porosity must be obtained from experiments rather than analytically. To do this under laboratory conditions, we calculate the bulk density of the material

$$\gamma_{bulk} = G_{bed}/V_{bed}.$$

from the known weight of material in the bed G and the measured volume of the bed V_{bed}.

Knowing the bulk density of the particles (see above), it is then simple to find the mean-bed porosity from the expression

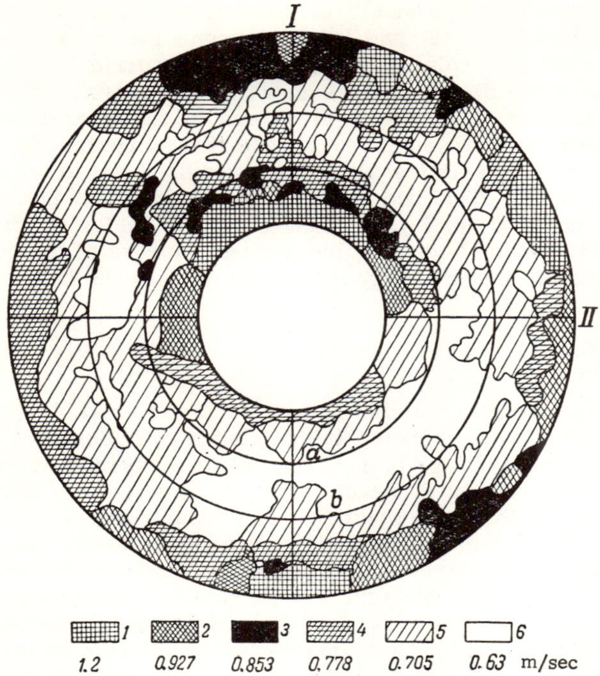

FIG. 1-4. Gas velocity distribution over the reactor cross section [6].

$$m = 1 - \gamma_{\text{bulk}} / \gamma_{\text{M}}. \tag{1-6}$$

It should be noted that in Ciborowski's monograph [677] the term "bulk weight" is taken to mean the weight of material per unit volume of the bed together with the medium included in the pores between the particles, rather than the weight of material per unit volume alone, so that the bed porosity is calculated as

$$\frac{\gamma_{\text{M}} - \gamma_{\text{bulk}}}{\gamma_{\text{M}} - \gamma_{\text{m}}},$$

where γ_{m} is the specific weight of the medium.

This cannot be regarded as a desirable approach, since it does not simplify the calculations or make them more precise, while it can be the source of misunderstandings and errors.

For approximate calculations one can use the numerical data of Ciborowski given in Table 1-2.

When large pressure drop or considerable change in the gas density is anticipated, it is necessary to take into account the compressibility of the gas in calculating the hydraulic resistance of a fixed bed (see Leva [988]).

Within the overall resistance of a fixed bed, the resistances at the inlet and outlet regions are of a different nature than the resistance through the center part of the bed, since they are not related to losses along the length of a "channel" but are typical local resistances [41, 729]. Borishanskiy [729] has isolated only an entry resistance, by means of which he explains the deviation from strict proportionality between the hydraulic resistance and the bed height, which is very noticeable for small numbers of layers of particles (over the bed height). In his measurements he did not find increased resistance of the last layers of particles along the stream. However, the process of fluidization of a fixed bed indicates (see below) that in many cases the resistance of these layers must obviously be higher than that of the same number of layers within the bed because of the practically complete loss of kinetic energy by the jets leaving the bed.

1.3 Fluid Mixing

In passing through a fixed bed of granular material, the fluid (gas) is mixed; this is usually termed fluid mixing or diffusion. The simplest scheme for representing fluid mixing is to suppose that the stream has a jetlike nature and that each jet branches and flows around various particles and then mixes with similar branches of neighboring jets [175, 744]. Thus it is noted in Chudnovskiy's monograph [177] that over a height of two layers of particles one-fourth of the gas stream "is interchanged with neighboring openings at a distance d apart" (the particle diameter). However, it is difficult to explain on the basis of such a representation why the transfer coefficient in the laminar region of flow varies according to a rather different law than in the transition region. Here it is useful to use another approach [9, 744], which was developed in detail by Aerov and Umnik [9]. They noted that turbulence in the stream between the particles occurs even at relatively low values of Re, and in the confined spaces adjacent to the particles a turbulent transfer mechanism predominates. The convective component of the diffusion coefficient in the bed is

$$D_c = D - D_0 = \Delta w l, \tag{1-7}$$

where D is the overall coefficient of diffusion; D_0 is the component of the overall coefficient of diffusion due to molecular transfer; and Δw is the change in the mean-stream velocity or the pulsation velocity over the length l in which these pulsations occur. The quantity l is proportional to the characteristic geometric dimension of the system: in the present case, to the equivalent diameter of the "channels" of the granular bed, $d_{e.h}$. If it is assumed that Δw is proportional to the fluid velocity, then

$$D_c = K w_f d_{e.h}. \tag{1-8}$$

A similar equation was proposed by Pshezhetskiy and Rubinshteyn [120], by Bakhurov and Boreskov [11] and by Baum [10]; the latter assumed that the whole diffusion coefficient was approximately proportional to Re. According to the data of Aerov and Umnik, the overall diffusion coefficient is practically proportional to w_f (i.e., molecular transfer can be neglected) from about $Re_{e.h} > 10$–20. Baum assumed that this proportionality became valid starting at Re = 5, but here Re refers to the particle dimension; i.e., the two assumptions are similar.

Aerov and Umnik assumed that at large Re the transfer in the granular bed was characterized by the same isotropy as at Re < 10. They experimentally demonstrated the isotropy of the effective thermal conductivity of the bed for the region Re < 10, where molecular transfer is important [7].

The numerical values of the proportionality constant K published in [9] differ depending on the particle shape, and in the case of a ceramic ring packing, K depends on the ring dimensions also (Table 1-3).

Table 1-3

Values of the Coefficient K in Granular Beds of Various Materials [9]

Material	Dimensions, mm	Range of Re	K	Notes
Glass spheres	7.42	11-470	0.0808	In all experiments $D_0 \approx 0.015$ m^2/hr
Steel spheres	7.15	11-470	0.0808	
Steel spheres	3.19	2-15	0.0808	
Shot	2.46	2-11	0.0808	
Catalyst lumps	Lumps of size 10	7-660	0.102	
Ceramic rings	8x8	5-370	0.135	
Ceramic rings	15x15	9-650	0.266	

Baum [10] found $K \approx 0.1$ for sand and for lumps of glass of irregular shape and various dimensions, but he used an equation which involved the actual granule size d instead of $d_{e.h}$. His data refer to the range $Re = w_f d/\nu$ from 5 to 2000. A much higher value of K for spheres than that found by Aerov and Umnik was also obtained in experiments of Bakhurov and Boreskov [11]. Aerov and Umnik explained this by the much higher roughness of the particles used in [11].

All of this indicates that fluid mixing has not been sufficiently studied. In particular, it is necessary to specify its mechanism and to make more precise the values of the proportionality factors (which are the reciprocals of the Péclet numbers for diffusion), as well as to investigate the isotropy of the mixing. The data of Rutherford and Amundson [703] and of Ebach and White [821] do not confirm isotropy of mixing, but indicate that the intensity of longitudinal mixing is usually higher than that of transverse mixing. However, reverse motion of the bulk of the medium can scarcely occur in a fixed bed. Obviously the phenomenon termed "longitudinal mixing" is related here to the nonuniformity of the velocity distribution over the cross section of the bed. Such nonuniformities, as shown in Chapter 5 for the case of heat transfer, lead to a reduction in the mean difference of the transfer potential in the bed, i.e., to an effect analogous to that observed in the intensive longitudinal mixing.

For a rough determination of the coefficient of radial fluid mixing (diffusion) within fixed beds of granular materials it is convenient to use Baum's equation:

$$D \simeq 0.1 d w_f. \qquad (1\text{-}9)$$

1.4 Moving Beds

In some parts of fluidized bed equipment we may have to deal with moving beds sliding under the action of gravity. For instance, this occurs in the hoppers under cyclones for recovering material entrained from a gas stream leaving a fluidized bed, in the circulating pipelines between various pieces of fluidized bed equipment, in bunkers above feed equipment, and in the hoppers below them. Sometimes it can even be convenient to carry out certain reactions or other processes within moving beds in such transport lines, and to operate with a gas permeating through such a bed in one direction or the other.

Here the main problems for designers are to determine the flow rate of material along tubes and the rate of efflux from openings under bunkers, to ensure uniformity of the efflux, to prevent the separation of the material into fractions, and, finally, in the case of gas flow, to determine the pressure drop through the moving bed. The pressure drop is not the same as the hydraulic resistance of a stationary dense bed not only because of the different porosity of the moving bed, but also because the moving bed acts as a pump, carrying the gas (fluid) along.

The gravitational motion of a dense bed of granular material has been studied by many investigators [2, 3, 12, 116, 52, 56, 80, 262, 292, 316, 344, 388, 398, 401, 475, 491, 507, 537, 554, 609, 647, 717, 725, 909, 1051].

Quite complete reviews of the foreign work in this field have been given by Zenz [621] and Leva [988]. Let us first consider moving beds without co- or countercurrent flow of a fluid medium.

1. Beds without a fluid medium (Bulk-solid flow)

Leva [988] notes that the porosity of a moving bed is very close to the free packing of a fixed bed. This is equivalent to saying that a bed of randomly packed particles is considerably expanded as a result of the transition to motion. However, the investigation of Luk'yanov, Gusev and Nikitina [647] has shown that the picture is really more complicated. These authors investigated the motion of various materials (Table 1-4) in a vertical tube 230 mm in diameter and 2500 mm long.

Table 1-4

No.	Material	Size analysis, wt. %				Bulk wt., kg/m
		1-3 mm	3-5 mm	5-7 mm	7-9 mm	
1	Spherical aluminosilicate catalyst.	7.36	89.22	3.22	0.2	790-830
2	Pelleted catalyst.	5.20	77.00	16.00	1.8	780-810
3	Firebrick particles	—	—	98.00	2.0	1100-1300
4	Activated carbon.	100.00	—	—	—	650-700

Layers of tagged particles were laid down at different levels in the upper part of the tube. The material flowed out through an outlet opening (D_0 = 30 mm) in the horizontal bottom of the tube. The velocity distributions of the material at cross sections more than two tube diameters above the opening were practically independent of the location of the opening. Up to this level the layers of tagged particles were deformed during bed motion, revealing a lag in a "boundary layer" of two to three layers of particles next to the wall, but otherwise symmetry was maintained [647] (Fig. 1-5). In the "boundary layer" the velocity of the material was about 10% lower than in the central part, i.e.,

$$w_{\text{wall}} = 0.9 w_{\text{center}}. \tag{1-10}$$

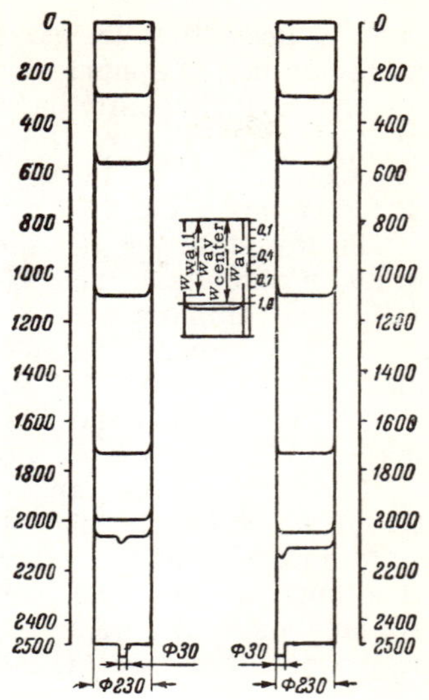

FIG. 1.5. Velocity distribution curves in the motion of finely ground activated carbon [647]. w_{av}—mean velocity of bed; w_{wall}—bed velocity near the wall; w_{center}—velocity in central part of bed.

These authors [647] also obtained velocity profiles for the efflux of material from a rectangular tube of section 232 times 100 mm. In order to obtain data indicating how the mean density of the bed changed with changes in its state of motion they carried out additional experiments with technically smooth tubes of diameters from 50 to 232 mm and length 3000 mm at moderate bed velocities (0.1-0.7 m/min). Under the conditions of these experiments the mean density of the moving beds differed from that of a loaded stationary bed by not more than 3-5%. Disintegration occurred over the whole lower part of the moving bed in the zone in which the opening had an effect; this had a height of not more than two opening diameters. Over the remainder of the tube there was only the "boundary-layer" effect mentioned above, while at the central (axial) region the moving bed could even become denser than the initial stationary bed; this has also been observed by the authors in experiments with pelletized activated carbon. In this case the mean density of the moving bed was equal to or somewhat greater than the mean density of the stationary bed.

It can be assumed that the increase in density of the central part of a moving bed will be encountered for other materials with particles of irregular shapes due to changes of their orientation as a result of the velocity gradients in the bed. Because of this distribution of porosity it can be expected that in the flow of a gas through a moving bed the velocity distribution of the fluid over the cross section of the bed will be even more nonuniform than in a stationary bed: the excess flow near the walls is increased.

According to experiments of Luk'yanov et al. [647], the disruption of a moving bed of particles of given shape depends directly on the particle diameter and inversely on the tube diameter and height of the moving bed (slightly).

The experimental results of Delaplaine [261] are in good agreement with Luk'yanov's data but disagree with the views of Leva; these results show that the porosity is close to 0.39 in the central part of a moving bed of spherical particles; i.e., it is close to the porosity of an ordinary randomly packed fixed bed. For moving beds of particles of irregular shapes, Delaplaine, like Luk'yanov, observed a slight increase in density in the central part of the bed and suggested that its porosity was about 0.02 smaller than that of a "static" bed.

In investigating the motion of material under gravity flow along tubes of diameters from 25 to 1270 mm, Gregory [316, 988] made the following comments:

1. The internal surface of the tube must be as smooth as possible, and must be at least 5-7 times larger in diameter than the diameter of the largest of the particles passing through it that are contained in the material in appreciable quantities.

2. Narrow size ranges move better under gravity than wide fractions, which have a tendency to hang.

A marked tendency toward hanging is also observed with narrow fractions of very fine particles ($d < 75\,\mu$) and with particles having surface moisture contents higher than 1-2%. Obviously this number represents the limiting permissible surface moisture only for a certain range of particle diameters.

It must be assumed that for very fine particles or particles covered with a film of moisture there are considerable forces of molecular interaction between them and the tube walls. Beilby [209] has shown that zinc, gold, silver and copper filings with particles smaller than 0.1 mm will adhere not only to the surface of any metal but also to glass, porcelain, etc. The density, surface conditions and electrostatic effect of the particles also affect hanging [988].

According to Gregory, there are two main forms of motion of a material under gravity: 1) continuous free flow, the velocity of which is proportional to the square root of the height of the column of material, and 2) discontinuous flow, which does not depend on the height of the column.

Other investigators [292, 388, 507, 537, 647] have not indicated the dependence of the efflux velocity on the height of the column of powdery material. Keneman, Zalogin, Vorob'yev and Antoshina [1051] have specified that the flow through an opening depends on the height of the column of material if this is smaller than the diameter of the opening. A "dynamic arch" zone extends over this height. In agreement with the "dynamic arch" hypothesis proposed by Pokrovskiy and Aref'yev [116], the authors assume that in the efflux of a free-flowing material the pressure due to the overlying column of material is not felt at the opening, and above the opening an arched bed structure is formed—the dynamic arch, which transmits the pressure of the overlying layers onto an annular area of the vessel bottom surrounding the opening.

As Leva [988] and Zenz [717] have pointed out, the widest and most systematic investigations of the gravity flow of beds have been carried out by Rausch [507] for materials which differed greatly in density, bulk weight, and particle size and shape (Table 1-5).

Rausch's data were obtained from work with tubes of diameters from 76 to 204 mm and for outlet openings of diameters from 1.6 to 51 mm. The ratio of the particle diameter to tube diameter varied from

Table 1-5

Characteristics of Granular Materials Used in Rausch's Experiments [717]

No.	Particles	d, mm	γ_{mi}, kg/m³	Porosity of stationary bed	Angle of repose on loading degrees
1.	Radish seed	2.49	730	0.41	38
2.	Soybeans	7.66	765	0.344	39
3.	Catalyst balls	4.31	730	0.435	35
4.	Aluminum oxide spheres	15.0	2218	0.385	43
5.	Aluminum oxide spheres	11.35	2040	0.331	38
6.	Aluminum oxide spheres	7.00	2070	0.384	34
7.	Steel spheres	9.76	5020	0.355	33
8.	Steel spheres	12.7	4830	0.38	37
9.	Lead shot	1.27	6750	0.375	23
10.	Lead shot	6.35	6590	0.43	33
11.	Lead shot	4.14	6600	0.418	30
12.	Sea sand.	0.50	1460	0.425	37
13.	Sea sand.	0.25	1370	0.447	36
14.	Sea sand.	0.18	1360	0.442	36
15.	Sea sand.	0.127	1360	0.442	35
16.	Iron powder	0.35	2380	0.48	42
17.	Iron powder	0.25	2440	0.481	41
18.	Iron powder	0.18	2250	0.48	40
19.	Iron powder	0.127	2218	0.48	40
20.	Glass spheres	0.29	1470	0.386	26
21.	Glass spheres	5.20	1363	0.421	32

about 2.5 to 250. The following observations were made during the experiments to determine the flow rate of material [988]:

1. The height of the column of material had practically no effect. According to Gregory, this would indicate that Rausch was dealing with discontinuous motion of the material.

2. The diameter of the feed pipeline had a small effect on the flow rate of material through the orifice.

3. The particle diameter d had little effect if the orifice opening D_0 was at least 25 times larger. When $D_0/d < 25$ the density of the bed of material obviously decreases during passage through the opening, thus affecting the flow rate of material. Obviously this is similar to the wall effect in cylindrical vessels containing fixed beds.

4. The angle θ of the conical part of the bunker above the feed pipe (Fig. 1-6) affects the flow rate of material. The effect of roughness of the feed pipe surface, which depends on θ and D_0/d, is not indicated in Fig. 1-6. It is important when the roughness of the pipeline and the particle diameter are of the same order of magnitude, but becomes insignificant as the ratio D_0/d increases.

5. It was also noted that a counterflow of air decreased the flow rate of material considerably.

All the experimental data are described by the following equation:

$$Q_{M.a} = CC_0 \frac{(D_0/d)^{2.70}\sqrt{g}\gamma_{bed}d^{2.5}}{\sqrt{\tan\beta_a}}, \qquad (1\text{-}11)$$

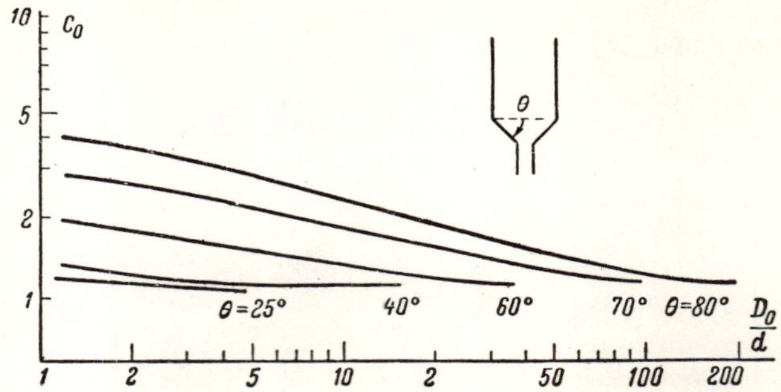

FIG. 1-6. Correction factor (C) for calculating the motion of granular materials under gravity as a function of the angle θ and the ratio of the outlet diameter (D_T) and particle diameter ($\bar{d}$) [988].

where $Q_{M.a}$ = flow rate of material during discharge into air under normal conditions, kg/sec;
β_a = angle of repose in air;
γ_{bed} = bulk weight of material in a stationary bed, kg/m^3;
g = gravitational acceleration, m/sec^2;
d = particle diameter, m;
D_0 = diameter of the opening or tube, m;
C = coefficient for the wall effect (Fig. 1-7);
C_0 = correction coefficient for taking into account the cone angle of the lower part of the bunker (see Fig. 1-6).

For polydisperse materials, the calculated particle diameter d is taken here as

$$d=\frac{\sum_1^n d_i^2}{\sum_1^n d_i}, \qquad (1-12)$$

where n = number of narrow size fractions;
$d_i = \sqrt{d_1 d_2}$ is the particle diameter of the narrow fraction, taken as the geometric mean of the screen openings between which the narrow fraction is retained.

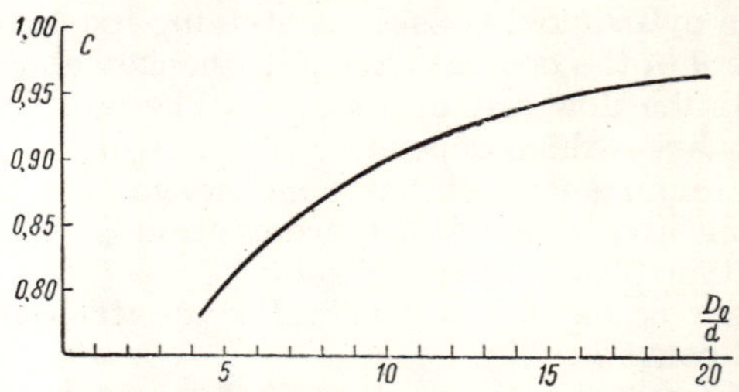

FIG. 1-7. Rausch's correction factor C for taking into account the wall effect on the motion of granular materials under gravity [988].
D_T—tube diameter; d—particle diameter.

The bulk weight of the material γ_{bed} is determined as described earlier (see p. 10-11).

The rate of efflux of material into any medium other than air under normal conditions is given as

$$Q_M = Q_{M.a}\sqrt{\frac{\tan\beta_a(1-\gamma_m/\gamma_M)}{\tan\beta}}, \qquad (1\text{-}13)$$

where β = angle of repose of the material in the given medium;
γ_m = specific weight of the medium, kg/m³;
γ_M = specific weight of the particle material, kg/m³.

Rausch's formula is suitable for calculating the flow rate of free-flowing materials from openings, vertical pipes and bunkers with conical bottoms. Figure 1-8 gives graphs corresponding to Eq. (1-11) which are rather more convenient to use in practical calculations than the formula. As the graphs show, Rausch's formula is in good agreement with the experimental data of other investigators: Delaplaine [717], Deming and Mehring [262], Franklin and Johanson [292] and Zenz [717]. The experiments of Borisevich [1025] showed that Rausch's formula is also applicable in the case of flow of materials along tubes of small diameter.

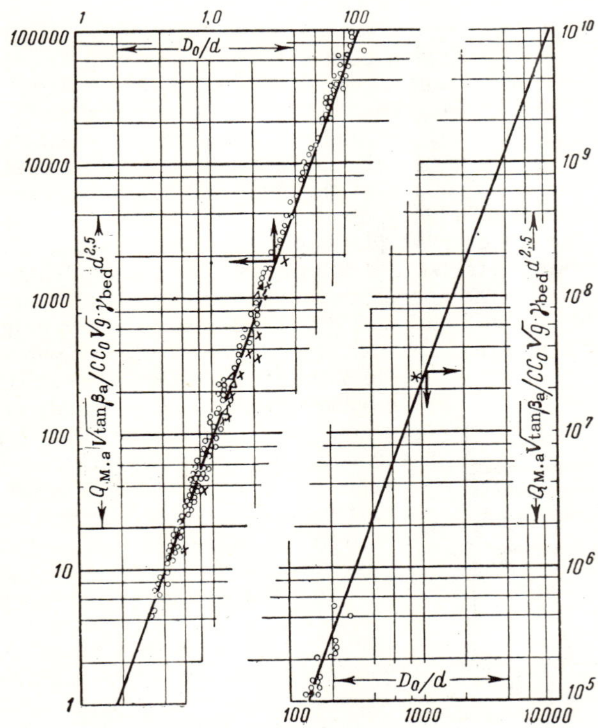

FIG. 1-8. Graphs of Eq. (1-11). Symbols as in text [717].

Some other equations for the efflux of free-flowing materials have been given in the monograph of Leva [988] and the work of Zenz [717]. In [988] a monograph is given which was constructed by Shirai [537] for the efflux of powders (d = 0.5–0.25 mm) through orifices.

The transverse velocity profile of the material in a bed sliding through a vertical tube is uniform, except for the "boundary layer" mentioned earlier [647], only at considerable heights above the opening. When the height is such that its ratio to the tube diameter is smaller than the tangent of the angle of internal friction of the free-flowing material, this uniformity of motion is lost [717].

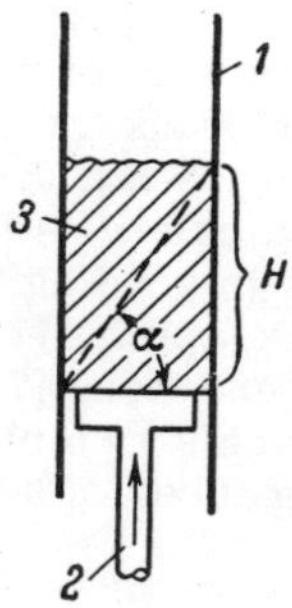

FIG. 1-9. Method for determining the angle of internal friction α. 1) tube; 2) piston; 3) bed; H) maximum bed height at which it is still possible to move the piston upward [717].

One of the methods for determining the angle of internal friction is shown in Fig. 1-9. The solid material is slowly poured into a tube in which there is a free-moving piston. It is found that if the ratio of the height of the bed above the piston to the tube diameter does not exceed some critical value, the bed can be raised by the piston without difficulty. However, if more free-flowing material is added and the experiment is repeated at larger values of the ratio $H/D_т$, it becomes impossible to move the piston upward. As Zenz has noted, it is easy to find the critical value of the ratio H/D, equal to $\tan \alpha$, by carrying out the experiment carefully. The tube must have a constant cross section, and the piston must move freely in it, but without allowing particles to fall into the gap between the walls and the piston. The angle of internal friction α is usually much larger than the natural angle of repose of the material β. Thus, for example, according to the data of [717], for sand (d = 0.48 mm), α = 63–65° and β = 35–37°; for one catalyst (d = 3mm), α = 71–73° and β = 35°; for a powdery cracking catalyst (d = 60 microns) the angle of internal friction reached 79°. The corresponding critical ratios of H/D for these materials were about 2, 3.3 and 5; i.e., the minimum bed height above which the motion of the material was comparatively uniform was from 2 to 5 diameters of the tube (bunker). Thus in bunkers, where large values of H/D are seldom used for constructional reasons, the velocity profiles as material flows out are usually nonuniform. In order to make this profile more uniform in equipment with small H/D, material feed is sometimes distributed through many openings or tubes over the cross section of the vessel bottom (Fig. 1-10).

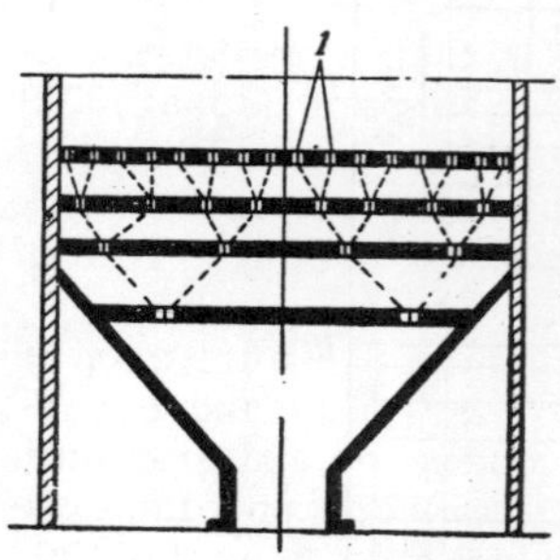

FIG. 1-10. Perforated grids for uniform efflux of free-flowing material from reactor [490]. 1) openings.

It must be noted that the actual rate of efflux of material will obviously be close to the calculated value only if the powdery material does not cake, i.e., either if the material does not have any tendency toward compaction and caking, or if the material stays in the equipment for so short a time that compaction does not have time to occur. Obviously in practice it is necessary to exclude other conditions that make possible the compaction of material, such as the use of vibrators for opening the outlets for the material, etc.

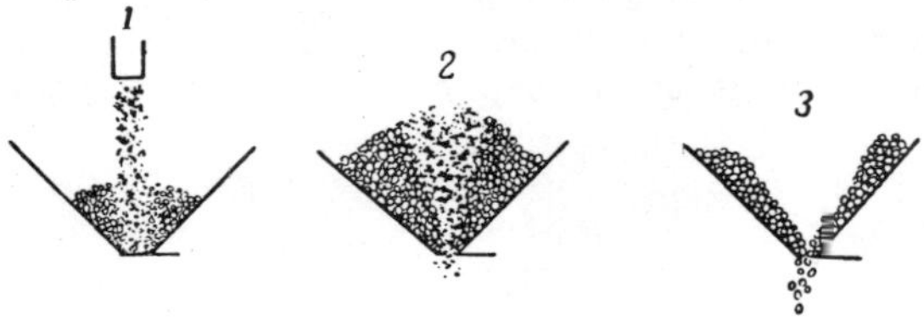

FIG. 1-11. Separation of coarse and fine particles during filling (1) and emptying (3) of a bunker [717].

As Zenz has noted [717], in many cases when working with dense beds of polydisperse materials it is desirable to avoid size separation of the particles, which can occur during loading of the material into the equipment. Thus, for example, in the localized supply of material into a bunker (Fig. 1-11) through a single central opening the finer particles on falling remain in the center, while the coarser particles roll down the slopes of the heap and separate along the periphery. As a result, the first portions on emptying the bunker will have an increased finer content, and when the bunker is almost empty the efflux will consist almost exclusively of coarser particles. Any scheme that prevents rolling of particles during the loading process (such as using a traveling feed point) will help to avoid separation.

2. Moving beds with gas flow

A number of investigators have dealt with the effect of a superimposed gas flow on the rate of efflux of a granular material [316, 988, 1051].

A counterflow of gas retards the efflux of material, while a cocurrent flow accelerates it. However, the corresponding quantitative relationships are not given, and obviously have not been determined.

There are few data in the literature on the hydraulic resistance of moving beds. We must refer here to the extensive work of Happel [332] and parts of the information from the theses of Kojabashian [395] and Shanahan [533] published in the chapter by Zenz in the collective work [490]; these, however, do not take into account the complete role of the motion of the material.

Happel carried out experiments to determine the pressure drop in moving beds with gas streams moving countercurrently and cocurrently to the material. The materials used included seeds and granular, spherical and pelletized catalysts (monofractional and polydisperse samples). Happel calculated the mean-particle diameter as $d = 100/\Sigma(X_i/d_i)$, where X_i is the weight percent of a narrow fraction and d_i is the diameter of the narrow fraction. The value of d varied from 0.38 to 7 mm. The vertical tubes in which the beds moved had diameters of 100 mm, and 50, 200 and 250 mm in a few experiments. Happel claims that as a result of experiments with the catalysts in laboratory models, it appears that the density of a moving bed is the same (to an error of less than 1%) as the charged weight of a freely packed stationary bed obtained by slowly inverting the measuring cylinder with the sample of material in it several times until the volume in it stabilizes. For correlating his experimental data, Happel proposed the equation

$$f_m = \frac{g\Delta P d}{2Hw_f^2\gamma_m(1-m)^3} = \psi\left[\frac{dw_f\gamma_m}{\mu}(1-m)\right], \quad (1\text{-}14)$$

where f_m = resistance coefficient found from the graph of the function ψ (Fig. 1-12);
d = mean-particle diameter;
m = porosity of the freely packed bed (obtained as outlined above).

The remaining symbols are as given previously.

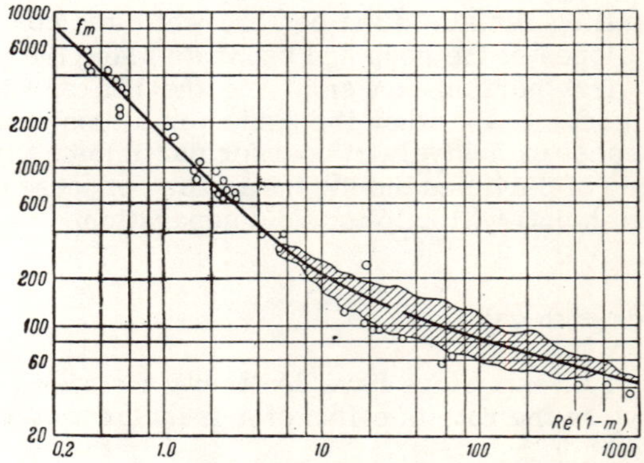

FIG. 1-12. Graph of Eq. (1-14) for determining the hydraulic resistance of a moving bed, according to Happel [332].

As can be seen from Eq. (1-14), Happel has not in any way taken into account the ejecting action of the moving bed. Leva [988] suggested that in order to take into account more completely the movement of the material, the fluid velocity in Eq. (1-14) should be replaced by the relative velocity of the gas and material, if the latter moves at more than 20% of the gas velocity. Obviously such a limit is not always valid, in particular for the transitional and turbulent regions of flow, when ΔP is very dependent on the relative velocity. How strongly the rate of motion of the material may sometimes affect ΔP is demonstrated well in Fig. 1-13, showing experimental data of Shanahan [533], who studied the pressure drop in moving beds of spherical catalyst particles with countercurrent and cocurrent air flow in a transparent tube of 25-mm diameter.

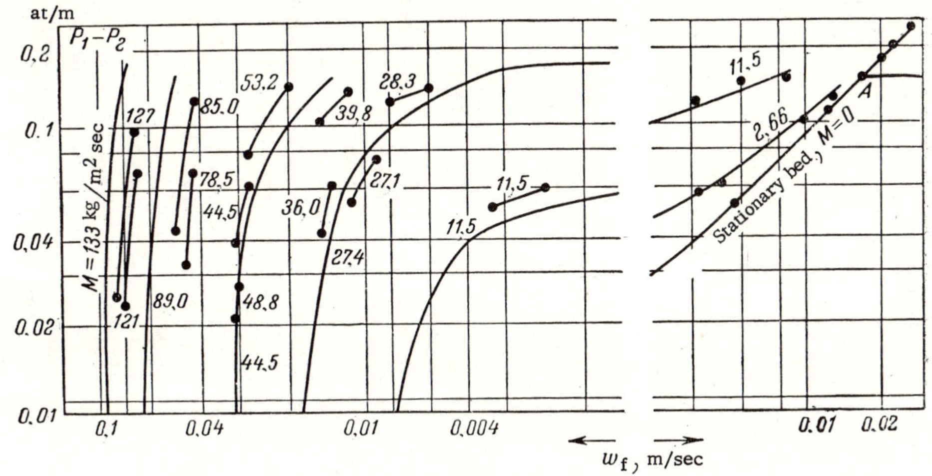

FIG. 1-13. Resistance of a dense bed as a function of its velocity in motion under gravity with a cocurrent (left) and countercurrent (right) air flow [490].

System: catalyst spheres–air. Notation on each line corresponds to the particle flow rate *M*, kg/m²·sec.

A—limit of stability for stationary bed.

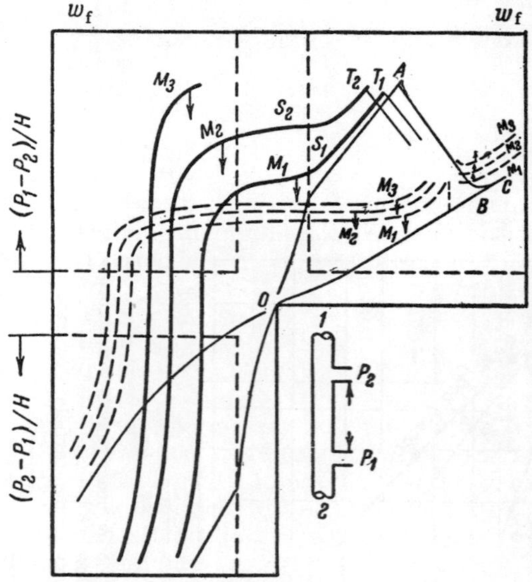

FIG. 1-14. Zenz phase diagram [850]. Hydraulic resistance (P_2—P_1) per unit height of bed (H) as a function of fluid velocity (w_f) for any medium –particle system.

The manner in which the pressure drop changes during fluid flow through a moving bed in cocurrent and countercurrent flow can be shown clearly on a Zenz phase diagram (Fig. 1-14).

In quadrant I, where the gas stream flows upward according to the conditions assumed in constructing the diagram, the line ST corresponds to countercurrent flow of the gas and material. The decreasing role of the velocity of the material at high fluid velocities can be seen clearly (the lines corresponding to various flow rates of material approach each other and the line OA for flow through a stationary bed). On the other hand, the lines diverge widely (the role of the material flow rate increases) as the fluid flow rate decreases. When the fluid velocity is equal to zero in the stationary bed, the pressure drop is also equal to zero (the line OA passes through the coordinate origin), while for a moving bed it has a considerable positive value (the line S_2T_2 has an intercept OS_2 on the ordinate axis), since the moving bed operates like a pump (leading to a pressure drop ΔP equivalent to the flow of gas at the same relative velocity through a stationary bed having the same structure as the moving bed). Changing the direction of the gas stream downward in cocurrent flow with the moving bed leads to lines on the phase diagram which extend the countercurrent lines and pass into the second and third quadrants. The very steep fall of the lines in Fig. 1-14 as the fluid velocity is increased is due not only to the schematic nature of the diagram, but also to the use of logarithmic scales. The lines for the various material flow rates approach each other and also the line for fluid flow through a stationary bed. This is quite natural at high fluid velocities, since then the flow rate of material constitutes a less important part of the total relative rate.

1.5 The Minimum Velocity for Fluidization

1. Description of the phenomenon

If a bed of granular material is fluidized for the first time, it has to be expanded before reaching the minimum fluidization velocity (the limit of stability). It expands from the initial porosity of the randomly packed dense bed (of the order of 0.39 for uniformly sized spheres) up to the maximum possible value (of the order of 0.5 for the same spheres),* after which limit the particles cease to be in direct contact with one another and assume the mobility familiar to fluidization. As the fluid velocity is decreased and the bed undergoes transition back to the stationary state, the line for this reverse process does not duplicate the line for the first expansion, but lies appreciably lower, since the bed has a certain degree of "irreversibility" of expansion with respect to the initial particle packing. The pressure-drop curve for the stationary bed before the initial fluidization forms a special loop with the line for the reverse process which is superficially reminiscent of a hysteresis loop (Fig. 1-15). When fluidization is repeated, the process occurs along the lower line for increasing and decreasing fluid velocity if the bed retains the maximum expansion for a fixed bed. "Incompressibility"—the ability to retain this expansion to the greatest degree—occurs in beds of rough particles of narrow size ranges, particularly if the resettled bed is not vibrated or jarred. Polydisperse beds of particles which do not separate into fractions do not retain their expansion as well [677].

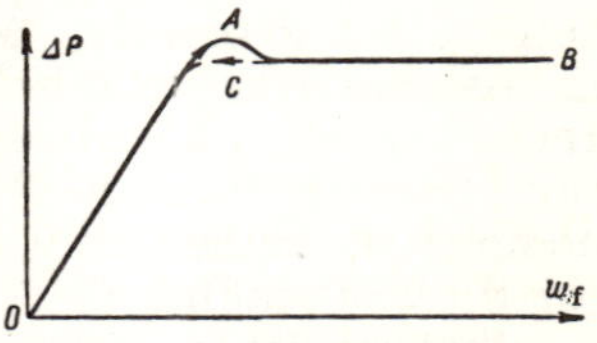

FIG. 1-15. Lines of initial fluidization (OAB) and the reverse process (BCO).

Thus the stage of preliminary bed expansion must be completed when fluid flow begins through a stationary bed before fluidization is repeated, even in the case of beds that are "incompressible" in practice. Because of these differences between initial and subsequent fluidization, we must be specific when we are speaking of initial fluidization, in order to avoid misunderstandings in evaluating the experimental data. The so-called pressure peak around the limit of stability is usually more clearly marked in initial fluidization; flow through the stationary bed continues until a pressure drop is reached exceeding the equivalent weight of the bed (the product of the bulk weight and the bed height), and then beyond the limit of stability the pressure falls to this theoretical equivalent.

Leva [988] explains the presence of the peak by the consumption of energy in unlocking the particles, Reboux [511] by the acceleration of the mass of material during the expansion before fluidization, and other

*According to recent data of Roberts, the expansion of the bed before fluidization is not so great.

authors by the need to overcome the "adhesion" of the particles at the walls. The reason put forward by Reboux can be important only for very rapid expansion of the bed, which does not usually occur in experiments for determining the limit of stability. In such experiments the fluid velocity is not increased abruptly, but very gradually. Regarding the adhesion of particles to one another and to the walls, both factors obviously exist. The rearrangement and expansion of a fixed bed before its initial fluidization cannot be completed until the adhesion is overcome. Obviously, when fluidization is repeated and repeated expansion is not necessary, the pressure peak is expressed only feebly. It must be noted that if the adhesion of the particles to one another and to the walls is small, it can be expected that partial fluidization of the bed will occur at pressure drops over the whole bed that are less than the weight of the column of material resting on a unit area.

This is because the resistance coefficient of the upper layers of particles of a dense bed that is uniform with respect to height is obviously larger than that of the middle layers, as mentioned earlier. Consequently, the hydraulic resistance force becomes equal to the weight of the particles earlier for the upper layers than for the rest of the bed. The suspension of the upper layer facilitates the fluidization of the succeeding layers with a certain increase in the flow velocity. As a result the bed is not fluidized all at once over its whole height, but gradually, beginning at the top. Thus gradual fluidization is possible even for a bed of uniform particles, and so it is more correct to speak of a critical region of velocities within which the transition from a fixed bed to a completely fluidized bed begins and is completed, rather than to speak of a critical velocity for fluidization (or limit of stability).

Conversely, in other cases fluidization commences below. This happens when the friction between the bed and the walls is small and, on reaching a velocity sufficient to suspend it, the whole bed is raised above the supporting screen like a piston. Due to random packing, there are always separate projecting particles and aggregations on the lower surface of the bed. The true velocity of flow around these particles will be much smaller than that around the particles within the bed. Particles and groups of particles will therefore begin to fall from the lower surface of the bed. Thus fluidization of the bed begins, and it assumes its free structure from the bottom upward.

The size of the pressure peak on initial fluidization can be of practical importance if it governs the choice of and rate of rotation of the air blowers. For piston equipment and similar types there are obviously no such relationships, since there is usually available an excess of pressure, over the working pressure, without having to change the speed of rotation of the equipment. There is usually also no difficulty in the use of centrifugal blowers, since they operate in the region of blower characteristics where it is possible to have an increased pressure at the decreased air flow rates sufficient for the onset of fluidization. It should be noted that with continuously operating equipment the problem of increased hydraulic bed resistance can arise only under the unusual conditions of restarting the operation after an unexpected shutdown of the air blast; the operation can be restarted with a small bed height, which is then gradually increased to the normal value.

2. Design relationships

A large number of relationships have been proposed for calculating the minimum fluidization velocity (limit of stability) $w_{1.s}$, [32, 82, 213, 511, 521, 570, 735, 863, 874, 930, 988, 1032, 1116, 1040, 1137, etc.].

Apart from setting up purely empirical equations, there are generally two routes for determining the types of design formulas. The first and more widely used method (see, for instance, the publications by Leva [988], Rykov and Shakhova [930] and Reboux [511]) reduces to determining the minimum velocity for complete fluidization from the resistance equation of a fixed bed as the fluid velocity at which the resistance force of this stationary but maximally expanded bed becomes equal to its apparent weight, i.e., its weight taking into account the Archimedes forces. This approach is equivalent to regarding the limit of stability as the limiting state of a still stationary bed.

The second approach is equivalent to a consideration of the transition process from the free fall of individual particles (a suspended dilute bed) to a state in which they are in relatively close proximity to one another (a fluidized bed) and finally to a state in which they settle out (formation of a stationary fixed bed). The limit of stability is then regarded as the limiting state in which the particles are still in suspension; the velocity of free fall of individual particles is one of the main factors appearing in the formulas. In this case, the effect of neighboring particles, which change the conditions of flow around particles more closely crowded together in a fluidized bed, is taken into account by some function of the porosity (see, for instance, the paper by Westfried and Cazacu [1137]). The porosity function is derived from the observed relationship between the expansion of a fluidized bed and the superficial fluid velocity. This relationship is also expressed as the factor giving the change in the resistance coefficient of the particles falling as a cloud compared with the resistance coefficient for the fall of individual particles in an infinite space.

In most cases the limit of stability lies within the region of laminar flow (Re < 10). Under these conditions, when the expression (1-3) for the bed resistance at the porosity corresponding to the limiting expansion of a stationary bed is equated to the resistance of the fluidized bed, Leva found that [988]

$$w_{1.s} = \frac{0\ 005 d^2 (\gamma_M - \gamma_m) m_{1.s}^3}{\mu \varphi^2 (1 - m_{1.s})}. \tag{1-15}$$

This formula was later simplified by Leva, Shirai and Wen [414]; they omitted the porosity and shape factor. The point of this is that the construct $m^3_{1.s}/\varphi^2(1-m_{1.s})$, denoted by C, changes very little in going from one type of particle to another. This is to be expected, since on changing from a bed of spheres, for example, to a material with particles of irregular shape, the shape factor increases, but at the same time so does $m_{1.s}$, since the particles of irregular shape form a stationary bed of considerably greater maximum porosity than do spheres. This is also indicated by experiments. For the region Re < 5 the construct C can be represented as the following function of Re [988]:

$$C = 0.0007\,\mathrm{Re}^{-0.063}. \qquad (1\text{-}16)$$

Figure 1-16 shows that Leva and co-workers have based this function on the experimental data of many investigators [189, 202, 276, 299, 338, 418, 422, 452, 604]. Substitution of this value for C into Eq. (1-15) gives

$$w_{1.s} = 1.58 \cdot 10^{-4} \frac{d^{1.82}}{\nu^{0.88}} \left(\frac{\gamma_M - \gamma_m}{\gamma_m}\right)^{0.94} \ \mathrm{m/sec}, \qquad (1\text{-}17)$$

where d is expressed in meters and ν is expressed in m^2/sec.

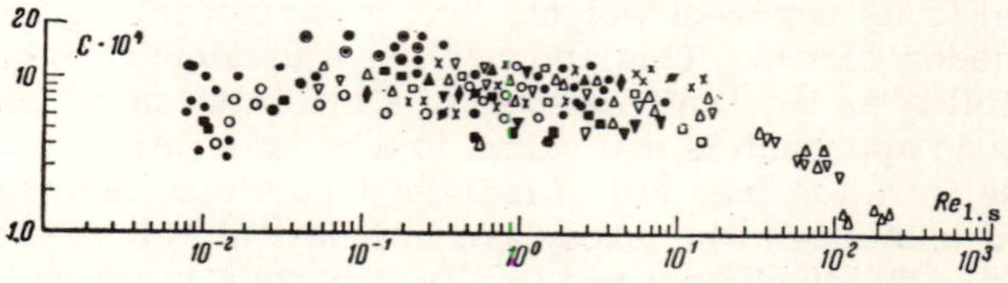

FIG. 1-16. Dependence of complex $C = m_{1.s}^3/\varphi^2(1 - m_{1.s})$ on Reynolds criterion |Re|[988].
$m_{1.s}$, porosity of layer at limit of stability; φ, form factor; ●, Leva [418]; ▲, Lewis [422]; ○, Miller and Logwinuk [452]; ×, Shirai [988]; ⊙, Van Heerden [338]; △, Wilhelm and Kwauk [604]; Agarwall and Storrow [189]; ▪, Baerg [202]; ▽, Ergun and Orning [276]; Furukawa et al. [299].

For Re = 10–300, the result calculated from (1-17) must be multiplied by the correction factor

$$K \approx 1.775\ \mathrm{Re}^{-0.272} \qquad (1\text{-}17a)$$

An approximate formula that is similar in form for the laminar flow region has been obtained by Graf [317] and is given in the paper by Landau and Franz [1116]:

$$w_{1.s} = A \frac{d^2(\gamma_M - \gamma_m)}{\mu}, \qquad (1\text{-}18)$$

where A is a constant coefficient.

Introduction of the kinematic viscosity instead of the dynamic viscosity yields

$$w_{1.s} = A_1 \frac{d^2}{\nu} \left(\frac{\gamma_M - \gamma_m}{\gamma_m}\right). \qquad (1\text{-}19)$$

In deriving this formula, Graf made an even greater simplification than Leva by assuming that

$$m_{1.s}^3/\varphi^2(1 - m_{1.s})$$

is completely constant.

It follows from Graf's approximate formula that in the laminar flow region the critical (or incipient) fluidization velocity cannot depend on the pressure except for compressible fluids (gases), at least not in the region of low pressures of the order of 1–10 atm abs. It is known that the effect of pressure on the dynamic viscosity can be neglected in this region [98]. The independence of $w_{1.s}$ on pressure (in the laminar region) has been confirmed by experiments of Sechenov and Al'tshuler [336] fluidizing aluminosilicate catalysts with nitrogen at pressures of 1 to 16 atm abs. In the so-called turbulent flow region, Sechenov and Al'tshuler noted that $w_{1.s}$ varied inversely with the square root of the gas density; i.e., it decreased somewhat as pressure increased.

Landau and Franz attempted to extend the correlation of Graf to the region of turbulent flow; they carried out experiments on the air fluidization of beds of roasted iron pyrites and low-temperature coke, and gave a nomogram for calculating $w_{1.s}$ for all regions of flow. The lines for each of the two materials they investigated are different in the turbulent regime.

Westfried and Cazacu [1137] also proposed a relationship for calculating the minimum fluidization velocity when the limit of stability lies in the turbulent regime; this relationship is given as

$$w_{1.s} = \sqrt{\frac{2g\Delta P d\Phi}{C_D \gamma_m H m_{1.s}(1 - m_{1.s})}}, \qquad (1\text{-}20)$$

where C_D = resistance coefficient for a sphere falling freely in space (Fig. 1-17);
ΔP = hydraulic resistance of the bed;
Φ = particle-shape factor.

The remaining symbols are as given previously.

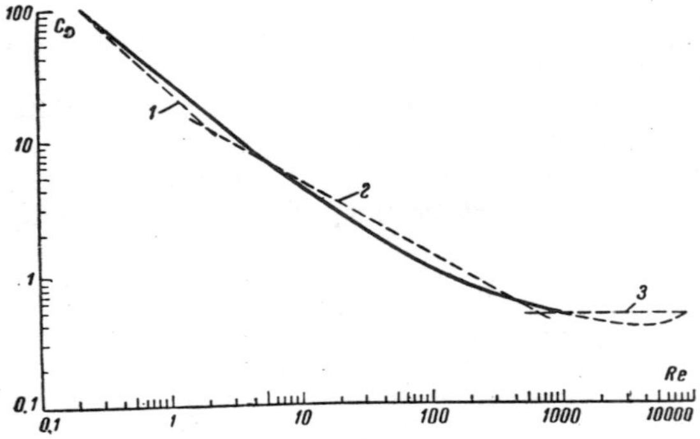

FIG. 1-17. Drag coefficient (C_D) of a sphere as a function of the Reynolds number (Re). Characteristic dimension is the sphere diameter. 1) Stokes's law, $C_D = 24/\mathrm{Re}$; 2) transition region, $C_D \sim 18.5\ \mathrm{Re}^{-0.6}$; 3) Newton's law, $C_D = 0.44$.

Westfried and Cazacu arrived at this formula from the data of Cohen de Lara [251].

Beranek and Sokol [874 and 959] proposed an expression of moderate complexity for calculating the minimum fluidization velocity over the range of Ω from 0.1 to 10^4:

$$w_{1.s} = 0.045 w_{fall} m_{1.s} \Omega^{0.2} \quad (1\text{-}21)$$

where $\Omega = \frac{w_{ff}^3}{g\nu} \cdot \frac{\gamma_m}{\gamma_M - \gamma_m}$, a dimensionless quantity usually known as the dynamic criterion of similarity, and

w_{ff} = free-fall velocity of the particles.

In contrast to the earlier correlations, Eq. (1-21) does not contain the particle diameter in explicit form, but it is necessary to know this diameter in order to evaluate w_{ff} unless it is determined experimentally.

Later, Beranek [1102] proposed modified expressions for calculating the minimum fluidization velocities for beds of uniform size particles of any shape:

$$w_{1.s}/w_{ff} = 0.019 \pm 0.003 \text{ for } B < 0.3; \quad (1\text{-}22a)$$

$$w_{1.s}/w_{ff} = 0.0213 B^{0.2} \text{ for } 0.3 < B < 10^3; \quad (1\text{-}22b)$$

$$w_{1.s}/w_{ff} = 0.090 \pm 0.005 \text{ for } B > 10^3, \quad (1\text{-}22c)$$

where B is the Ω criterion for spherical particles, determined from the equation

$$B = 6.494 \cdot 10^{-4} Ar^2 \text{ for } \mathrm{Ar} < 4; \quad (1\text{-}23a)$$

$$\log B = -8.502 + [211.350 - \log(\mathrm{Ar} - 13.605)^2]^{1/2} \text{ for } 4 < \mathrm{Ar} < 10^7; \quad (1\text{-}23b)$$

$$B = 9.809\, \mathrm{Ar}^{1/2} \text{ for } \mathrm{Ar} > 10^7, \quad (1\text{-}23c)$$

where $\mathrm{Ar} = Zg/\gamma_m \nu^2$; and

Z = apparent bed weight (the weight taking into account the Archimedes forces).

Using the two-term Ergun formula [276] for the resistance of a stationary bed and the usual theoretical expression for the resistance of a fluidized bed, Todes, Goroshko and Rozenbaum [735] obtained an interpolation formula for determining the critical velocity for onset of fluidization:

$$\mathrm{Re}_{1.s} = \frac{\mathrm{Ar}}{150 \frac{1 - m_{1.s}}{\tilde{m}_{1.s}^3} + \sqrt{\frac{1.75}{m_{1.s}^3} \mathrm{Ar}}}, \quad (1\text{-}24)$$

where

$$\mathrm{Ar} = \frac{gd^3}{\nu^2} \cdot \frac{\gamma_M - \gamma_m}{\gamma_m}, \quad \mathrm{Re}_{1.s} = \frac{w_{1.s} d}{\nu}.$$

For nonspherical particles, Zhavoronkov's shape factor must be introduced in the Reynolds number $\mathrm{Re}_{1.s}$:

$$\varphi_1 = (0.207 F / V^{2/3})^{1/2}.$$

Assuming that the particles in industrial reactors with fluidized beds are rounded, so that $\varphi_1 \approx 1$ and $m_{1.s} \simeq 0.4$, these authors finally arrived at the following equation for approximate calculations (with an accuracy of the order of ±20%):

$$\mathrm{Re}_{1.s} = \frac{\mathrm{Ar}}{1.400 + 5.22\sqrt{\mathrm{Ar}}}. \tag{1-25}$$

Volkov [1032] compared the formula of Fedorov, Re = f(Fe), which he recalculated graphically for the conditions at the limit of stability, with the interpolation formula (1-25), where $\mathrm{Fe} = d\left(\frac{4g}{3\nu^2} \cdot \frac{\gamma_M - \gamma_m}{\gamma_m}\right)^{1/3}$. Using the formula obtained by Fedorov [170] for the conditions at the onset of intensive mixing in a fluidized bed:

$$\mathrm{Re}_{im} = 0.19\,\mathrm{Fe}^{1.56}, \tag{1-26}$$

where $\mathrm{Re}_{im} = w_{im}d/\nu$, and w_{im} is the fluid velocity at the onset of intensive mixing, Volkov assumed that $w_{1.s} \simeq 0.5\, w_{im}$, so that

$$\mathrm{Re}_{1.s} \approx 0.5\mathrm{Re}_{im} = 0.095\mathrm{Fe}^{1.56}. \tag{1-27}$$

Graphical comparison of Eqs. (1-27) and (1-25) on a plot with coordinates Fe and $\mathrm{Sch} = w_f\left[\frac{3}{4} \cdot \frac{\gamma_m}{g\nu(\gamma_M - \gamma_m)}\right]^{1/3}$ shows that they agree well over the range Fe = 40–200.

When polydisperse materials are fluidized it is known [930, 1157, 1236] that there is no single critical velocity for the limit of stability, but rather a transitional region of velocities within which the resistance law deviates from the equations for flow through a fixed bed, but is still not in agreement with the resistance law for a fluidized bed. The reason is the simultaneous occurrence of two processes: fixed bed flow through the coarse fractions and fluidization of the fine fractions [1157]. Thus in the case of polydisperse materials it is necessary to distinguish between the velocities corresponding to the onset of the transitional region and to its end, which is complete fluidization. Rychkov and Shakhova [930] and later Winterstein and Biesz [1236] proposed finding the minimum velocity for complete fluidization of a polydisperse bed for any temperature and composition of medium by using single experiments carried out under simple laboratory conditions using air at room temperature.

The introduction of a stream of fluid into a fluidized bed changes its temperature sharply. Due to mixing of the particles and intensive heat transfer between the particles and the stream, this change occurs over a very short distance (of the order of ten rows of particles).

It is of considerable interest to estimate how the linear velocity and mass flow rate for minimum fluidization as well as the fluidization index (the ratio of the velocity to the minimum velocity for fluidization) change as a result of this nonisothermal behavior. Consider the laminar region only (Re < 10), for which simple relationships can be obtained. Using the approximate formula of Graf,

$$w_{1.s} \approx A_1 \frac{d^2}{\nu} \frac{\gamma_M - \gamma_m}{\gamma_m} = A \frac{d^2 (\gamma_M - \gamma_m)}{\mu}. \tag{1-19a}$$

the ratio of the linear velocities for minimum fluidization at t_1 and t_2 (°C) will then be (neglecting the change of γ_m)

$$(w_{1.s})_2/(w_{1.s})_1 = \mu_{t_1}/\mu_{t_2}. \tag{1-28}$$

The mass flow rate for minimum fluidization is

$$G_{1.s} = w_{1.s}\gamma_m \approx A_1 \frac{d^2}{\nu}(\gamma_M - \gamma_m).$$

The ratio of the mass flow rates at minimum fluidization for different temperatures (neglecting changes in $\gamma_M - \gamma_m$) is therefore

$$(G_{1.s})_2/(G_{1.s})_1 \approx \nu_{t_1}/\nu_{t_2}. \tag{1-29}$$

If we use the familiar Sutherland formula for the relationship between the dynamic viscosity of a gas and its temperature:

$$\mu_t = \mu_0 \frac{1 + C/273}{1 + C/(273+t)} \sqrt{\frac{t+273}{273}},$$

it is easy to obtain the following modifications of Eqs. (1-28) and (1-29) for gases:

$$(w_{1.s})_2/(w_{1.s})_1 = \frac{1 + C/(273 + t_2)}{1 + C/(273 + t_1)} \sqrt{\frac{273 + t_1}{273 + t_2}}; \tag{1-30}$$

$$(G_{1.s})_2/(G_{1.s})_1 = \frac{273 + t_2 + C}{273 + t_1 + C} \left(\frac{273 + t_2}{273 + t_1} \right)^{-2,5}. \tag{1-31}$$

The fluidization index is

$$N = \frac{w_f}{w_{1.s}} = \frac{w_f \gamma_m}{w_{1.s}\gamma_m} = \frac{G_f}{G_{1.s}}.$$

As the temperature of a given gas stream is changed (G_f = const), the fluidization index varies as

$$N_2/N_1 = \frac{G_f/(G_{1.s})_2}{G_f/(G_{1.s})_1} = \frac{(G_{1.s})_1}{(G_{1.s})_2} = \nu_{t_2}/\nu_{t_1}. \tag{1-32}$$

Thus the fluidization index varies directly with the change in the kinematic viscosity of the stream. This means that in the region of high gas temperatures the fluidization index will be larger. In liquid fluidization, on the other hand, the fluidization index will be lower at higher stream temperatures.

As regards the choice of a final design correlation it should be pointed out that even in the case of so-called uniform particle size beds there actually exists a narrow range of sizes, and hence there must also exist a critical region of fluidization velocities rather than just one critical velocity; obviously a formula assuming the presence of a strictly defined critical velocity for a given bed cannot therefore be exact. In addition, as Todes and co-workers [735] have pointed out, such factors as the size and shape distributions of the particles must have an effect on the onset of fluidization, and in design practice it is not possible to allow for such factors. Therefore it is quite permissible and convenient in practice to use simple approximate relationships for calculating $w_{1.s}$, in particular, the interpolation formula (1-25) of Todes and co-workers and the simplified formula (1-17) of Leva et al. (in the laminar region).

1.6 The Onset of Fluidization in Special Cases

1. Fluidization of moving beds

In the general case the onset of fluidization is determined by the velocity of gas flow relative to the bed particles rather than relative to the vessel walls. Thus the limit of stability will be reached for moving beds at the same value of the relative velocity as in the case of stationary beds. However, for moving beds the critical velocity $w_{1.s.m.b}$ is usually calculated with respect to the stationary coordinate axis. Consequently, $w_{1.s.m.b}$ differs from the critical fluidization velocity of a stationary bed $w_{1.s}$ by the value of the vertical velocity of the bed w_M:

$$w_{1.s.m.b} = w_{1.s} \pm w_M. \tag{1-33}$$

Here the plus sign in front of the arithmetic value of the bed velocity w_M refers to ascending motion of the fluidized bed, and the minus sign refers to descending motion. An ascending fluidized bed can be visualized as shown in Fig. 1-18a; the material is supplied continuously through a sealed feeder into the bottom of the bed on the gas distributing screen and allowed to overflow freely from the upper end of the tube. A descending fluidized bed is shown schematically in Fig. 1-18b; it can be visualized as a feed of material at the top and removal of it through a discharge part from the lower part of the bed into a sealed space.

Various aspects of moving fluidized systems have been considered in many papers and in the patent literature [350, 490, 511, 677, 692, 707, 832, 847, 1000].

The efficiency of raising material by using an ascending fluidized bed and its hydraulic resistance are of practical interest.

It is known that in principle there is no significant difference between raising material in a fluidized bed and pneumatic conveying (a suspended bed): in both cases the particles of material are suspended. The boundary between the two forms of transport can be logically traced to changes in the nature of fall of the particles, and it is assumed that the fluidized state ends and simple pneumatic transport begins when the transition from dense phase fall to practically free

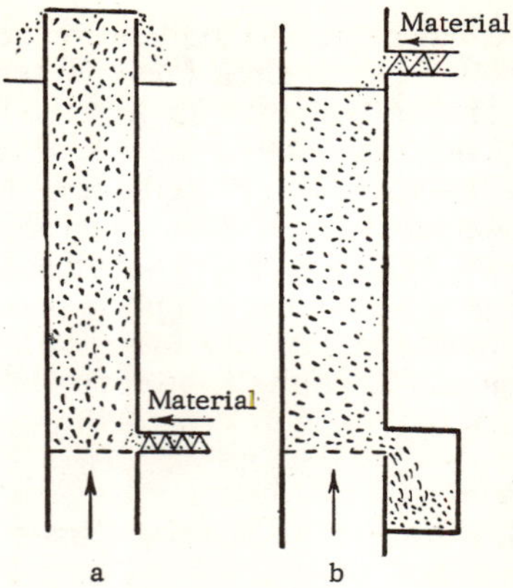

FIG. 1.18. Arrangements for moving fluidized beds.
a) ascending; b) descending.

fall occurs, i.e., at a fluid velocity equal to or exceeding the velocity of free fall of individual particles in an infinitely large space. It must be noted, however, that in gas fluidization there is no smooth transition from a dense ascending fluidized bed to pneumatic transport, since porosities of about 0.7-0.95 correspond to a zone of unstable fluidization (pulsation).

From energy considerations, the use of fluidization for lifting material vertically is not always very efficient compared with pneumatic transport because of the rather high resistances of stable fluidized beds at high particle concentrations. Ciborowski [677] noted that vertical transport in fluidized beds is most suitable for raising large quantities of material through small distances. This was considered in greater detail by Reboux [511], who also derived an expression for the efficiency of this type of transport. However, the expression given by Reboux

$$\eta = w_M / w_f, \tag{1-34}$$

where w_M is the linear velocity of the material, is fundamentally incorrect. Reboux made some simplifying assumptions in the derivation, such as neglecting energy loss by friction of material and gas at the walls, and energy loss in accelerating the material, but in the expression for the pressure drop in a fluidized bed moving upward, it is also necessary to introduce a term corresponding to the work done in lifting. This was not done by Reboux or other investigators, such as Struve, Lapidus, and Elgin [847], who have also made special studies of vertically moving fluidized systems. The pressure drop required for suspending particles in a fluidized bed does not include the energy required for raising them during the bulk upward motion of the bed as a whole.

The expression sometimes used for the "consumption of energy in suspension" is not applicable since this is essentially, as mentioned before, the consumption of energy in overcoming the resistance to gas flow in the "channels" between the particles (or to the gas flow around the particles). The validity of having to take into account the additional pressure drop ΔP_{lift} in raising the material in an ascending fluidized bed is quite clear for the case when the fluidization of the material is poorly developed and there is still no appreciable agitation of the particles. In principle this case is the most economical for transporting material using a fluidized bed, since only a small flow rate of the fluidizing medium is required.

The situation is much more complicated in the case of high velocities, when intensive mixing and particle circulation occur in the fluidized bed. However, even in this case, which is less suited for the transport of material, there is no basis for assuming that the pressure drop through a bed which is not moving upward will be practically the same as that in an ascending fluidized bed for any rate of lifting of the material. The experimental data supposedly confirming such an equality [847] refer to laboratory investigations in which there are obviously comparatively small rates of upward movement of the material, so that the lifting work is insignificant compared to the work of fluidization. This also means that the efficiency of such transport is very small under the experimental conditions.

Consider an expression for the efficiency of transport in an ascending fluidized bed. The power usefully consumed in lifting the material is

$$N_{lift} = F_{sect} w_M \gamma_{bed} l, \tag{1-35}$$

where F_{sect} = cross-sectional area of bed;
l = height of lift, equal, for example, to the bed height (see Fig. 1-18a);
w_M = linear velocity of rising material;
γ_{bed} = bulk density of bed.

The power required for fluidizing a bed of height l is

$$N_{fl} = F_{sect} w_f \gamma_{bed} l. \tag{1-36}$$

Neglecting, as is customary [511, 847], the consumption of energy due to friction of the gas and material against the walls and the energy loss due to acceleration, the efficiency of raising a material by means of a fluidized bed will be

$$\eta = \frac{N_{lift}}{N_{fl} + N_{lift}} = \frac{w_M}{w_f + w_M}. \tag{1-37}$$

Obviously the expression $\eta = w_M / w_f$ derived by Reboux is valid only in the special case when $w_M \ll w_f$, which occurs when $\eta \ll 1$, i.e., when transport by this means is energetically uneconomic.

Thus transport by means of an ascending fluidized bed is more economically feasible at high velocities of material along tubes of relatively small diameter. In calculating the efficiency at particularly high material velocities it may become necessary to allow for the loss in kinetic energy of the material when it leaves (overflows from) the transport line.

This loss of power is given as:

$$N_M = \frac{G_{M.\,sec} w_M^2}{2g} = \frac{w_M^3 F_{sect} \gamma_{bed}}{2g} \tag{1-38}$$

and hence an exact expression for the efficiency is

$$\eta = \frac{N_{lift}}{N_{fl} + N_{lift} + N_M} = \frac{w_M}{w_f + w_M + w_M^3/2gl}. \tag{1-39}$$

The velocity of an ascending fluidized bed is controlled by varying the solids feed rate through the lower feed device (see Fig. 1-18a). According to Reboux [511] it is interesting that in the absence of a grid in the tube below the feeder, it would generally be impossible to obtain an ascending fluidized bed; the material supplied by the feeder would simply fall, though at a somewhat reduced rate due to the retardation caused by the countercurrent, ascending gas stream. The feed device cannot supply material in the form of a dense, flat-bottomed piston filling the whole cross section of the tube. Thus the addition of a screen, which in principle does not experience the pressure of the material in the working section, leads to a radical change in the nature of the motion of the material.

Solids transport in the form of ascending fluidized beds as opposed to pneumatic conveying is used in certain cases because of such other advantages as the small gas flow rate, less erosion of the transport line, lower attrition and lower entrainment of the particulate material. In addition, it is possible to carry out various reactions with ascending fluidized beds in the transport line itself; i.e., it is possible to build reactors in the form of transport lines [832].

On a phase diagram the line for an ascending fluidized bed falls in quadrant I. The pressure drop $\Delta P/H$ per unit height of an ascending fluidized bed decreases in the usual way as the velocity increases, since the bed porosity simultaneously increases.

The pressure drop measured experimentally in ascending fluidized beds (not the calculated value determined from the bulk weight of the bed) is an overall value, which includes such items as the pressure drop for lifting and accelerating the material, as well as the opposing effect of gas ejection by the ascending stream of material.

At a constant fluid velocity w_f (referred to the stationary walls of the tube) the density of the bed increases as the feed rate of material increases (i.e., the velocity of the ascending fluidized bed will increase), since the gas velocity relative to the bed is thereby decreased.

2. Conical equipment

When fluidization occurs in a conical apparatus tapering downward, there is a considerable pressure peak at the limit of stability [1033]. This peak is much larger than at the onset of fluidization in an apparatus with a constant cross section and is due to the conical form of the bed. In some experiments of Gel'perin et al. [1033] the pressure drop before the onset of fluidization was two to three times greater than the value established after the limit of stability. The authors explained this phenomenon by the observation that in a conical apparatus the bed passes into the fluidized state only after fluidization of its upper portions. Since at this time the gas velocities in the lower and middle parts of the bed are considerably greater than the critical fluidization velocities, the pressure drop in the stationary bed portions increases rapidly compared with that at the limit of stability in a cylindrical bed of the same height. Immediately beyond the limit of stability the pressure drop falls to approximately the usual theoretical value, equal to the product of the bed density and its height, regardless of the bed configuration, in the same way that the hydrostatic pressure on the bottom of a vessel does not depend on its shape.

As the fluid velocity is further increased the pressure drop does not remain constant but, in contrast to the cylindrical bed, begins to fall. Gel'perin et al. noted that this is because in a conical apparatus, as the porosity increases, the bed height increases more slowly than its volume, so that the quantity $H(1-m)$ decreases. A similar movement of the lower limit of a conical fluidized bed occurs as the fluid velocity increases.

In deriving a theoretical formula for the pressure peak $\Delta\pi$ (Fig. 1-19), the authors of [1033] made the assumption that the fluidizing gas was uniformly distributed over the bed cross section at every height in the bed, and that the motion of the gas in the bed was laminar. They arrived at the expression

$$\Delta\pi_{\text{theor}} = \frac{d_0}{2\tan\frac{\alpha}{2}}\left(\frac{D}{d_0} - 1\right)\left(\frac{Kw_{1.s}D}{d_0} - \gamma_{\text{sta}}\right), \quad (1\text{-}40)$$

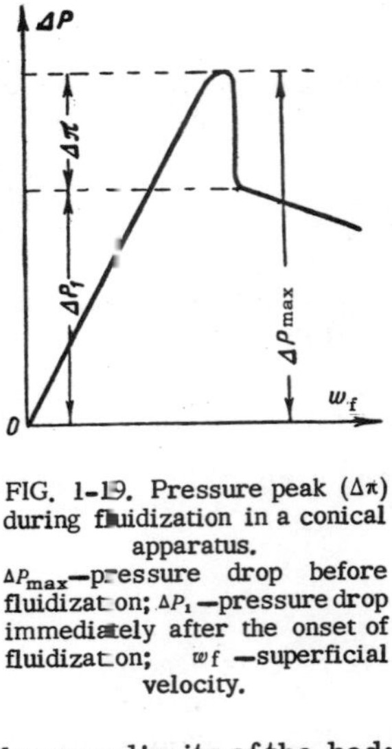

FIG. 1-19. Pressure peak ($\Delta\pi$) during fluidization in a conical apparatus. ΔP_{max}—pressure drop before fluidization; ΔP_1—pressure drop immediately after the onset of fluidization; w_f—superficial velocity.

where K = proportionality coefficient in the expression relating the resistance of the fixed bed and the fluid velocity, $\Delta P = K w_f H_{sta}$;

H_{sta} and γ_{sta} = height and bulk density of the stationary bed;

d_0 and D = diameters of the lower and upper limits of the bed;

α = cone apex angle.

Considering that the assumption of uniform distribution of the gas over the conical bed, made in the derivation of Eq. (1-40), is far from fulfilled in practice, the authors have not proposed this as a design equation, but only as a basis for choosing the type of empirical formula. For this purpose they proposed

$$\Delta\pi = f\left(D/d_0, \tan\frac{\alpha}{2}\right). \quad (1\text{-}41)$$

It follows from Eq. (1-40) that $\Delta\pi$ can vary even when D/d_0 and α are unchanged due to differences in the value of K. It is also probable that the empirical formula

$$\Delta\pi_{\text{exptl}} = 2{,}53\left(\frac{D}{d_0}\right)^{2,55}\left(\tan\frac{\alpha}{2}\right)^{-1,19}, \tag{1-42}$$

which was proposed by these authors, and confirmed by them over the ranges α = 10-60° and D/d_0 = 1-30-6.77, is valid only within the conditions under which the experiments were carried out (with sand).

A more general expression must be used as the basis for setting up an empirical formula. Dividing both sides of (1-40) by ΔP (i.e., expressing the size of the pressure peak in relative units, comparing it with the theoretical pressure drop in a fluidized bed of any configuration immediately after the stability limit) and taking the following substitutions in the right-hand part of (1-40):

$$D = 2H_{\text{sta}}\tan\frac{\alpha}{2} + d_0;\quad \Delta P = \gamma_{\text{sta}} H_{\text{sta}};$$

yields

$$(\Delta\pi/\Delta P)_{\text{theor}} = \left(2H_{\text{sta}}\tan\frac{\alpha}{2}\right)\Big/d_0, \tag{1-43}$$

and the relative value of the maximum pressure drop in a conical bed before fluidization as

$$\left(\frac{\Delta P_{\max}}{\Delta P}\right)_{\text{theor}} = \frac{\Delta\pi + \Delta P}{\Delta P} = \frac{2H_{\text{sta}}\tan\frac{\alpha}{2}}{d_0} + 1. \tag{1-44}$$

Accepting the assumption of Graf that for a bed of fine particles fluidized by gas

$$w_{1.s} \sim d_{\gamma m}^2,$$

and also taking into account that $K = C\mu/2d$, we get

$$\left(\frac{\Delta P_{\max}}{\Delta P}\right)_{\text{theor}} \sim \frac{C\mu D}{d_0}$$

where C is a coefficient taking into account the shape of the particles.

From these considerations, the following statement can serve as the basis for an empirical formula:

$$\frac{\Delta P_{\max}}{\Delta P} = f(\alpha,\ d_0,\ H_{\text{sta}}). \tag{1-45}$$

Note that in almost all cases the actual values of $\Delta\pi$, and hence of $\Delta P_{\max}/\Delta P$, must be lower than the theoretical, since nonuniformity of the gas distribution over the conical bed will lead to the fluidization of a more or less narrow column of material before the mean velocity at the exit cross section reaches the critical value $w_{1.s}$, i.e., at smaller hydrodynamic resistances of the fluidized bed. A reservation must be made concerning the opposite state of affairs when in practice the bed is close to the simple cylindrical bed. For this case it must be expected that the experimental $\Delta P_{max}/\Delta P$ will be larger than $(\Delta P_{max}/\Delta P)_{theor}$(close to unity), since in the theoretical formula no allowance is made for the pressure peak caused by the adherence of the particles to the walls, which is observed even in a true cylindrical bed (and which is also noticeable for repeated fluidization of materials that shrink). It is therefore possible to justify the presence of tan $(\alpha/2)$ in (1-41), and to introduce this argument into the right-hand side of (1-45):

$$\frac{\Delta P_{max}}{\Delta P} = f(C, \mu, D/d_0, \tan\frac{\alpha}{2}). \qquad (1\text{-}45a)$$

In their experiments on fluidization in conical beds, Gel'perin et al. observed that the free surface of the bed assumes a concave shape; the surface of the material in the peripheral stagnant zone is higher than in the "fluidized" core. They attribute this to the fact that in a conical apparatus the pressure drop is smaller than that calculated from the bulk density and bed height at the walls. Obviously it is possible to form gas channels lower down in the central part of the bed.

3. Fluidization with poor initial gas distribution. Grid requirements

The transition from a stationary bed to the fluidized state can be complicated by poor initial gas distribution. For instance, if in a simple cylindrical bed the gas passes through a small number of large openings in a grid with a small free area, it is obvious that near the openings the velocity field will be unable to equalize itself, and the velocity above the openings will be much greater than the mean value based on the whole cross section of the bed. As a result (just as in the case of the conical bed) the minimum fluidization velocity in the upper part of the bed will be reached at pressure drops in the lower part of the bed that are larger than $\Delta P = (\gamma_M - \gamma_m)(1 - m)H$. As a result of the nonuniformity of the gas distribution, fluidization will occur only within narrow cones. As soon as this occurs, and still at the increased pressure, the particles from the central zones of the cones will be thrown upward to the free bed surface, forming little mounts; the pressure will then fall abruptly, since the gas will pass freely through the "channels" formed in this way. In contrast to the conical equipment, the mounds of ejected material will not subside rapidly, forming circulation patterns, since here they are resting firmly on a foundation of stationary particles not in the immediate neighborhood of the walls.

Thus the rapid attainment of the limit of stability in the narrow zones above the openings leads to entrainment of a certain part of the material and formation of channels, but not to fluidization of the whole bed. Such channels are not formed, or do not persist, in deep beds. They pass only through the lower parts of the bed, while the upper parts are fluidized satisfactorily. Generally speaking, in the latter case ΔP is again lower than the theoretical value, since the lower part of the bed is penetrated by channels. Channels do not usually develop at small values of the ratio γ_M/γ_m of the densities of the material and the flowing medium (for instance, in fluidization by liquids). However, an important group of fluidized bed processes are carried out at large values of γ_m/γ_m, and so, bearing in mind the consumption of energy in overcoming the hydraulic resistance of the bed, it is advantageous to operate with comparatively shallow beds (i.e., beds particularly subject to channel formation). As a result, the choice of the gas-distributing device is of particular practical importance.

Many investigators, for instance, Wicke and Hedden [601] and Fedorov [170], have paid special attention to the problem of gas distribution in fluidized bed equipment. Wicke and Hedden assumed that in order to obtain a satisfactorily uniform gas distribution the grain size of the material used for making the porous inlet gas-distributing device (where the gas passes through the pores) should be only about 1/10 of the diameter of the particles making up the fluidized bed. They based this conclusion on laboratory experiments, in which they used filter papers as the screen supporting the fluidized beds. However, experiments with milled peat have shown that this requirement is excessive if the "degree of suspension of the bed" given by

$$R = \frac{\Delta P}{(\gamma_M - \gamma_m)(1-m)H} = 0.85\text{–}1.0^*$$

(where ΔP is the hydraulic resistance of the fluidized bed, kgf/m^2) is satisfied.

Fedorov [170] mentions that "if the (gas-distributing) screen resistance is equal to or greater than the resistance of the bed of material, fluidization occurs uniformly regardless of the thickness of the bed, the particle size, or the area of the vessel." However, this condition is not always practical. In the case of sufficiently small openings in the distribution grid and sufficiently large initial heights of the fluidized bed, the requirement of such a large screen resistance is quite excessive. Thus, Wicke and Hedden [601] observed uniform fluidization, without channels and bubbles, with grinding powder ($d = 50\mu$) on a "grid" of filter paper. The bed had a height of 20 cm and a resistance of 330 mm water, while the pressure drop across the "grid" was only 1% of this value. In another case, using a screen with a small number of comparatively large openings, its resistance was higher than the bed resistance due to the small free cross section, but this did not ensure satisfactory fluidization, since with the necessary flow rate of gas the velocity in the openings was very large and the resulting gas jets lead directly to channels in the bed. Obviously here channel formation sets in very easily in the case of thin beds.

Wicke and Hedden proposed the condition that "the dimensions of the particles and filaments of the material" used for constructing the inlet-distribution device should be about 1/10 of the diameter of the particles of the fluidized bed, but this is incorrect. For example, if we take peat particles (d = 5–7 mm) and make a bed on a wire screen with openings of 0.5 mm, the Wicke-Hedden condition would be satisfied, but fluidization is poor. Here blow-through of the gas is reduced, since for each particle that forms a "ceiling" gap between the bed particles and the screen, there will be a large number of openings, which are small compared with the particle, but very densely distributed; as a result the force of the total pressure of all these jets

*The significance of such conditions will be dealt with later.

on the "ceiling" should be sufficient to expand the gap. Such enlarged gaps—imperfect places in the bed—continuously appear above the grid supporting the fluidized bed (or above any other initial gas distributing device). The particles immediately above the grid openings experience greater pressure forces from the jets of gas than the neighboring particles lying above blind spots, and as a result of the familiar mobility possessed by the bed after passing the limit of stability, they will ascend in the bed, increasing the cavities.

Zabrodsky [642] came to the conclusion that the correct approach to the construction of a grid is to ensure its limiting action on the gas jets that pass through into randomly exposed positions above the grid, or into places where random initial packing caused defects in the bed. In doing this, two important conditions must be satisfied simultaneously.

The first condition, opposing channel formation, requires a device with such a fine grid that there are not only openings opposite to each particle lying on it, but also opposite an average sample of the blank surface. In this way, no single defective position, even having a width equal only to the particle diameter, can exist under conditions favorable to channel formation. Thus the spacing of the openings should not exceed the diameter of the particles lying directly on the grid. Further decrease in the spacing and opening diameter is obviously unnecessary.

The second condition is that the resistance of the grid at the operating fluid velocity should be sufficiently large that uncovering parts of it cannot lead to a sharp decrease in the gas velocity in the parts covered with material and reduce it below the minimum necessary for fluidization. There will then be no stable "channels," since their walls will "erode."

Consider the values of the screen resistance that also satisfies the first condition, which is necessary in order to ensure that when almost the whole grid is uncovered (for example, all the openings except for one) the material above the openings covered by it will remain in the fluidized state and spread over the grid.

The parallel nature of the gas-flow paths through the exposed openings and those closed by material leads to the expression

$$\Delta P'_{\text{grid}} = \Delta P''_{\text{grid}} + \Delta P_{\text{fb}} , \tag{1-46}$$

where $\Delta P'_{\text{grid}}$ = hydraulic resistance of the screen in the exposed positions;
$\Delta P''_{\text{grid}}$ = hydraulic resistance of the screen in the positions covered by material;
ΔP_{fb} = hydraulic resistance of the fluidized bed as such above the openings covered by the material.

The rate of efflux of the gas from the free (uncovered) openings is

$$w' = \varphi_1 \sqrt{\frac{2g\Delta P'_{\text{grid}}}{\gamma_g \xi'}} , \tag{1-47}$$

where φ_1 = discharge coefficient;
γ_g = density of the gas;
ξ' = resistance coefficient.

Analogously, for the efflux of gas from the openings covered by the bed:

$$w'' = \varphi_2 \sqrt{\frac{2g\Delta P''_{grid}}{\gamma_g \xi''}}. \tag{1-47a}$$

The ratio of the gas velocities through the covered and exposed parts of the grid is

$$\frac{w_f''}{w'_f} = \frac{w''}{w'} = \frac{\varphi_2}{\varphi_1}\sqrt{\frac{\xi'}{\xi''}\left(\frac{\Delta P'_{grid} - \Delta P_{fb}}{\Delta P_{grid}}\right)},$$

or

$$\frac{w''_f}{w'_f} = \frac{\varphi_2}{\varphi_1}\sqrt{\frac{\xi'}{\xi''}\left(1 - \frac{\Delta P_{fb}}{\Delta P'_{grid}}\right)}. \tag{1-48}$$

The velocity increases n times as the grid is uncovered, i.e.,

$$w'_f = nw_f, \tag{1-49}$$

where $n > 1$.*

Hence, calculating the grid resistance by quadrature,

$$\Delta P'_{grid} = n^2 \Delta P_{grid} \tag{1-50}$$

where ΔP_{grid} is the grid resistance at the normal operating velocity.

Denoting the ratio of the grid resistance at the operating velocity to the resistance of the fluidized bed itself as $\Delta P_{grid}/\Delta P_{fb} = x$, consider the limiting case for the the existence of fluidization, i.e., the condition when $w_f'' = w_{1.s}$, where $w_{1.s}$ is the velocity at the limit of stability.

Thus for the case of interest

$$\frac{w_{1.s}}{nw_f} = \frac{\varphi_2}{\varphi_1}\sqrt{\left(1 - \frac{1}{n^2 x}\right)\frac{\xi'}{\xi''}},$$

whence

$$x = \frac{1}{n^2 - \left(\frac{w_{1.s}\varphi_1}{w_f \varphi_2}\right)^2 \frac{\xi''}{\xi'}}. \tag{1-51}$$

In the case of thin fluidized beds, where the bed resistance is only a small fraction of the total resistance of the system,

$$w_f \approx w'_f, \text{ i.e., } n \approx 1.$$

Furthermore, in the fluidization of fine particles with fluid velocity $w_f \gg w_{1.s}$, fluidization above the openings can only be achieved when

*For the centrifugal types of air blowers usually used for equipment with fluidized beds. In the case of piston-type air blowers and similar types, $n = 1$.

$x \approx 1$; i.e., the hydraulic resistance of the screen should be no smaller than the resistance of the fluidized bed itself. Consequently, at $R = 1$

$$\Delta P_{\text{grid}} \geqslant (\gamma_{\text{M}} - \gamma_{\text{m}})(1 - m)H. \tag{1-52}$$

This condition ensures spreading of the material over a fine grid as long as the angle of repose is smaller than 90°. Finally, in order to calculate the minimum value of ΔP_{grid} from Eq. (1-52), it is necessary to substitute not any value of H, but some value $H_{\min}$ at which the condition of constrained flow around the particles is still observed, as a result of which actually $w_{1.s} \ll w_{\text{f}}$. The quantity $H_{\min}$ must still be defined more accurately, but according to experiments described in [642], it can be taken to be of the order of 40 mm for air-dried milled peat. In this case, the required working resistance of a fine grid is

$$\Delta P_{\text{grid}} = 16 \text{ mm water.}$$

Note that with centrifugal blowers the required value of x decreases in the case of thick layers of granular materials, particularly when the resistance of the fluidized bed itself makes up an important part of the total resistance of the system. In this case there is an increase in the velocity when the screen is exposed, i.e., $n > 1$.

For instance, when $n = 1.4$ and $w_{1.s} \ll w_{\text{f}}$, $x = 0.5$ rather than 1. It is possible that under the conditions of Wicke and Hedden's experiments the values of n were even higher, which would lead to the requirement of very small values of x for homogeneous fluidization.

Equation (1-51) indicates that for any given fluidized bed a change to smaller values of the operating velocity w_{f} leads to an increase in the relative value of the grid resistance x, since w_{f} tends toward $w_{1.s}$.

Experiments have shown that it is possible to have satisfactory fluidization quality ($R > 0.8$) even when there are considerable deviations from satisfying the first condition (the requirement that the spacing of the openings should not exceed the particle dimensions).

The proposal that the main function of the grid should, and must be, to limit any random local breakthrough of the gas at small operating resistances of the grid has served as the basis for attempts to obtain satisfactory fluidization of milled peat on such grids with small operating resistances and has given favorable results [642, 1045].

Chapter 2

The Development of Homogeneous (Particulate) and Nonhomogeneous (Aggregative) Fluidization

2.1 Homogeneous and "Two-Phase" Fluidization

As the superficial fluid velocity increases the fluidized bed expands. It may behave like an ideal fluidized bed, in which the distribution of material in the stream is always uniform at all superficial velocities, from that for minimum fluidization to that at which particles are entrained. The actual velocity of flow around a suspended particle in an expanded bed cannot remain constant but will increase. The fact is that the velocity of fall of a particle in a fluidized bed is a variable quantity [50], which, as we know, varies from the velocity corresponding to the limit of bed stability to the velocity of free fall of individual particles in an open tube. This upper limit is reached at a porosity of unity, i.e., when the fluidized bed ceases to exist and particles are carried off by the fluid stream. Fluidization with a uniform particle distribution is termed homogeneous, particulate or uniform. In practice, uniform fluidization can be observed experimentally with a liquid, for instance water, as the fluidizing medium.

Usually there is quite a different picture in the gas fluidization of solid particles; part of the gas breaks through the bed in the form of channels or bubbles, almost depriving the bulk of the solid particles of gas. This is termed nonhomogeneous, aggregative or nonuniform fluidization. Visual observations, motion-picture filming and other methods of measuring the size of bubbles even remote from the tube walls show that the bubbles increase in size as they rise through the bed. On passing through the surface of the fluidized bed, the bubbles raise a thin layer of particles and "burst," scattering these particles and demonstrating that there is an excess pressure within the bubbles [966].

Many investigations have been concerned with the study of nonuniform fluidization, particularly those by Orochko [107], Wilhelm and Kwauk [604], Morse [459], Wicke and Hedden [604], Toomey and Johnstone [567], Trawinski [573], Davidson et al. [966], Baumgarten and Pigford [1099], and Wace and Burnett [1234]. Observations of gas-fluidized beds have been updated by Obryadchikov [106] and Orochko [107]. Later, in more developed form, Morse [459] and Toomey and Johnstone [567] proposed the hypothesis of two-phase fluidization.

According to this hypothesis, a fluidized bed consists of two different parts or "phases." The first phase—the "dense," "continuous" or "aggregative" phase—is formed of particles of material that are in a state of minimal fluidization and that are uniformly distributed in the fraction of the gas stream supporting them. The second—"discontinuous"—phase consists of pure gas, and exists in the form of bubbles or channels, or sometimes pistons with diameters equal to the bed diameter. Thus, according to this hypothesis, all the "excess' gas above the minimum required for recipient fluidization inevitably breaks through the bed, hardly contacting the particles. Since the inlet gas-distributing device plays no part in this hypothesis, certain doubts have arisen as to its usefulness [744]. However, if we consider a sufficiently deep fluidized bed and disregard its lower region (i.e., the region where the screen has an effect), the two-phase theory gives a valid representation of the tendency to develop nonuniform fluidization, which is inherent in a bed with a random initial particle packing.

Recent experimental data generally confirm the two-phase theory of gas-fluidized beds. Thus, for example, Baumgarten and Pigford [1099] measured the instantaneous local densities of fluidized beds of glass spheres (diameters 24–234 μ) by means of γ-ray absorption and found that at a considerable height above the inlet gas-distributing device the dense phase reached a concentration corresponding to that at the minimum fluidization conditions. Relative to the density of a settled bed, the concentration of the dense phase of the fluidized bed of glass spheres was 0.92–1.0. In experiments with spherical aluminosilicate cracking catalyst (diameter 41 μ) much poorer agreement was obtained (0.60–0.84). The authors mention that this discrepancy could be partly due to the presence of fine bubbles ($d <$ 18 mm) which were not discernible in the equipment with the catalyst bed; the accuracy of measurement was much greater in the beds of glass spheres. Yasui [716] indicated that within certain limitations his experimental data confirm the theory of two-phase fluidization.

Dotson [970] determined the effects of various factors on fluctuations of the local density (uniformity) of fluidized beds of mixtures of silicon and copper powders (93:7 by weight) in a column of 100-mm diameter by means of a complex statistical experiment using a capacitance probe. A mixture of nitrogen and carbon dioxide (18:15 by volume) was used as the fluidizing agent. The effects of five factors were investigated: 1) the particle diameter (fractions with $d < 250\,\mu$ and $d < 74\,\mu$); 2) the type of gas-distributing device (porous plate or grid with openings of $\varnothing$ = 1.8 mm and an open area cross section of 23%); 3) the overall bed height (1520 and 2740 mm); 4) the height above the grid at which the bed density was measured (457 and 1220 mm); and 5) the superficial fluidizing velocity (w_f = 1152, 3.04, 9.12, 15.2, 21.3 cm/sec). The superficial velocity had the largest effect on the uniformity of fluidization; it appeared that a large part of the gas above that required for minimum fluidization passed through the bed in the form of bubbles. The type of inlet gas-distributing device had an important effect only in the lower part of the bed (up to 450 mm from the screen).

No clear picture has as yet been proposed for the mechanism by which fluidized beds tend toward nonuniformity or two-phase fluidization with the formation of dense phases having porosities close to the porosity of a bed during minimum fluidization. Toomey and Johnstone [567], for example, have not even offered to explain such fundamental facts as the presence of two-phase fluidization in beds fluidized with gases and of single-phase fluidization of the same materials when liquids are used.

The work of Morse [459] is quite different; this is an early but very detailed investigation of nonuniformity in fluidization. Morse analyzed the differences between fluidization with liquids and gases and arrived at the valid conclusion that the tendency toward nonuniform fluidization increases with increase in the ratio $(\rho_M - \rho_m)/\mu$, where ρ_M is the density of the material, and ρ_m and μ are the density and dynamic viscosity of the medium. Morse did not give any conclusive physical explanation of why this kind of relationship should prevail but derived it from a rather formal application of the Carman-Kozeny equation (for flow through a fixed bed) to the determination of the "velocity of separation of liquid from the particles," which is a rather nebulous concept.

Wilhelm and Kwauk [604] assumed that nonuniform fluidization is caused by the appearance of hydrodynamic Bernoulli forces but showed only an empirically straight-line relationship between the degree of nonuniformity and the value of the Froude criterion $Fr = w_f^2/gd$. In considering the development of nonuniformities in fluidization, Wicke and Hedden [601] paid particular attention to the action of the dynamic pressure of the jets flowing into the forming bubbles and channels, the "defective spots" of the bed, i.e., cavities of dimensions larger than the spaces between neighboring particles. The presence of defective spots is explained by nonuniformities in the initial packing of the material. The dynamic pressure of the jets lifts the particles in the "ceiling" of the cavity and thus elongates the cavity. Wicke and Hedden mention that for a cavity to persist and grow the mobility of the particles must be somewhat retarded, or the nonuniformity will be filled in rapidly (particularly in beds fluidized with liquids of high lubricating power rather than gases). However, these authors made the dubious assumption that particles fill the cavities under the action of simple hydrodynamic drainage, whereas there is usually an excess static pressure in the bubble.

Wicke and Hedden's view that particles entrained from the bottom of the cavity bombard the walls of the bubble also appears to be unfounded. Such phenomena can be observed only in certain cases, as in open channels through which the gas moves with a high velocity, greater than the mean-superficial velocity and greater than the free-fall velocity of the particles.

As a characteristic for the tendency toward formation of fluidization nonuniformities, Wicke and Hedden introduced the coefficient φ, which represents a measure of the ratio of the dynamic pressure on the particles at the limit of stability of the bed to the gravity pressure (the apparent weight of the particles divided by their cross-sectional area). Bed uniformity should deteriorate as φ increases. However, when Wicke and Hedden calculated φ for the fluidization of the same particles with air and with water, they found that the difference was small ($\varphi_{air}/\varphi_{water} \approx 1.55$), and mentioned that the interactions between the particles, which are not taken into account in deriving φ, may be of great importance in beds fluidized by gases. These authors also observed cohesion between particles and formation of channels, even in beds fluidized by water, when the particles were sharp-angled (splinters). Thus the coefficient φ retains its usefulness as a measure of the tendency towards nonuniformity only when smooth particles are distributed in a liquid stream.

To examine the causes of fluidization nonuniformity, Trawinski [573] took into consideration the frictional forces between particles, starting with the concept that bubbles appear in the bed as soon as the fluid velocity is high enough to cause the interparticle forces—the surface friction forces—to hinder the uniform expansion of the bed and the required quantity of fluidizing medium can no longer pass through the interstices between the particles. In Trawinski's opinion, bubbles have a stable shell of particles and float in the fluidized bed flowing past this shell, just as in a viscous liquid; the bubbles move under the action of Archimedean forces and obey Stokes's law. According to Trawinski, the bubble dimensions are determined by the residence time at the point of formation (on the gas-distributing grid), which in turn depends on the friction forces that must be overcome when the bubble separates from the grid. The excess static pressure (according to Bernoulli's law) inside the bubble causes it to expand as long as it remains on the grid and is fed through the grid openings. The bubbles break away when they acquire sufficient buoyancy to overcome the friction between the particles.

Some of Trawinski's assumptions have not been confirmed by the experimental data of other investigators. For example, bubbles not only are "fed" with medium at the point of formation on the grid, but continue to grow while rising through the bed. The bubbles rise not only due to Archimedes forces but also as a result of the collapse of the "ceiling" of the bubble to the "floor." Going to the other extreme, Yasui and Johanson [856] and Solchani [1177] assume that the rate of collapse of the particles surrounding the bubble is the main factor determining its rate of rise.

Parker and Stevens [999] and Deemter [1196] emphasize that in a fluidized bed there is a continuous process of formation and breakup of particle aggregates. Under stable fluidization conditions a dynamic equilibrium is set up in which the rates of aggregation and breakup are equal, and a certain fraction of the material exists in the form of aggregates and the rest in the form of separate particles.

It can be assumed that there are two opposite and competing tendencies in fluidized beds: first, the tendency toward aggregation under the action of hydrodynamic forces and the forces of attraction between the particles (molecular, electrostatic, etc.); and, second, the tendency toward filling in the cavities formed as a result of particle movement. The hydrodynamic factors, which lead to an expansion of the initial "defective spots," include the dynamic pressure of the jet entering such a spot and the excess static pressure arising within the bubble as a result of the decrease in the velocity of the medium and the transformation of the dynamic pressure into a static pressure. The dynamic pressure of the jet can play an extremely important part in the development of nonuniformities only in the case of a poor gas-distributing device (for instance, a perforated grid with a small open cross section), when the velocities of the gas jets leaving the grid openings may be many times larger than the superficial fluidizing velocity, and will lead to the development of channels (cavities extending upward, penetrating right through the whole fluidized bed or only its lower part), for instance, by the scheme of Wicke and Hedden [601]. In most cases the most important role of the excess static pressure in the cavities, which have expandable, movable walls, is to increase the size of the cavities and the density of the surrounding parts of the bed.

The recent experimental data of Wace and Burnett [1234] appear at first to contradict the view that there is excess pressure inside the bubbles. Experiments have shown that streaks of tracer gas deviate into the side of passing bubbles (Fig. 2-1), which would indicate a negative pressure inside the bubble. However, a more detailed consideration of the data shows that such a deflection can be caused by the existence of lower pressure in the wake below the bubble. Higher up, on a level with the bubble walls, the deflection is reversed—away from the bubble. As visual observations have shown, the tracer gas [1234] enters the bubble only through the "floor," while at the walls and "ceiling" it passes out into the dense phase, which is in complete agreement with the concept of the transformation of the dynamic pressure of the gas jets entering the bubble into a static pressure, so that there exists an excess pressure inside the bubble.

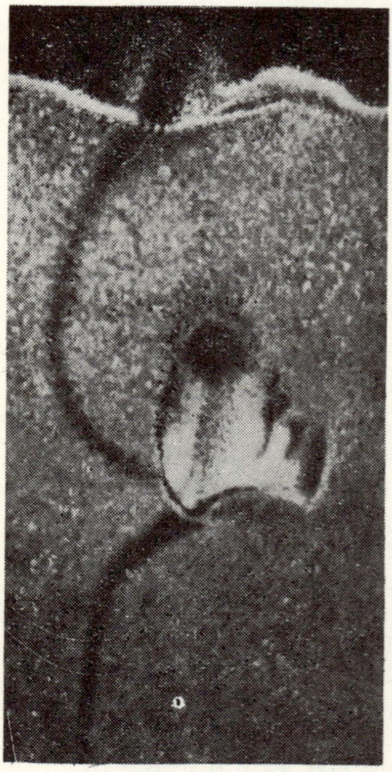

FIG. 2-1. Deflection of a colored gas jet by a bubble rising in a fluidized bed [1234].

The random initial particle distribution, the nonuniformity of the flow around the particles, and the growth and rise of bubbles cause a random and pulsating motion of the particles in a fluidized bed and intensive mixing of the particles. Pulsations in the stream of medium can equally well lead to such motion of the particles. The particle mixing caused mainly by nonuniformities in the bed produces an opposite tendency toward the breakup of aggregates and the establishment of uniform fluidization, since in ideal mixing all the particles are uniformly distributed in the bed like molecules in a gas.

"Gravitational mixing" is also important; this involves the collapse of particles from the upper surfaces and walls of the bubbles. When particles or their aggregates fall away for any distance inside a bubble, the flow around them ceases to be as constrained as in the main mass of the fluidized bed. Correspondingly, the suspension velocity of such particles is larger than the superficial fluidizing velocity, and hence they fall.

Obviously mixing will be facilitated when the particles are highly mobile and there is no friction or cohesion between them.

In principle the dynamic equilibrium between the processes of particle aggregation and breakup (or local increases in bed density) can occur at various levels depending on the degree to which the first or second tendency predominates. Hasset [1204] has even stated that uniform and nonuniform fluidization may be encountered simultaneously in a bed, and only the proportion between them determines the overall type of a given fluidized bed. However, a tendency toward a limiting state of breakthrough (gas eruption) of all the excess quantity of fluidizing medium above that required for minimum fluidization is the usual observation.

A natural limit to the compression of the "dense phase" occurs under the squeezing action of the expanding bubbles; the aggregates reach a constant degree of closeness between neighboring particles. The small fraction of particles moving through the large bubbles at any given time can be neglected.

Most cases of gas fluidization of deep beds are examples of this "pure two-phase" state. This limiting state is not achieved immediately above the gas-distributing device in the lower part of a fluidized bed or in a shallow bed but is approached in the upper parts of deep fluidized beds. In the experiments of Baumgarten and Pigford [1099] the beds were comparatively uniform immediately above the gas distributor, but at a considerable height, significant nonuniformity was observed and the "dense phase" reached concentrations corresponding to that a minimum fluidization. Dotson [970] also mentioned an increasing fluidization nonuniformity at large distances above the grid.

The force F expanding a defective spot of given dimensions at a given degree of conversion of the dynamic pressure into static pressure can be regarded as proportional to the dynamic pressure of the incoming jet. Thus, at superficial velocities close to the limit of stability, or at similar fluidization indices, the aggregating forces F will be proportional to $w^2_{\ell.s}\gamma_m$:

$$F \sim w^2_{\ell.s}\gamma_m.$$

Substitution of Leva's approximate formula (1-17) for the minimum fluidization velocity yields

$$F \sim \frac{d^{3.64}(\gamma_M - \gamma_m)^{1.88}}{\nu^{1.76}\gamma_m^{0.88}} = \frac{d^{3.64}(\gamma_M - \gamma_m)^{1.88}\gamma_m^{0.88}}{\mu^{1.76}}. \tag{2-1}$$

Hence, with other factors equal (d, γ_M), in gas-fluidized beds F will be several times larger than in liquid-fluidized beds. For instance, consider fluidization of a material (γ_M = 2600 kg/m^3) in one case with water and in the other with air at room temperature; the calculated ratio of the F values corresponding to water (F_1) and to air (F_2) is simply

$$F_1/F_2 = \left(\frac{\gamma_{air}}{\gamma_{H_2O}}\right)^{0.88}\left(\frac{\nu_{air}}{\nu_{H_2O}}\right)^{1.76}\left(\frac{\gamma_M - \gamma_{H_2O}}{\gamma_M - \gamma_{air}}\right)^{1.88} \approx 0.127,$$

i.e., the hydrodynamic forces will be many times smaller in a bed fluidized with water than in a bed fluidized with air. The reasons for this are the small apparent weight of the material in water and the low water velocity necessary for fluidizing the bed.

However, the principal cause promoting bed uniformity during water fluidization is the high lubricating ability of water compared with that of a gas; defective spots have no time to develop, but are filled in as a result of the rapid transfer of the highly mobile particles and the incompressible liquid "back-pressure" from other defective spots. Actually, as Wicke [601] has mentioned, if special pains are taken to ensure that the particles will cohere in water (e.g., by causing a bed of needle-shaped particles) it is still possible to have very marked fluidization nonuniformities.

Finally, even gases have some lubricating action. It is evident from the hydrodynamic theory of lubrication that very thin lubricant layers are able to sustain high pressures and prevent solid surfaces from coming into direct contact if the surfaces have no protuberances exceeding the necessary thickness of the lubricant layer. Hence, if it is assumed that the absolute height of the characteristic roughness of the surfaces of the particles decreases as the particles are made finer, the medium can be expected to have a lubricating effect on very fine particles, even in the case of gases. Finally, for very small particle dimensions, aggregation occurs due to the forces of molecular interaction.

This is also the explanation for other experimental observations, such as the way in which the fluidization nonuniformity changes with increase in superficial velocity, γ_M/γ_m and with particle diameter.

As the superficial velocity is increased, the amount of gas passing through in the form of bubbles increases. If in the bed near the limit of stability another form of gas breakthrough occurs, namely, channel formation, this usually changes to bubble formation as the superficial velocity increases. This has been noted by a number of investigators (see, for instance, [601], [744]). This change is portrayed by the following argument: when the gas velocity increases, the walls of the channels collapse and a greater amount of particles is thrown upward, forming mounds on the surface of the bed. The "hydrostatic" pressure of these mounds causes mixing of the bed; during the mixing the channels are broken up and converted into bubbles.* When the first bubbles appear, mixing is increased still further, and the transition to this type of gas breakthrough is stabilized. The fact that increasing bed mixing does not lead to its uniformity is, at least qualitatively, the explanation for the sumultaneous increase in the aggregating forces F.

It is not possible to establish from published experimental data whether there is a change in the phase density (and in which direction) as the superficial velocity is increased, or whether the density remains constant and close to the value of the voidage at minimum fluidization.

It follows from very simple consideration that the nonuniformity of fluidization and the passage of bubbles and slugs cannot be maintained over the entire range of superficial velocities up to the entrainment velocity. The fact is that the rate of bubble rise is limited and, at best, is the sum of two components: the velocity with which the bubbles rise due to Archimedes forces and the rate of their translation due to the

*The mechanism of the mixing of material in a "normal" fluidized bed with bubbles is considered more rigorously in [1250, 1245].

continuous collapse of particles from the "ceiling" to the "floor" of the bubbles. Consequently, at high superficial velocities exceeding the rate of bubble ascent it becomes impossible in principle for all the "excess" gas to pass in the form of bubbles. Obviously passage in the form of bubbles will cease even earlier—as soon as the impossibility of further gas passing in the form of bubbles leads to filling of almost the entire bed with gas bubbles.

Figure 2-2a illustrates the case when the mean-bed voidage $m \approx 0.9$. It is obvious that such a state is hardly likely to be stable, since there is insufficient material in the region of the gas distributor to form "shells" around the continuously supplied gas. The conversion of the velocity head of the gas into an excess static pressure falls off, meaning that the tendency toward aggregation also falls off. The formation of pistons of material, as shown in Fig. 2-2b, also cannot continue, since the particles falling from the lower surface of the piston cannot accumulate in the lower parts of the bed, filling the whole cross-sectional area with a dense phase (with $m \approx m_{\ell.s}$). Long before such a dense phase could be achieved, the particles would be picked up by the fluid stream, the velocity of which is close to the free-fall velocity of individual particles.

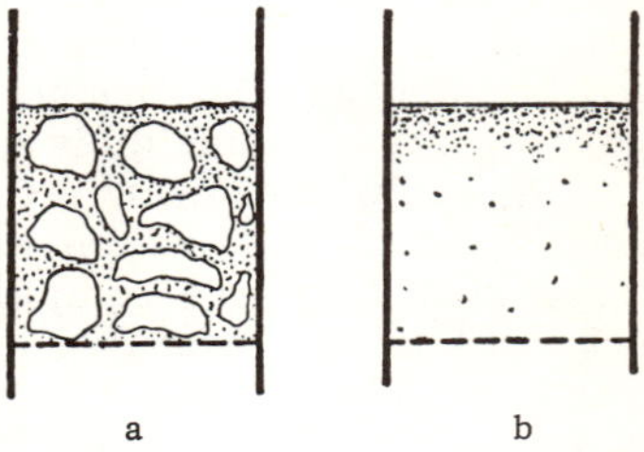

FIG. 2-2. Unstable states of nonuniform fluidization at high bed voidages.

According to the data of Cairns and Prausnitz [1105], the dense phase becomes the discontinuous phase instead of the fluid even at $m \approx 0.7$. This change can be regarded as preliminary to the succeeding state of a uniform rarefied fluidized bed.

A reappearance of uniform fluidization can therefore be expected at high mean-bed voidages. According to the evidence of Zenz [490], uniform fluidization actually reappears when the concentration of material in the bed is 0-10% of the concentration in the dense phase. This corresponds approximately to a bed voidage of $m = 0.94$. There is a view that the maximum concentration of the dispersed phase for the system to remain uniform can be calculated from the value of the allowable distance between particles without interference with the motion of separate particles. This minimum distance obviously corresponds to the length of the particle wake. There exists a fairly wide zone of small bed porosities (large particle concentrations) corresponding to unstable regimes with clearly marked pulsation and slugging of the gas passing through. Even in this unstable zone the dense phase ceases to be continuous and parts of it become separated by gas layers.

As γ_M/γ_m decreases the aggregating force F should decrease because of the decreasing apparent weight of the material and the lower gas velocity required for fluidization. The friction forces between the particles also decrease; i.e., the particle mobility increases. Hence uniform gas fluidization can be expected for particles of very low bulk density. In air fluidization of beds of hollow paper cubes, Happel and Grosselfinger [490] observed uniform bed expansion over the whole range of porosities from about 0.4 to 1. Zenz [490] assumes that the value of γ_M/γ_m should not exceed 2-3 for uniform fluidization. Obviously this value corresponds to conditions of fluidization by liquids (see above).

As the particle diameter increases, the aggregation force F must also increase because the transition to larger diameters implies a transition to larger fluidization velocities. However, many investigators have observed an increase in the tendency toward nonuniform fluidization, and the formation of larger gas cores (bubbles) and slugs, as the particle diameter is increased [459, 490, 601, 856, 970, etc.].

The above fundamentally general arguments as to the reasons for, and forms of, fluidization nonuniformities are presented without pretense that they are complete or final; they are useful in analyzing actual cases of appearance of these nonuniformities as encountered in practice. There exist a variety of views concerning the development of nonuniformities, observed and described by various investigators [459, 573, 601], because the specific conditions in the experiments have not always been the same.

The circumstances under which cavities appear in beds of smooth spherical particles fluidized with liquids are still not clear. Such cavities have been observed in Hassett's experiments [1204] and those of other investigators [1105]. Hassett claims that in the water fluidization of beds of glass spheres a little coarser than 1 mm, horizontal mushroom-shaped streaks with a reduced concentration of spheres were observed; these moved upward and disappeared at the bed surface. With glass spheres of 2-mm diameter the streaks appeared to be free of particles, reached 1-5 cm in width, and rose at 5 cm/sec when the superficial water velocity was about 2 cm/sec. Hassett proposed that such cavities could have an eddy nature.

Without defining the nature of the nonuniformities that developed, Jackson [1244] recently showed that the state of uniform fluidization is unstable for any system, since disturbances, which always occur, are certain to grow in the upper portions of the bed. Calculations have shown [1243, 1244] that the growth of disturbances in beds fluidized with liquids is many times slower than in beds fluidized with gases, which could explain the observed uniformity of many shallow water-fluidized beds. At the same time, the development of nonuniformity at sufficiently large distances above the grid of water-fluidized beds of glass spheres was predicted and observed.

Harrison, Davidson, Rowe, and others [1249, 1245, 1247, 1248, 1246] did not pursue the problem of initiation and growth of nonuniformities in fluidized beds, but rather investigated the passage of single bubbles through fluidized beds, including the question of their stability, the flow of the medium through and around the bubbles, and the motion of the bed material caused by the bubbles.

The flow of solid particles around the bubbles in a fluidized bed closely resembles the motion of elements of nonviscous fluids around solid bodies rising through them, although the analogy is not complete [1249, 1245] and does not extend to the mechanisms of the two phenomena. The rising bubbles carry a considerable quantity of material with them and are the main reason for particle mixing in beds fluidized with gases.

Harrison and Davidson [1249, 1248] claim that the resulting gas velocity field in a fluidized bed with bubbles is obtained by the summation of the velocity calculated by D'Arcy's law (without regard to the bubbles) and the volume of gas entrained by the particles flowing downward along the sides of the spherical bubbles. For these calculations it is necessary to know in particular the shape of the bubbles and the ratio of the bubble velocity to the velocity of the gas in the interstices between the particles, i.e., $w_b/(w_{\ell.s}/m_{\ell.s})$. When $w_b > w_{\ell.s}/m_{\ell.s}$, as usually the case for deep beds of fine particles, all the gas in the bubble circulates through the bubble and around it (in the immediate vicinity) [1249]. Visual observations made by introducing colored tracer gases into the bubbles have confirmed that there is a "cloud" of gas around a rising bubble which circulates from the bubble, and vice versa, and which does not mix with the remainder of the gas stream. However, calculations of the amount of contact of the gas with the material, carried out on the assumption of piston flow [1245] in the continuous ("emulsion") phase and closed circulation in the "cloud," gave values of the conversion rate which were smaller than observed experimental values. This implies that the actual gas transfer between the bubbles and the emulsion phase is improved due to "diffusion" of the cloud, particularly in the wake zone of the bubble (the tail). Neglecting to take into account contact with the particles passing through the bubble [1245] could also lead to a difference in experimental and calculated values of the degree of conversion.

Nevertheless, from measured conversions [1245] it is quite clear that gas transfer between the bubbles and the emulsion phase is very poor in normal fluidized beds with bubbles, and to intensity the processes of heat and mass transfer between particles and gases it is desirable to avoid the formation of bubbles, even at the expense of making the apparatus more complicated (devices for retarding bubble formation, imposing vibrations, operating with thin beds, etc.), as discussed in Chapters 5, 8 and 11. In order to demonstrate the important adverse effect of micrononuniformities in the gas distribution on heat transfer to particles, numerical calculations were carried out for a very simple physical model of a nonuniform fluidized bed (see Chapter 8) with aggregates of the minimum thickness (one layer of particles), followed by complete mixing of the gas.

By assuming under these conditions "models for gas microeruptions" with different (larger) numbers of layers of particles in the aggregates, it was possible to characterize quantitatively any large degree of bed nonuniformity.

Rowe assumed [1252] that the presence of bubbles in gas-fluidized beds imparts a high degree of order to the motion of the gas and solids. However, elements of randomness remain, since, for example, the course of formation and the residence time of the bubbles in the bed are random under real conditions. There is as yet insufficient evidence, despite calculations and experiments with artificially injected single bubbles, to substantiate a high degree of order in nonuniform beds, and hence the transition to real conditions must add many elements of randomness, since here not one but a multitude of bubbles and "excess gas" move simultaneously through the bed; these enter not through a single nozzle but through a screen with numerous openings in close proximity.

So far there is no correlation available for calculating nonuniformity. It is still not possible to take into account quantitatively all the effects mentioned above (shape and surface characteristics of the particles, γ_M/γ_m, type of gas, type of gas distributor, distance of a particular bed zone above the gas distributor, particle diameter, superficial velocity, etc.). The problem is complicated by the importance of knowing a great deal about the mechanism of the development of nonuniformities in fluidized beds in order to find the most efficient ways of limiting these nonuniformities in various practical instances. It is therefore necessary to carry out detailed and systematic investigations of fluidization nonuniformities.

2.2 The Initial Region of Fluidization

1. Incipiently fluidized beds

Intensive particle motion does not occur immediately after passing the limit of stability. The bed is comparatively homogeneous if there is no channel formation such as is caused, for instance, by a poorly designed gas distributor. Channel formation is favored by small particle diameters and low bed heights, which assist in the formation of continuous channels or "rat holes" due to the action of the dynamic pressure of the jets rising from the gas-distributor openings. There are also indications in the literature [988] that the tendency to form channels increases as the bed diameter increases, i.e., on going from a small laboratory apparatus to large-scale industrial equipment. To some extent this is explained by nonuniformity of the gas supply below large grids.

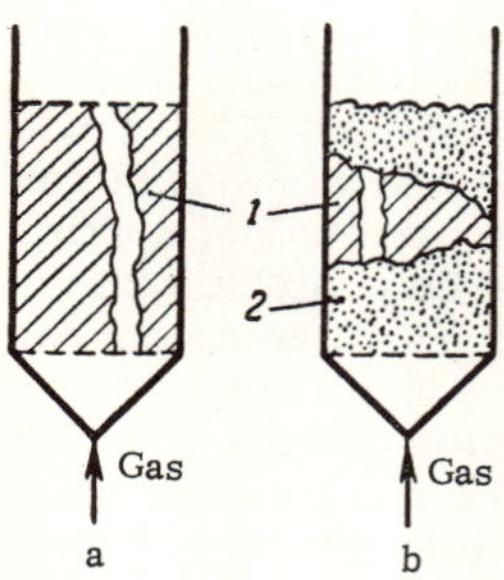

FIG. 2-3. Two types of channel formation in particle beds [988]: a) continuous; b) internal; 1) zone of stationary dense beds; 2) fluidized zone.

Leva mentions [988] that channels sometimes appear not in the lower part of the bed above the grid but in the middle part of the bed (Fig. 2-3); he does not explain the reasons for such a phenomenon. It can be assumed that channel formation in the middle part of a bed is natural for materials that have a poor or variable (across the bed) flowability and are able to create an arch in the middle part of the bed above any large cavities. These cavities become positions for the concentrated passage of large quantities of gas, and well-developed channels are usually found above them. A more probable path for the development of cavities or rarefied areas in the bed is the initial existence of channels penetrating the whole bed and the ejection of particles from the lower parts of the bed along this length without any filling in from the top, due to the presence of the arch.

Operation in the initial region of fluidization, where there is no intensive mixing of the particles of material, has both advantages and disadvantages. The low velocity and flow rate of the fluidizing medium in this region is an advantage if fluidization is carried out with the object of transporting material or separating it into fractions; it is a disadvantage if the bed exists simply as a heat transfer medium or reagent, since then the small flow rate of the medium corresponds to a small productive capacity of the aggregate. The absence of intensive particle mixing and the horizontal motion of the fluidized bed as a whole from the loading point to the takeoff point, as will be shown below, are favorable to heat transfer between the gas and the particles in the cross stream as well as improvement of the utilization of gaseous heat transfer media, but reduce the heat transfer coefficient between the bed and heating surfaces surrounding it and also reduce the effective thermal conductivity of the bed.

Another interesting hydrodynamic phenomenon, predicted by Yavorskiy [184] and investigated by Sharlovskaya, is observed in the initial region of fluidization, immediately beyond the limit of stability. It is usually assumed that the mean value of the true velocity of flow around the particles increases monotonically as the superficial velocity increases. In determining experimentally the dependence of the bed voidage on the superficial velocity near the limit of stability, Sharlovskaya showed [793, 952, 953] that this is not entirely correct. Within a transition zone between the limit of stability and the zone of intensive fluidization, the true mean velocity of flow around the particles decreases as the superficial velocity increases. This corresponds to a change in the nature of the flow around the particles in this region such that the resistance coefficient does not decrease but rather increases as the bed voidage increases. According to Sharlovskaya the width of this transition zone (up to the minimum true mean velocity w) (Fig. 2-4) increases as the particle diameter increases, but at the same time the absolute value of the fall in the actual velocity decreases; i.e., the transition zone is less sharply marked. In the opinion of Sharlovskaya, the transition zone is less efficient from the point of view of heat intensity [793].

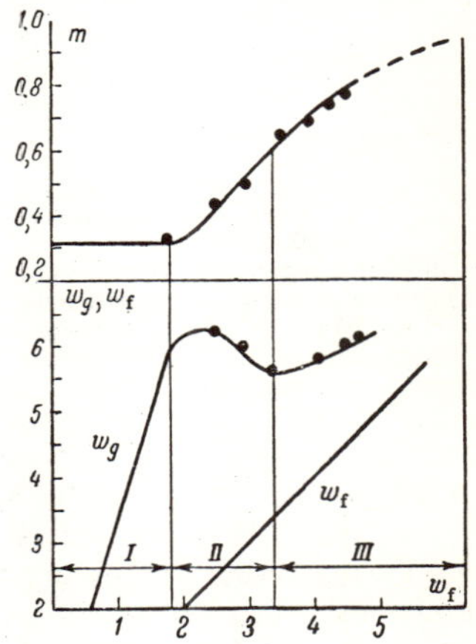

FIG. 2-4. Change of voidage (m) and actual velocity of flow around the particles (w_g) in a bed of porcelain spheres, d = 4.63 mm, initial bed height 60 mm [793]: I) stationary dense bed; II) transition zone of fluidized bed; III) zone of intensive "boiling"; w_f) fluidization velocity.

Pneumatic conveying (Fig. 2-5) [26], which has long been used for the horizontal transport of free-flowing materials, is an example of the use of fluidized beds with poorly developed particle mixing. The pneumatic conveyors used for transporting cement dust (particle size 20–40 μ) have slopes of the order of 4° and carry material in a layer 50–60 mm thick at a velocity of about 1 m/sec with an air fluidization velocity of 0.12 m/sec [26]. Pneumatic conveyors have reached lengths of 50 m, but even this is not the limit.

An arrangement was once described for drying during the controlled flow of a material in a fluidized state, i.e., at incipiently developed

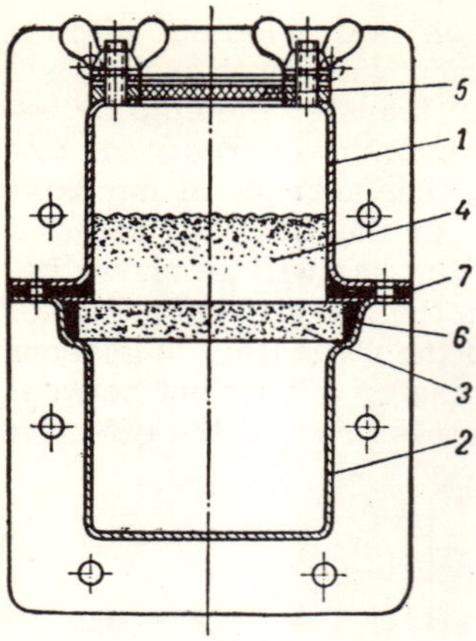

FIG. 2-5. Cross section of fluidized conveyor.
1) upper trough; 2) lower trough; 3) porous plate; 4) material transported; 5) fabric retaining filter; 6) sealing cement; 7) rubber or felt gasket.

fluidization [82]. Various modifications of such dryers were proposed later. Przhetslavskiy and Solov'yev [777] proposed that the drying of coal and other powdery materials be carried out in fluidized conveyors in which the slope could be varied to control the degree of drying and the solids throughout. Information is available on the drying of coal (of particle diameter 0-13 mm) concentrates from floatation processes, residues and sands in models with a grid cross-sectional area of 0.5 m^2 [717].

The lower part of a bed of polydisperse material is in a state of incipient fluidization in the dryer developed by the Power Institute of the Belorussian SSR Academy of Sciences for drying milled peat in thin layers by low-temperature flue gases. Zabrodskiy, Solomonov and Begun have demonstrated the successful operation of such dryers [1045]. The bed thickness was automatically maintained at a level such that the total operating resistance of the combined grid and bed came to about 50 kgf/m^2. The grid was kept horizontal.

Fraser, Yancey and Przhetslavskiy [942] proposed a process for enriching coal by means of a fluidized bed of an auxiliary granular material. Przhetslavskiy's process has been tested industrially. The operation of this process is based on the phenomenon of segregation of particles, which differ over the bulk of the bed. The particles of greater bulk weight than that of the fluidized bed of auxiliary material sink in the latter, whereas the coal particles "float" on the surface.

Sorting polydisperse materials according to particle size by using incipiently fluidized beds has also been proposed, since a tendency toward particle segregation is noted in these beds; small particles move to the surface and coarse ones sink [511, 988].

The phenomenon of particle segregation in fluidized beds has long been recognized but never studied in detail. However, some explanations can be given for segregation according to particle density (bulk weight) and dimension.

2. Particle segregation

Segregation by density. The pressure within a fluidized bed is approximately proportional to the total bulk weight of the column of material and fluid above the level considered. Hence when a body is submerged in the bed (in the ideal case when the fluidization is not disturbed) there is a buoyant force on the body equal to the volume of bed displaced by the body, i.e., a force analogous to an Archimedes force. However, Reboux [511] mentions that the actual buoyant force in a fluidized bed is smaller than this theoretical value. In spite of

Reboux's view, it seems quite likely that the difference is caused by the disturbance of the fluidization directly above the submerged body, as a result of which the bed settles down onto the body, becomes more dense, and decreases the net buoyant force. However, this assumption requires experimental confirmation.

In testing the previously mentioned Przhetslavskiy process it was found that optimum results were obtained when the fluidization index was 2–3. At larger fluidization indexes the mixing of the bed was intensified, leading to poor segregation, and at smaller fluidization indexes there was a large effective bed viscosity, which slowed down the floating of the coal particles and sinking of the gangue particle. At fluidization indexes of 2–3 the effective viscosity of the fluidized bed of auxiliary material (barytes of particle size d = 0.17–0.3 mm) was only 1.2–1.4 times greater than the viscosity of water [942].

Particle size segregation. The phenomenon of particle segregation according to size in water-fluidized beds has long been known from observations of bed settling in sand filters. Andrieu [988] carried out a systematic investigation of this type of segregation. Practically total particle segregation according to size was achieved in experiments of Neužil [1124] and of Hoffman, Lapidus and Elgin [1114] in water-fluidized beds; this can be explained by a reduction of the tendency toward particle mixing during uniform fluidization. Sinking of large particles and concentration of fine particles of the same material in the upper parts of the fluidized bed also occur in cases of gas fluidization, though again not in intensively mixed beds.

Leva [988] proposed the following mechanism for the size segregation of particles: small channels appear in the initially uniformly fluidized bed and the fine particles are carried upward through them; the segregation process begins in this way. As they pass up through the lower parts of bed, there is a gradual depletion of the fine particles causing fluidization to cease in these parts. The lower parts of the bed then "solidify" somewhat and serve only as a coarse-particle gas distributor for the upper portion, which in most cases is briskly fluidized and exhibits distinct segregation. In this description one detail is rather questionable: how is it possible to retain distinct segregation in the intensively fluidized upper portions of the bed if they are not quite dilute? (See section below on segregation in dilute fluidized beds.)

Leva maintains that the same mechanism applies to the segregation of particles according to density. This is incorrect (see above section on density segregation).

It is also impossible to explain segregation during fluidization with liquids by means of Leva's mechanism, since in this case channel formation is not usually observed. The segregation mechanism of Leva corresponds to only one special case in which the fines are carried to the upper part of the bed.

It can be shown that the tendency toward segregation by particle size can exist in very different cases of fluidization (in homogeneous fluidized beds, in beds with gas passing in the form of bubbles with unstable channels). Consider first a homogeneous fluidized bed of wide particle size distribution. The particles of various sizes will have the same velocity of fall, corresponding to the overall superficial velocity, only when there are different interstitial bed voidages around the fine and coarse particles. In general, the voidage around the small particles must be larger or they will be entrained upward. Obviously systems with particles of different sizes occurring at the same level are to some extent unstable, since this requires maintenance of definite but different voidages at different points in the same horizontal cross section of the bed. Furthermore, even in uniform fluidization there are small rapid fluctuations in the local density (voidage) of the bed. Thus during temporary reduction in voidage around the fine particles, they will be able to move partially upward under the action of the increasing drag, despite their inertia. Under analogous circumstances the coarse particles have a smaller chance of rising because of their larger inertia and the larger gap width that they need for passing between the particles lying above. Sinking of the fine particles when the voidage around them is increased occurs more infrequently, since the counterflow of fluidizing medium contributes to holding the fine particles back in the regions downstream of the large particles below them.

In nonuniform fluidized beds with breakthrough of the gas in the form of bubbles, particles collapse within the bubbles. The collapse must be accompanied by segregation, since the fine particles fall more slowly than the coarse ones in an ascending stream of pure gas. Thus in a bed of initially uniformly mixed coarse and fine particles, the repeated passage of bubbles will lead to a preferential net movement of the large particles downward and the fine ones upward as long as the mixing in the bed does not overwhelm this tendency. Placing a horizontal screen barrier in a fluidized bed (so-called bubble retardation) makes it more difficult for particle mixing to occur [988, 1065] and actually facilitates particle segregation. In order to effect particle segregation in ordinary unrestrained fluidized beds of constant cross section, Leva [988] suggested that the conditions are optimum when the superficial velocity is slightly greater than the minimum velocity required for fluidizing the largest particles in the bed.

Finally, variation of the bed cross section over the height of the equipment also has an important effect on segregation of particles by size. In such cases the superficial velocity varies over the bed height. For example, in a conical apparatus expanding upward it will increase from top to bottom. Although conical beds are fluidized relatively uniformly (liquid or gas fluidization at small cone angles and small fluidization indexes), the tendency for the small particles to rise and the large ones to sink is much more pronounced than in a bed of uniform cross section. Conical beds also operate under conditions in which the coarse particles cannot in general be fluidized in the upper portions of the bed, even at the minimum bed voidage corresponding to the limit of stability. They therefore fall to the level where the superficial velocity is equal to their minimum fluidization velocity.

2.3 Development of Fluidized Beds

1. Uniform bed expansion

A fluidized bed expands monotonically as the superficial velocity is increased. The law of expansion is quite complicated even for very simple, uniformly fluidized beds because of the complex nature of the relationship between the velocity of hindered fall of the particles and the bed voidage. If this rate of fall were constant and equal to the true velocity of the medium in the interstices between the particles, the conditions for existence of a fluidized bed could be written as

$$w_{\text{fall}} = w_{\text{true medium}} = \frac{w_f}{m} = w_{ff} = \text{const},$$

since the limit at $m = 1$ corresponds to free fall of the particles. This equality represents a so-called voidage function (a relationship between the bed voidage and the superficial velocity). The form obtained for the voidage function

$$w_f / w_{ff} = m,$$

is equivalent to stating that the velocity at the limit of stability $w_{\ell.s} = m_{\ell.s} w_{ff} \approx 0.4 w_{ff}$ is incorrect, since the particle drag in hindered fall is actually much larger than in free fall, and $w_{\ell.s}/w_{ff} < 0.4$. Hence in an actual voidage function of the type $w_f/w_{ff} = m^n$ the exponent on m must be larger than unity.

Leva et al. [411], assuming that the relationship between the resistance of a particle around which flow occurs and the bed voidage was the same in a fluidized bed as in a fixed bed, derived the voidage function

$$w_f / w_{ff} = 0.09 m^3/(1 - m). \tag{2-2}$$

However, experiments have shown that the actual resistance coefficients of particles in fluidized beds are 20-50% lower than they would be in an equivalent fixed bed of the same voidage. This could obviously be the result of alignment of the "channels" between the particles in a fluidized bed. It is generally impossible to justify attempts at extending the laws of fluid flow in channels to the high-voidage region of a fluidized bed. The flow in an expanded fluidized bed can be regarded as a typical problem of internal flow in bodies.

Brötz [221, 810] proposed a different type of voidage function for determining the expansion of fluidized beds:

$$w_f / w_{\ell.s} = \left(\frac{m}{m_{\ell.s}}\right)^z, \tag{2-3}$$

in which the exponent z is a function of the particle diameter given by the graph in Fig. 2-6 for the laminar regime (Re < 10).

Assuming that the linear law of resistance is replaced in the turbulent flow regime by a quadratic law, the calculated exponent should be equal to $z/2$.

The height of the expanded bed H is given by the equation

$$H/H_{\ell.s} = (1 - m_{\ell.s})/(1 - m). \tag{2-4}$$

Mints [95] proposed an empirical formula for the expansion of a fluidized bed that satisfies the upper limiting condition ($m = 1$) but not the lower (the limit of stability).

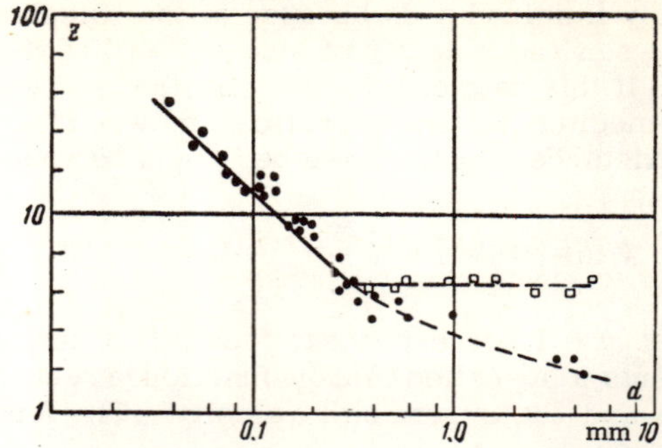

FIG. 2-6. Relationship between the exponent z in Eq. (2-3) and the particle diameter d according to Brötz [810].
□—bed fluidized with water; ●—bed fluidized with gas.

Yerkova and Smirnov [33, 34] gave a relationship for the expansion of beds of spherical particles fluidized with a liquid which took the form of a single-term empirical relationship between the voidage m,

$$\mathrm{Re} = \frac{w_\mathrm{r}\, d}{\nu} \text{ and } \mathrm{Ar} = \frac{gd^3}{\nu^2} \cdot \frac{\gamma_\mathrm{M} - \gamma_\mathrm{m}}{\gamma_\mathrm{m}}:$$

$$\mathrm{Re} = 0.472\ \mathrm{Ar} \cdot m^5 \text{ for } \mathrm{Ar} \cdot m^5 < 60 \text{ (laminar region);} \tag{2-5}$$

$$\mathrm{Re} = 0.110\ (\mathrm{Ar} \cdot m^5)^{0.78} \text{ for } \mathrm{Ar} \cdot m^5 = 60\text{–}5000 \text{ (transition region);} \tag{2-6}$$

$$\mathrm{Re} = 1.15\ (\mathrm{Ar} \cdot m^5)^{0.5} \text{ for } \mathrm{Ar} \cdot m^5 > 5000 \text{ (turbulent region).} \tag{2-7}$$

Yerkova and Smirnov found that within a given hydrodynamic regime the equations for particles of different shapes differed only in the values of the coefficients.

Table 2-1

Comparison of the slopes n of the bed expansion lines calculated from Richardson and Zaki's formulas with slopes determined experimentally.

Flow regime	No. of line in Fig. 2-7	Slope		Mean value of Re
		calculated	experimental	
Laminar	3 2 4 1	4.9 4.8 4.5 4.4	5.1 5.1 4.3 4.0	0.02 0.06 0.3 0.5
Transitional	5 6 7	4.1 3.6 3.3	3.8 3.8 3.1	2.0 8.0 20.0
Turbulent	8 9	2.5 2.39	2.7 2.5	350.0 1000.0

Richardson and Zaki [518, 988] proposed describing the expansion of a fluidized bed by a porosity function:

$$w_f / w_{ff} \sim m^n. \qquad (2\text{-}8)$$

They found that the exponent n depended mainly on the flow regime characterized by Re, and gave the following empirical equations:

$$n = (4.35 + 17.5 d/D_T)\,\mathrm{Re}^{-0.03} \quad \text{for Re} = 0.2\text{–}1.0; \qquad (2\text{-}9)$$

$$n = (4.35 + 18 d/D_T)\,\mathrm{Re}^{-0.1} \quad \text{for Re} = 1.0\text{–}200; \qquad (2\text{-}10)$$

$$n = 4.45\,\mathrm{Re}^{-0.1} \quad \text{for Re} = 200\text{–}500; \qquad (2\text{-}11)$$

$$n = 2.39 \text{ for Re} > 500. \qquad (2\text{-}12)$$

The correlation of Richardson and Zaki has been checked [988] by comparison with the experimental data of numerous investigators [335, 373, 422, 604, 623] (Fig. 2-7 and Table 2-1).

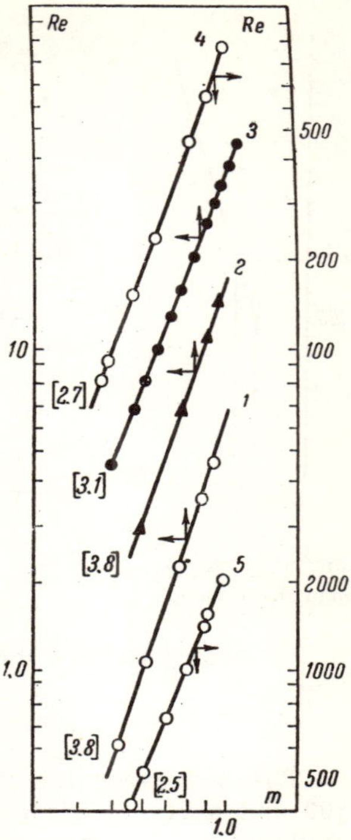

FIG. 2-7. Data on the expansion of a fluidized system of liquid and solid particles in the transition and turbulent regimes evaluated by the method of Richardson and Zaki [988].
1) Lewis et al. [422], d = 0.284 mm; 2) Young [623], d = 0.411 mm; 3) Wilhelm and Kwauk [604], d = 0.556 mm; 4) Wilhelm and Kwauk [604], d = 1.283 mm; 5) Wilhelm and Kwauk [604], d = 5.21 mm; numbers in brackets indicate the slopes of the lines.

Richardson and Zaki's procedure for the graphical analytical calculation of the expansion of a fluidized bed is as follows [1118]: the calculated minimum fluidization velocity $w_{\ell.s}$ and the known bed voidage at the limit of stability $m_{\ell.s}$ establish one point on the expansion line on plot of m vs. w. By calculating Re at the limit of stability and at a given superficial velocity the exponent on the mean value of Re is found from expressions (2-9) to (2-11). A line with slope n is drawn through the limit of stability on a logarithmic plot of m vs. w or Re. The voidage of the expanded bed (from which the bed height is calculable) at the required value of the superficial velocity (or of Re) is then read from this graph.

Example 1. A spherical catalyst ($d = 4.42 \times 10^{-3}$ m) is to be fluidized with water at 21°C in a column of 0.15 m diameter. The bulk density of the catalyst particles $\gamma_M = 1600$ kg/m³. The settled height of the bed $H_{\ell.s} = 0.71$ m, and its voidage $m_{\ell.s} = 0.37$. Find the height of the expanded bed if the superficial water velocity $w_f = 10.4 \times 10^{-2}$ m/sec.

Solution: The minimum fluidization velocity is

$$w_{\ell.s} = 9.35 \cdot 10^{-3} \frac{d^{1.82}}{\nu^{0.88}} \left(\frac{\gamma_M - \gamma_m}{\gamma_m} \right)^{0.94} =$$

$$= 9.35 \cdot 10^{-3} \frac{(4.42 \cdot 10^{-3})^{1.82}}{(9.8 \cdot 10^{-7})^{0.88}} \left(\frac{1\,600 - 998}{998} \right)^{0.94} = 5.87 \cdot 10^{-2};$$

$$Re_{\ell.s} = \frac{w_{\ell.s} d}{\nu} = \frac{5.87 \cdot 10^{-2} \cdot 4.42 \cdot 10^{-3}}{9.8 \cdot 10^{-7}} = 265.$$

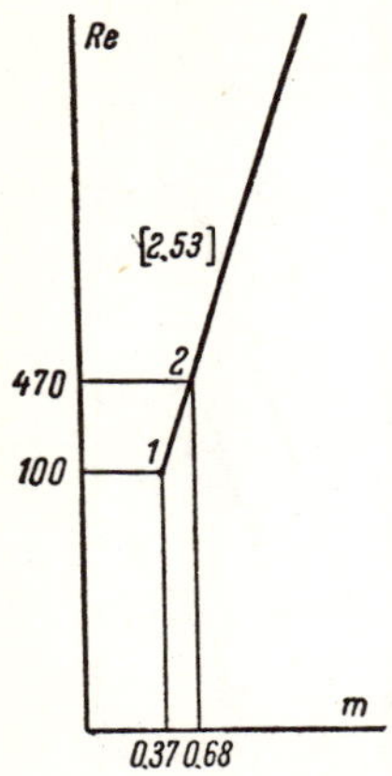

FIG. 2-8. Bed expansion line (Example 1).

From (1-17a) the correction factor is 0.38, i.e., finally $w_{1.s} = 0.38 \times 5.87 \times 10^{-2} = 2.23 \times 10^{-2}$ m/sec, and $Re_{1.s} = 0.38 \times 265 = 100$.

The Reynolds number at the operating superficial velocity is

$$Re_f = \frac{10.4 \cdot 10^{-2} \cdot 4.42 \cdot 10^{-2}}{9.8 \cdot 10^{-7}} = 470.$$

The mean (calculated) Reynolds number is

$$Re = \frac{470 + 100}{2} = 285.$$

According to the correlation of Richardson and Zaki,

$$n \doteq 4.45 \cdot 285^{-0.1} = 4.45 \cdot 0.569 = 2.53.$$

The bed expansion line passing through the point m = 0.37, Re = 100 and having a slope of 2.53 can now be drawn (Fig. 2-8) on a logarithmic plot. At Re = 470 this line passes through the point for which m = 0.68. Hence the height of the expanded bed is

$$H = H_{\ell.s} \frac{1 - m_{\ell.s}}{1 - m} = 0.71 \frac{1 - 0.37}{1 - 0.68} = 1.4 \text{ m}.$$

Note: In Leva's monograph [988] a similar example is solved incorrectly; the Reynolds number of the expanded bed and the slope of the expansion line are in error.

2. Expansion of a nonhomogeneous fluidized bed

A considerable body of information has been published on the expansion of nonhomogeneous gas-fluidized beds [221, 276, 322, 335, 373, 418, 422, 433, 482, 492, 604, 657, 752, 810, 813, 865, 1097]. Some typical experimental data are shown in Fig. 2-9, taken from Leva [988], in the form of curves for the voidage function $(1-m)/m^3$ vs. the superficial mass velocity plotted in logarithmic coordinates. In this system of coordinates uniform fluidization is expressed by a straight line of slope $n = 1$. Lines with a gentle slope (line 1) correspond to flow expansion of the bed as the superficial velocity increases.

It can be seen from Fig. 2-9 that the more rapid the expansion, the more homogeneous it is (slopes $n > -1$) for beds of coarse particles. Under otherwise similar conditions, beds of fine particles (line 4) expand more slowly than in uniformly fluidized beds. Also as the superficial velocity is increased above some value the slope of the expansion line decreases; i.e., more rapid expansion commences in nonuniformly fluidized beds. The experimental points on line 3 do not show this tendency, possibly because of the limited range of superficial velocities used in this instance.

For practical purposes the experimental data on the expansion of nonuniform beds can be approximated in most cases by a straight line on a logarithmic plot. It is permissible to assume that the slope *n* depends in a single-valued manner on the particle dimensions (see Fig. 2-10, which was compiled by Leva [988] from the experimental data of several investigators).

In an attempt to explain the experimental data on nonhomogeneous expansion of fluidized beds, Leva proposed that there exist two opposing factors in gas-fluidized beds. The first is particle aggregation, which presumably always tends to increase the bed expansion. The second factor is particle motion. It will be shown later that Leva's concept that the effect of aggregation (the two-phase nature of fluidization) is always the same is not correct. This seems to be unfounded, as is Leva's assertion that particle motion always leads to a retardation of bed expansion. Frequently beds expand more slowly in nonuniform fluidization

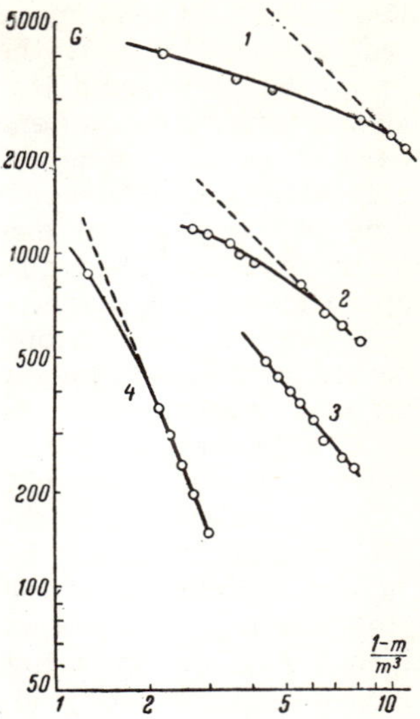

FIG. 2-9. Data on the expansion of gas-solid fluidized systems [988]: 1) gas spheres, d = 1.006 mm [433]; 2) spherical sand particles, d = 0.384 mm [418]; 3) glass spheres, d = 0.249 mm [314]; 4) iron Fischer-Tropsch synthesis catalyst, d = 0.178 mm [418]; G—superficial mass flow rate, kg/m² · hr; m—bed voidage.

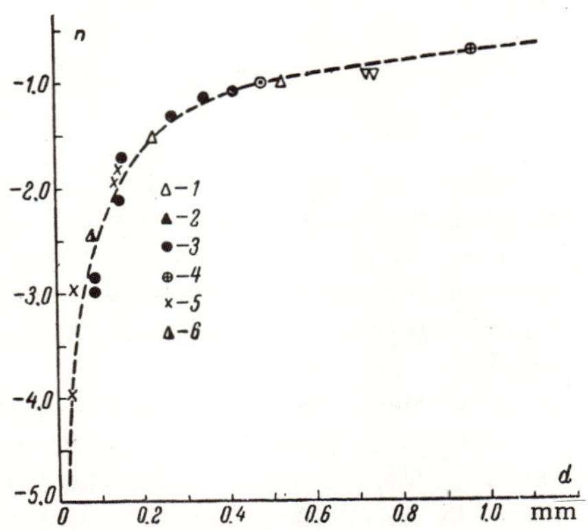

FIG. 2-10. Slopes n of the tangents to the expansion curves during gas fluidization as functions of the particle diameter d.
1) Ergun and Orning [276]; 2) Graf [314]; 3) Lewis, Gilliland and Bauer [422]; Massimilla and Bracale [433]; 5) Olin and Dead [483]; 6) Parent, Yagol and Steiner [492].

though the particles move many times more intensively than in uniform fluidization. At the same time the hydraulic resistances are almost the same for uniform and nonuniform beds. Thus, at the same overall consumption of energy, nonuniform fluidization (with the larger energy consumption necessary to maintain a high kinetic energy of the particles) should be accompanied by a smaller expansion (a smaller value of the potential energy of the bed). Such an argument is inconclusive because the hydraulic resistances in the two cases are only approximately the same.

Toomey and Johnstone [567] showed that under otherwise similar conditions the resistance of a nonuniformly fluidized bed is larger than that of a uniform bed, and they assume that this difference goes into the kinetic energy of the particles. This concept is not exact since it does not take into account the loss of energy due to friction at the walls.

Furthermore, it is doubtful if the work of bed expansion, as Leva [988] considers it, can be carried out continuously in a fluidized bed. The bed is expanded at the instant that the given fluidization regime is established, and after that no consumption of energy for bed expansion is required. Bed pulsations are not so large that repeated bed expansion is required over and over again. Obviously it is more correct to regard the energy continuously consumed in fluidizing ("suspending") the particles as being an energy dissipation mainly for overcoming the total resistance to flow around the particles. There is no accompanying detectable increase in the bed temperature, since the heat liberated in the degradation of energy is carried off by the fluidizing agent.

As a result, the characteristic that Leva introduced for the intensity of particle motion in a fluidized bed—the "fluidization efficiency"—assumes a rather arbitrary nature. Leva uses the term "fluidization efficiency" for the fraction of the total energy consumed in fluidization, W_{fb}, that contributes to imparting motion to the particles, with the assumption that the only other important component is the "energy for bed expansion" W_{exp}:

$$\eta = \frac{W_{fb} - W_{exp}}{W_{fb}}.$$

In estimating the effect of particle motion on bed expansion it is obvious that it is not the consumption of energy in maintaining motion that is important, but the effect of the latter on aggregation. Particle mixing, which hinders aggregation, will impede bed expansion only when the velocity with which the so-called discontinuous bubble phase passes through it remains quite small.

It is interesting to note that Leva's data [988] on the monotonic increase of η with the fluidization index are not in agreement with those of Todes and Bondareva [662], which indicate the presence of a maximum in the relationship between the particle velocity and the fluidization index.

By using the concepts of the two-phase theory of fluidization, attempts have been made to establish a relationship between the expansion of a nonuniform fluidized bed and a fundamental variable such as the superficial velocity w_f, the minimum velocity of fluidization $w_{1.s}$, and the velocity with which the "discontinuous phase" passes through the bed w_d. From the concept that the excess quantity of gas passes in the form of bubbles, it follows that

$$Q_{excess} = w_f F - w_{\ell.s} F = (w_f - w_{\ell.s}) F \text{ m}^3/\text{sec} \tag{2-13}$$

where F is the cross section of the bed, m².

The volume of all bubbles simultaneously existing in the bed is designated Σv_b. These bubbles, which pass through the bed height H with a linear velocity w_b, carry the gas volume Q_{excess}; hence

$$Q_{excess} = \Sigma v_b \frac{w_b}{H} = \frac{(\Sigma v_b) w_b F}{V_{bed}}, \tag{2-14}$$

where V_{bed} is the volume of a bed of constant cross section.

However, the volume of bubbles is equal to the total volume of the bed minus the volume occupied by the dense phase, i.e., by the aggregates that have a porosity equal to that at the limit of stability, $m_{\ell.s}$:

$$\Sigma v_b = V_{bed} - V_{dense\ phase}. \tag{2-15}$$

Equating (2-13) and (2-14) and substituting (2-15):

$$w_f - w_{\ell.s} = w_b \left(1 - \frac{V_{dense\ phase}}{V_{bed}}\right). \tag{2-16}$$

However, $V_{dense\ phase}$ is obviously equal to the volume of the bed at the limit of stability:

$$V_{dense\ phase}/V_{bed} = V_{\ell.s}/V_{bed} = H_{\ell.s}/H.$$

Thus

$$w_f - w_{\ell.s} = w_b \left(1 - \frac{H_{\ell.s}}{H}\right) \tag{2-17}$$

or the relative expansion of the nonhomogeneous bed is

$$H/H_{\ell.s} = w_b/(w_b - w_f + w_{\ell.s}). \tag{2-17'}$$

It follows from Eq. (2-17') that when the bubbles rise very rapidly, so that $w_b \gg (w_f - w_{\ell.s})$, the relative expansion of the bed is small. As $w_b \to \infty$, $H/H_{\ell.s} \to 1$. However, the velocity of bubble rise is actually not infinitely large but comparatively small. Usually $w_b < 1$ m/sec. Only a few investigators [21] have reported bubble velocities as large as 1-2 m/sec.

Turning to the possible mechanism of bubble rise in fluidized beds, it becomes apparent why this velocity is limited and quite unrelated to the velocity of the gas; i.e., in some cases it is smaller than the superficial velocity, and in other cases larger.

The physical picture of the rise of a bubble in a fluidized bed is considerably more complicated and diverse than is the course of a solid body rising to the surface or of a gas bubbling through a Newtonian liquid. During the rise of a bubble in a fluidized bed its gas exchanges with the surrounding two-phase "medium"—the particle aggregates; one bubble may merge with another, so that the bubble increases in size as it rises, and hence the Archimedes force on it may increase. In contrast to the viscosity of a Newtonian liquid, the effective viscosity of a fluidized bed may change quite markedly depending on the prevailing hydrodynamic conditions and the velocity of the fluidizing gas. Moreover, the viscosity of a fluidized bed is not the same over its entire height at a given superficial velocity. Nevertheless, the main factors determining the rate of rise of a bubble are obviously the same as in the case of a solid rising through a liquid: the viscosity of the medium (in this case the effective viscosity of the fluidized bed) and the Archimedes forces [in this $v_b(\gamma_{bed} - \gamma_g)$].

The maximum bubble size is limited. Where the two-phase theory of fluidization is valid, the difference between the specific gravities of bed and bubble is constant; where this is not the case, it decreases as the superficial velocity increases. The effective viscosity of the bed is far from zero. Hence the rate of bubble rise is again limited. Since the Archimedes forces increase with the third power of the bubble diameter while the cross section of the bubble is proportional to the second power, large bubbles will rise more rapidly than small ones of the same shape (other conditions being equal); this is observed in practice [966].

It is worth noting the relationship between the effective viscosity of a fluidized bed and the superficial velocity. As the superficial velocity is increased above the limit of stability, there is a sharp decrease in the viscosity, then something of an increase, and finally it falls again, but slowly [1130].

Many details of the passage of gas bubbles through fluidized beds are still unclear. For instance, it is not clear how closely Trawinski's concept [573] of stable bubble "shells" consisting of particles surrounding the bubbles corresponds to reality. The growth of bubbles and the frequently observed collapse of particles from the "ceiling" of the bubbles, falling through them as showers of particles, are in poor agreement with this concept. The collapse of particles* is a reason for a certain additional increase in the rate of bubble rise. However, in spite of the views of Yasui and Johanson, this increase cannot be very great, even if it is assumed that a dense continuous curtain of particles falls through the bubble (the limit at which it ceases to be a bubble).

If, within the accuracy of the calculation, the falling resistance of the particles is neglected in this limiting case, the time of fall from the upper to the lower edges of the bubble will be $\tau = (2h_b/g)^{1/2}$, which is valid for uniformly accelerated motion with zero-initial velocity. Here h_b is the diameter of the bubble. The mean-particle velocity in this time will be $w = h_b/\tau = (gh_b/2)^{1/2}$. It can be seen that the rate of movement of the upper bubble boundary (the rate of rise of the bubble) due to the collapse of particles cannot exceed w. Thus, for quite a large

*If it actually occurs inside the bubbles, and not just around the tube walls.

bubble diameter $h_b = 2$ cm, this component of the bubble rise velocity is less than 0.3 m/sec.

Due to the limited velocity of bubble rise, a large expansion can be expected in a nonuniform bed at high superficial velocities, since as the latter increases it begins to approach the velocity of bubble rise. For instance, it follows from Eq. (2-17') that the relative expansion of the bed will be around two when the superficial velocity reaches half the linear bubble velocity. Actually, in this case

$$H/H_{\ell.s} = 2w_f/(2w_f - w_f + w_{\ell.s}) = 2 - \frac{2w_{\ell.s}}{w_f + w_{\ell.s}} \approx 2,$$

since usually $w_{\ell.s} \ll w_f$ in the region of high superficial velocities. However, it should be noted that in practice gas-fluidized beds are rarely operated with $H/H_{\ell.s} > 1.15$–1.2 in order to avoid entrainment of material during the vigorous oscillations of the bed level due to the increasing nonuniformity of the bed. Assuming that also in this case $w_f \gg w_{\ell.s}$, i.e., neglecting $w_{\ell.s}$ in Eq. (2-17'), a relative bed expansion of $H/H_{\ell.s} = 1.2$ will correspond to $w_f \approx w_b/6$. Ratios of the superficial gas velocity to the bubble velocity close to this value were used in experiments of Orochko et al. [21]. The superficial velocity was 0.2–0.5 m/sec and the rate of bubble rise 1–2 m/sec.

The general tendency toward accelerated expansion at high superficial velocities (or, more exactly, at high-fluidization indexes) has also been confirmed by experimental data (see, for example, Fig. 2-9). Possibly the expansion lines in Fig. 2-9 begin to deviate downward as soon as the increase in the mean-bubble velocity with increasing superficial velocity starts to slow down.

What should be the relationship between the relative expansion of a nonhomogeneous bed and the minimum fluidization velocity $w_{\ell.s}$? According to Eq. (2-17'), the bed expansion decreases as $w_{\ell.s}$ increases at a given (constant) superficial velocity. This is to be expected, since in this case the operating regime approximates the limit of stability. If at a constant fluidization index $N = w_f/w_{\ell.s}$ the previous rate of bubble rise is maintained, the relative bed expansion must accelerate according to (2-17') as $w_{\ell.s}$ is increased (for instance, by going to particles of larger diameter). This can be demonstrated more easily if Eq. (2-17') is rewritten as

$$H/H_{\ell.s} = \frac{w_b}{w_b - (N-1)\,w_{\ell.s}}. \qquad (2\text{-}17'')$$

This accelerated expansion of nonuniform beds of coarse particles has been confirmed by experimental data (see Fig. 2-9).

Solchani [1177] also considered the expansion of a nonhomogeneous fluidized bed as a function of the rate at which the excess fraction of the fluidizing medium passes through in the form of bubbles. However, like Yasui and Johanson [586], he holds the quite unfounded view that the bubbles move upward mainly as a result of the fall of particles through them.

The rate of expansion of a nonuniform (two-phase) fluidized bed is governed by a larger number of factors than is the expansion of a uniform bed. As we have shown, there is nothing to prevent it from being larger than, smaller than, or equal to the rate of expansion of a uniform bed under various conditions. The fact that these expansions might happen to be equal at the same superficial velocity does not by any means indicate that a gas-fluidized bed is homogeneous, but only that the decrease in expansion due to the high density of the aggregates is compensated for by the increase in expansion due to the presence of a large number of bubbles moving quite slowly.

In some cases the difference between the expansion values for homogeneous and nonhomogeneous beds is quite small. According to Brötz [810], the same expansion law [Eq. (2-3)] can be assumed for a bed of particles finer than 300 μ fluidized with either liquids or gases in the laminar region (Re < 10) (see Fig. 2-6). Beds of particles coarser than 300 μ fluidized with gases (nonhomogeneous) expanded to a greater degree than uniform beds. This follows from Fig. 2-6, considering that

$$m/m_{\ell.s} = (w_f/w_{\ell.s})^{1/2}.$$

Brötz made the reservation that for gas-fluidized beds this correlation is useful only for relative expansions $H/H_{\ell.s} = (1-m_{\ell.s})/(1-m) \leqslant 1.15$-$1.20$. When the superficial velocity is further increased, vigorous pulsations of the bed level set in, and the upper boundary of the bed becomes indefinite.

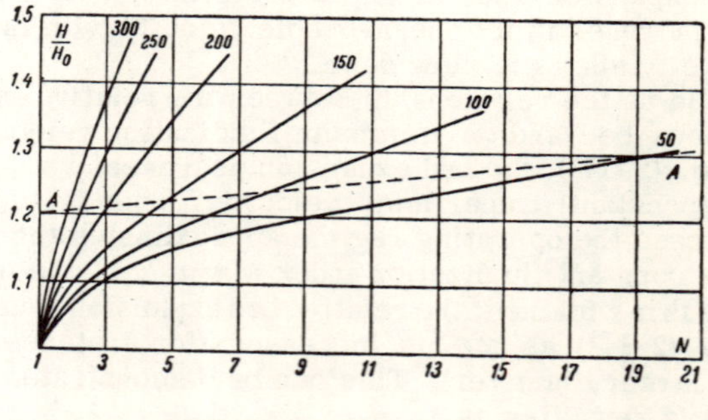

FIG. 2-11. Relative bed expansion H/H_0 as a function of the fluidization index N according to Leva [988]. The dotted line A–A indicates the recommended limit of expansion for narrow size fractions. The parameter is the particle diameter (mm) (numbers alongside the curves).

Leva [988] presented a graph for calculation purposes (Fig. 2-11) in which the relative expansion of a gas-fluidized bed is given as a function of the fluidization index N, with the particle diameter (d = 50-300 μ) as a parameter.

Figure 2-12 shows Leva's plot of the limiting attainable fluidization index versus particle diameter. In establishing these graphs, Leva assumed that poor fluidization (piston formation, violent fluctuations of the bed level) starts approximately where the expansion line in

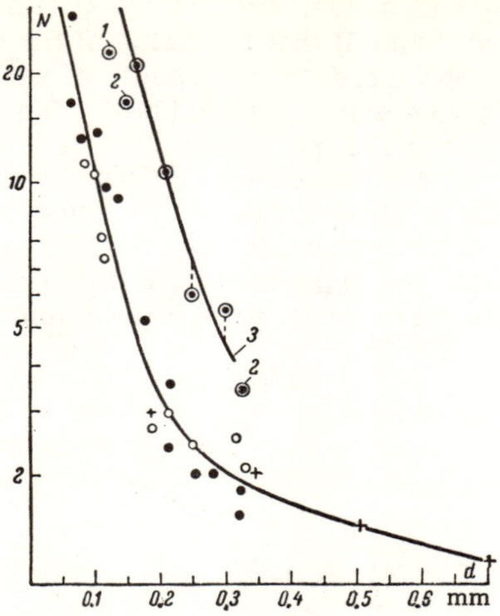

FIG. 2-12. Effect of particle properties on the limiting attainable value of the fluidization index N for gas-solid systems [988]: Narrow particle size fractions: ●—quartz sand in tube, D_T = 63.5 mm [418]; ○—quartz sand, D_T = 101 mm [418]; +—coal, D_T = 41.4 mm [189]. Wide particle size fractions (heterogeneity index > 2.5) in tube (D_T = 101 mm) [418]: 1) Fe_3O_4; 2) quartz sand; 3) anthracite.

logarithmic coordinates (w_f vs. $(1-m)/m^3$) begins to curve. The limiting fluidization index decreases as the particle diameter increases. According to Fig. 2-12, the limiting fluidization indexes are greater for fluidization of a wide particle size fraction than for a narrow fraction of the same mean-particle size, but with a nonhomogeneity coefficient $C_0 > 2.5$.

Leva defines the nonhomogeneity coefficient as the ratio d_{60}/d_{10}, where d_{60} and d_{10} are the screen openings that allow 60% and 10% of the bed material to pass. Leva states that, in general, beds of wide particle size distribution are better fluidized than uniform size beds, since they have a smaller tendency toward piston formation and aggregation. However, the numerical data on the limiting fluidization indexes (Fig. 2-12) for materials of wide size distribution cannot be regarded as reliable because of the small number of experimental points, the indefinite nature of the upper limit of the nonhomogeneity coefficient, and the inaccuracy in determining the limit of stability for such beds.

Generally speaking, there is no basis for assuming the same limiting attainable value of H/H_0 = 1.2–1.3 (Fig. 2-11) or of the fluidization index (Fig. 2-12) for all fluidized beds. For example, in the heat exchangers proposed by Zabrodsky et al. [1251], pulsations of the fluidized bed are used for organizing circulation of the material between the chambers of the heat exchanger, and it is beneficial to operate with much larger fluidization indexes and bed expansions than are recommended by Brötz and Leva.

It is surprising and convenient for calculation purposes (though results need to be confirmed) that the value of the expansion of a gas-fluidized bed does not depend on its breadth of particle size distribution, as found by Davies and Robinson [1107]. They investigated beds of glass spheres and coke particles fluidized with air in a tube of 57-mm diameter. The air passed into the bed through a porous bronze plate. Narrow fractions of particles in the range 74-912 μ were fluidized, as well as wide fractions with 5-, 10-, 25- and 100-fold ranges of diameters but with the same mean-particle diameter ($d_{av} = 271\ \mu$). Electrostatic effects were reduced to a minimum by slightly humidifying the air.

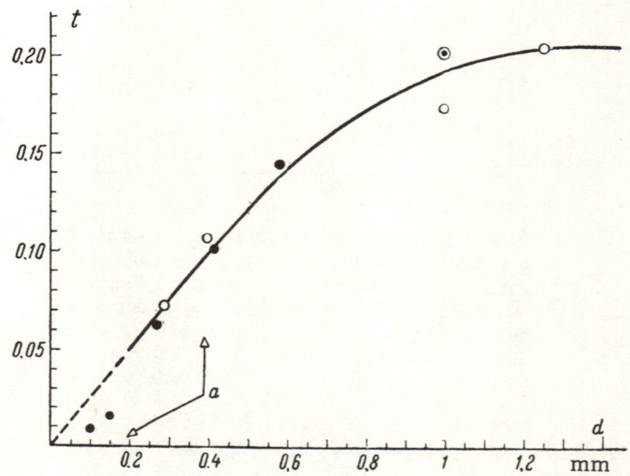

FIG. 2-13. Value of t in Eq. (2-18) as a function of the particle diameter d [988]:
●—Lewis et al. [422]; ○—Wilhelm and Kwauk [604]; ⊙—Massimilla and Bracale [433]; a) mixture of glass spheres, Lewis et al. [422].

As the superficial velocity is increased the surface fluctuations of a gas-fluidized bed become more violent. Leva [988] attempted to represent the relative oscillation of the bed level in the form of the function

$$r = e^{t(N-1)}, \qquad (2\text{-}18)$$

where $r = H_{max}/H_{min}$ is the ratio of the maximum instantaneous bed level to the minimum value at the given fluidization index N and $t = f(d)$ is an empirical function of the particle diameter, shown in Fig. 2-13. The form of the function $r = f(t, N)$ is shown in Fig. 2-14, where the dotted line marks the approximate lower limit of the unstable regime. At a given fluidization index the level fluctuations are more marked for large particles. From Fig. 2-13, the fluctuations are smaller for beds of wide particle size than for passes into the region of the unstable regime without a break. There is also no noticeable relationship between the relative fluctuation r and the ratio of the bed height to its diameter (Fig. 2-14). This is not in agreement with the usually observed effect of the ratio of bed dimensions on piston

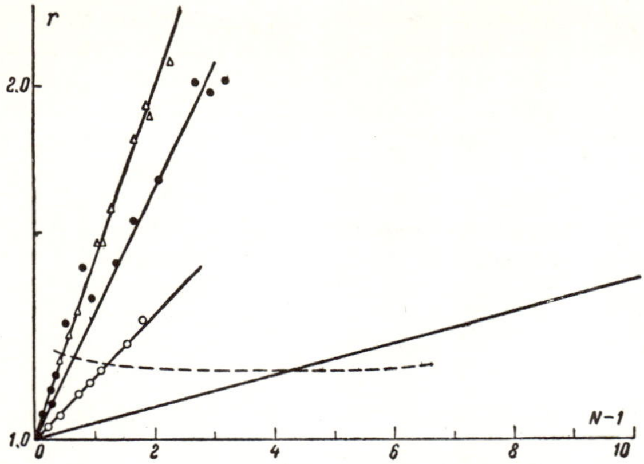

FIG. 2-14. Relative fluctuation of the bed level as a function of the fluidization index N and the value of t for narrow size fractions of glass spheres [988].
---- approximate limit of the unstable regime for narrow size fractions.

Symbol	d, mm	H_0/D_T	t
△	0.569	4.25	0.142
●	0.452	4.6, 5.3, 10.0	0.108
○	0.284	2.6, 10.0	0.060
Line without points	0.155	2.7, 9.8	0.015

formation; the subject of level fluctuations in fluidized beds requires further investigation.

Leva's correlation, which indicates a monotonic increase in the level fluctuations of a gas-fluidized bed, is obviously inapplicable over the region of large relative bed expansions $H/H_{\ell.s} > 3$–4, since there the nonuniformity of the bed decreases, and experimental data [657, 942] give evidence that a definite upper boundary of the fluidized bed reappears.

3. Interpolation formulas

The following interpolation formula of Todes, Goroshko and Rozenbaum is recommended for preliminary calculations of the expansion of beds fluidized with liquids and gases as a function of the superficial velocity (apart from the piston-flow regime and generally the region of vigorous level fluctuations):

$$\mathrm{Re} = \frac{\mathrm{Ar} \cdot m^{4.75}}{18 + 0.6\sqrt{\mathrm{Ar} \cdot m^{4.75}}}. \tag{2-19}$$

To solve the reverse problem of finding the superficial velocity required for a given bed expansion, it is convenient to invert Eq. (2-19):

$$m = \left(\frac{18\,\mathrm{Re} + 0.36\mathrm{Re}^2}{\mathrm{Ar}}\right)^{0.21}. \tag{2-19'}$$

Todes et al., derived the interpolation formula from the limit of stability of a bed of randomly packed circular particles with $m_{\ell.s} \approx 0.4$:

$$\mathrm{Re}_{\ell.s} = \frac{\mathrm{Ar}}{1\,400 + 5.22\sqrt{\mathrm{Ar}}} \tag{1-25}$$

(see above, p. 31) and the free-fall or terminal velocity of individual spherical particles [656];

$$\mathrm{Re}_{\mathrm{entr}} = \frac{\mathrm{Ar}}{18 + 0.61\sqrt{\mathrm{Ar}}}. \tag{2-20}$$

By simple algebraic rearrangement, Eq. (1-25) can be written as

$$\mathrm{Re}_{\ell.s} = \frac{0.0128\mathrm{Ar}}{18 + 0.59\sqrt{0.0128\mathrm{Ar}}}. \tag{2-21}$$

Neglecting the difference in the coefficients in front of the square-root signs in (2-20) and (2-21), and regarding the coefficients of Ar as some function of the voidage, the whole range over which a fluidized bed can exist, from $m_{\ell.s} = 0.4$ (the limit of stability) to $m = 1$ (entrainment), can be represented by a single interpolation formula:

$$\mathrm{Re} = \frac{\mathrm{Ar}\,\Phi(m)}{18 + 0.6\sqrt{\mathrm{Ar}\,\Phi(m)}}. \tag{2-22}$$

By approximating $\Phi(m)$ as a power function m^n from the two limiting values $\Phi(0.4) = 0.0128$ and $\Phi(1) = 1$, Todes, Goroshko and Rozenbaum obtained $n = 4.75$, and hence arrived at Eq. (2-19).

Equations (1-25) and (2-20) also make it possible to determine the range of fluidization indexes within which the fluidized bed can exist:

$$N_{\max} = w_{\mathrm{entr}}/w_{\ell.s} = \mathrm{Re}_{\mathrm{entr}}/\mathrm{Re}_{\ell.s}. \tag{2-23}$$

For very fine particles or large viscosities of the fluidizing medium (Ar $<$ 20):

$$N_{\max} \approx \frac{1}{18}\mathrm{Ar}\Big/\frac{1}{1\,400}\mathrm{Ar} \approx 78. \tag{2-23'}$$

For coarse particles or small viscosities of the medium (Ar $> 10^4$):

$$N_{\max} \approx \frac{1}{61}\sqrt{\mathrm{Ar}}\Big/\frac{1}{5.22}\sqrt{\mathrm{Ar}} \approx 8.5. \tag{2-23''}$$

Pinchbeck and Popper [501] determined earlier the rates of minimum fluidization and entrainment for beds of regular and irregular particles in nitrogen and arrived at very similar values for the limiting operating fluidization indexes:

$N_{\max} = 76$ for fine particles;

$N_{\max} = 10$ for coarse particles.

Thus fluidized beds can exist over a wider range of fluidization indexes with fine particles than with coarse. However, the range of absolute superficial velocities $w_{entr} - w_{\ell.s}$ [rather than dimensionless values (N)] over which a fluidized bed can exist is much wider for coarse particles than for fines, since the value of $w_{\ell.s}$ for coarse particles is many times larger than for fines.

Todes et al. [735], noted that the range over which a fluidized bed of wide particle size distribution can exist appears to be narrower than the value given by Eqs. (2-23') and (2-23") when the mean-particle diameter is used in the calculations. The disintegration of a fluidized bed of wide particle size must be governed by the carry-over of the finest particles in the bed, i.e., at some value of Ar other than that at the limit of stability.

Figure 2-15 (taken from [735]) compares Eq. (2-19) with the experimental data of these authors on the fluidization and free fall of particles of various materials in water and more viscous media, as well as with the data of several other investigators.

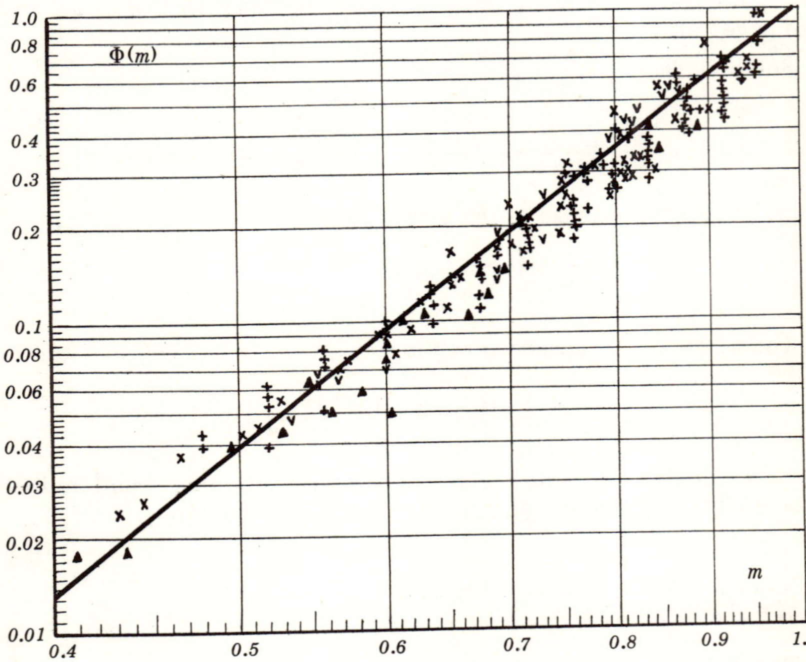

FIG. 2-15. Function $\Phi(m)$ of the bed voidage for Eq. (2-22) by Goroshko, Rozenbaum and Todes [735].
●—data of Goroshko, Rozenbaum and Todes [735]; ∨—data of Yerkova [32]; ×—data of Lewis et al. [744]; ▲—data of Wicke.

Recently, Zabrodsky and Shlapkova modified the interpolation formulas (2-19) and (2-19') due to Todes et al., in order to describe the expansion of fluidized beds under moderate vacuum (down to about 1 mm Hg), when flow with slip occurs in the bed. They proposed a more general empirical correlation:

$$\mathrm{Re}_{\ell.\mathrm{s}} = \frac{\mathrm{Ar}(1 + 10\,\mathrm{Kn}_{\ell.\mathrm{s}})^{1.18}}{1400 + 5.22\sqrt{\mathrm{Ar}}} \quad (2\text{-}24)$$

$$\mathrm{Re} = \frac{\mathrm{Ar}\,m^{4.75}(1 + 10^4\,\mathrm{Kn})^{0.054}}{18 + 0.6\sqrt{\mathrm{Ar}\,m^{4.75}(1 + 10^4\,\mathrm{Kn})^{0.054}}} \quad (2\text{-}24')$$

$$m = \frac{18\,\mathrm{Re} + 0.36\,\mathrm{Re}^2}{\mathrm{Ar}(1 + 10^4\,\mathrm{Kn})^{0.257}}^{\,0.21} \quad (2\text{-}24'')$$

where $\mathrm{Kn} = \bar{\Lambda}/d_h$, the Knudsen number;
$\bar{\Lambda}$ = mean-free path of the molecules;
d_h = mean-hydraulic diameter of the "channels" between the particles in the bed;
$\mathrm{Kn}_{\ell.\mathrm{s}}$ = value of Kn at the limit of stability of the bed.

Equations (2-24) to (2-24") might need to be improved when new data on fluidization under vacuum become available.

Example 2. Using the Todes interpolation formulas, find the height of a water-fluidized catalyst bed, using the conditions of example 1 (see page 61).

Solution. $\mathrm{Re} = \frac{wd}{\nu} = \frac{0.104 \cdot 4.42 \cdot 10^{-3}}{0.98 \cdot 10^{-6}} = 470;$

$$\mathrm{Ar} = \frac{gd^3}{\nu^2} \cdot \frac{\gamma_{\mathrm{M}} - \gamma_{\mathrm{m}}}{\gamma_{\mathrm{m}}} = \frac{9.81\,(4.42 \cdot 10^{-3})^3 (1\,600 - 998)}{(0.98 \cdot 10^{-6})^2\; 998} = 5.34 \cdot 10^5.$$

From (2-19') the voidage of the expanded bed is

$$m = \left(\frac{18 \cdot 470 + 0.36 \cdot 470^2}{5.34 \cdot 10^5}\right)^{0.21} = 0.685.$$

The height of the expanded bed is

$$H = H_{\ell.\mathrm{s}} \frac{1 - m_{\ell.\mathrm{s}}}{1 - m} = 0.71\,\frac{1 - 0.37}{1 - 0.685} = 1.42 \text{ m}.$$

This simpler procedure gives practically the same result as that obtained by the graphical analytical method (example 1).

4. Voidage distribution over the height of a fluidized bed

The solution of heat transfer problems requires not merely knowledge of the mean voidage of the bed (which governs its overall height), but also detailed information on the voidage distribution within the bed. The previous sections dealt only with such problems in fluidized bed structure as particle separation by weight and size; it has been shown that due to particle separation by size, an increase in voidage occurs over the height of any bed of wide particle size distribution. The vast majority of fluidized beds in industry and even in laboratory practice are not strictly of uniform particle size. The bed appearance is quite different if size separation is prevented in intensively pulsating nonuniform fluidized beds up to the piston-flow regime. Beyond the unstable regime, size separation will again be amplified in dilute-phase fluidized beds. Pulsations of dense aggregates actually facilitate the entrainment of the fine particles adjacent to the coarse ones, including those below. However, in the comparatively uniform fluidization of a dilute bed, where the particles remain far apart from one another, such interaction becomes negligible and separation by particle size increases again, confirmed experimentally.

Marked separation into different size fractions is observed at large bed expansions, when $H/H_{\ell.s} > 3$. Syromyatnikov [154] proposed another theory for the nonuniformity of particle concentration over the height of very dilute fluidized beds ($H/H_{\ell.s} > 3$); this is also valid for beds of uniform particle size to a first approximation. In a system of particles moving randomly in a gravity field it is characteristic that their concentration decreases exponentially with height (compare the so-called barometric formula of the molecular-kinetic theory of gases). According to the barometric formula the atmospheric pressure at a height h is

$$p_h = p_0 e^{-\frac{3gh}{c^2}}, \tag{2-24}$$

where c is the mean-square velocity of the randomly moving molecules.

By analogy, the pressure at a height h in a fluidized bed due to the random motion of particles in a gravity field (not the total pressure, however) is equal to

$$p'_h = p'_0 e^{-\frac{3gh}{u^2}} = p'_0 e^{-Bh}, \tag{2-25}$$

where u is the mean-square velocity of the random motion of the particles in a dilute-phase fluidized bed. The quantity B is constant for a given bed and given operating conditions (superficial velocity). Syromyatnikov has shown experimentally that the total pressure (the excess above atmospheric pressure) in a dilute phase fluidized bed also varies exponentially:

$$p_h = p_0 e^{-Ah}. \tag{2-26}$$

The voidage distribution is related to A as follows:

$$\frac{1-m}{1-m_0} = \frac{\gamma_{bed}}{\gamma_{0bed}} = e^{-Ah}, \tag{2-27}$$

where m and γ_{bed} are the voidage and bulk density of the bed at height h, and m_0 and γ_{0bed} are the voidage and bulk density at zero level.

Syromyatnikov's experiments [942] were carried out with beds of sand (d = 1 mm), coke (fractions of 1.25-5 mm) and chamotte (fractions of 2.5-5 mm) fluidized with air in cylindrical and conical tubes. The data obtained in a cylindrical tube of 52-mm diameter are given in Table 2-2.

Figure 2-16 shows the relationship between p_0/p and h in semilogarithmic coordinates for the experiments with coke. The points fall on straight lines with constant slopes A, in agreement with Eq. (2-26).

Table 2-2

Experimental data of Syromyatnikov [942].

No.	Material and particle size investigated	w_f, m/sec	$H_{l.s}$, mm	H, mm	P_0, kgf/m²	G_M/F, kgf/m²	A, m⁻¹
1.	Coke, 1.25-5 mm	2.22	32	150	24.0	23.6	21.2
2.	Coke, 1.25-5 mm	2.75	32	200	24.8	23.6	16.0
3.	Coke, 1.25-5 mm	2.95	32	225	26.0	23.6	14.3
4.	Coke, 1.25-5 mm	2.00	110	375	72.0	71.5	11.2
5.	Coke, 1.25-5 mm	2.78	110	575	75.0	71.5	7.5
6.	Coke, 1.25-5 mm	2.96	110	650	76.0	71.5	6.5
7.	River sand, 1 mm	1.62	25	100	50.0	47.2	40.0
8.	River sand, 1 mm	1.62	50	200	98.0	95.0	23.3
9.	River sand, 1 mm	2.45	100	500	205.0	189.0	10.5
10.	Chamotte, 2.5-5 mm	2.77	175	900	200.0	189.0	6.0

It is impossible to calculate the mean-particle velocity in a fluidized bed from Syromyatnikov's experimental data, using Eq. (2-27), since $u \neq \sqrt{3g/A}$. The numerical values of A are determined from changes in the total pressure; they reflect the change in particle concentration over the bed height due to all causes, not only to the random motion of the particles in the gravity field. As was shown earlier, the separation of particles according to size is also facilitated by the reduction in concentration of the material in the upper part of the bed; i.e., there are two effects operating on the voidage distribution which act in the same direction. Hence, calculating $\bar{u}$ as $\sqrt{3g/A}$ would be equivalent to ascribing to particle motion the effect producing separation, i.e., to knowingly exaggerate the value of the actual mean velocity of the random particle motion if there is separation.

Syromyatnikov's data (see Table 2-2) indicate that separation occurred in all the experiments. The coke and chamotte were known to

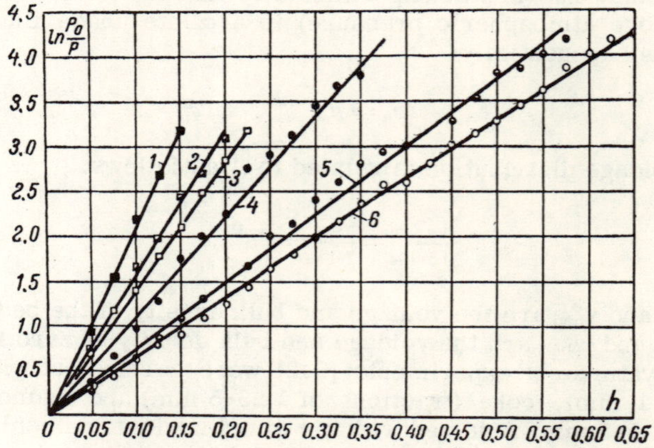

FIG. 2-16. Pressure distribution over the height of a fluidized bed of coke fines [942].

P_0/P—ratio of bed pressure at the grid level to the pressure at a distance h (m) above the grid. Numbers beside the lines indicate the experimental run numbers (see Table 2-2).

be of wide size distribution. Comparison of experiments 7 and 8 shows that the nominally uniform-size sand also had some distribution. In fact, at the same superficial velocity the value of A decreased from 40 to 23.3 only because of the different initial bed heights. It is doubtful whether a simple increase in the quantity of material in a bed of truly uniform particle size could lead to this sharp decrease in A, i.e., to such a marked increase in the particle velocities. It is more probable that there were different sizes of sand grains, leading to separation, and that with increased quantity of material the size range was more pronounced.

The expression proposed in [942]:

$$m = 1 - A \frac{g}{F_{gr} \gamma_M} e^{-Ah}, \tag{2-28}$$

makes it possible to determine the voidage at any height in a dilute fluidized bed if we know its weight g, the grid area F_{gr}, the bulk density of the particles γ_M, and the coefficient A, though the latter must be determined experimentally. However, even for the same bed, A varies with the superficial velocity, and no analytical relationship exists for this dependence. In addition, A must depend on the fractional composition of the bed. This all makes the practical application of Eq. (2-28) very limited.

Omae and Furukawa [484] incorrectly applied the barometric formula to describe the pressure distribution in a sand bed of wide size range fluidized with water, where there is generally no intensive particle motion. Eventually, Omae and Furukawa and also Andrieu [199, 672] approximated the experimental data for wide size range beds fluidized with liquids by a simple linear relationship between the pressure gradient and height.

Syromyatnikov [154] noted that over the range of bed heights from the grid to the initial level of filling ($H_0 \approx H_{\ell.s}$) the pressure distribution obeyed a linear law for $H/H_{\ell.s} < 2$, indicating a constant mean voidage over this range. Bakker and Heertjes [803, 1097] also obtained results which in many respects confirmed the data of Syromyatnikov on the pressure distributions in dense and dilute fluidized beds. They carried out fluidization experiments with narrow size fractions of glass spheres (d = 100-385 μ), silica gel, and polystyrene, using air in a 90-mm-diameter tube and a grid with 80-μ openings, above which they placed a layer of Raschig rings. The superficial velocity was 0.08-0.98 m/sec. The time-average local voidage was measured by means of a capacitance probe. At each point in the bed the voidage varied in time from $m_{\ell.s}$, corresponding to the limit of stability, to unity during the passage of a bubble. However, they found a regular distribution for the time-average value of the voidage.

When not in the piston regime of fluidization and in the absence of wall effects (beds of large diameter or the axial zone of a small bed), it was possible to distinguish three zones in the bed. The first was the zone in which the gas-distributing grid had an effect. The height of the first zone and the voidage distribution within it depended on the type of grid. The second was the zone of constant voidage ($m_{p.\ell}$) and lay above the first zone up to the initial bed height $H_{\ell.s}$. The third, the zone in which the voidage increased exponentially, lay above $H_{\ell.s}$.

The simultaneous existence of a dense phase with a dilute phase above it is explained by the ejection of particles by the bubbles. Here the dilute phase represents a zone of gas separation and falling of material back into the bed. There is a considerable quantity of material above the initial bed level, corresponding to the total volume of the bubbles that are simultaneously present in the bed below this level. On the basis of this model of the structure of a nonuniformly fluidized bed, Bakker and Heertjes proposed an equation with empirical coefficients for determining the voidage in the third zone (dilute fluidized bed) at a height h above the grid:

$$1 - m = (1 - m_{p.\ell}) \exp[-C_2(h - H_{\ell.s})^{C_3}], \quad (2\text{-}29)$$

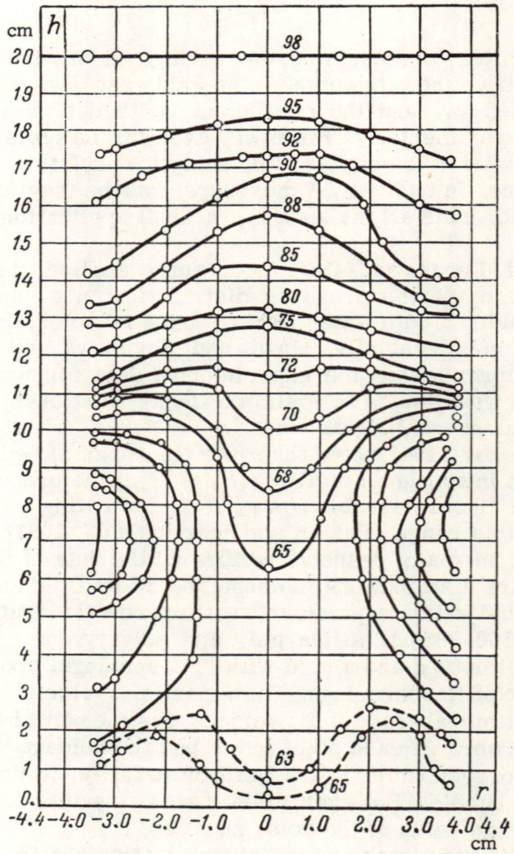

FIG. 2-17. Lines of constant voidage ("isopores") in a diametric cross section of a fluidized bed of glass spheres in the experiments of Bakker and Heertjes [1097].

h—height above grid, cm; r—radial distance from bed axis, cm;

Numbers of lines are the local bed voidages in percentages. Particle diameter 175–210μ, weight of bed 1 kg, superficial velocity 0.596 m/sec.

where $m_{p.\ell}$ is the voidage in the second zone:

$$m_{p.\ell} = m_{\ell.s} + \frac{w_f - w_{\ell.s}}{\overline{w}_b - w_{\ell.s}}(1 - m_{\ell.s}),$$

and $\overline{w}_b$ is the mean velocity of rise of the bubbles.

Bakker and Heertjes did not specify theoretically the boundary between the zones of the different laws of pressure distribution in the

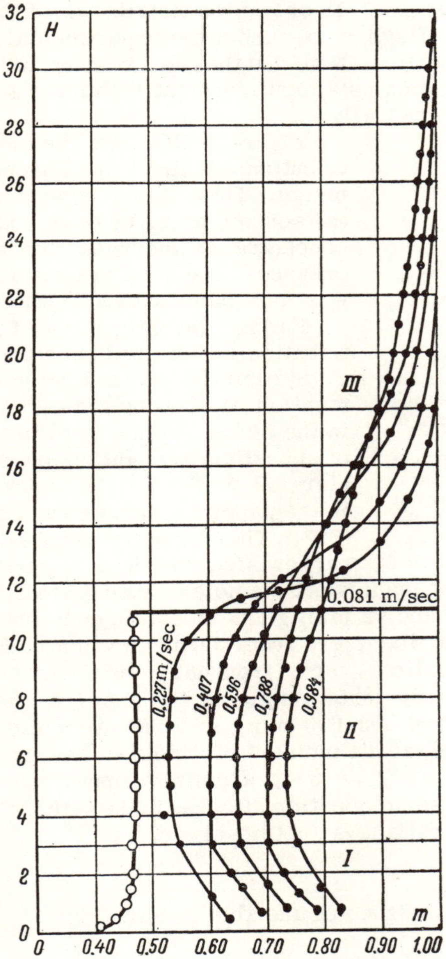

FIG. 2-18. Distribution of time-averaged voidage m along the axis of a fluidized bed of glass spheres at various superficial velocities w_f; minimum fluidization velocity $w_{\ell.s}$ = 0.08 m/sec; d = 175–210μ [1097].

I, II, III: zones of the fluidized bed according to the classification proposed by Bakker and Heertjes.

fluidized bed (between the second and third zones), and hence the experimental data (see Fig. 2-17) must be regarded as preliminary. It is important that Bakker and Heertjes drew attention to the existence of the first zone—the zone in which the gas-distributing device had an effect—though the measured data on the mean voidage over the bed height for this zone and the others are only of a preliminary, exploratory nature. For instance, little could be said previously concerning the first zone; also it was made clear that the type of grid used could have an important effect on the height of this zone and on the voidage variation in it. Figure 2-18 shows the distribution of the mean voidage along the axis of a fluidized bed under the experimental conditions used by Bakker and Heertjes. Note that the mean voidage initially decreases somewhat, then remains almost constant within the second zone, and finally increases gradually.

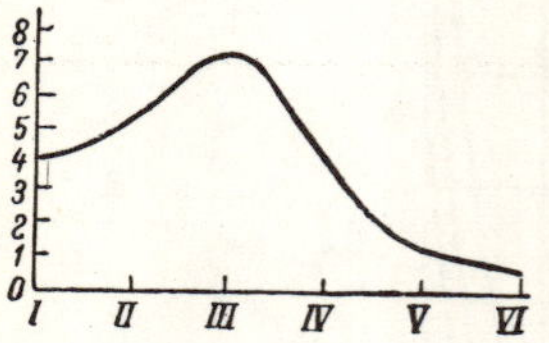

FIG. 2-19. Particle concentration distribution over the height of a fluidized bed according to the experiments of Chechetkin [942]. Ordinate: relative particle concentration. Abscissa: height above the bottom of the bed (figures refer to the number of the section).

Figure 2-19 shows the concentration distribution obtained in Chechetkin's experiments. This agrees approximately with the the scheme given by Bakker and Heertjes: the concentration increases initially and then decreases. The zone of constant concentration was not pronounced in these experiments.

Further investigations of the voidage distribution in fluidized beds are required.

Syromyatnikov and Volkov [942] made no mention of passing through an unstable zone in the piston regime despite their wide range of superficial velocities and small bed diameter (52 mm). Similar information is given in the paper by Romankov, Lepilin and Nemet [657]. They investigated air-fluidized beds of narrow size fractions of particles of irregular shape and mean-equivalent diameters from 0.5 to 3.5 mm in small-diameter (32, 49 and 52 mm) glass tubes. In each fraction the deviation of the particle diameters from the mean was ± 0.25 mm. The bulk densities of the particles varied from 482 to 1080 kg/m^3, and the initial bed heights from 50 to 500 mm. In the piston regime the bed level fluctuations reached 500-600 mm, and the hydraulic resistance increased by 40-60% at the moment of "slug" formation. When the bed had expanded to $H/H_{\ell.s} = 3$-4, a definite upper bed boundary reappeared. There is some question attached to the authors' statement that particle size separation was not observed.

5. Boundaries of the zone of unstable fluidization

In the vast majority of cases the zone occupied by the unstable regime of nonhomogeneous fluidization—the zone of vigorous fluctuations or even of the absence of a definite upper limit to the bed—cannot be used in practice because of the large carry-over of the material from the bed. It is therefore of practical importance to know the limits of this zone and the limiting values of the fluidization index or of the relative bed expansion.

Table 2-3

Limits of the unstable regime [717].

No.	Porosity corresponding to the upper limit of the unstable regime	Porosity corresponding to the lower limit of the unstable regime	$d \times 10^3$, m	γ_M, kg/m³	γ_m, kg/m³	$\frac{\gamma_M}{\gamma_m}$	Investigator
1.	No information	0.6	5.2	2,340	1.22	1920.0	Kwauk
2.	Stable bed		5.2	2,340	1000.0	2.34	Kwauk
3.	Stable bed		4.41	1,602	1000.0	1.60	Kwauk
4.	Unstable bed		3.81-8.38	14,000	1000.0	14.0	Kwauk
5.	Stable bed		3.17	147	5.6	26.2	Grosselfinger
6.	0.945	No information	1.68	1,090	1.22	892.0	Zenz
7.	No information	0.7	1.00	2,630	1.22	2150.0	Kwauk
8.	0.987	No information	0.93	2,640	1.22	2165.0	Zenz
9.	0.983	No information	0.59	2,480	1.22	2040.0	Zenz
10.	No information	0.7	0.51	2,500	1.22	2050.0	Kwauk
11.	No information	0.8	0.28	2,500	1.22	2050.0	Kwauk
12.	0.978	No information	0.168	2,100	1.22	1720.0	Zenz

The lower boundary of the unstable regime can conveniently be regarded as the fluidization index, voidage, or relative bed expansion when piston formation begins (in beds of small diameter) or when intolerably large level fluctuations begin (in beds of any diameter). As the upper boundary it is convenient to use the fluidization index, voidage, or relative bed expansion at which a definite surface (now of the dilute fluidized bed) reappears. It is also possible that limiting attainable bed level fluctuations could be agreed upon for the upper limit of the unstable regime. As already mentioned, the limiting superficial velocity, corresponding to the upper limit of the unstable zone (the lower limit for the existence of a homogeneous dilute fluidized bed), is lower than the choking velocity of a vertical pneumatic transport tube (see Chapter 3).

Unfortunately there are still few data on the boundaries of the unstable regime. Table 2-3 gives data published by Zenz [717] derived from his own investigations and the work of Kwauk and Grosselfinger on the limits of slugging. They did not indicate the tube diameter, and it is obvious that the term "slugging" implies the unstable regime in general. The tabulated data are in quite good qualitative agreement with the preceding discussion on the dependence of the bed nonuniformity on the type of medium, the ratio of the densities of the material and medium, and the particle diameter. For example, with finer particles (d = 0.279 mm) the unstable fluidization regime commences later (at m = 0.8 rather than 0.7) than with coarser particles (d = 0.51 mm); i.e., the unstable regime is narrower.

If we approximate the bed voidage at the limit of stability as 0.4 for all the experiments indicated in the table, the lower limits of the unstable regime correspond to a relative expansion of $H/H_{\ell.s}$ = 115-3, while the upper limits correspond to 11-46.

This is not in agreement with the statement of Brötz [810] that the upper boundary of a gas-fluidized bed is lost at relative expansions of 1.15-1.20.

There is also a lack of agreement with the data of Romankov, Lepilin and Nemet [657] on the upper limit of the unstable zone, in that a definite interface appears at a relative bed expansion of the order of 4, rather than 10 or more.

To a certain extent these discrepancies can be attributed to the lack of a clear definition of the unstable fluidization regime. It is necessary to establish a single definitive terminology. The main problem is to carry out further experimental investigations that will be more systematic and complete than those made to date. In order to approximate the lower limit of the unstable fluidization regime the graphs in Fig. 2-14 for fine particles ($d < 0.3$ mm) are applied by specifying an arbitrary permissible fluctuation of the upper level of the fluidized bed.

6. Discharge of fluidized beds

Massimilla [1211] investigated the free efflux of gas and solids through openings in the walls of tubes containing gas-fluidized beds of quartz sand, iron ore, coke, sawdust, glass granules and limestone

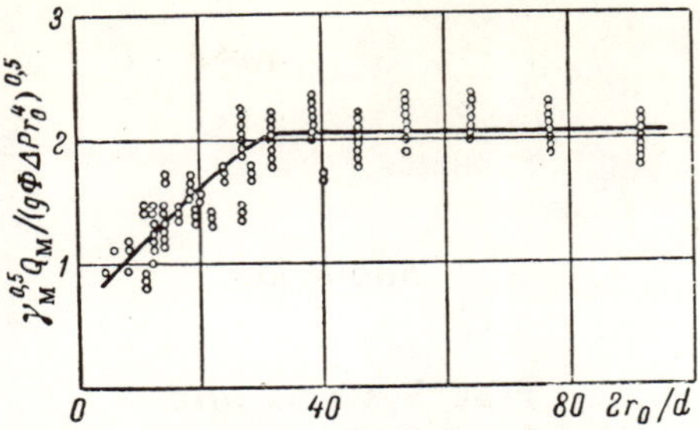

FIG. 2-20. Empirical relationship (2-31) for determining the volumetric flow rate of material Q_M (m^3/sec) through an opening of diameter $2r_0$ [1211].

pellets. The particle diameters varied from 220 to 800 μ and the pressure at the point of efflux from 60 to 5000 kgf/m^2.

One series of experiments was carried out in a tube 900 mm in diameter and 1200 mm high with openings 3-7 mm in diameter. A second series of experiments was carried out in a tube 90 mm in diameter and 2700 mm high, using openings 7-20 mm in diameter.

Massimilla obtained a relationship (shown in Fig. 2-20) between the dimensionless parameters that satisfactorily described his experimental data:

$$\frac{\gamma_M^{0.5} Q_M}{(g\Phi\Delta P r_0^4)^{0.5}} = f\left(\frac{2r_0}{d}\right), \qquad (2\text{-}31)$$

Figure 2-20 permits determination of the bulk flow rate of material Q_M (m^3/sec) through an opening of radius r_0 (m). In the above equation Φ is the Carman shape factor and ΔP is the efflux pressure, kgf/m^2.

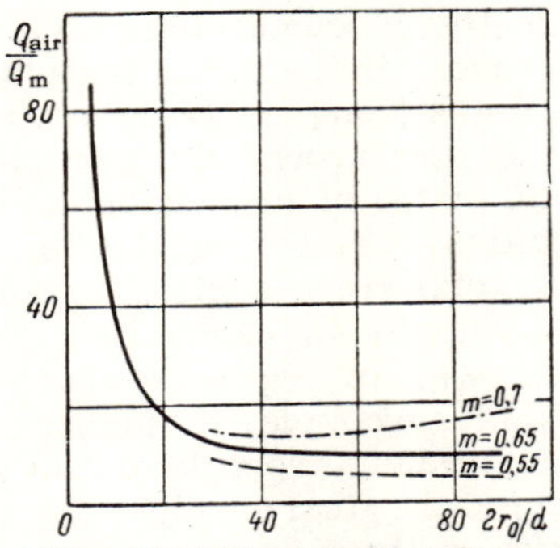

FIG. 2-21. Voidage (Q_{air}/Q_M) of a two-phase jet flowing from a fluidized bed through an opening of diameter $2r_0$. Parameter: voidage of fluidized bed (m).

The volumetric flow rate of gas, Q_{air} (m^3/sec), through the opening is found from Fig. 2-21, where the upper correlation curve corresponds to a fluidized bed voidage of m = 0.70, the middle curve to m = 0.65, and the lower curve to m = 0.55.

Chapter 3

Free Systems and Particle Transport

3.1 Classification of Systems of Suspended Particles

Particles of solid material may exist in a number of quite different states of hindered fall: in a fluidized bed (ascending or descending), in vertical pneumatic transport (with the material moving upward or downward), or in a settling bed with a fluid medium at rest. The velocity of hindered fall depends on the population density of the particles; in other words, on the bed voidage.

An ideal homogeneous bed of given voidage, consisting of uniform particles equidistant from one another, must be fluidized by the same relative gas velocity throughout the bed in all the states mentioned above. This follows directly from the principle of relativity in classical mechanics. In fact, such a bed's hydraulic resistance, which balances the particle weight, depends only on the relative velocity of the medium and is independent of whether the medium moves through the bed, the medium is at rest and the bed moves, or the medium and bed both move cocurrently or countercurrently. Exceptions occur due to the effects of differences in the initial turbulence of the medium in each of the various states (compare Du Buat's paradox [43]).* Thus, the relative velocity of flow around the particles at a given voidage must be the same in all states of suspension. Consequently, the relative superficial velocities are also equal, i.e., the nominal relative velocities based on the total cross-sectional area of the bed.

The mean-relative velocity $\overline{w}_{rel}$ cannot be expected to equal the velocity of hindered fall, w_{fall}, for regions of accelerated or decelerated motion of the material, for example, the pickup zone, where the mean velocity of the material varies from the initial velocity with which it is fed into the system to the final steady-state velocity. Equilibrium conditions do not exist in these regions. In real transport systems with suspended materials $\overline{w}_{rel} \neq w_{fall}$ even in regions where a constant material velocity has been established. In an ascending stream of gas $\overline{w}_{rel} > w_{fall}$, and in a descending stream $\overline{w}_{rel} < w_{fall}$. This is attributable to retardation of the moving particles by impingement on the tube walls [170]; essentially the same reason for the difference between $\overline{w}_{rel}$ and w_{fall} holds in an entry section, namely, acceleration (retardation)

*Discussed, for example, by Rouse and Ince, History of Hydraulics, 1957, p. 134.

of the particles. The difference between $\overline{w}_{rel}$ and w_{fall} increases as the stream velocity and concentration of material increase and as the tube diameter decreases [170].

The positions of the feed and discharge points of the suspending medium and material, the presence and positions of screens that limit the motion of the material, and the feed rates of material and medium all play an important part in determining which state will exist in the suspended material (fluidized bed, suspended bed, etc.). Lapidus and Elgin [692] introduced the approximate classification of "free" and "mechanically restrained" for vertical systems involving suspended material. Free systems are those in which the stream of material is not restrained by any special devices within the tubes (columns) and the supply of material is regulated externally (with a flow rate of fluid sufficient for suspension). Systems with mechanical restraint of the material are those in which the particle paths in the tubes are limited by screens, and the flow rate of material through the system is regulated at the discharge point. Cocurrent and countercurrent systems differ, as usual, in the mutual directions of motion of the material and fluid, and, depending on the direction of motion of the material itself, systems are differentiated into ascending and descending.

At sufficiently high superficial velocities, a contraction in the cross section of the apparatus can be used instead of a grid for restraining the fall of the particles. Thus it is possible to fluidize materials without grids in conical equipment contracting downward [170].

Free countercurrent system. Particles and flowing medium are fed in at opposite ends of a vertical tube. If the particle density of the material γ_M is greater than the density of the medium γ_m, the particles are fed in at the top and fall downward (a countercurrent "falling bed" in Chuckhanov's terminology). The settling of a bed is a special case of a free countercurrent system in which the solids fall through a fluid at rest. If $\gamma_M < \gamma_m$ the material is fed into the free countercurrent system at the bottom and the flowing medium at the top.

Free cocurrent system with descending flow. Here γ_M must be larger than γ_m. The particles and fluid are fed in at the upper end of the tube and move downward (a cocurrent "falling bed").

Free cocurrent system with ascending flow. Particles and fluid are fed in at the bottom and move upward (conventional pneumatic transport).

System without supply of material, restrained below. Here $\gamma_M > \gamma_m$; the fluid is supplied at the lower end of the tube, below a grid, above which there is a charge of material (batch-operating fluidized system). If the grid is removed, the system degenerates into an unstable free system which finally collapses due to the fall of particles projecting from the lower surface of the bed. The free lower bed surface can never be stable because of particle movement, which is characteristic of true suspensions.

System restrained below with supply of material at top. The fluid is fed in at the bottom, below a screen; material is supplied at the top and removed at the bottom of the bed (descending fluidized bed).

System restrained below with supply of material at bottom. With a screen at the lower end, material ($\gamma_M > \gamma_m$) is fed in at the bottom and removed at the top of the tube (ascending fluidized bed).

Lapidus and Elgin also included in their classification the rare case of a system restrained above and below and, incorrectly, two systems with descending motion of the material, in which the material is not suspended, strictly speaking, and which are concerned with downward-moving dense beds. Their classification does not cover all possibilities. For instance, in a single vertical tube an ascending suspended bed of fine particles and a falling bed of coarse particles can exist simultaneously [46].

Wilhelm and Valentine [605] investigated a free system with the material supplied halfway along the tube height. Such a system cannot be classified as either purely countercurrent or purely cocurrent. It operates in either manner depending on the velocity of the stream of fluid.

The countercurrent falling bed with artificially retarded particle fall, applied by Fedoseyev [1084] in grain dryers, deserves special mention; it does not differ greatly from systems with suspended materials. Retardation of the grain fall is achieved by placing horizontal screens (openings about 16 × 16 or 20 × 20 mm) in the column.

3.2 Free Systems and the "Entrainment Velocity"

In restrained systems, such as fluidized beds, there is one less degree of freedom than in free systems, and a single-valued relationship between the bed voidage and the superficial velocity. Knowing the stream velocity, an additional condition is necessary in free systems in order to calculate the bed voidage. Even the simple problem of the rate of settling of a bed of particles of one given dimension in a given quiescent medium has a multivalued solution when only the single condition is given that the relative velocity is equal to the velocity of hindered fall; in general, an infinite number of solutions are possible.

Suppose that a bed consisting of a definite quantity of material lies on a grid in an infinitely long vertical tube filled with a fluid. When the screen is removed instantaneously, hindered fall (settling) of the bed occurs. It is obvious that if the particles in the very lowest layer of the bed fall more rapidly than those above, the bed will expand during the time of fall. When the upper particles experience practically the same hydrodynamic conditions as the lower particles an equilibrium state is established with high voidage. The situation will be quite different if at any voidage the upper particles overtake the lower ones during their fall. In such a case a low voidage is established during settling, close to the voidage of a dense bed.

Additional information describing the actual situation is obtained from experimental observations of the upper boundary of the bed, which is initially uniformly mixed, during settling in a vertical tube (vessel). As shown experimentally [380], this surface settles initially at a constant velocity (Fig. 3-1) that depends inversely on the concentration of the material. The subsequent slowing of settling of the surface is explained by the bed becoming denser, which occurs because of the limited height of the vessel.

Todes et al. [735], determined the rate of settling from the subsidence of the upper surface of a fluidized bed after cutting off the supply of fluidizing agent. The experiments were carried out in a vertical glass tube 17 mm in diameter and 1400 mm high. The velocities of fluidization and hindered fall were measured for suspensions of coal particles 0.45-0.5 mm in diameter, sand particles with mean diameters from 0.23 to 1.54 mm, steel spheres (d = 1 mm) and lead spheres (d = 2.14 mm) in water and more viscous mixtures of glycerine and calcium chloride solution. The voidage varied from 0.96 to 0.52. The Archimedes number for the various series of experiments varied from 0.2 to 10^6; i.e., it extended over a very wide range of laminar, transitional and turbulent flow conditions. The experiments confirmed that the settling rate determined in this way was equal to the superficial fluid velocity during fluidization. Thus the particles in the upper part of a settling suspension actually fell with the same velocity, and during this fall the bed retained its initial voidage, at least in its upper parts, during the period of observation.

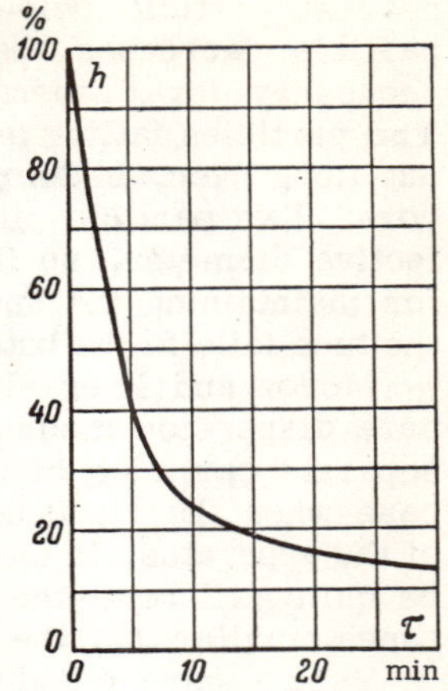

FIG. 3-1. Settling of the upper boundary of a typical suspension, according to Kalinske [380].
h—height of column of suspension, as percentage of initial height; τ—time from the beginning of settling.

On the other hand, investigations of free systems with ascending suspended material (hydrotransport and pneumatic transport) and descending material (falling beds) showed that there is a characteristic velocity of the medium at which the stream is saturated with material and at which disturbances and stoppages in the operation of the system set in—the material ceases to be transported in the given direction. This is the so-called "choking velocity" for vertical pneumatic conveyors. For example, decreasing the velocity of the ascending stream of medium in the case of vertical pneumatic transport while maintaining the same material feed rate leads to an increase in the true material concentration in the suspended layer [47], i.e., a decrease in the bed voidage m similar to the decrease in m in a fluidized bed when the superficial velocity is reduced. Ultimately, as the velocity of the medium is further reduced, particles begin to drop out and pneumatic transport ceases, even though this velocity (the choking velocity) is many times larger than the minimum fluidization velocity for the same material in a system with a screen; i.e., it would otherwise appear that dropout of the material is impossible even at velocities much smaller than the choking velocity.

Zenz [717] gives a graphic and valid qualitative explanation of the choking phenomenon in pneumatic transport. In suspensions with very low solids concentrations in the ascending gas stream the particles could be at distances of, say, about 100 particle diameters from one another; each particle leads to the formation of an eddy zone 20 particle diameters long behind (above) itself. When the fluid velocity is decreased somewhat, the concentration of particles in the suspension increases so that the mean distance between particles becomes less than 20 diameters. Each particle then falls in the eddy trail of the next nearest particle below. Usually turbulence in the stream around a particle decreases the drag coefficient; i.e., a higher velocity is necessary for suspending the particles in the turbulent eddy zone. The particles falling into the turbulent trails therefore begin to settle out along them. In doing this, they come into contact with their neighbors. Two particles in contact one above the other have a larger effective diameter, so that the fluid velocity is obviously insufficient for maintaining the suspension, and the solid material contained in the tube falls to the bottom.

Moroz and Frenkel' [99] indicate that a "cloud of suspension in a pure dispersion medium falls many times more rapidly than do the separate particles of the suspension. Finally, we must consider the case when the cloud does not fill the whole transverse cross section of the apparatus. In the opposite case the effect of collective fall would be quite different—the rate of hindered fall of the bed would be many times smaller than the rate of fall of individual particles.

Zenz's scheme and the experimental data of Gasparyan and Zaminyan [733] are in agreement. In their experiments on the fall of two adjacent steel spheres along the axis of a tube filled with glycerine, the velocity of simultaneous fall was about 1.5 times larger than the rate of fall of a single sphere in the same tube. Measurements carried out by Rowe and Henwood [1224] on a hydraulic model of a particle system also confirmed that the resistance of particles in the wakes around other particles falls sharply.

A relationship must be found between the choking velocity w_{choke} and the nominal concentration of material. On the basis of Zenz's model of the choking mechanism it can be assumed that choking corresponds to some definite value of the true material bulk concentration. Hence, at a given density of medium and a given particle density, the true weight concentration at choking is therefore also a constant quantity:

$$\mu_{\text{true choke}} = \text{const} = C_1.$$

The true material concentration in vertical pneumatic transport in steady-state motion [47] depends on the nominal concentration μ, the rate of fall of the particles w_{fall}, and the gas velocity w_{gas} in the following way:

$$\mu_{\text{true}} = \mu/(1 - w_{\text{fall}}/w_{\text{gas}});$$

where $\mu = Q_M/Q_m$ is the ratio of the weight flow rates of material and conveying fluid.

At choking conditions,

$$\mu_{\text{true choke}} = \mu/(1 - w_{\text{fall}}/w_{\text{gas}}). \tag{3-1}$$

For a suspended bed we can write $w_{\text{fall}} = \text{const} = C_2$, and solving (3-1) for w_{choke}, we get

$$w_{\text{choke}} = C_1/(1 - C_2\mu). \tag{3-2}$$

Thus in vertical pneumatic transport there is a direct relationship between the choking velocity and the nominal concentration μ obtained from the flow rates. If μ is large, choking begins even at high velocities of the transporting fluid.

Experiments show that the choking velocity is higher than the velocity of transition between a dilute fluidized bed and the unstable regime in the same system when restrained by a screen. This is to be expected, since it follows from what has been said above that the settling out of material during choking does not require the formation of definite aggregates; it is sufficient for the particles simply to approach each other. The tendency for both of two adjacent particles lying one above the other to settle out is also present in systems restrained with grids (fluidized beds), as long as the particles are in the necessary orientation. However, the material cannot settle into the lower part of the tube because of the screen. In a fluidized bed such behavior leads only to the development of pulsations. The screen itself is obviously never completely unloaded, and receives short-term impulses from the bed. The total pressure of the fluidized bed on the screen cannot be large, since the individual impulses are small and displaced in time.

It is most efficient to operate pneumatic conveyors with a minimum velocity of the conveying medium so as to have the least consumption of energy for transport and to reduce erosion of the tube and attrition of the material. It is therefore best to keep close to the limiting minimum velocity—the choking velocity. With this in mind, the industrial practice is sometimes to use pneumatic transport lines that expand upward. The system initially operates at a high gas velocity, considerably in excess of the choking velocity, and then the velocity of the transporting agent decreases gradually, approaching the economic regime [717]. The purpose of the expansion is to ensure that the velocity at the upper end of the tube is smaller than at the lower end, compensating for the expansion of the transporting medium (gas, vapor) due to the pressure drop along the tube, and in some cases by heating. As a result, choking will be initiated at the upper end of the tube. This leads to slugging and pressure fluctuations, but the transport system does not become clogged with material since the velocity lower down is larger than the choking velocity. The first pressure pulsations at the upper expanded end of the tube signal the approach to choking conditions, and warn that the gas velocity should not be lowered further.

Zenz [717] noted validly that if the pressure drop is not limiting, it is entirely possible to operate a vertical transport line of small diameter even in the piston-flow regime, i.e., at velocities below the choking point. This can cause serious vibration of the transport line if its diameter is large, since there will then be large pressure fluctuations as massive quantities of material slug through the pipe. However, in relatively long pipes of small diameter, the transition from material transport in the dilute or disperse phase to the piston regime might be scarcely noticeable [717]. Obviously the fluctuations in the total pressure drop are small in the case of long lines and small slugs, since the individual changes do not occur at the same time. Zenz makes no mention of means for preventing the material from falling back to the lower part of the tube. In this connection it is obviously useful to constrict the lower part, or simply fit a grid, i.e., convert the free system into a restrained system (transport in an ascending fluidized bed in the piston regime).

Table 3-1

Summary of experimental data on the limiting transportable load in the dilute phase and on the velocities of the medium at the onset of saltation and choking.

Investigators	Transport medium	d, mm	γ_M, $\frac{kg}{m^3}$	Size fraction	w_{choke} or w_{salt} m/sec	G_{Mmax} $\frac{kg}{m^2 \cdot sec}$	μ_{max}	$\frac{w^2_{choke} \text{ or } w^2_{salt}}{gd\gamma^2_M} \cdot 10^{-3}$, $m^6 \cdot kg^{-2}$	D_T, mm	Direction of stream
Smith and Carruthers [546]	Water	0.240	2595	Narrow	0.824	16.8	0.0204	0.0204	76	Horizontal
	Water	0.240	2595	Narrow	1.22	97.6	0.0799	0.0934	76	Horizontal
	Water	0.240	2595	Narrow	1.37	273.8	0.201	0.117	76	Horizontal
Howard [351]	Water	0.378	2595	Wide	2.29	468.5	—	—	102	Horizontal
Yufin [181]	Water	0.25	2595	Narrow	1.20	24.4	0.0204	0.0922	250	Horizontal
	Water	0.25	2595	Narrow	1.40	28.3	0.0204	0.118	300	Horizontal
	Water	0.25	2595	Narrow	1.50	429.4	0.286	0.137	250	Horizontal
	Water	0.25	2595	Narrow	1.70	844.2	0.50	0.175	250	Horizontal
	Water	0.25	2595	Narrow	2.00	1000	0.50	0.243	300	Horizontal
	Water	0.32	2595	Narrow	1.25	62.9	0.0505	0.0739	250	Horizontal
	Water	0.32	2595	Narrow	1.60	159.1	0.10	0.119	300	Horizontal
	Water	0.32	2595	Narrow	2.70	2689	1.0	0.345	300	Horizontal
Settle, Perkins [546]	Water	0.059	2003	Wide	0.275	30.5	—	—	102	Horizontal
	Water	0.079	2003	Wide	0.406	82.5	—	—	89	Horizontal
Blatch [217]	Water	0.584	2595	Narrow	1.56	205	0.132	0.0603	25.4	Horizontal
	Water	0.584	2595	Narrow	2.04	536.8	0.263	0.104	25.4	Horizontal
	Water	0.584	2595	Narrow	2.26	878.4	0.391	0.127	25.4	Horizontal
	Water	0.584	2595	Narrow	2.50	1308	0.524	0.156	25.4	Horizontal
	Water	0.584	2595	Narrow	2.56	1664	0.65	0.164	25.4	Horizontal
	Water	0.584	2595	Narrow	2.75	2147	0.785	0.188	25.4	Horizontal
	Water	0.584	2595	Narrow	2.81	2538	0.906	0.197	25.4	Horizontal
Gregory [317]	Water	0.012	2563	Wide	0.702	151.3	—	—	102	Horizontal
	Water	0.012	2563	Wide	1.10	307.5	—	—	102	Horizontal
	Water	0.012	2563	Wide	1.59	556.3	—	—	102	Horizontal
	Water	0.012	2563	Wide	1.98	805.2	—	—	102	Horizontal
	Water	0.012	2563	Wide	2.41	1113	—	—	102	Horizontal

Table 3-1 (continued)

Investigators	Transport medium	d, mm	γ_M, kg/m³	Size fraction	w_{choke} or w_{salt}, m/sec	G_{Mmax} kg/m²·sec	μ_{max}	$\frac{w^2_{choke} \text{ or } w^2_{salt}}{gd\gamma^2_M \cdot 10^{-3}}$, m⁶·kg⁻²	D_T, mm	Direction of stream
Ambrose [193]	Water	0.582	2643	Narrow	1.14	2.37	0.00207	0.0325	141	Horizontal
	Water	0.582	2643	Narrow	1.03	0.942	0.00092	0.0263	141	Horizontal
	Water	0.582	2643	Narrow	0.899	0.478	0.000531	0.0202	141	Horizontal
	Water	0.582	2643	Narrow	0.772	0.239	0.00031	0.0149	141	Horizontal
	Water	0.582	2643	Narrow	0.659	0.094	0.0001435	0.0109	141	Horizontal
	Water	0.582	2643	Narrow	0.567	0.0478	0.0000831	0.008	141	Horizontal
	Water	0.25	2643	Narrow	0.839	0.488	0.000581	0.0409	141	Horizontal
	Water	1.62	2643	Narrow	0.854	0.473	0.000555	0.00654	141	Horizontal
Zenz [625, 717]	Air	1.68	1089	Narrow	3.05	19.0	5.1	0.475	45	Horizontal and vertical
	Air	1.68	1089	Narrow	3.42	39.5	9.45	0.595	45	Same
	Air	1.68	1089	Narrow	3.89	68.3	14.33	0.770	45	Same
	Air	1.68	1089	Narrow	4.49	104.4	19.0	1.031	45	Same
	Air	1.68	1089	Narrow	5.11	141.5	22.7	1.330	45	Same
	Air	0.929	2643	Narrow	3.57	6.54	1.62	0.199	45	Same
	Air	0.929	2643	Narrow	5.19	239	4.78	0.261	45	Same
	Air	0.929	2643	Narrow	4.76	46.9	8.05	0.354	45	Same
	Air	0.929	2643	Narrow	6.16	94.7	12.54	0.593	45	Same
	Air	0.929	2643	Narrow	7.99	155.2	15.88	0.9997	45	Same
	Air	0.929	2643	Narrow	10.3	234.2	18.8	1.646	45	Same
	Air	0.587	2483	Narrow	3.43	13.2	3.14	0.331	45	Same
	Air	0.587	2483	Narrow	3.97	34.2	7.05	0.442	45	Same
	Air	0.587	2483	Narrow	4.58	58.6	10.45	0.588	45	Same
	Air	0.587	2483	Narrow	6.16	126.9	16.83	1.066	45	Same
	Air	0.168	2099	Wide	4.27	11.1	—	—	45	Horizontal
	Air	0.168	2099	Wide	7.32	24.5	—	—	45	Horizontal
	Air	0.168	2099	Wide	11.6	40.5	—	—	45	Horizontal
	Air	0.168	2099	Wide	1.75	11.1	5.18	0.422	45	Vertical
	Air	0.168	2099	Wide	2.38	40.5	13.9	0.782	45	Vertical
	Air	0.168	2099	Wide	2.93	69.3	19.3	1.178	45	Vertical

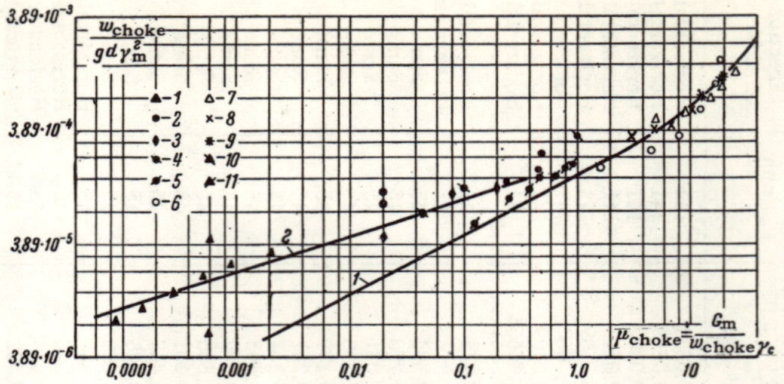

FIG. 3-2. Zenz's correlation for determining the choking velocity, w_{choke}, or saltation velocity for very narrow material fractions, w_{salt}, during transport by gases (line 1) or by liquids (line 2) as functions of the overall material concentration of the stream (μ_{choke}). The letter symbols are as given in the main list of symbols. The graph is also suitable for determining w_{salt}, since for uniform size material $w_{salt} = w_{choke}$ according to Zenz using the overall material concentration at the beginning of saltation, μ_{salt}, instead of μ_{choke}.

No.	D_T, mm	d, mm	γ_M, kg/m3	Fluid	Ref.
1.	168	0.58	2640	Water	193
2.	254-305	0.245	2600	Water	181
3.	76.2	0.241	2600	Water	546
4.	254-305	0.32	2600	Water	181
5.	25.4	0.585	2600	Water	217
6.	44.5	0.93	2640	Air	625
7.	44.5	1.675	1090	Air	625
8.	44.5	0.59	2480	Air	625
9.	44.5	0.168	2100	Air	625
10.	168	0.245	2640	Water	193
11.	168	1.62	2640	Water	193

It is obviously of great practical interest to make design calculations for the choking velocity. Zenz [717] has proposed a correlation, shown in Fig. 3-2, based on the experimental data given in Table 3-1. Note that some of the experimental data refer to the velocities at which fallout (saltation) of material begins during transport by gases or liquids in horizontal tubes. The correlation assumes that the choking velocity and the saltation velocity are equal for uniform particle size materials. Zenz noted the very approximate nature of his proposed correlation, in particular as regards the data on hydraulic transport, in which no clear effect of the particle diameter was noted.

In Fig. 3-3, taken from Leva [988], Zenz's correlation is compared with the data of Lewis, Gilliland and Bauer [422] and Culgan [256]. Lewis, Gilliland and Bauer [422] investigated the behavior of glass spheres in an air stream in a vertical system with material recirculation. Culgan investigated horizontal transport. The data of Lewis, Gilliland and Bauer are quite close to the values given by Zenz, while Culgan's data indicate a choking velocity that is many times larger. Leva assumes that the deviation between the data of Culgan and those of the others is connected with the great length of the horizontal pipeline used in Culgan's experiments (about 10 m), while the others used short tubes, in which it was not possible to eliminate the

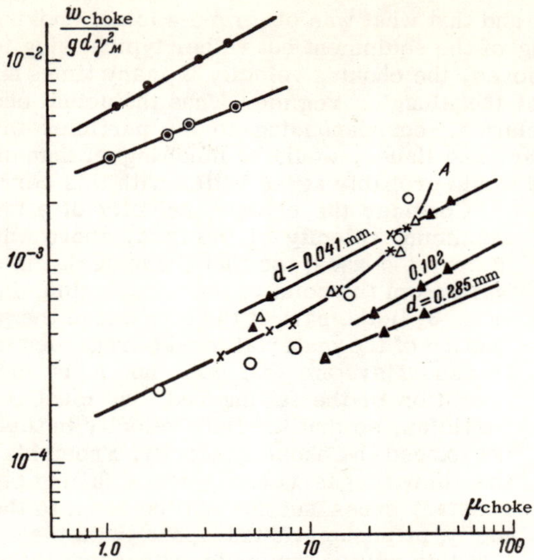

FIG. 3-3. Relationship for transport of a solid phase proposed by Zenz [717] compared with data of other authors [988].
Dimensions of quantities on ordinate axis: w_{choke}, m/sec; d, m; γ_M, kg/m³; g, m/sec².
A—Zenz's correlation [717]; D_T = 44.5 mm

Symbol	d, mm	γ_M, kg/m3
△	1.678	1090
○	0.930	2640
×	0.587	2480
*	0.168	2100
Data of Culgan [256]; D_T = 77.7 mm		
●	3.05	1133
⊙	6.35	1170

▲—Data of Lewis et al. [422]; D_T = 63.5 mm, glass spheres, γ_M = 2480 kg/m³.

effects of acceleration of the material on the measured choking velocity. Such an explanation does not inspire confidence. The fact is that in vertical pneumatic transport the uncompleted acceleration of the material must be facilitated by the ready onset of material fallout (at large gas velocities), i.e., by increase in the choking velocity, which was not observed (Fig. 3-3, data for short and long tubes). There is also some doubt as to the validity of Leva's comparison of the data of Lewis, Gilliland and Bauer (Fig. 3-3) with Zenz's correlation. It is not clear from the description of the experimental equipment (with material recirculation) given in the original paper by these authors [422] whether the equipment represented a free system, where choking could appear. Furthermore, from the description of the way in which the unstable operating regime in this apparatus developed as the rate of supply of material was increased, it can be concluded that the system was not free and that what was observed was not choking with subsequent clogging of the equipment but rather typical slug formation. As already mentioned, the choking velocity is many times higher than the upper limit of the slugging regime. Thus the actual choking point in Zenz's correlation, corresponding to the particles investigated by Lewis, Gilliland and Bauer, would be much higher than shown by Leva (Fig. 3-3) and would probably agree better with this correlation.

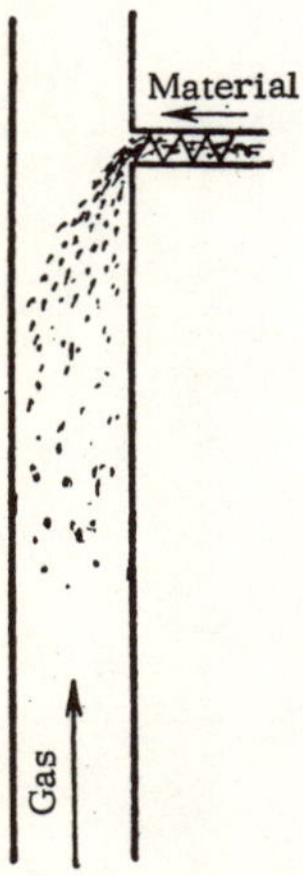

FIG. 3-4. Falling bed.

Consider the choking velocity of a falling bed—the maximum velocity of the fluid, above which fall of the material ceases, so that the material is entrained upward from the point of its introduction. In a falling bed (Fig. 3-4) the part of the tube above the point of introduction of the material must be regarded as a pneumatic transport system that does not function under normal operation of the falling bed, but must be free of solid particles, so that the fluid velocity in this section cannot exceed the choking velocity. From this point of view, the limiting gas velocity for a falling bed in a tube of constant cross section will be equal to the choking velocity in pneumatic transport, with the sole difference that this will be the upper rather than the lower operating limit of the fluid velocity. This is also the limiting gas velocity for the operation of a descending fluidized bed, since in this case the part of the tube above the material supply point, which must be free of solid particles, represents a cocurrent free system.

However, the actual breakdown in operation of a falling bed does not occur as an instantaneous transition to pneumatic transport at w_{choke} but at considerably smaller velocities of the medium. This is confirmed by Wilhelm and Valentine [605], who investigated a free system with the material supplied at the middle of a vertical tube; as a result, there was a falling bed in the lower part of their tube at low gas-flow rates and pneumatic transport in the upper part of the tube at high velocities. The accuracy of Wilhelm and Valentine's data should not be overestimated, since there were large errors in measuring the limits of the unstable region. These errors explain the rather unlikely results of the measurements, such as the apparent constancy of the choking velocity in pneumatic transport regardless of the feed rate of material.

However, the onset of unstable operation of a falling bed at velocities smaller than the choking velocity in pneumatic transport for the same particles and the same material flow rate can be regarded as quite definitely established. This disruption of falling-bed operation begins at bed voidages that are much smaller (i.e., at true particle concentrations that are much larger) than at choking conditions in pneumatic transport. Thus falling beds can be operated stably at larger material saturations of the stream than can suspended beds. For example, in the experiments of Wilhelm and Valentine the value of $(1-m)$ did not exceed 0.01 for a suspended bed, while for a falling bed at the same feed rate it reached almost 0.07.

The disruption of a falling bed appears in the form of abrupt fluctuations in the bed density and, just as in the case of choking during pneumatic transport, is related to the close approach of the particles to one another. However, the approach of the particles to one another in the vertical direction cannot in itself cause an interruption in the motion of the material in a falling bed, but it helps to accelerate the fall of the material, increasing the flow capacity of the bed. At large particle concentrations the close horizontal approach of the particles to one another will begin to have an effect. Gasparyan and Ziminyan [733] investigated the fall of groups of steel spheres in glycerine and showed that close proximity of the spheres to one another had a twofold effect: fall was accelerated as the spheres approached each other in the vertical plane, but as they approached each other in the horizontal plane it was retarded. This has also been confirmed by the work of Rowe and Henwood [1224]. In a falling bed, unstable complexes of particles of various and varying configurations fall with different velocities, leading to temporary increases in concentration at various points in a horizontal plane, i.e., to retardation and increase in density of the bed. However, the complexes are mobile and rearrange themselves. The vertical proximity of the particles within them increase again, and the horizontal proximity is reduced, so that the fall of the complexes accelerates and the density of the falling bed decreases. Powerful fluctuations in the density of the falling bed result, and its motion becomes unstable.

In conclusion, we must mention the efficiency of pneumatic raising of material, since this problem is not always treated correctly.

The advantage of pneumatic transport is the simplicity of the equipment required, and its drawback is its low mechanical efficiency.

The hydraulic resistance of a pneumatic transport line is given by

$$\Delta P = \frac{\gamma_M (1-m)(w_{M2}^2 - w_{M1}^2)}{2g} + \lambda \frac{H}{D_T} \cdot \frac{\gamma_m w_{\text{stream}}^2}{2g} (1 + k\mu) + \gamma_M (1-m) H, \tag{3-3}$$

where w_{M1} and w_{M2} = the velocities of the material particles at the beginning and end, respectively, of the lift line, m/sec;

H = height to which the material is lifted (height of the pipeline), m;

μ = nominal weight concentration of material (from the flow rate per unit time);

k = Gasterstadt coefficient [24, 127];

λ = friction factor for motion of gas in the tube;

w_{stream} = fluid-stream velocity, m/sec.

The first term on the right-hand side of Eq. (3-3) corresponds to the pressure drop in accelerating the particles of material, the second, to the pressure drop due to friction, and the third, to the pressure drop in lifting the material to the height H. Only the lifting work ΔP_{lift} is useful. Consequently, the efficiency of the pneumatic transport (lift) of material is

$$\eta = \frac{\Delta P_{\text{lift}}}{\Delta P} = \left(\frac{w_{2M}^2 - w_{1M}^2}{2gH} + \frac{\lambda \gamma_m w_{\text{stream}}^2 (1 + k\mu)}{2g D_T \gamma_M (1-m)} + 1 \right)^{-1}. \tag{3-4}$$

At very low material concentrations, i.e., when $k_\mu \to 0$ and $(1-m) \to 0$, the pneumatic transport efficiency tends toward zero, as can be seen from Eq. (3-4). This is to be expected, since then the useful work is negligible, while there is still a loss of energy in overcoming the gas friction against the tube walls. Thus the view, sometimes expressed, that the efficiency should tend to a maximum at very small concentration is wrong. Sometimes it is better from an energy point of view to raise material by means of an ascending fluidized bed than by pneumatic transport (see page 35), to say nothing of simple mechanical lifting (for example, by bucket elevator).

Hariu and Molstad [334] showed experimentally that for most materials the mean-equilibrium particle velocity w_{2M} is approximately equal to half the velocity of the transporting medium. The accuracy of these data is not high, but they can be used for an approximate estimation of the pressure drop in accelerating the material.

3.3 Representation of Vertical Systems on the Phase Diagram

Figure 3-5 (taken from [988]) shows the curves for vertical pneumatic transport, falling beds and nonhomogeneous fluidized beds schematically on a phase diagram $\left[\frac{\Delta p}{H}=f(w_{\text{stream}})\right]$. The line for a nonhomogeneous fluidized bed $DSst$ has a break between the points S and s; this is the region of the unstable regime of nonhomogeneous fluidization, with pressure fluctuations or piston formation. Naturally, processes in this region cannot be represented by any line on the phase diagram, since the pressure drop fluctuates greatly with time, so that there is no single-valued relationship between $\Delta P/H$ and the superficial velocity.

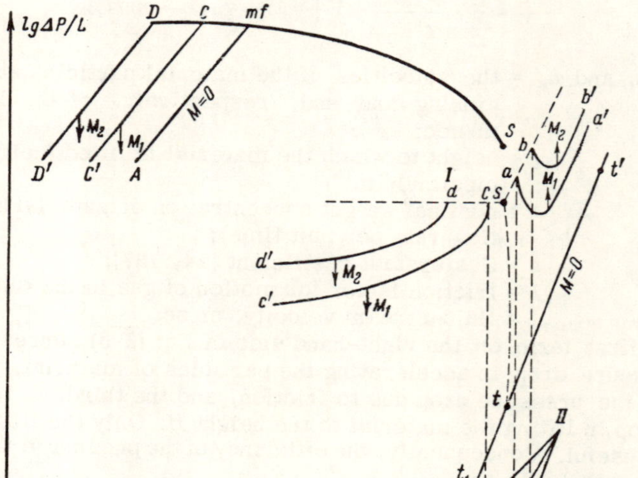

FIG. 3-5. Schematic Zenz phase diagram for vertical pneumatic transport, falling beds and nonhomogeneous fluidized beds [988]. I) zone of unstable regime; II) choking velocities.

The section st represents a dilute fluidized bed. The point t corresponds to the limit of expansion of a fluidized bed, i.e., to $m = 1$, and therefore lies on the line t_1t' representing the relationship between the hydraulic resistance of 1 m of tube and the stream velocity for the flow of the pure medium (not containing particles of solid material). The abscissa of the point t is the superficial stream velocity at which a comparatively uniform and very dilute fluidized bed appears.

The lines for a falling bed are shown as $d'd$ and $c'c$. The first correponds to the larger material flow rate ($M_2 > M_1$). The points d and c, where the lines for the falling bed break, correspond to the onset of the unstable operating regime of the falling bed due to the high true bulk concentration of the material that occurs at high velocities of the ascending stream of the medium. This limiting concentration occurs earlier (at smaller velocities of the ascending stream) in the case of the larger material feed rate ($M_2 > M_1$). The fact that all the break points for the falling bed lines for given particles in a given medium fall along a single straight line on the phase diagram parallel to the abscissa, i.e., that the unstable regime begins at the same value of $\Delta P/H$, was assumed by Zenz without experimental verification.

The lines for vertical pneumatic transport are aa' and bb'. The second corresponds to the higher material flow rate (larger pressure drop per unit length of tube). The right-hand branches of these curves go off to infinity, tending toward the line tt' for a stream of pure medium (at $w_{\text{stream}} \rightarrow \infty$), since as $w_{\text{stream}} \rightarrow \infty$ the true and even the nominal material concentrations in the stream tend to zero and have no effect on the total pressure drop per unit length of tube. The left-hand branches of the pneumatic transport curves show breaks at the choking points (a and b). The abscissas of these points are the choking velocities. These velocities increase with the material flow rates.

The curves for vertical pneumatic transport have a characteristic minimum in the value of $\Delta P/H$ not far from the choking point. The presence of this minimum value of $\Delta P/H$ is to be expected, since one of the terms in Eq. (3-3) for the hydraulic resistance during pneumatic transport in a vertical tube depends directly on the stream velocity (ΔP_{fr} = pressure drop due to friction), while the other varies inversely (ΔP_{lift} = pressure drop due to lifting the material). Thus, reducing w_{lift}, i.e., the gas feed rate, while retaining the same material feed rate M accomplishes the same work of lifting the material with a smaller quantity of gas, leading to a larger change in the gas pressure, i.e., to larger ΔP_{lift}. In the region of high-stream velocities and small true material concentrations $(1-m)$, the friction loss ΔP_{fr} inevitably predominates, while reduction of w_{stream} leads to a monotonic decrease in the overall pressure drop ΔP and in $\Delta P/H$, since the value of ΔP_{lift} remains small. However, near the choking point, where $(1-m)$ is large and w_{stream} is small, the situation is reversed: $\Delta P_{\text{lift}}/H$ predominates and the total $\Delta P/H$ will vary accordingly; i.e., it will decrease as the stream velocity increases.

3.4 Horizontal Transport in the Dilute Phase

In auxiliary fluidized bed equipment conditions frequently occur of horizontal transport, transport in inclined tubes, and horizontal transport at high material concentrations. A detailed description of the phenomena occurring in such devices lies outside the scope of the present book. A limited amount of information is given below on these forms of transport.

The horizontal transport of granular materials through pipelines by gases or liquids has been studied by many investigators [24, 181, 193, 217, 317, 334, 351, 546]. Zenz [717] and Leva [988] give reviews of much of this work.

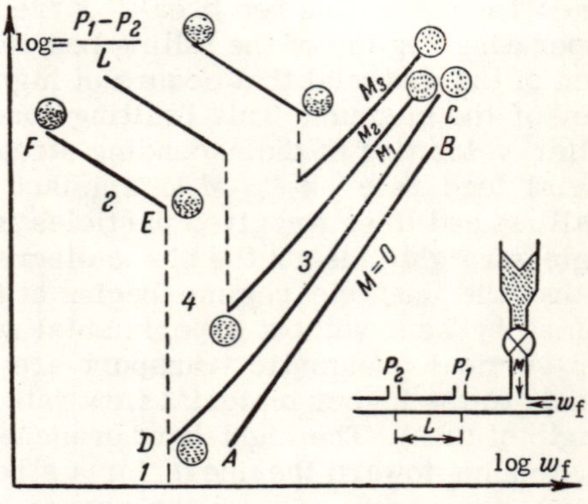

FIG. 3-6. Schematic Zenz phase diagram for horizontal pneumatic or hydraulic transport [717].
1) saltation velocity w_{salt} for given flow rate of solid material M_1; 2) transport above a salted layer; 3) material completely dispersed in stream; 4) saltation.
$M_3>M_2>M_1>0$.

Figure 3-6 gives a schematic representation of horizontal transport in the dilute phase on a Zenz phase diagram [717]. The line *CDEF* corresponds to some flow rate of the solid material M_1. The section *CD* corresponds to transport in the dispersed state (without saltation), with a gradual increase in the true concentration of the material as the velocity of the transporting agent is reduced. Finally, at the point *D* the concentration is so large that material begins to settle out all along the bottom of the tube. This is an unstable or transient regime with a continuously increasing $\Delta P/L$, in spite of the constant feed rate and overall velocity of the transporting medium. $\Delta P/L$ increases continuously since the particles undergo saltation and thus constrict the free cross section available for the passage of the suspension. This unsteady-state transition regime is shown by the dotted lines in Fig. 306. After some interval of time, sometimes several hours [717], a steady state is again reached at the same velocity at which saltation begins, w_{salt}. This is point *E* on the diagram, where further increase in the salted layer of particles ceases. A well-dispersed suspension flows above the settled material. As the velocity of the transport medium is further decreased (going from point *E* to point *F*), the thickness of the equilibrium salted layer increases, and again some time is required for attaining a new equilibrium state.

Zenz [717] mentions that for fine particles ($d < 2.8$ mm) in tubes of diameter $D_T > 37$ mm, and for still smaller particles ($d < 0.152$ mm) in tubes of $D_T > 13$ mm, about half of the cross section of the tube is filled with precipitated particles immediately after reaching the equilibrium state (point E). For coarser particles (1.5 mm or more) in tubes of very small diameter ($D_T \leqslant 13$ mm), the area of the free cross section above the precipitated particles is usually so small that immediately after the beginning of saltation the two-phase stream changes to slug flow. The saltation process for narrow particle fractions in horizontal tubes is reversible; as the fluid flow rate is increased, the salted layer gradually becomes thinner, until at point E an unsteady-state period occurs, culminating in the total disappearance of the salted layer.

The velocity of the medium when saltation of the material begins, w_{salt}, is shown by Zenz's data [717] to be practically equal to the choking velocity for materials of uniform particle size and can be calculated from the curves in Fig. 3-2 for w_{choke}. However, for materials of wide particle-size distribution, $w_{salt}/w_{choke} > 1$, and for horizontal transport of the same material the required velocity is several times larger than in vertical transport. This ratio may reach ten if the material feed rate is high. Figure 3-7 gives Zenz's graphs; by using these and Fig. 3-2 it is possible to determine the velocity of initial saltation w_{salt} in horizontal transport by the method of successive approximations. The use of the graphs can best be demonstrated by means of an example.

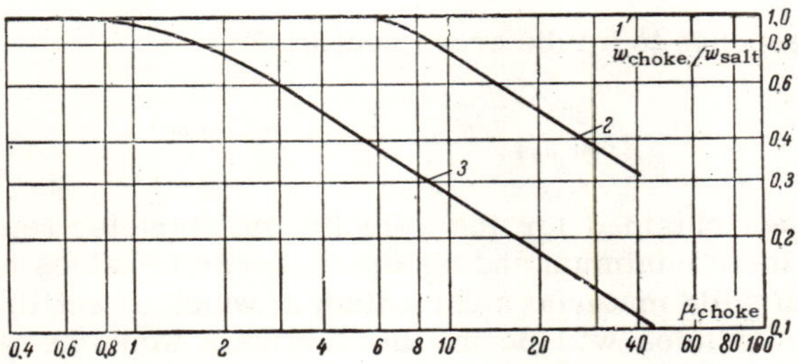

FIG. 3-7. Effect of particle-size distribution on the ratio of the critical choking and saltation velocities (w_{choke} w_{salt}) at various overall concentrations (μ_{choke}) which are limiting in vertical pneumatic transport [717].
1) uniform particles; 2) narrow fractions ($d_{max}:d_{min} = 3–4$); 3) wide fractions ($d_{max}:d_{min} = 6–12$).

Example. Calculate the velocity w_{salt} (and tube diameter) for horizontal transport of a material of wide size analysis, given the following data: mean-particle diameter 60 μ (6×10^{-5} m); bulk weight of particles 1762 kg/m³; material feed rate 181,000 kg/hr; gas-flow rate 2260 m³/hr; density of gas 24.8 kg/m³.

Solution. The nominal weight concentration for which the saltation velocity is to be calculated is

$$\mu_{salt} = \frac{181 \cdot 10^3}{2.26 \cdot 10^3 \cdot 24.8} = 3.2.$$

As a preliminary step, assume that the choking velocity is equal to 1 m/sec:

$$\frac{w^2_{\text{choke}}}{gd\gamma^2_M} = \frac{6}{9.81 \cdot 6 \cdot 10^{-5} \cdot 1\,762^2} = 0.545 \cdot 10^{-3}\,\text{m}^6\text{kg}^{-2}.$$

Then from Fig. 3-2,

$$\mu_{\text{choke}} \approx 8.$$

According to Fig. 3-7,

$$\mu_{\text{salt}} : \mu_{\text{choke}} = 0.3.$$

Consequently, $\mu_{\text{salt}} = 0.3 \times 8 = 2.4$, which does not agree with $\mu_{\text{salt}} = 3.2$, corresponding to the initial data. Obviously, too small a choking velocity was assumed; 1 m/sec would be satisfactory only at $\mu = 2.4$, i.e., at a material feed rate smaller than the one given. As a second approximation, take $w_{\text{choke}} = 1.5$ m/sec. Again calculate $w^2_{\text{choke}}/gd\gamma^2_M$ and determine μ_{choke} and $\mu_{\text{salt}}:\mu_{\text{choke}}$, arriving at $\mu_{\text{salt}} \approx 3.2$, which agrees with the value calculated from the initial data. This means that the saltation velocity is

$$w_{\text{salt}} = \frac{w_{\text{choke}}}{\mu_{\text{salt}}/\mu_{\text{choke}}} = \frac{1.5}{0.16} = 9.4\ \text{m/sec},$$

which corresponds to a tube cross section

$$F = \frac{2\,260}{3\,600 \cdot 9.4} = 0.067\ \text{m}^2, \quad \text{or} \quad D_T = 0.292\ \text{m}.$$

The values obtained for the velocity and diameter represent, respectively, their minimum and maximum possible values for the given flow rates of solid material and medium at which no settling or saltation of the particles will occur. In practice it would be necessary to select a somewhat smaller tube diameter, which would then ensure a higher and safer velocity.

Leva [988] recommends Hinkle's formula for determining the pressure drop in horizontal pneumatic transport in the dilute phase. Hinkle [345] investigated horizontal pneumatic transport of various materials (Table 3-2) in glass tubes of diameter 50 and 76 mm and

Table 3-2

Properties of materials investigated by Hinkle [988].

Material property	Tenite	Alundum	Catalin	Polystyrene
Particle density, kg/m^3	1293	1810	1116	1049
Weight "as filled," kg/m^3	729	1137	785	641
Particle diameter, m	Wide fraction	—	$6.35 \cdot 10$	—
Mean-particle diameter, m	Wide fraction	$8.38 \cdot 10$	—	$(2.03 \div 3.56) \cdot 10$
Maximum-particle diameter, m	$3.05 \cdot 10$	$9.14 \cdot 10$	—	$8.89 \cdot 10$
Shape factor (sphericity)	0.93	1.00	1.00	0.96

about 9 m long, in which the pressure was measured every 1.5 m. The particle velocities were determined by high-speed motion-picture filming. The pressure drop in a tube of circular cross section was given as

$$\Delta P = 5.17\cdot 10^{-2}\,\frac{f_{g'}\,w^2_{\text{stream}}L}{D_T}\left(1+\frac{f^*_M w^*_M}{f_{g'}w_{\text{stream}}}\right), \tag{3-5}$$

where f_g = friction coefficient for flow of pure gas in the tube, dimensionless (Fig. 3-8);
f_M = coefficient of friction of the material, dimensionless;
w_{stream} = overall gas velocity, m/sec;
w_M = velocity of the solid particles, m/sec;
L and D_T = length and diameter of the pipeline, m;
* = denotes equilibrium (after acceleration).

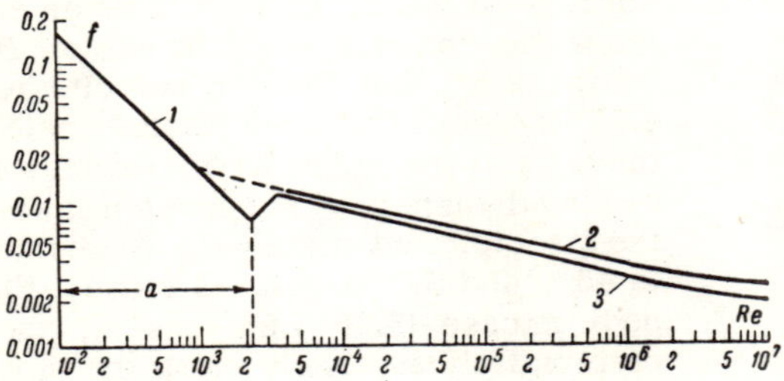

FIG. 3-8. Friction factor f as a function of Reynolds number Re for isothermal flow of liquids and gases in circular tubes [988].
a—region of laminar flow; 1) laminar stream in circular tube, f = 16/Re; 2) industrial steel and cast iron pipes; 3) smooth tubes, such as glass or copper.

The friction coefficient of the material was

$$f^*_M = \frac{3\gamma_{g'}C_D D_T}{2d\gamma_M}\left(\frac{w_{\text{stream}} - w^*_M}{w^*_M}\right)^2, \tag{3-6}$$

where C_D = drag coefficients for a sphere (see Fig. 1-17). The equilibrium particle velocity was

$$w^*_M = w_{\text{stream}}\,[1 - 2.02d^{0.3}(\gamma_M/998)^{0.5}]. \tag{3-7}$$

3.5 Pneumatic Transport in Inclined Tubes

Pneumatic transport of granular materials in inclined tubes has not been extensively studied [42, 717]. Zenz [717] gives some experimental data on the effect of tube inclination on the conditions of pneumatic transport of salt, magnetite powder and rape seed, but makes no attempt to correlate the data. It is mentioned that when the material moves along an inclined tube there is a reduction in the pressure drop caused by the sharp turn of the stream through 90° (from a horizontal to a vertical tube). The abrasive action of the solid material is also reduced. However, an appreciably higher stream velocity is required, which in some cases avoids the appearance of the piston regime and the accompanying undesirable vibration of the pipeline.

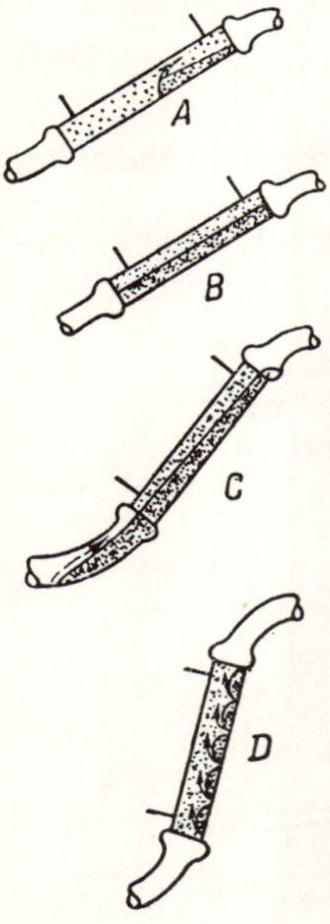

FIG. 3-9. Ways in which solid particles (rape seed) settle out at various pipeline slopes and stream velocities [717].

There are various types of material movement in inclined tubes. Some examples in Fig. 3-9, taken from Zenz [717], show schematically the nature of the precipitation of solid particles at different tube slopes and stream velocities; the arrows show the direction of motion of the precipitated material. Sketch *A* corresponds to a slope of 22.5° and an air velocity of 5.2-5.5 m/sec; particles were precipitated only in the upper half of the tube; according to Zenz, this could be an entrance effect from the lower elbow. In sketch *B* the slope is again 22.5°, but the air velocity is lower (3.05-4.57 m/sec); the precipitated material does not move, and transport occurs exclusively in the form of a well-dispersed suspension in the space above the precipitated material. In sketch *C*, the slope is 45°, and the velocity is close to that in the preceding case (3.36-4.26 m/sec); the precipitated material slides slowly along the tube and is picked up by the air at the bottom and returned to the dispersed stream. In sketch *D* the slope is 67.5° and the air velocity is almost the same as in case *A* (51.8 m/sec); aggregates of heaps of precipitated material are formed; at lower air velocities slug flow begins at these positions.

3.6 "Dune" Transport

At low velocities of the transporting medium (considerably below the saltation velocity w_{stream}) horizontal transport in the disperse phase ceases, but so-called "dune" transport is possible. The granular material forms mounds (dunes) in the horizontal pipeline, and is carried along, flowing from one dune to the next along the path of the transporting medium. A decrease in the velocity w_{stream} of the latter leads to an increase in the frequency and height of the dunes, until finally at very small values of w'_{stream} the line becomes plugged. In dune transport it is possible to operate with very small flow rates of the transporting medium and very high values of the nominal material concentration μ. In Wen's experiments [988] values of $\mu = 850$ were attained, while in normal horizontal pneumatic transport there is a nominal weight concentration of the order of 0.5-5.0 [988]. However, in dune transport the hydraulic resistance is considerably higher than in pneumatic transport, since the work in transporting the material along the pipeline in dunes is derived from a smaller quantity of gas. If the granular material is supplied to the pipeline by a feed device which aerates (fluidizes) the material rather than by a mechanical feeder, transport will occur like flow in a fluidized bed in the initial section (near the feed point). The material then settles out, and over the greater part of the pipeline there is the formation and drifting of dunes.

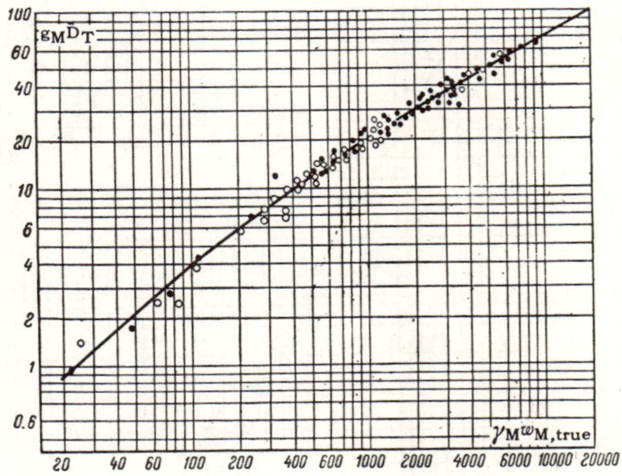

FIG. 3-10. Relationship for determining the mean-true particle velocity $w_{M,true}$ in dune transport [988].

The paper of Koble, Johnes and Koehler [394] compared horizontal pneumatic transport and dune transport. In a review of several papers on dune transport [192, 203, 394, 428, 594] Leva [988] notes that the most detailed investigation of this problem is given in Wen's thesis [594]. Wen carried out experiments in an apparatus consisting of a horizontal tube, a bunker and feed device supplying it with material, and a separator for removing the material after leaving the tube. Measurements were made in horizontal glass sections of the pipeline of length 3.05 m and I. D. 25.4, 19 and 12.7 mm and also in a horizontal steel tube of diameter 9.2 mm. Rapid-acting slide valves were located at both ends of the sections investigated. The mean-weight concentration per unit volume of pipeline was determined by closing both slide valves simultaneously and collecting the material in the pipe between them. The pressure drop over the section was recorded. The transport air was dried in advance by passage through a bed of activated silica gel. The granular materials used were coal dust (d = 730, 500 and 112 μ, γ_M = 1300 kg/m^3) and glass spheres (d = 280, 147 and 71 μ, γ_M = 2500 kg/m^3). The air velocity varied from 0.15 to 15.2 m/sec and the nominal weight concentration from 50 to 850 kg/kg. Wen gave empirical correlations for his data in the form of graphs (Figs. 3-10 and 3-11). At a given tube diameter D_T and weight-flow rate of material g_M (kg/sec·m^2) referred to the total cross section of the tube, Fig. 3-10 gives the mean-true particle velocity of the material along the tube, $\overline{w}_{M,true}$. At this value of $\overline{w}_{M,true}$, Fig. 3-11 gives the modified friction factor for the material in dune transport:

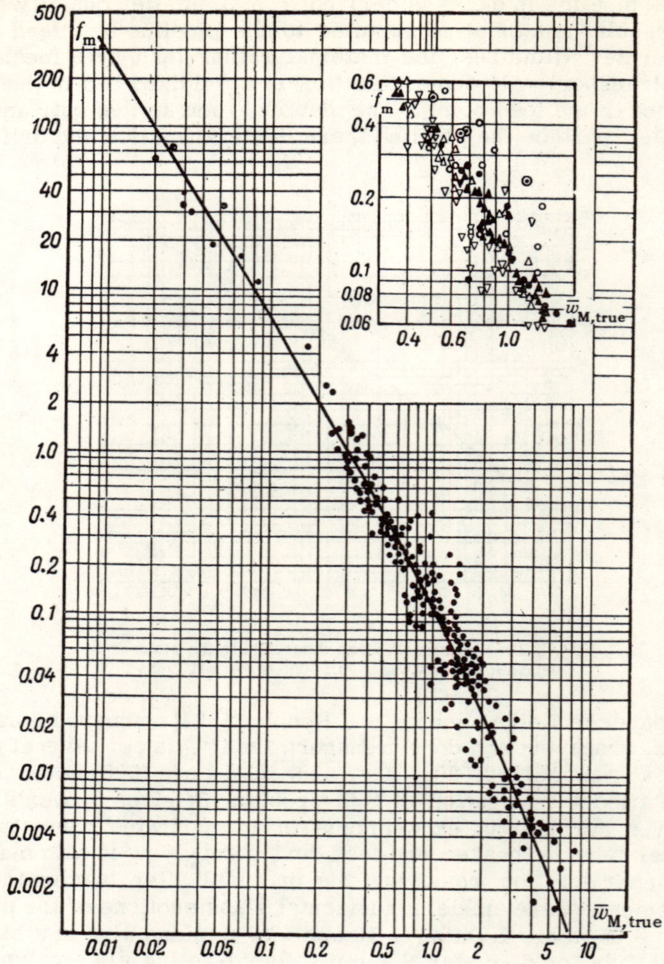

FIG. 3-11. Modified friction factor as a function of particle velocity $\overline{w}_{M,true}$ in dune transport [988].

$$f_M = \frac{9.81 \Delta P D_T}{2 G_M L \overline{w}_{M,true}},$$

where ΔP = hydraulic resistance of the pipeline in the type of transport for which the numerical value of f_M is defined.

Wen showed that the mean-true velocity of the air above the dunes was approximately twice as true as $\overline{w}_{M,true}$:

$$w_{g,true} \approx 2\overline{w}_{M,true}. \tag{3-8}$$

In order to go from the mean-true velocities of the gas and material along the tube to the overall values (calculated from the flow rates and total cross section of the tube), Wen used the concept of apparent densities of the dispersed gas and material, $\gamma_{d.g}$ and $\gamma_{d.M}$. Here $\gamma_{d.M}$ is the ratio of the weight of material contained in the pipeline at any moment, and nominally dispersed in the gas, to the volume of the pipeline, while $\gamma_{d.g}$ is the ratio of the weight of gas contained in the tube (dispersed with a nominally uniform distribution of material) to the tube volume. It can be shown that $\gamma_{d.g}$ and $\gamma_{d.M}$ are, respectively, equal to $\gamma_M(1-m)$ and $\gamma_g m$ for a uniform bed.

Wen used the relationships

$$\gamma_{d.M} = \frac{G_M}{\bar{w}_{m.\,true}}; \tag{3-9}$$

$$\gamma_{d.g} = (1 - \gamma_{d.M}/\gamma_M)\,\gamma_g; \tag{3-10}$$

$$G_g = \bar{w}_{g,true}\,\gamma_{d.g}, \tag{3-11}$$

where G_g is the overall weight flow rate of gas, referred to the total pipeline cross section.

A numerical example of a dune transport calculation will illustrate the use of Wen's data.

Example. A powdered catalyst ($d = 147\,\mu$, $\gamma_M = 1600$ kg/m³) is to be transported in a horizontal direction through a glass pipe ($D_T = 25.4$ mm) for a distance of 15 m at the rate of 446 kg/hr. The pressure at the discharge end of the pipeline is atmospheric, and the temperature is 20°C.

For dune transport, find:

1) the air pressure required at the inlet of the pipeline;
2) the nominal material concentration;
3) the required air-flow rate.

Solution. The mean-true material velocity is determined from the quantity $G_M D_T^{0.7}$.

The weight rate of material is

$$G_M = M/3\,600F = 446/3\,600\cdot 5.07\cdot 10^{-4} = 244\ \text{kg/m}^2\cdot\text{sec};$$

$$G_M D_T^{0.7} = 244\,(2.54\cdot 10^{-2})^{0.7} = 18.65.$$

From Fig. 3-10:

$$\gamma_M \bar{w}_{M,true} = 975,$$

whence

$$\bar{w}_{M,true} = 975/1\,600 = 0.61\ \text{m/sec}.$$

From Fig. 3-11, at $\bar{w}_{M,true} = 0.61$, the material friction factor is

$$f_M = 0.27.$$

Hence

$$\Delta P = \frac{2\cdot 15\cdot 244\cdot 0.61\cdot 0.27}{9.81\cdot 2.54\cdot 10^{-2}} = 4\,840\ \text{kg}_f/\text{m}^2.$$

Thus the air pressure at the inlet of the pipeline is

$$P_1 = 0.484 \approx 0.5 \text{ atmospheres gauge.}$$

The weight of material per m^3 of pipeline is

$$\gamma_{d.M} = G_M / w_{M,true} = 244/0.61 = 400 \text{ kg/m.}$$

The weight of gas per m^3 of pipeline is

$$\gamma_{d.g} = (1 - \gamma_{d.M}/\gamma_M)\gamma_g = (1 - 400/1\,600)\,1.2 = 0.9 \text{ kg/m}^3.$$

The mean-true gas velocity is

$$\overline{w}_{g,true} \approx 2\overline{w}_{M,true} = 2 \cdot 0.61 = 1.22 \text{ m/sec.}$$

The weight flow rate of air is

$$G_g = \overline{w}_{g,true} \cdot \gamma_{d.g} = 1.22 \cdot 0.9 = 1.1 \text{ kg/m}^2 \cdot \text{sec.}$$

Hence the nominal concentration of material is

$$\mu = G_M/G_g = 244/1.1 = 222 \text{ kg/kg.}$$

The weight flow of air is

$$Q_g = 3\,600 G_g F = 3\,600 \cdot 1.1 \cdot 5.07 \cdot 10^{-4} \approx 2 \text{ kg/hr,}$$

and at 20°C and 1 atm abs., this corresponds to

$$2/1.2 = 1.67 \text{ m}^3/\text{hr.}$$

According to Wen's correlation, ΔP and $\overline{w}_{M,true}$ appear to be independent of the size, shape and bulk weight of the particles. Leva contends [988] that in dune transport the material moves more or less as a mass, and so the effect of moving the individual particles is suppressed.

The investigations of Barker et al. [203], and Longhouse et al. [428], have shown that there is no advantage in using high pressure of the transporting gas. Barker et al., investigated dune transport at gas pressures up to 11.5 atm abs.

As Leva has noted [988], dune transport has still not been sufficiently investigated. For example, the data of the earlier investigators [192, 203, 394] are not well correlated by Wen's method. Leva suggests that factors not taken into account by Wen, such as the gas viscosity and roughness of the pipeline, must have an effect on dune transport. However, it is doubtful if the roughness of the tube would have an important effect on the creeping of the dunes. No investigations have been made of the feasibility of dune transport in inclined or large-diameter horizontal pipes.

Chapter 4

Spouting Beds

4.1 Survey

Beds are said to "spout" when there is a jet of fluid passing ("fountain") up through their central (axial) parts, the solid particles in the jet move upward as in a pneumatic transport regime, falling back and settling in an annular zone near the walls after the jet leaves the bed. The particle concentration in this annular zone is high. Particle circulation in a spouting bed occurring in this manner—rise of the particles in the central part of the bed and descent around the periphery—is more intensive than that encountered in ordinary fluidized beds. Most of the equipment used in studying spouting beds has been cylindrical in shape with conical inlet sections (Fig. 4-1). In some cases the gas inlet opening has been covered with a screen.

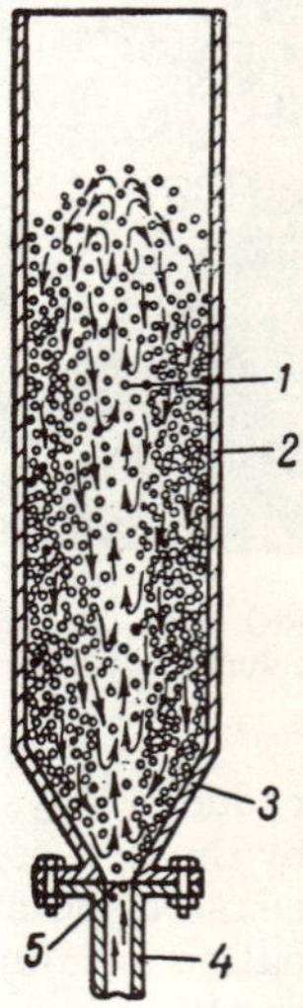

FIG. 4-1. Sketch of spouting bed [436]: 1) central spout; 2) vessel; 3) cone; 4) gas (air) inlet tube; 5) screen.

Mathur and Gishler [436] studied spouting beds of wheat in half-cylindrical tubes of 0.15- and 0.61-m diameter; the fluid velocities were 0.9-1.5 m/sec. They used transparent columns of clear plastic with semicircular cross sections and diameters of 0.15 m (Fig. 4-2) to permit visual observation and high-speed motion-picture filming of the particle motion. Air was passed into the beds through a semicircular opening of 12.7-mm diameter. The transparent "half-round" made it possible to observe flow patterns similar to those actually occurring in the diametrical cross section of a full cylindrical column. The films showed that the upward particle velocity reached a maximum value of 7.5 m/sec at 0.2 m above the air inlet for a spouting bed 0.6 m high. The descending particle velocity was greater at the upper end of the column than at the lower end, indicating a radial movement of particles from the annular peripheral space to the central fountain over the whole height of the bed. However, most of the particles passed through the complete circulation path without such short-circuiting. The mean-descending particle velocity near the walls was about 25 mm/sec. In a bed 1.2 m high the particles went around a complete circulation path in less than one minute. The air jet expanded upward (Fig. 4-2); in the

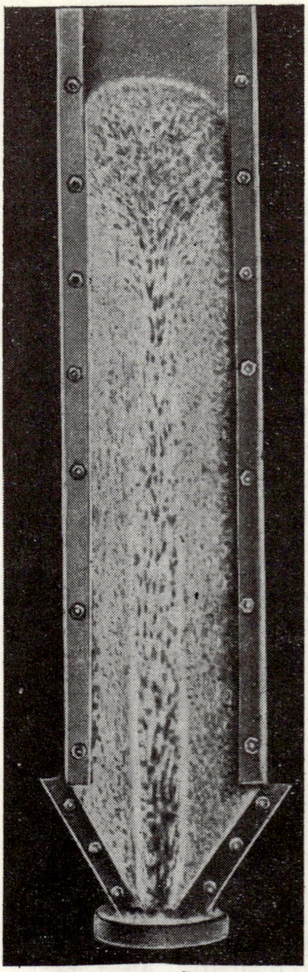

FIG. 4-2. Transparent "semicolumn" with spouting bed [436].

upper part of the bed about 85% of the air passed through the part of the bed bounded by the peripheral annulus (where the particles move downward). Under other conditions (column diameter $D_T = 0.15$ m, bed height $H = 0.635$ m, and mean air velocity $w_f = 1.14$ m/sec), only 65% of the air passed through this region (Fig. 4-3).

More recent experiments of Thorley et al. [1011], showed that the distribution of air between the fountain and the peripheral annulus depends on w_f and the geometry of the equipment. More of the gas flows through the peripheral annulus as the diameters of the apparatus and gas inlet are made larger, or as the air velocity and angle of the conical inlet section are reduced. Figure 4-4 shows typical curves for the relationship between the bed resistance and w_f in a conical-cylindrical apparatus both before and after the beginning of spouting. Leva [988] explains the nature of these curves as follows: Initially, the fluid permeates through the bed in the ordinary manner, after which a short channel is visualized as forming in the lower portion; as a result of this channel the bed resistance begins to fall (section *bc*) when the fluid velocity is further increased, since the channel then grows longer. The bed than begins to expand and the bed resistance first stabilizes (section *cd*) under the influence of two opposing forces—the increase in resistance due to the growth in bed height and the decrease in ΔP due to development of the channel. When the channel finally extends through the whole bed, there is another sharp fall in ΔP (section *de*) and a second stabilization of the overall bed resistance as spouting sets in. If the air flow rate were now reduced "hysteresis" would be noted due to the freer packing acquired by the bed compared with the initial state. Spouting therefore does not cease at point e in Fig. 4-4 but at point $e_{m.s}$, correspoinding to the so-called minimum spouting velocity. Photographs of the development of spouting beds in transparent semicolumns in the paper of Madonna and Lama [1120] confirm that there is actually a partly developed channel at the bottom of the bed in a conical-cylindrical apparatus when the bed resistance is maximum, and that this channel then breaks through and forms the fountain or spout.

Within the spout the voidage decreases with height; this occurs particularly in small-diameter columns. Leva [988] mentions that it can be expected that the spout will disappear at some height at which the velocity of the medium approximates the settling velocity, but the situation is complicated by the fact that the spout is not the same as a strictly defined pipeline.

According to Mathur and Gishler [437], the minimum velocity at which spouting begins, $w'_{m.s}$, increases linearly with the column diameter D_T and the cone angle α. Increase of the air inlet diameter d_0 requires a larger $w'_{m.s}$ (referred to the cross section of the cylindrical column) because of the increase in the air flow rate required to produce the same jet velocity as in the previous narrow cross section.

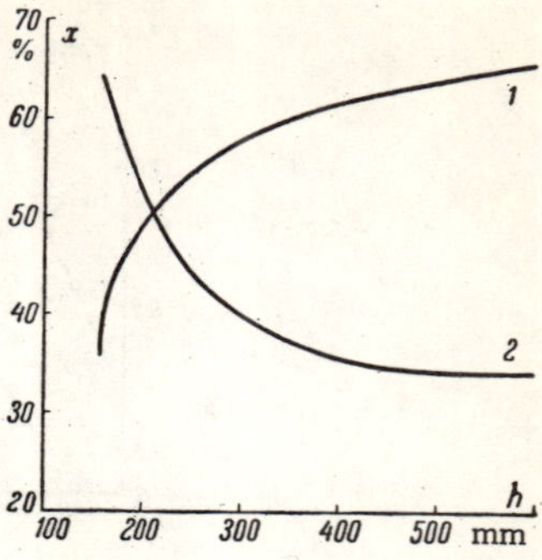

FIG. 4-3. Variation, over the bed height h, of the percentage x of the total air stream passing through the peripheral annular region and through the central part of the bed [988]. Column diameter 152.5 mm, wheat bed 635 mm deep, mean-air velocity 1.135 m/sec.
1) annular region; 2) core (spout).

Leva explains the dependence of the velocity $w'_{m.s}$ on the column diameter by reasoning that the air distribution over a large cross section is not as good as over a small cross section, and so spouting could begin at a lower value of the average velocity $w'_{m.s}$ over the cross section. This explanation is more probable than that of Thorley et al. [1011], who attributed the effect to change in the ratio of the particle diameter to the tube (column) diameter. The attenuation in the velocity distribution of the fluid medium over the cross section can also explain the effect of increasing the cone angle α. Madonna, Lama and Brisson [1210] have shown that the maximum pressure drop recorded before the onset of spouting is always smaller than that observed during fluidization. It was their opinion that the pressure drop in a spouting bed was determined not by the

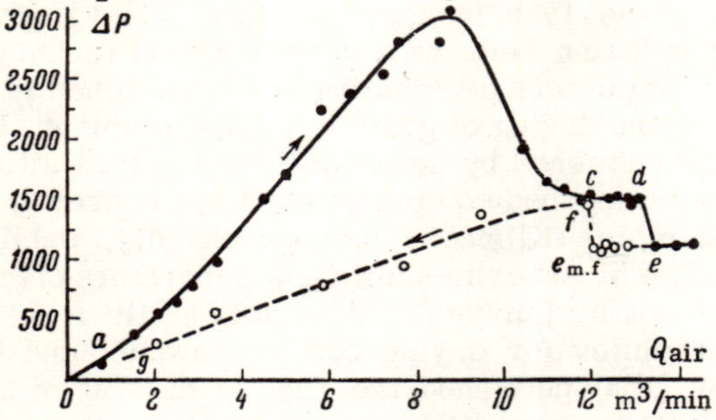

FIG. 4-4. Typical curves of bed resistance (ΔP) as a function of the superficial fluid velocity (w_f) in a conical-cylindrical apparatus before and after the onset of spouting [1011].
Apparatus diameter 610 mm, bed of wheat 1790 mm deep, cone angle 45°, air inlet tube diameter 101.6 mm. Bulk density of wheat = 800 kg/m³.

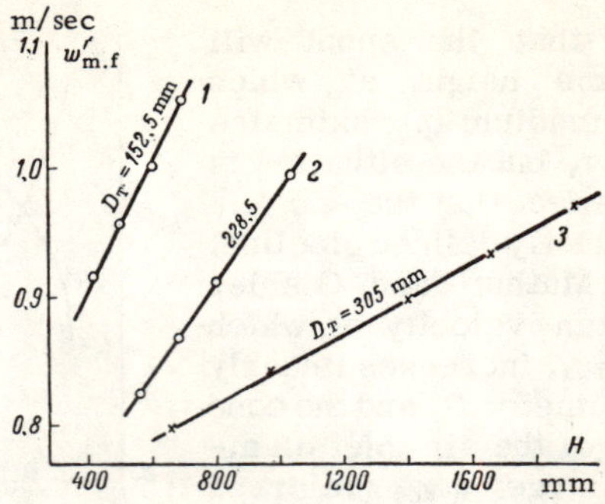

FIG. 4-5. Effect of bed height H on the value of the minimum velocity at the onset of spouting ($w'_{m.s}$) according to Mathur and Gishler [437].
Experiments with wheat. $D_T/d_0 = 6$; d_0 = diameter of air inlet opening.

force required to suspend the bed but by the maximum loss of energy expended by the jet in forcing its way through the bed.

Some of the observations and relationships mentioned above for spouting beds are purely qualitative; others are valid only for special cases. The empirical correlations for calculating the minimum spouting velocity and hydraulic resistance of a spouting bed that are given at the end of this section are not generalized or reliable, since the hydrodynamics of spouting beds have still not been sufficiently studied. The majority of investigators have studied spouting beds for drying, gasification or other process applications and have been satisfied to find the overall characteristics and operability of these processes in a particular apparatus without concern for the hydrodynamic structure of the beds.

Spouting bed equipment appeared much earlier in the Soviet Union than abroad. In his 1959 monograph, Leva [988] was quite wrong in implying that spouting beds were a very recent technique and stating that the only application of spouting beds mentioned in the literature up to then was the drying of grain in pilot equipment. Leva mentions only the papers published by Canadian investigators after 1955. There are data on spouting bed equipment even in the prewar Soviet literature. For example, Klimov's monograph [61], published in 1939, gives information on experiments in gas generators of the air-fountain type. The textbook by Lur'ye [82] describes an air-fountain dryer proposed by Voroshilov for drying cotton, sawdust and grain. In 1951 Syromyatnikov [153] published the results of tests of an air-fountain type combustion chamber. There is information on the use of spouting beds for the thermal treatment of peat in the paper by Bogdanov et al. [875]. Romankov and his associates mentioned the drying of dyestuffs and pasty materials in air-fountain equipment [659, 779, 1069], while Sazhin [1176] dealt with the operation of such equipment in a Soviet plant. A commercially tested air-fountain dryer is described in the booklet by Beker [634].

4.2 Analogy With Channel Formation

To demonstrate the feasibility of spouting beds and to find their limiting efficiences, it is first necessary to investigate their aerodynamics. It would seem obvious to compare, or contrast, a spouting bed with an "ordinary" fluidized bed during channel formation. Spouting is a special case of channel formation and is an example of using this more or less undesirable phenomenon to ensure contact of the medium and particles by stable and intense circulation of the granular material.

There are a number of instances in bed operation wherein some form of channeling takes place:

1) the formation of open channels (craters), while the main part of the bed remains in the unsuspended state (fixed bed);

2) the formation of internal channels in the lower and middle parts of the bed, with the upper part of the bed above the channels fluidized;

3) a pouting bed with "organized" particle circulation in the case of a single inlet for the fluid stream and sufficiently narrow columns;

4) a second channel-formation regime occurring in an ordinary fluidized bed after passage of gas in the form of bubbles (cores) and replacing the first channel-formation regime (1 and 2).

The second onset of channel formation is possible because at very high superficial velocities the dynamic pressure of the gas jets leaving the openings of the gas distributor prevents aggregates of particles from crossing the jets and "clogging" the channels.

Leva [988] mentions the "apparent similarity" between spouting and ordinary channel formation in beds. Both are facilitated by shallow beds and large particle diameters. Leva also noted the "striking similarity" between the minimum velocities for fluidization and spouting.

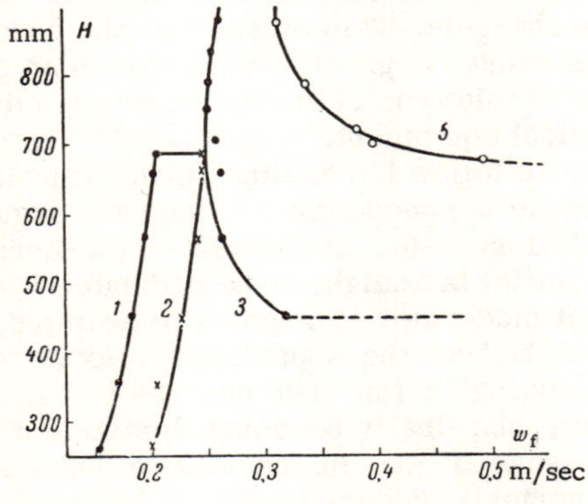

FIG. 4-6. Mathur and Gishler phase diagram [437] for a bed of Ottawa sand. Gas enters through a tube of i.d. d_0 = 12.6 mm; column diameter D_T = 152.5 mm; diameter of sand particles d = 0.42–0.84 mm.
1) fixed bed; 2) good spouting; 3) poor spouting (fountain scarcely noticeable); 4) fluidization; 5) unstable regime (slugging); H — bed height, mm; w_f — superficial fluid velocity, m/sec.

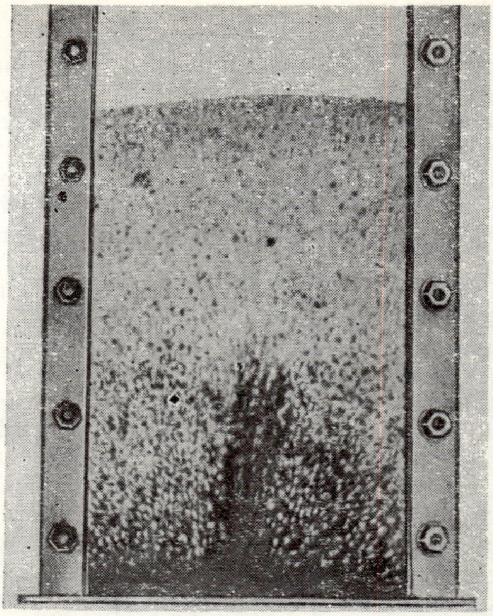

FIG. 4-7. Channel development in the lower part of a bed before the onset of spouting [1120].

The difference between spouting and other types of channel formation is due to the intensive material circulation, which is aided by the proximity of the walls. The friction against the the channel walls is small compared with the internal friction through the bed forming the walls of ordinary channels. This means that the concentration of material in a spout must be greater than that in an ordinary channel. Spouting consists of this type of channel formation during fluidization in conical-cylindrical equipment.

It is convenient to follow the analogy between spouting and ordinary channel formation on a special phase diagram suggested by Mathur and Gishler [437] (Fig. 4-6). In contrast to the Zenz phase diagram, this one uses the initial bed height as the ordinate. It can be seen from the diagram that at moderate bed heights (for example, 0.5 m) spouting replaces the fixed bed as the superficial velocity of the fluid is increased ("good spouting"); this then changes to "poor spouting" (with a diffuse fountain), and finally becomes fluidization without spouting, i.e., ordinary passage of gas in the form of bubbles. A similar sequence of events normally occurs in simple fluidized beds of moderate height with channels replaced by bubble motion. In very shallow beds it is usually easy to form channels, which are retained even at large superficial velocities, without overreaching a state of good fluidization. This also appears to be true for shallow spouting fluidized beds (Fig. 4-6). Top-to-bottom channels do not occur at very large initial bed heights; after fluidization with the passage of bubbles, slug formation begins. Similarly, spouting does not occur in deep beds (see the phase diagram), and there follows the same transition to the regime of ordinary fluidization. The phase diagram given in Fig. 4-6 does not show the region of high fluid velocities where the zone of dilute-phase fluidization and the second onset of channel formation occur. Also the boundary between fluidized and spouting beds shown on the diagram does not correspond to the limit of stability but to the rather indefinite moment when the fountain disappears and loses its sharp boundaries. The boundary between fixed beds and spouting beds shown on the phase diagram is provisional. As can be seen from a typical curve for $\Delta P = f(w_f)$ for a spouting bed (see Fig. 4-4), the critical velocity for the transition from a fixed bed to the spouting state is higher than the velocity for the reverse-order change. Strictly speaking, the transition from a fixed to a spouting bed (or to channel formation) passes through a stage in which some columns of the material are transiently fluidized, since it is necessary for the particles to acquire a certain minimum mobility in these zones before the spout can break through. Of course, judging from photographs (Fig. 4-7), the development of channels in the lower part of the bed can also lead to fluidization in the upper parts due to the pressure of the jets and the sufficiently free initial packing of irregularly shaped particles in the bed. It is possible that such behavior occurred in the experiments of Gel'perin et al. [1033] on fluidization in conical-cylindrical equipment, although they did not consider this possibility in their discussion.

4.3 Empirical Correlations

Several investigators have proposed empirical correlations for approximate calculation of the minimum superficial fluid velocity at the transition from fixed to spouting beds as well as correlations for the hydraulic resistance and the maximum height at which a bed may spout.

Mathur and Gishler [437] carried out experiments in columns of 0.15-, 0.23- and 0.30-m diameter and proposed the following empirical equation for calculating the minimum spouting velocity:

$$w_{m.s} = \frac{d}{D_T}\left(\frac{d_0}{D_T}\right)^{1/3}\left[2gH\left(\frac{\gamma_M-\gamma_m}{\gamma_M}\right)\right], \qquad (4\text{-}1)$$

where d, d_0 and D_T are, respectively, the diameters of the particles, the air inlet, and the cylindrical part of the apparatus.

Thorlety et al. [1011], pointed out that the value of $w_{m.s}$ for a spouting bed of wheat in a large-diameter column ($D_T = 0.61$ m) calculated by this formula was about 10% lower than that obtained experimentally (Table 4-1).

Leva [988] gives the formula

$$w_{m.s.} = \frac{d}{D_T}\left(\frac{d_0}{D_T}\right)^{n}\left[2gH\left(\frac{\gamma_M-\gamma_m}{\gamma_m}\right)\right]^{1/2}, \qquad (4\text{-}2)$$

where the exponent n is 1/3 for a column of $D_T = 0.15$ m and is independent of the cone angle over the range 45-85°. For a column of diameter $D_T = 0.61$ m the exponent does depend on the cone angle: $n =$

Table 4-1

Minimum Spouting Velocity $w_{m.s}$ for Wheat in a Column of Diameter D_T = 0.61 m [1011]

Cone angle α°	$d_0 \cdot 10^3$, m	H, m	$w_{m.s}$, m/sec	
			experimental	calculated
45	102	1.80	0.683	0.660
		1.52	0.656	0.607
		1.25	0.595	0.550
		0.92	0.510	0.472
60	102	1.83	0.660	0.665
		1.52	0.625	0.607
		1.40	0.574	0.543
85	102	1.86	0.756	0.671
		1.59	0.710	0.620
		1.25	0.662	0.550

0.23 for d_0 = 48-102 mm and $\alpha = 45°$, and $n = 0.13$ for $\alpha = 85°$. This formula is based on experimental data in which spouting was caused by air or water. The following ranges of the variables were studied:

$$d = 0.585\text{-}6.35 \text{ mm}; \quad \gamma_M = 1105\text{-}7430 \text{ kg/m}^3;$$
$$d_0 = 1.59\text{-}102 \text{ mm}; \quad \gamma_m = 1.18\text{-}1000 \text{ kg/m}^3;$$
$$D_T = 76\text{-}610 \text{ mm}; \quad H = 0.089\text{-}2.69 \text{ m}.$$

On the basis of experiments with spouting beds of wheat, barley, oats, vetch, peas and soya beans in columns with D_T = 0.1-0.15 m, Madonna and Lama [1120] proposed the following formula for calculating the maximum hydraulic bed resistance before spouting:

$$\Delta P_{\max} = \frac{17\,950 HG \mu \varphi^2 (1-m)^2}{d^2 \gamma_m m^3} \text{ kg/m}^2, \tag{4-3}$$

where μ = dynamic viscosity, kg · sec/m²;
G = mass flow rate of medium, kg/m² · sec;
$\varphi = 0.205 S/V^{2/3}$ = shape factor;
S = surface area of particle, m²;
V = volume of particle, m³;
m = voidage of settled bed after spouting.

These authors [1011] also gave an equation for calculating the bed resistance after the onset of spouting:

$$\Delta P_s = 1\,185 HG^{1.25} \frac{\mu^{0.75}}{g^{0.25}\gamma_m} \left(\frac{\varphi}{d}\right)^{1.75} \frac{(1-m)^{1.75}}{m^3} \text{ kg/m}^2, \tag{4-4}$$

where g = gravitational acceleration, m/sec².

Madonna and Lama's equations were obtained for columns of diameter D_T = 0.15 m and inlets of d_0 = 12.7 mm, and are valid within the following ranges of the variables: d = 2-6.35 mm; m = 0.358-0.525, and fluid viscosities close to that of air at standard conditions.

Becker [1190] carried out experimental investigations of spouting in the passage of air through beds of narrow particle size distributions from 0.763 to 6.93 mm (sand, turnip seed, flax seed, wheat, barley, oats, maize) in five sizes of column (diameter of cylindrical section D_T = 152-610 mm, angle of inlet cone β = 90-180°).

Becker's approach to correlation was first to establish the maximum value of the minimum spouting velocity, $(w_{m.s})_{max}$, corresponding to the deepest bed that can still spout. He suggests that $(w_{m.s})_{max}$ can be calculated as

$$(w_{m.s})_{max} \approx 1.25 w_{\ell.s}, \tag{4-5}$$

where $w_{\ell.s}$ is the minimum fluidization velocity of a similar but cylindrical bed. The maximum spouting height H_{max} is then calculated from the relationship

$$\Omega \left(\frac{12.2 d_0}{D_T}\right)^n \left(22 + \frac{2\,600}{\mathrm{Re}_{max}}\right) \Phi^{2/3} \mathrm{Re}_{max}^{1/3} = 42, \tag{4-6}$$

where $\Omega = (H_{max}/d)(d/D_T)^{1.76}$;
$n = 1.6 \exp(-0.0072\ \mathrm{Re}_{max})$;
$\mathrm{Re}_{max} = (w_{m.s})_{max} d/\nu$;
Φ = experimental shape factor;
d_0 = diameter of inlet.

The expression for calculating H_{max} is simpler for beds of spherical particles for $\mathrm{Re}_{max} < 100$:

$$\Omega^{-1} = 4.4 + 1.6 \cdot 10^4/\mathrm{Re}_0, \tag{4-7}$$

where $\mathrm{Re}_0 = w_0 d/\nu$;
$w_0 = (w_{m.s})_{max}(D_T/d_0)^2$.

The minimum spouting velocity, $w_{m.s}$ of a bed of given height H can then be found from the expression

$$w_{m.s}/(w_{m.s})_{max} = 1 + 0.0071\,(D_T/d_0)\,\Phi^{2/3} \mathrm{Re}_{max}^{0.295} \ln(H/H_{max}), \tag{4-8}$$

where H is the height of the settled bed.

On the basis of experiments carried out in glass columns of 152-mm diameter with β = 180°, Becker [1190] proposed the following expression for calculating $w_{m.s}$ for very shallow beds ($H/D_T < 1$):

$$w_{m.s}/(w_{m.s})_{max} \approx 1.2\,(H/D_T) \exp(-0.058 D_T/d_0). \tag{4-9}$$

Chapter 5

Mixing in Fluidized Beds

5.1 Particle Mixing

Intensive mixing of the solid material and fluid can occur in fluidized beds. Mixing assists some processes and hinders others; it has very important effects on heat and mass transfer and on chemical reactions. The study of mixing is therefore of considerable practical interest and has occupied the attention of many researchers, but because of its complexity the subject has still only been superficially investigated.

There is a definite relationship between the motions of the particles and the fluid in a fluidized bed. As mentioned earlier, the reason for particle motion within the bed is mainly the nonuniformity of the flow around the particles and the nonuniformities caused by the passage of bubbles. A bed becomes fluidized and particle motion commences only after reaching a sufficiently large fluid velocity (in particular, according to Rowe's data [2147], the appearance of bubbles is noticeable only at fluidization indices $N \geqslant 1.3$). In this sense the fluid motion is of primary importance. However, moving particles, particularly, aggregates of particles, can entrain some fluid with themselves and thus lead to mixing of it. For example, the material circulation sketched in Fig. 5-1, with downward motion of the particles around the walls, causes the fluid (gas) at the walls to move in a direction opposite to that of the main stream. Because of this, Leva [988] considers fluid mixing in a fluidized bed to be a secondary phenomenon caused by the particle mixing. On the contrary, Reboux [511] neglects transport of the fluid by the particles and assumes that this is negligible in the absence of adsorption. Ioffe and Pis'men [1050] also assume that particle motion can only affect gas transport through an adsorption-desorption mechanism. However, experiments of Gilliland and Mason [307] and Cairns and Prausnitz [963] have again indicated the opposite. In spite of an insignificant adsorption helium, a descending stream of nonporous

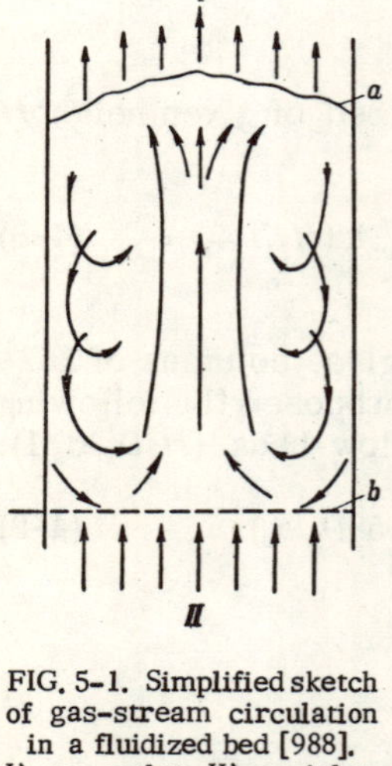

FIG. 5-1. Simplified sketch of gas-stream circulation in a fluidized bed [988]. I) gas outlet; II) gas inlet; a) upper boundary of fluidized bed; b) gas-distributing screen.

glass particles entrained an appreciable quantity of this tracer gas. The gas transported in this way could be trapped between the particles of a moving aggregate. A similar "ejecting" action was encountered in examination of the pressure drop in moving dense beds. By analogy, an enhanced ejecting action can be expected on going to smaller particles, all other conditions remaining unchanged. Gilliland and Mason [307] determined the gas mixing by analyzing for the concentration of the tracer gas (helium) at various points in the bed. The reverse motion of a colored tracer gas could scarcely be noticed in experiments of Rowe and Wace [1126], which were carried out by a mixing method not very well suited to quantitative determinations. They made visual observations and took motion pictures of the gas-stream structure in air-fluidized beds of glass spheres (d = 460 μ) to which a nozzle supplied a thin stream of the tracer. The fluidized bed had an unusual flat shape (cross section 305 × 25 mm).

In concluding these introductory remarks, it must be noted that Leva's view [988] that fluid mixing in a fluidized bed is purely secondary is also incorrect, since even in a fixed bed of particles there is mixing of the medium due to passage between the particles, and longitudinal mixing again occurs due to effects caused by nonuniformities in the velocity profiles.

1. Visual Observations

Many papers have been published on the mixing of particles, the attendant effective bed viscosity, and methods for decreasing the mixing of particles in fluidized beds [107, 264, 307, 410, 511, 515, 573, 652, 696, 699, 824, 706, 903, 961, 988, 993, 1024, 1065, 1119, 1123, 1130].

In describing the usual particle circulation in small-diameter fluidized beds (see Fig. 5-1), Leva [988] noted that the particles could be seen through the transparent walls of a tube to move downward and then suddenly away from the tube walls into the bed. At the same point or a little lower, other particles appeared from within the bed and after some time also moved downward. These active zones ended at some finite distance from the bottom of the bed. If the lower part of the bed were conical such active zones appeared closer to the screen than in a completely cylindrical bed, in which the height of the inactive zone could be of the order of the bed radius. However, this height, as well as the whole picture of the particle motion, depends on the bed height. In shallow fluidized beds the descending particle motion around the walls is feebly developed, and mixing in the bed appears to be more uniform. Bed structure depends to a considerable extent on the type of gas distributor, primarily in shallow fluidized beds and in the lower parts of deeper beds. Leva also noted [988] that at similar values of the fluidization index, close to the limit of stability, there is a much more intensive particle motion in gas-fluidized than in liquid-fluidized beds. Referring to a private communication from Sackman, Leva stated that the picture of particle motion is altered drastically if the bed axis deviates only a few degrees from the vertical.

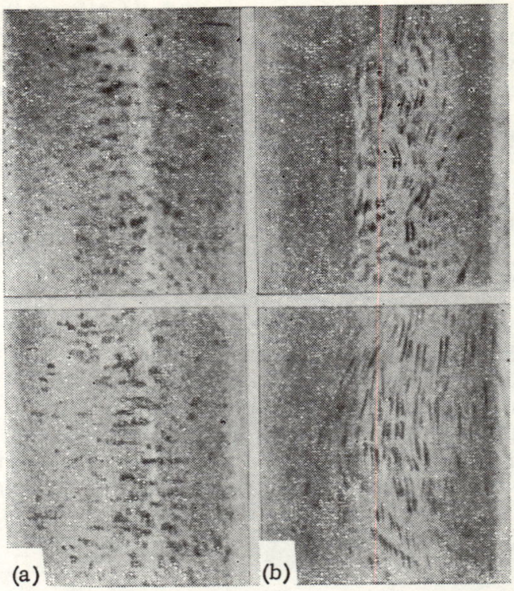

FIG. 5-2. Photographs of very dilute fluidized beds of steel spheres [1024].

In their study of particle motion in fluidized beds, Todes and Bondareva [662, 728, and 1024] drew an analogy between this motion and turbulent pulsations of fluids. Groups of particles with similar velocities move through the bed (like "eddies" in a turbulent fluid); individual particles gradually drop out of the groups and are combined in other groups. The mean distance L between these groups is the "mixing length" or "scale of turbulence." Motion-picture photographs of highly expanded fluidized beds of steel spheres (d = 1 mm) in a glass tube (D_T = 28 mm) confirmed the presence of pulsating particle motion [1024]. Two frames (Fig. 5-2) show the almost parallel motion of groups of neighboring particles along the walls in many places. Bondareva and Todes explained that each sphere gives two blips on the frame which are slightly separated in the direction of motion due to the finite time of exposure (1/250 sec). The projection of the particle pulsation velocity in the plane of the photograph can be measured from the length of these streaks. This varies (down to zero) from frame to frame; after a certain number of frames the flow picture recurs. Knowing the rate of filming (64 frames/sec) it is possible to determine the pulsation period T and hence the scale of turbulence $L = UT/2$, where U is the pulsation velocity. The results of such measurements are given in Table 5-1.

The table shows that as the bed expands the scale of the pulsations increases from a value rather larger than the bed radius to a value greater than the bed diameter.

Massimilla and Westwater [1121] also observed particle pulsations in fluidized beds. They took high-speed motion pictures (200 frames

Table 5-1

Particle Motion in an Air-Fluidized Bed of Steel Spheres (d = 1 mm) in a Tube of D_T = 28 mm [1024]

H/H_0	w_f, m/sec	Mean-pulsation velocity of particles, U, m/sec		Scale of turbulence, $L = UT/2$, mm
		from films	from turbulimetry	
2.00	3.44	0.141	—	17.6
2.40	4.75	—	0.151	
3.33	5.42	0.135	—	
4.00	5.62	—	0.345	
5.00	5.85	0.194	—	36.5
6.00	6.25	—	0.268	
6.67	6.32	0.170	—	
8.00	6.40	—	0.205	
8.30	6.44	0.275	—	
8.80	—	0.272	—	43.5

per second) of the movement of glass spheres (d = 710 μ) and alumina particles (d = 74-120 μ) in beds of 95-mm diameter fluidized by air. The gas-distributing screen had 74-μ openings. An alternating rapid and slow movement of the particles and aggregates up and down the walls could be clearly observed. Rotation of individual particles was also observed.

No change of sign of the vertical particle velocity component along the wall was reported in Toomey and Johnstone's fluidized bed investigations [568], which were carried out by the same method of high-speed filming. Their bed diameter was 120 mm, and the initial bed height about 0.61 m. Glass spheres of 376-μ diameter were fluidized with air, and particle velocities measured 0.38 m above the porous-plate gas distribution. Figure 5-3 shows typical examples of Toomey and Johnstone's data on particle velocities in fluidized beds. Leva [988] mentions the following results of these experiments:

1) the periodically appearing maxima give an explanation of the tendency toward the formation of the piston regime;

2) if we consider the mean value rather than the peak, then the particle velocity is lower than the superficial fluid velocity;

3) the particle velocity increases as the superficial fluid velocity increases.

The first point raises some doubts: if the peak values were actually caused by the slugging motion, both ascending and descending particle motions ought to be observed, with discontinuous circulation.

In Toomey and Johnstone's experiments the relationship between the particle velocity and the distance above the gas distributor was fixed (Fig. 5-4). Particle velocities reached as high as 6-61 cm/sec; these values are not in agreement with the experimental data of Leva and Grummer [410] for quartz sand and silica gel fluidized in tubes of 33.5- and 51.6-mm diameter, where particle velocities reached only 0.9-3.8 cm/sec. Leva [988] explains this difference by the effects of particle shape and bed diameter. Leva and Grummer's experiments

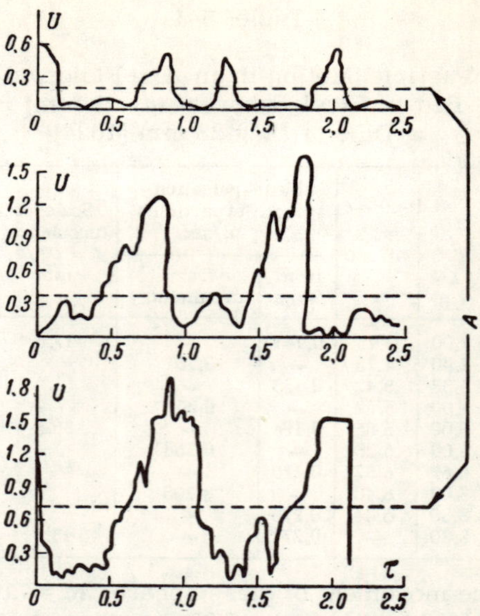

FIG. 5-3. Time changes of particle velocities in a bed of glass spheres (d = 376 μ) at a bed height of 584-610 mm, according to data of Toomey and Johnstone [568]. Velocities were determined at a distance 381 mm from the bottom of the bed. u = particle velocity, m/sec; τ = time, sec; A = superficial fluid velocity, m/sec.

were carried out in small-diameter beds (large retarding effect of the walls), and the particles were of irregular shapes, while Toomey and Johnstone worked with more mobile glass spheres in a large-diameter bed. Values of 10-30 cm/sec were reached in Bondareva's experiments [728] with steel spheres.

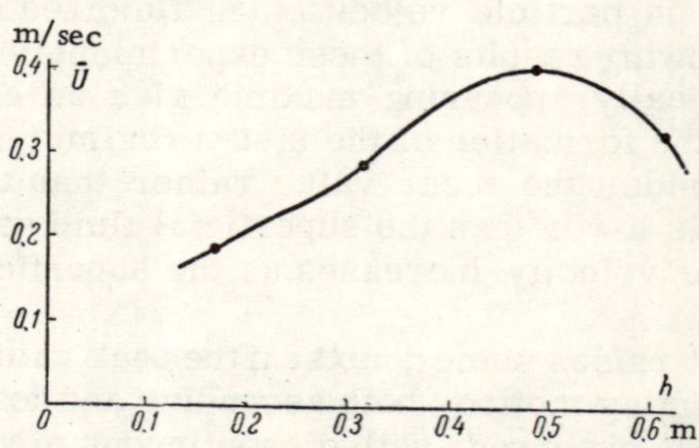

FIG. 5-4. Particle velocity at the walls of the apparatus at different cross sections of the bed, according to data of Toomey and Johnstone [568]. These data are characteristic of the case investigated; they refer to glass spheres (d = 376 μ) fluidized with air at a weight flow rate G = 1320 kg/m²·hr; different results can be obtained for other particles fluidized under varied conditions.
h = distance from bottom of bed, m.

In addition to determining the vertical components of the pulsating velocity by means of motion-picture films, Bondareva and Todes also attempted to calculate them from the oscillations of a metallic sphere suspended from a spring and submerged in the bed. This sphere, which was larger and heavier than the bed particles, was connected rigidly through a rod to a flat horizontal spring fitted with an electric tensiometer forming part of an unbalanced bridge. The alternating e.m.f. appearing in the diagonal of the bridge was amplified and read on a loop oscilloscope. The pulsating particle velocity was determined from the mean velocity of the spherical turbulence detector, assuming that the sphere diameter was much smaller than the fundamental scale of the pulsation. Equations of Landau and Lifshits relate the pulsation velocity of the sphere to the pulsation velocity of an ideal fluid surrounding it:

$$U = u_s \frac{\rho + 2\rho_s}{3\rho}, \tag{5-1}$$

where U and ρ are, respectively, the pulsation velocity and, in this instance, the bed density, and u_s and ρ_s are, respectively, the pulsation velocity and density of the sphere.

The pulsation velocities calculated in this way are also shown in Table 5-1; note that in one case the particle pulsation velocities obtained from the films and from the turbulence measurements are similar, while in the other cases they are quite different. In addition, the pulsation velocities calculated from the turbulence measurements initially increase as the superficial fluid velocity increases and then begin to decrease rapidly, while the cinematographic measurements indicate no such tendency over the range of fluid velocities investigated. The difference can be explained either by a difference in the pulsations at the periphery and in the middle of the bed, or by an error in the turbulence experiments. Bondareva and Todes feel that the decrease in the particle velocity as the superficial fluid velocity increases is quite definitely established at sufficiently large values of the fluid velocity [728, 1024]. To explain this theoretically, they proposed that at very large fluidization indices the bed voidage has increased appreciably, while the hydrodynamic attraction forces decrease approximately inversely with the fourth power of the distance between the particles. However, the hydrodynamic effect can be quite different in a dilute fluidized bed, leading to intensive particle motion, such as the well-known Magnus effect. To explain the reliability of the results obtained from the turbulence measurements, it would be necessary to have more information on the particle concentration distribution in the bed under the experimental conditions. If the local density of the fluidized bed during its expansion became appreciably lower than the mean-calculated value at the position of the spherical turbulence probe, the calculated particle velocity would be lower than the true value.

The presence of a maximum in the particle velocity–fluid velocity relationship agrees with the data of Cairns and Prausnitz [1105]. They fluidized beds (about 250 mm high) of lead and glass spheres (d = 3 mm) with water in a 101-mm-diameter glass tube. Under these conditions they visually observed nonhomogeneous two-phase fluidization and a maximum mixing velocity of the particles at $m \approx 0.7$, when the dense phase, instead of the liquid, became the continuous phase. On the basis of these experiments, however, it would be more natural to relate the appearance of a maximum particle velocity to the previously mentioned second onset of more uniform fluidization after the unstable regime rather than to a simple decrease of the hydrodynamic attraction force.

According to the data of Brötz [221], who investigated particle mixing over the whole volume of a water-fluidized bed, there was a monotonic increase in particle velocity as the superficial fluid velocity increased. Brötz fluidized beds of glass beads with water; initially the top part of the beds was colored. Brötz failed to reach fluidization indices as high as those in the investigation of Bondareva and Todes.

Bondareva [728] indicated that the particle pulsation velocity does not vary monotonically with the Archimedes number, i.e., with the particle diameter. The absolute values of the pulsation velocity for sand particles increased when the particle diameter was increased from 100 μ (Ar = 100) to 233 μ (Ar = 1260), but when the particle diameter was increased further from 233 to 477 μ (Ar = 11,000) it decreased. Bondareva explains this by the change in the hydrodynamic regime of flow around the particles.

Gilliland and Mason [307] and Singer, Todd and Guinn [705] also investigated particle mixing in fluidized beds. Gilliland and Mason demonstrated the presence of particle mixing by measuring longitudinal temperature profiles along the axis of a tube, the lower part of which was cooled with water and the upper part heated electrically. They compared the temperature profiles obtained with and without a fluidized bed of particles within the tube. Singer et al. [705], determined the residence time of cracking catalyst particles tagged with radioactive scandium-46 and cerium-144 in industrial-scale equipment, and concluded that ideal mixing of the catalyst in each chamber could be assumed under the operating conditions of such equipment.

2. Effective Viscosity

Particle mixing is a manifestation of the effective viscosity of fluidized beds. It is known that when a body, for instance, the vane of a Stormer viscosimeter, moves in a fluidized bed it experiences friction in the bed. At relatively small bed expansions the transfer of momentum is accomplished mainly by surface friction between neighboring particles in contact rather than by mixing and elastic collisions between particles of different energies. By analogy with the viscosity of gases, the effective viscosity determining the mixing of particles in sufficiently dilute beds, or regions of beds, can be written as [537]

$$\mu^* = K\rho_{\text{bed}} ua, \tag{5-2}$$

where K = proportionality factor;
ρ_{bed} = bed density;
u = mean-particle velocity;
a = length of mean-free path.

Trawinski proposed this formula, then abandoned it, assuming that it was incorrect, since it involved the assumption that the viscosity increased as the bed expanded. However, the investigation of Diekman and Forsyth [264], published in the same year, and the more recent paper by Shuster and Haas [1130] showed that over some regimes there is actually an increase in the effective viscosity of a fluidized bed as the superficial fluid velocity increases.

For the effective viscosity Bondareva and Todes [728] proposed a similar equation based on the analogy between mixing in a fluidized bed and turbulent mixing of a fluid:

$$\eta_{eff} = K\rho_{bed}uR, \tag{5-3}$$

where R = scale of motion (tube radius).

These authors mentioned later that it is possible to have a relationship between u and R at a given superficial fluid velocity and bed expansion [1024].

Equation (5-3) was confirmed by experiments of Bondareva [728] for large bed expansions, when there is no friction between one particle and another. The coefficient K appeared to be close to 100. The measurements were carried out with the same sand and in the same tubes in which the pulsating sand velocities were determined by turbulence measurements. Viscosities were determined both from the damping of torsional oscillations of a suspended hollow cylinder and with a rotating viscosimeter. The first method could be used for measuring viscosities of sands of various diameters only at relatively large expansions $H/H_0 > 10$. Measurements with the rotating viscosimeter, fitted with a cylindrical perforated attachment suspended in the fluidized bed, were possible over the range $2 < H/H_0 < 10$. Under the experimental conditions there was initially an abrupt decrease in the effective viscosity as the superficial fluid velocity was increased, followed by a gradual change approaching the viscosity of air.

In these investigations, just as in those of Matheson, Herbst and Holt [434] and Furukawa and Omae [824], only one value is reported for the viscosity at each fluid velocity in a given fluidized bed, while actually the effective viscosity is different at different points within the fluidized bed. Kramers [396] was one of the first to demonstrate this by measuring vertical viscosity profiles in a fluidized bed. Liu and Orr [1119] determined the local values of the effective viscosity at various points along the vertical axis of an air-fluidized bed of glass spheres (d = 44, 88, 99 and 123 μ; γ_M = 2470 kg/m^3) and of aluminosilicate cracking catalyst particles (d = 45 μs, γ_M = 2120 kg/m^3); the relative humidity of the air was 15%. The effective viscosity was measured with a viscosimeter having a rotor provided with two concentric cylinders fixed to a shaft. The diameter of the outer cylinder was 19 mm, and its height was 12.7 mm. Effective viscosities were not measured closer than 50 mm from the surface of the fluidized bed because of the violent particle motion in this region. External vibrations were avoided, since they have an important effect on the effective viscosities of beds of fine particles. In the lower part of each fluidized bed the effective viscosity decreased as the superficial fluid velocity w_f increased; in the middle of the bed the viscosity changes were small and could be in either direction; and at the top of the bed the effective viscosity increased with w_f. The authors assumed that as w_f increased, more and more of the particles were transferred from the lower to the upper parts of the fluidized bed, thus decreasing the effective viscosity at the bottom and increasing it at the top of the bed. This explanation is scarcely valid.

Shuster and Haas [1130] made more detailed measurements of viscosities in fluidized beds. They determined the local viscosities of air-fluidized beds of glass spheres (d = 76.5-1020 μ) in transparent 146- and 70-mm-diameter columns, and found the dependence of the effective viscosity on the superficial fluid velocity w_f up to relative bed expansions of 1.5-2.3; initial bed heights varied from 38 to 160 mm. Effective viscosities were determined with a modified Stormer viscosimeter having a very small measuring head (19 × 19 × 116 mm), which could be moved to any point in the bed by means of a positioning device. The main measurements were carried out at one rate of rotation of the viscosimeter shaft (n = 200 r/min). The absolute value of the effective viscosity measured by a Stormer viscosimeter is a function of the size and angular velocity of the measuring head, the particle density, the particle-size distribution, the type of gas distributor, and the superficial fluid velocity. The effective viscosity is a linear function of the bed density and the vertical particle velocity component. Shuster and Haas found that the shape of the curve giving the effective viscosity as a function of the fluid velocity did not change with the angular velocity of the viscosimeter shaft and could be divided into three parts, in agreement with theoretical considerations. At very small superficial fluid velocities the effective viscosity was very large and decreased rapidly as w_f increased (Fig. 5-5, region *I*, with negligible particle motion). At moderate values of w_f the curve began to rise and passed through a maximum (region *II*, with rapidly increasing particle velocity and initially slowly decreasing bed density). At larger values of w_f the effective viscosity again fell to some equilibrium value (region *III*, with increasing particle velocity but rapidly decreasing bed density). Compared with a screen with 75-μ openings, a porous gas-distributing plate gave a smaller value of the effective viscosity and greater uniformity of the latter over a horizontal cross section of the bed. Mixtures with spheres of two sizes ($d_1:d_2$ = 0.41 and 0.273) gave effective viscosities that were larger than those for either component in region *II*, while in region *III*, just as in the case of Furukawa and Omae's experiments [824],

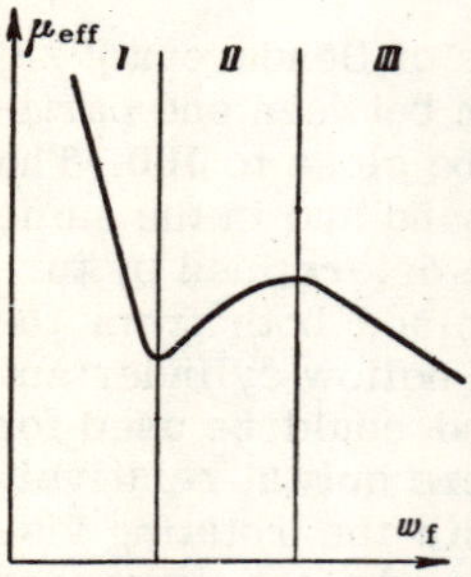

FIG. 5-5. Nature of relationship between the local effective viscosity of a fluidized bed and the superficial fluid velocity, according to Shuster and Haas [1130].

$$\lg \mu_{mixt} = y_a \lg \mu_a + y_b \lg \mu_b, \tag{5-4}$$

where y_a and y_b are the weight fractions of components a and b in the mixture; μ_{mixt}, μ_a and μ_b are the effective viscosities of the mixture and the components.

Thus the effect on the viscosity of the fluidized bed of adding fine particles to coarse ones varies in sign as well as in magnitude for the different zones of fluidization. It is known [573, 717] that Trawinski's model concepts explaining the decrease in the bed viscosity when fine particles are added to a coarse bed are confirmed in the region of small bed expansions (obviously corresponding to region *I* in the notation of Shuster and Haas), where friction effects between neighboring particles predominate. Trawinski assumes that the viscosity is caused primarily by friction between the large particles. The introduction of fines between the large granules therefore acts as a lubricant, decreasing friction and leading to a decrease in viscosity. According to this concept, the minimum fines concentration at which there is an appreciable viscosity reduction corresponds to a quantity of fine particles sufficient to cover each coarse particle with a monolayer of fine particles (Fig. 5-6). The weight fraction of fine particles necessary to form a monolayer is

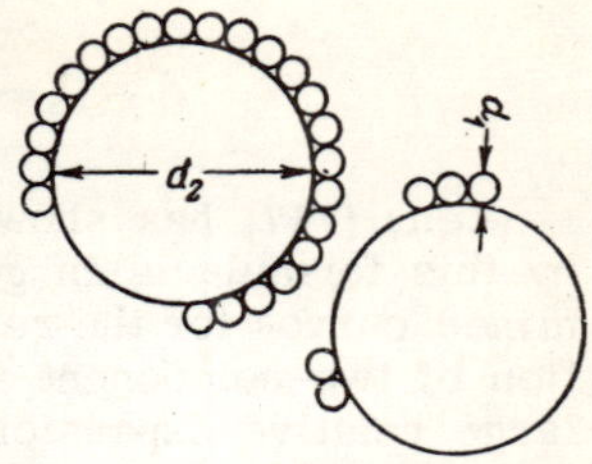

FIG. 5-6. Model assumed by Trawinski for explaining the effect of the size analysis of the solid on the effective viscosity of a fluidized bed [717].

$$W_{d_1} = 1 - 1/(A + 1), \tag{5-5}$$

where

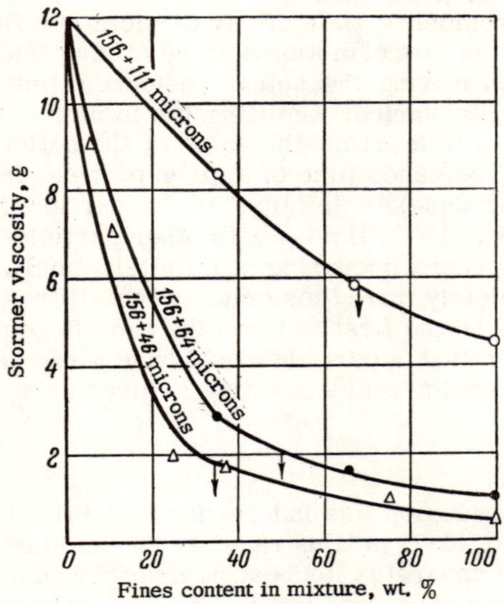

FIG. 5-7. Data of Matheson, Herbst and Holt [434] on viscosities of fluidized two-component mixtures.
The numbers on the lines refer to the sizes of the coarse and fine particles. The arrows indicate the minimum fines concentration calculated from Eq. (5-5) [771].

$$A=(1-m)\left\{3\left[\frac{d_1}{d_2}+\left(\frac{d_1}{d_2}\right)^2\right]-\frac{H-H_{l.s}}{H_{l.s}}\right\}.$$

Zenz [717] has shown that the minimum fines fraction calculated by this formula is in good agreement with the experimentally determined curves for the relationship between the viscosity and composition of two-component mixtures (Fig. 5-7). However, in the zone of large relative expansions of a fluidized bed, when the particles are separated from one another, Trawinski's model for the lubricating action of the fines ceases to be valid; the experiments of Furukawa and Omae, and later Shuster and Haas, showed that the viscosities of two-component mixtures varied quite differently with composition. The optimum size analysis of the bed for minimizing the viscosity differs in the various regions of fluidization.

3. Particle Residence Times

In a fluidized bed with continuous supply and removal of material the deviation in the residence times of individual particles from the mean-residence time of the material in the bed serves as a measure of the particle mixing; the mean-residence time is calculable as

$$\tau_0=G_{\text{bed}}/Q_{\text{M}}, \tag{5-6}$$

where G_{bed} = bed weight, kg and Q_{M} = weight rate of feed or removal of material, kg/hr (kg/sec).

The actual residence time of any particle in a fluidized bed would be equal to τ_0 in the case of motion with no mixing (piston flow regime).

In a bed with mixing the actual residence times of the individual particles are quite varied. Omitting the minimum time required for passage of a particle from the inlet to the outlet in the absence of collisions, the residence time of a given particle can vary from zero (in this limiting case) to infinity, if the particle is prevented from reaching the outlet by collisions with other particles. The probability of these limiting cases occurring is infinitely small. For the simplest case of a completely mixed material, when all the particles have the same chance to be the first to leave the bed, Reboux [511] calculated the probability x that a particle would have a residence time greater than τ, when the mean-residence time is given as τ_0:

$$x=e^{-\tau/\tau_0}. \tag{5-7}$$

A similar expression was independently obtained by Fedorov [170].

The use of Eq. (5-7) is illustrated by an example.

Example. An apparatus for heating material in a thin fluidized bed (H_0 = 0.1 m) has a cross section F = 3.47 m². The material bulk density γ_{bulk} = 1600 kg/m³ and the throughput Q_{M} = 20,000 kg/hr. Find the mean-residence time of the material in the bed and the distribution of actual heating times for the case of complete mixing.

Solution. The bed weight G_{bed} = 3.47 × 0.1 × 1600 = 555 kg. The mean-residence time of material in the bed $\tau_0 = G_{\text{bed}}/Q_{\text{M}}$ = 555 × 3600/20,000 = 100 sec.

The probability x that any particle will have a residence time larger than τ is equal to $e^{-\tau/\tau_0}$.

τ, sec	x	τ, sec	x
1	0.991	50	0.61
10	0.905	100	0.37
20	0.82	500	0.007

Thus, in spite of the fact that the mean-residence time is 100 sec, 0.9% of the particles stay in the bed for less than 1 sec, 37% for more than 100 sec, and 0.7% for more than 500 sec.

4. Diffusivity

At present no correlation exists for calculating the mixing of material in a fluidized bed. Some researchers have attempted to evaluate the experimental data on the basis of an analogy with molecular or turbulent diffusion, and effective diffusivities have been calculated [205 and 696]. However, the information obtained in this way is very fragmentary. According to data of Bart [205] published by Leva [988], the particle diffusivity in a fluidized bed is approximately proportional to the superficial fluid velocity. Bart studied gas-fluidized beds of cracking catalyst particles (d = 115 μ) in a 32-mm-diameter tube. Massimilla and Bracale [696] studied mixing in an air-fluidized bed of glass spheres (d = 0.7 mm) in a 100-mm-diameter tube by continuously adding colored spheres at the top and measuring their concentrations in samples taken at different points along the bed axis. In order to maintain a constant bed height as many particles were removed at the bottom as were added at the top of the bed. Under Massimilla and Bracale's experimental conditions the material diffusivity D^* was not proportional to the superficial fluid velocity w_f:

w_f, cm/sec	D^*, cm²/sec
43	0.3
50.3	0.5

For experiments carried out in his laboratory Todes calculated the order of magnitude of the proportionality factor φ in the expression for the turbulent diffusivity of fluidized beds:

$$D^* = \varphi u l, \qquad (5\text{-}8)$$

where u = pulsating particle velocity, determined from turbulence measurements; l = scale of motion.

The quantity φ appeared to fall in the range 1/10 to 1/20.

Practical calculations using effective diffusivities are still far in the future. Information is required on the effects of particle properties, size analysis and diameter of the fluidized bed on the effective diffusivity, as well as for taking particle circulation in the bed into account.

5.2 Fluid Mixing

1. Homogeneous Beds

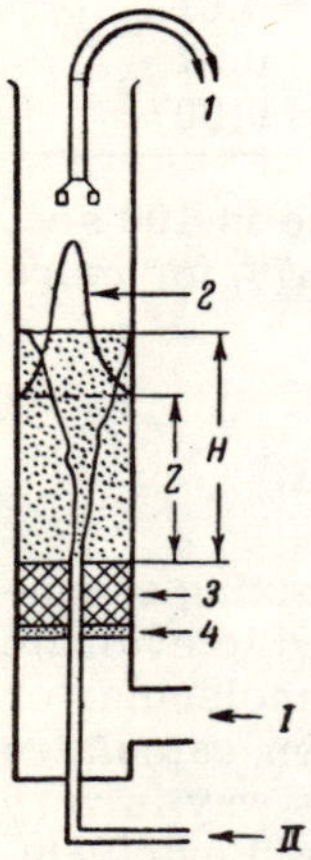

FIG. 5-8. Sketch of experimental equipment used by Wicke and Trawinski [602]. I) main fluidizing liquid (water); II) tracer liquid (HCl or hot water); 1) probe with leads to bridge device; probe can be moved vertically and horizontally; 2) radial concentration profile; 3) packed zone for damping stream; 4) porous-distributing plate; H = bed height; z = level at measuring point.

Wicke and Trawinski [602], Hanratty, Latinen and Wilhelm [331], Blickle and Kaldi [961] and Cairns and Prausnitz [1104] have investigated fluid mixing in fluidized beds of solid particles.

Figure 5-8 is a sketch of the main part of the experimental equipment used by Wicke and Trawinski [602]. They fluidized beds of glass spheres and alumina and plastic particles with water in tubes of 90-mm diameter. The particle diameters varied from 0.9 to 12 mm. Tracer liquid (3 N hydrochloric acid solution or hot water) was fed in continuously through a tube of 4- or 6-mm diameter placed axially with its outlet in the plane of the bed support screen. The concentrations of hydrochloric acid at various points were determined by electroconductivity probes. In those cases where hot water was used as tracer liquid, thermocouples were used as the probes. Typical curves for the tracer concentration distribution obtained in the experiments of Wicke and Trawinski are shown in Fig. 5-9. The experimental data agree approximately with the Gaussian distribution function:

$$C(r,Z)=k\frac{w_f}{Z}\,e^{-w_f r^2/4D^*Z}, \tag{5-9}$$

where D^* = effective diffusivity.

This equation corresponds to the assumption that the distribution of the tracer in the main stream (the mixing of the medium) occurs analogously to molecular diffusion. The initial equation for steady-state one-dimensional radial diffusion in a moving stream;

$$\frac{\partial C}{\partial Z}=\frac{D^*}{w_f}\cdot\frac{\partial^2 C}{\partial r^2} \tag{5-10}$$

can be obtained from the similar equation for unsteady-state one-dimensional diffusion in a stationary medium.

Wicke and Trawinski found that the effective diffusivity increased linearly with the superficial fluid velocity (Fig. 5-10), and that the linearity was not broken on passing through the limit of stability. This is equivalent to saying that mixing of the medium is not reduced during fluidization compared with mixing in a fixed bed, though the increase of voidage (expansion) of the bed was small. According to [602], the effective diffusivity increases as $d^{0.85}$:

$$D^*=2.76\cdot10^{-2} w_f\, d^{0.85}\quad \mathrm{m^2/sec}, \tag{5-11}$$

where w_f is expressed in m/sec and d in meters.

The height of the fluidized bed and the distance from the point of tracer injection appeared to have little effect on D^*. The effect of the nature of the material was not established. The results obtained with HCl and hot water as tracer were in agreement.

Hanratty, Latinen and Wilhelm [331] investigated the mixing of the fluid stream in a water-fluidized bed of glass spheres (d = 3 mm) in a 54-mm-diameter tube. A tracer (aqueous solution of methylene blue dye) was fed in axially. Liquid samples were taken at various points in the bed and analyzed electrophotometrically. This investigation covered a much wider range of bed expansions than that of Wicke and Trawinski; maximum mixing was observed at a bed voidage of $m \approx 0.7$.

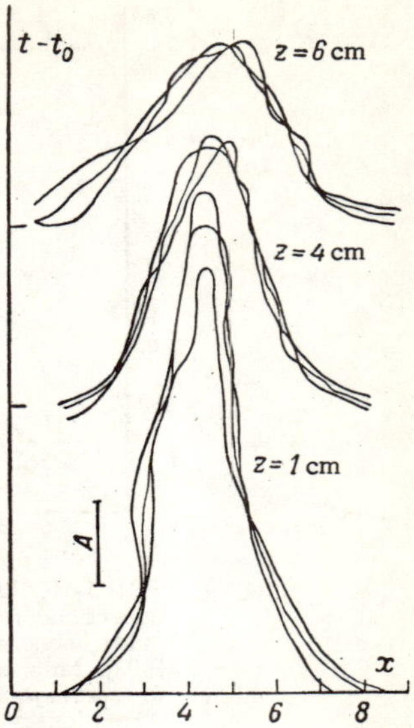

FIG. 5-9. Typical data for fluid mixing in a fluidized bed of plastic spheres (d = 9.8 mm), according to Wicke and Trawinski [602]. Hot water as tracer liquid. $t-t_0$ = excess of temperature over the value corresponding to complete mixing; A = interval corresponding to 1°C on the ordinate axis; x = distance from one wall along a diameter, cm; z = distance from the bottom of the bed.

The more recent investigations of Blickle and Kaldi [961] and Cairns and Prausnitz [1104] have confirmed the presense of such a maximum. Blickle and Kaldi used a noncontinuous tracer feed (NaCl solution) and determined the change in concentration at the bed outlet. The mixing coefficient increased initially with increase in the superficial fluid velocity but then began to fall off. The authors explained that one reason for this could be the marked decrease in particle concentration in the bed at large fluid velocities.

Cairns and Prausnitz [1104] studied the longitudinal mixing of water in fluidized beds of glass (d = 3.2 mm) and lead (d = 113 and 3.0 mm) spheres in 50- and 100-mm-diameter along the axis at intervals of five bed diameters from the electrolyte tracer feed point (sodium nitrate solution). The electrolyte was introduced simultaneously at 156 points over the cross section; even at an axial distance of five particle diameters the nonuniformity of the concentration profile did not exceed 9% when the electrolyte was supplied continuously. By using a by-pass line with a rapid-acting solenoid reversing valve it was possible to cut off the electrolyte supply very rapidly. Radial profiles of the electrical conductivity were obtained by using small 3-mm-diameter probes to measure the electrical conductivity of a volume of the order of 1 mm . The electrolyte concentration was assumed proportional to the electrical conductivity. The bed voidage had an important effect on the intensity of longitudinal mixing; maximum mixing was observed at $m \approx 0.7$. The effective longitudinal turbulent diffusivity depended directly on the bulk density of the particles and the ratio of bed diameter to particle diameter, D_T/d. The turbulent diffusivity is a function of the product of a characteristic length and a characteristic velocity; a nonuniform profile of the longitudinal diffusivity. According

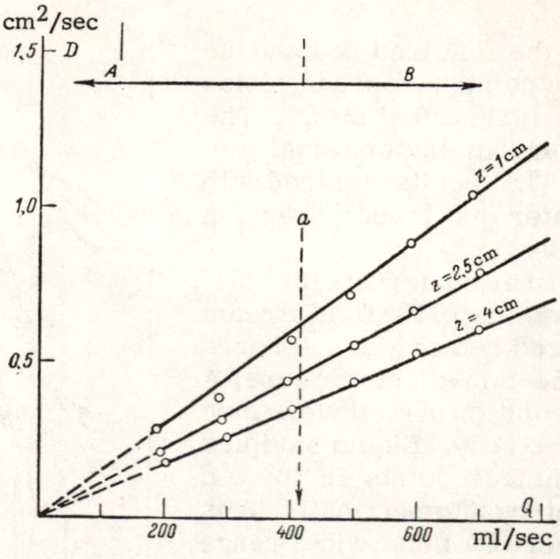

FIG. 5-10. Linear change of the diffusivity D with superficial fluid velocity in beds fluidized with liquids, according to data of Wicke and Trawinski [602]. No break in the lines is observed at the point a, corresponding to the onset of fluidization [988]. A) stationary dense bed; B) fluidized bed; z = distance from bottom of bed; Q = water flow rate, ml/sec.

to the data of Cairns and Prausnitz, the ratio of the mean-longitudinal turbulent diffusivity to the radial diffusivity was 20-30 under conditions of maximum nonuniformity of the fluid velocity profile of ± 20%.

2. Nonhomogeneous Beds

Mixing of gases in fluidized beds has been the subject of a large number of papers [72, 201, 307, 308, 309, 515, 683, 893, 1050, 1126, 1196].

Gilliland and Mason [307-309] studied the mixing of a gas in fluidized beds of glass spheres (155 μ), catalyst microspheres (75- to 210-μ fraction) and filter-aid particles in 25- and 76-mm-diameter tubes. Helium and, occasionally, carbon dioxide were used as tracer gases. Adsorption phenomena were reduced to a minimum by the use of helium. The tracer stream was fed axially through a 5-mm-diameter glass tube. Gas samples were taken at various points in the fluidized bed and analyzed for helium by a weight method. The gas-distributing grid had openings of 75-μ diameter. Figures 5-11 and 5-12 show typical longitudinal and transverse tracer concentration profiles in the bed; C is the tracer concentration at the given point; C_0 is the tracer concentration for the case of total mixing; $C_0 = v_{tr}/v_g$, where v_{tr} and v_g are the volumetric flow rates of tracer and fluidizing gas.

The presence of tracer gas below its feed point (Fig. 5-11) along the axis of the fluidized bed clearly indicates that there is reverse (longitudinal) gas mixing. In contrast to the analogous profiles for homogeneous beds (fluidized with liquids), the radial concentration

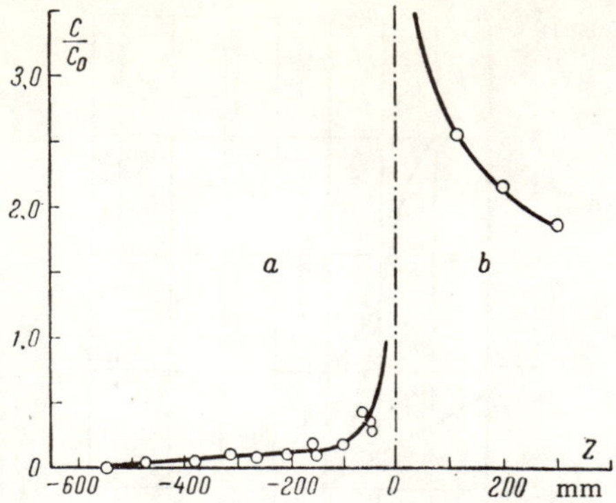

FIG. 5-11. Results of analyses of gas samples taken along the axis of the apparatus for glass spheres fluidized with air, using helium as the tracer gas [307, 988].
Superficial fluid velocity = 0.354 m/sec; z = distance from helium feed point; *a*) zone below feed point; *b*) zone above feed point.

profiles in Fig. 5-12 show a nonmonotonic radial concentration change with a minimum above the tracer feed point. Leva [988], who did not report this result quite correctly, explains the presence of the minimum by downward motion of gas between the core and peripheral annulus of the fluidized bed. Obviously the descending motion of particles and gas is always most marked along the walls, i.e., in the peripheral zone. Therefore, above the cross section where the tracer concentration profile levels out, a great deal of tracer is carried off, and its concentration along the walls increases. This is the cause of the minimum; without this effect there would be a monotonic fall in tracer concentration right up to the wall caused by the symmetrical axial position of the tracer feed point. Leva [988] conjectures why the measured tracer concentrations on some transverse profiles appear higher at all points than the mean value obtained during ideal mixing; his "anomaly" he feels is caused by the passage of bubbles containing hardly any tracer gas through the bed. The bubbles pass so rapidly that scarcely any of the gas from them is collected by the sampling device.

Gilliland and Mason attempted to give a correlation for the back mixing of the gas based on the equation of turbulent diffusion:

$$\frac{dm}{d\tau}=D^*\frac{dC}{dZ}, \tag{5-12}$$

where m = quantity of mass transferred across unit area;
τ = time;
Z = vertical distance from the tracer feed point;
D^* = turbulent diffusivity.

To integrate this equation, the authors made the far from valid assumptions that the diffusivity and the gas velocity were uniform at all points over the cross section obtaining

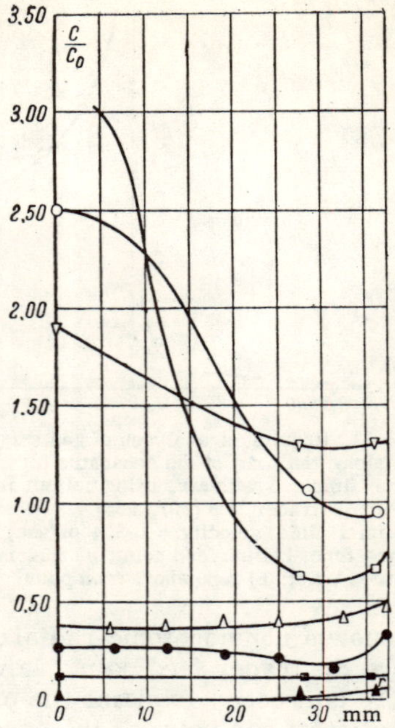

FIG. 5-12. Radial concentration profiles above and below tracer gas feed points, according to data of Gilliland and Mason [307, 988]. Superficial fluid velocity = 0.354 m/sec.

Symbol	Level in bed	
▽	305.0 mm	below helium feed point
○	152.5 mm	
□	50.8 mm	
△	50.8 mm	below helium feed point
●	152.5 mm	
■	254.0 mm	
▲	457.0 mm	

$$\ln \frac{C}{C_0} = -\frac{w_f}{D^*} Z + b. \tag{5-13}$$

If the experimental points are plotted as C/C_0 versus Z in semilogarithmic coordinates (Fig. 5-13), the slope of the line fitting the points gives the approximate value of w_f/D^*, from which D^* can be determined.

Leva [988] gives a graphical correlation obtained by Reman [515] for the data of Gilliland and Mason and unpublished data of Stemerding in the form of a relationship between the "mixing length" (D^*/w_f) and the superficial fluid velocity w_f (Fig. 5-14). This figure shows that the mixing length increases with the fluid velocity and density of the fluidizing gas. Most of the lines in Fig. 5-14 are almost horizontal, indicating that here D^* is approximately proportional to w_f. It can also be seen from Fig. 5-14 that, in general, gas mixing becomes more pronounced as tube diameter increases, as was also noted by another investigator [201]. Handlos, Kunstman and Schissler [683] indicate that there is vigorous gas mixing in large-diameter fluidized beds; they carried out investigations in regenerators of industrial dimensions.

Ioffe and Grigorov [893] measured the reverse (longitudinal) mixing of a tracer (carbon dioxide) in fluidized beds of porous and nonporous materials (silica gel and sand) (see Table 5-2).

In Table 5-2 the circulation multiplicity factor r represents the ratio of the recycled volume of carbon dioxide, CO_{bottom}, to the carbon dioxide supplied, CO_{2top}—$CO_{2bottom}$:

$$r = \frac{CO_{2bottom}}{CO_{2top} - CO_{2bottom}}.$$

The tabulated data show that under Ioffe and Grigorov's experimental conditions an adsorption-desorption mechanism of gas mixing prevails, while effects not related to adsorption of the ejected CO_2 entrained in the particle aggregates lie within the range of the measurement error.

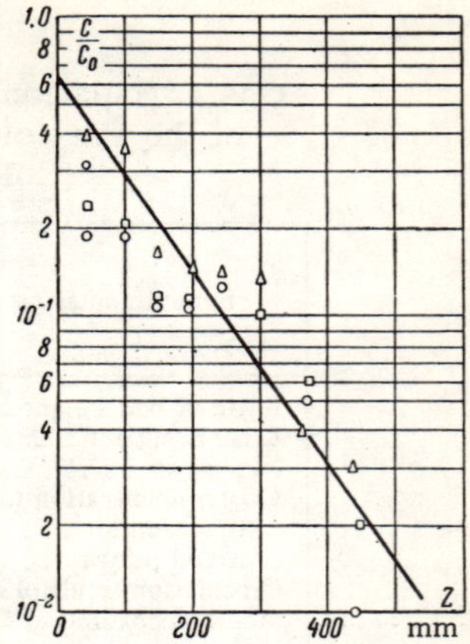

FIG. 5-13. Correlation of Gilliland and Mason's data on longitudinal gas mixing in fluidized beds [307, 308].
Z = distance below helium feed point, mm.

Symbol	Radius, mm
○	0.0
□	27.9
△	36.3

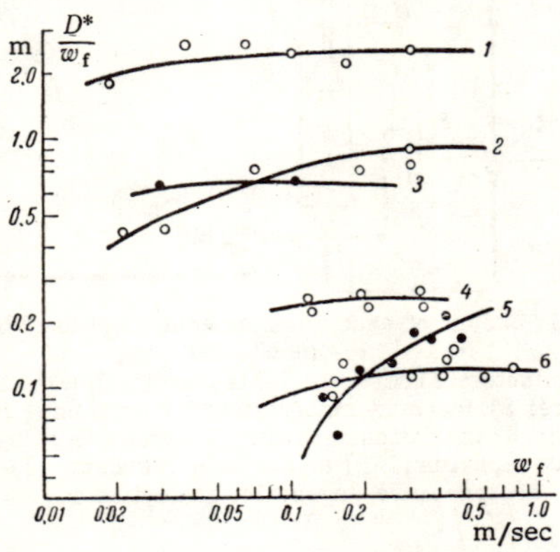

FIG. 5-14. Relationship between turbulent diffusivity D^* and superficial gas velocity according to data of Rema [515, 988].
1) D_T = 152.5 mm; air, CO ; cracking catalyst; 2) D_T = 152.5 mm; air, H ; cracking catalyst; 3) D_T = 101.6 mm; air, CO ; cracking catalyst; 4) D_T = 76.2 mm; air, He; cracking catalyst; 5) D_T = 76.2 mm, air, He; glass spheres; 6) D_T = 25.4 mm, air, CO ; filter-aid particles.

Table 5-2

Gas Circulation Multiplicity Factor as a Function of the Particle Properties in a Fluidized Bed (Column Diameter 58 mm)

Experimental parameter	Material				
	Silica gel (50-250 μ)			Sand (40-250 μ)	
Ratio of bed height to diameter	4.5	4.5	7.4	5.7	7.4
Linear superficial fluid velocity, m/sec.	0.17	0.125	0.11	0.23	0.15
CO_2 concentration in gas:					
at bed outlet	8.5	11.8	13.6	9.6	8.7
at bed inlet	3.7	4.2	1.9	0.0	0.0
Circulation multiplicity factor factor according to Orochko . .	0.77	0.55	0.16	0.0	0.0

The interesting paper by Ioffe and Pis'men [1050] analyzing the macrokinetics of chemical processes in fluidized catalyst beds investigated in particular detail the mixing of gases, using experimental residence time distribution functions for a tracer gas fed into the beds in short bursts.

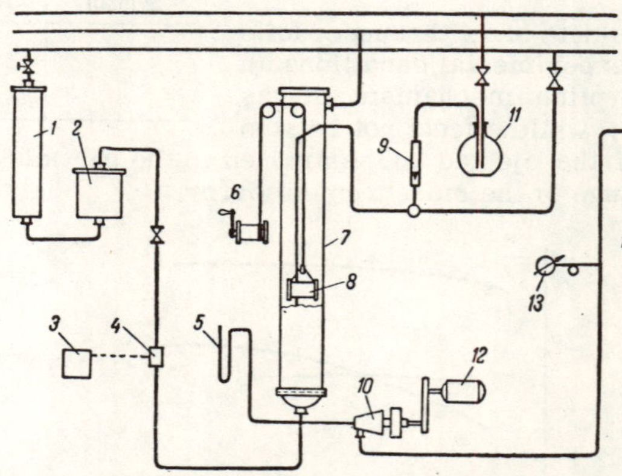

FIG. 5-15. Sketch of experimental equipment used by Ioffe and Pis'men [1050].
1) oil eliminator; 2) reservoir; 3) recording flow meter; 4) double orifice; 5) mercury manometer; 6) positioning mechanism; 7) column; 8) ionization chamber; 9) rotameter; 10) reversing valve; 11) capacitor; 12) motor with reducer; 13) membrane pressure gages; 14) gas cylinder.

Figure 5-15 is a sketch of the equipment used in these experiments [1050]. The 2-m-high column had an i.d. of 150 mm. A pulse of tracer gas (helium) was supplied by a rotating valve driven by an electric motor. The pulse feed point was immediately below the column. A radioionization method was used for measuring the tracer gas concentration at the outlet. The gas leaving the bed passed through one of the cells of a differential ionization chamber; the other cell of the chamber, which was filled with air, was used to compensate for the

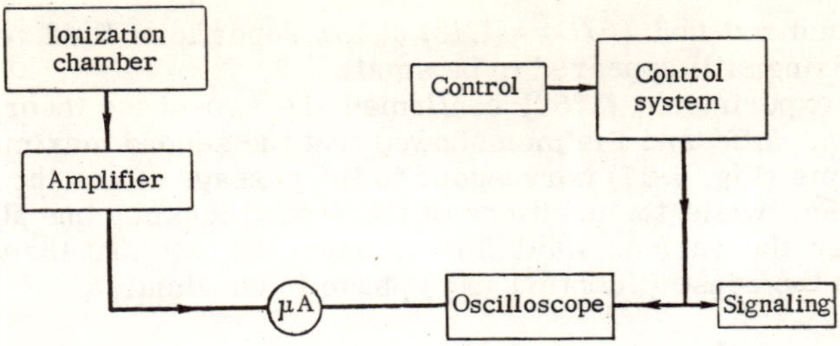

FIG. 5-16. Electrical scheme of Ioffe and Pis'men's apparatus [1050].

ionization current flowing at zero concentration of the tracer gas. The gas ionization in the two cells was excited by promethium b-radioactive sources of equal strength. The difference between the ionization currents from the two cells of the chamber (proportional to the concentration of tracer gas) varied over the range $0\text{-}10^{-11}$ amp. This was amplified electrometrically and read on a loop oscillograph. Helium was chosen as the tracer gas since its relative ionization is furthest from that of air. The detector was suspended in the column immediately above the upper boundary of the bed. About 50 cm^3 of helium was released in the column at each pulse. The air flow rate was measured by paired orifices. The electrical arrangement (Fig. 5-16) ensured synchronization between the pulses and the photographing of the oscillograms.

Experimental distribution functions were obtained for the empty column, a stationary bed, and a fluidized bed of quartz sand of 120-μ mean-particle diameter. The bed height and superficial fluid velocity were varied. The experimental curves obtained for the same regime were averaged. Examples of the curves obtained for the case of the fluidized bed are shown in Fig. 5-17.

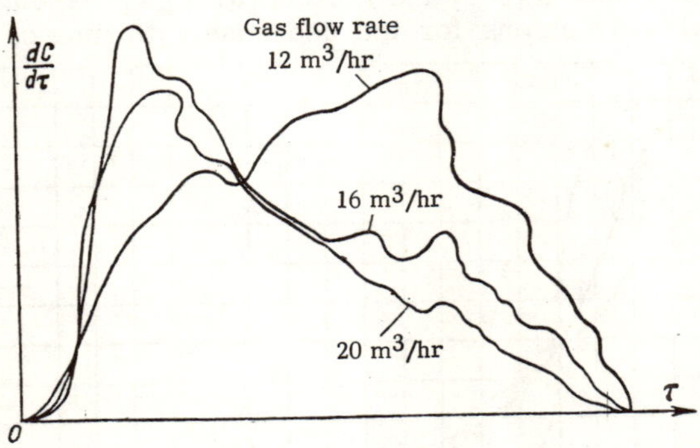

FIG. 5-17. Experimental gas residence time distribution functions in fluidized bed [1050].
Ratio of initial bed height to diameter, H/D_T = 3.4; C = helium concentration; τ = time.

Under Ioffe and Pis'men's experimental conditions, the fixed bed operated close to the regime of ideal displacement, and even in a shallow fluidized bed (H/D_T = 1.13) at low superficial fluid velocities the gas mixing still appeared to be small.

These experiments [1050] confirmed the two-phase theory of gas fluidization. Ioffe and Pis'men showed that the second maxima on the oscillograms (Fig. 5-17) correspond to the passage of gas through the dense phase, while the positions of the second maxima one above the other under the various fluid flow regimes indicate that the true gas velocity in the dense ("continuous") phase is constant.

3. Residence Time of the Fluid Medium in the Bed

Gilliland and Mason [308] and Reboux [511] derived analytical expressions for the gas residence time in a fluidized bed for the limiting case of complete longitudinal mixing of the gas.

It is known that in the simple case of motion of the medium through the bed—the regime of total displacement without any longitudinal mixing—each element of the fluid volume is in the bed for the same length of time:

$$\tau_0 = H/w_{\text{true}} = mH/w_f, \qquad (5\text{-}14)$$

where H = bed height and w_{true} = true mean-gas stream velocity.

According to Reboux's expression [511], the gas molecules lose all individuality in the case of total mixing of the medium. Each of them has the same chance of leaving the fluidized bed at a given moment, which may be the moment of entering the bed. Thus the actual residence times of given molecules lie within the very wide range from zero to infinity. The probability that a molecule will stay in the bed for a time τ becomes smaller the more τ differs from τ_0, given by mH/w_f.

Gilliland and Mason [308] and Reboux [511] gave essentially identical analytical expressions for the decrease with time of the relative

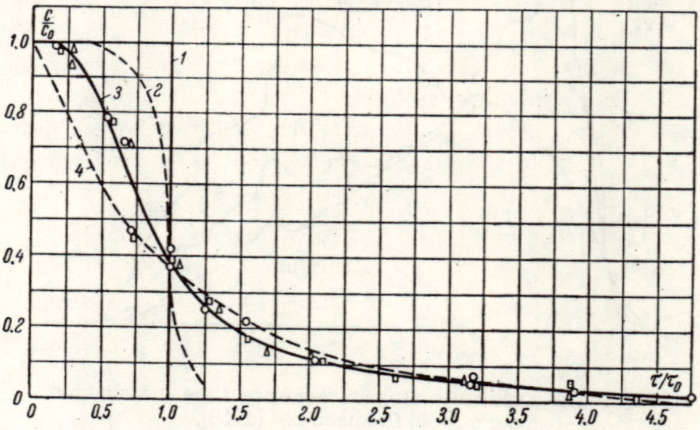

FIG. 5-18. Gas residence times in bed for various limiting conditions compared with experimental data of Gilliland and Mason [308] for a fluidized bed of glass microspheres.
1) line for ideal displacement; 2) fixed bed; 3) fluidized bed, from experiments [308]; 4) motion with complete mixing.

tracer concentration (after its supply is stopped) in the main stream of medium leaving the fluidized bed:

$$\frac{C}{C_0} = e^{-\tau/\tau_0}, \tag{5-15}$$

where τ is the time after the tracer feed has been cut off. This expression makes it possible to use experimental data to estimate how closely the effective longitudinal mixing of the medium in the fluidized bed approximates the ideal case. Figure 5-18 shows the experimental line for a dense bed, as well as the lines for the limiting cases of complete mixing and total displacement. For the regime of total displacement the relative tracer concentration in the discharge gas remains constant and equal to unity for a period of time τ_0 after the tracer supply is cut off, after which all the tracer is displaced from the bed, corresponding to the cut-off point on the graph. Beyond this point C/C_0 falls directly to zero. It can be seen from Fig. 5-18 that the line for a fixed bed is quite close to that for complete displacement, in agreement with the data of Ioffe and Pis'men. However, the data for fluidized beds do not show such agreement, since Fig. 5-18 shows that under the experimental conditions of Gilliland and Mason there was apparently close to complete longitudinal mixing of the gas (helium) in the fluidized bed, in spite of the small degree of adsorption of the tracer by the particles. This can be seen from Fig. 5-19, where the experimental points and the theoretical line for complete mixing are

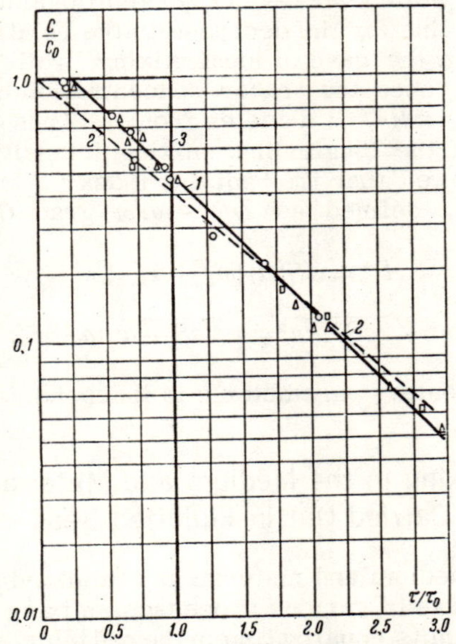

FIG. 5-19. Gas residence time in fluidized bed of glass microspheres, according to data of Gilliland and Mason [309]. Superficial fluid velocities: ○ 0.134 m/sec; △ 0.186 m/sec; □ 0.305 m/sec; 1) experimental line; 2) line for complete mixing; 3) line for total displacement.

plotted in semilogarithmic coordinates. However, such an explanation of Gilliland and Mason's experimental data [508] is essentially incorrect. The fact is that the extended time for which an appreciable tracer concentration is retained after the tracer feed is cut off (or the long duration of the discharge pulse compared with the initial pulse), which at first sight indicates a considerable longitudinal (reverse) mixing of the gas in the fluidized bed, can be explained equally well by the nonuniformity of the gas velocity profiles in a nonuniform bed with a part of the gas passing in the form of bubbles. Ioffe and Pis'men [1050] have taken this into account to give a valid explanation of the presence of two peaks in the discharge pulse. The first is explained by the rapid passage of tracer in the bubbles, and the second by tracer moving through the continuous phase. It can be seen from Fig. 5-17 how the first peak increases in size and moves to the left as the gas flow rate (superficial fluid velocity) increases through a given bed; this corresponds to an increase in the fraction of the gas passing in the form of bubbles and to the formation of larger and faster bubbles. The confusing impression of the value of the longitudinal gas mixing caused by velocity nonuniformity in a nonhomogeneous fluidized bed was subsequently also noted by Rowe [1246]. He pointed out that the experiments of Gilliland and Mason [307, 308], Reman [515], and Bracale, Cabella and Massimilla [808] were not designed in such a way as to permit the nonuniform gas velocity profile effect to be separated from true reverse mixing.

When C/C_0 is plotted versus τ/τ_0 in semilogarithmic coordinates, the slopes of the line for the decrease in the relative tracer concentration equals 1 in the case of ideal mixing, while for complete displacement, $S = \infty$ once the tracer concentration begins to fall, i.e., once the line $C/C_0 = f(\tau/\tau_0)$ deviates from 1. Thus S is a useful numerical measure of the longitudinal mixing intensity. Lavrovskiy and Rozental' [72] showed that the "mixing index" S is related to the effective diffusivity, defined as $D^* = q/m$ grad C, by the equations

$$D^* = (H/2a)(w_f - w_{\ell.s})/m; \tag{5-16}$$

$$\tan\sqrt{2aS - a^2} = \sqrt{2aS - a^2}/(a - S), \tag{5-17}$$

where the dimensionless parameter a is independent of w_f.

5.3 Effects of Mixing in the Medium and Material on Processes Carried Out in Fluidized Beds

Mixing of the medium and material in a fluidized bed has important effects on the course of various processes. Mixing of the material in a fluidized bed permits equalization of the bed temperatures even when it is used for carrying out a process with a large thermal effect; it also leads to an increase in the heat transfer rate between a fluidized bed and heated or cooled walls surrounding it. This makes it possible to maintain suitable temperatures in fluidized bed equipment. Boreskov and Slin'ko [636] noted that gas mixing had the advantage of allowing components of a reacting mixture to be introduced separately into a catalyst bed. This is particularly useful if the mixture of the initial components is explosive over a wide range of compositions.

However, in other respects the longitudinal mixing of the material and medium is disadvantageous; for instance, in carrying out heat transfer between the medium and material or in heterogeneous chemical reactions. The mean value of the potential difference (for instance, the temperature driving force in heat transfer) is reduced as a result of the longitudinal mixing, and the possibility of thermal counterflow is lost.

There are additional difficulties in determining the effects of medium and material mixing by comparing the kinetics of processes in fluidized beds and in systems without mixing; it is necessary to estimate the degree and importance of gas break-through in nonhomogeneous fluidization [893].

Contradictions are encountered in estimating the degree of approximation of the behavior of a fluidized bed to the limiting cases of total displacement and complete mixing; often these are connected not only with actual differences in the operating conditions and degree of mixing, but also with the use of different methods for determining the degree of mixing, such as the mixing index, the effective diffusivity, the circulation multiplicity coefficient (Orochko), etc. The various methods of estimating mixing differ in their accuracy in determining deviations from a given limiting regime of total displacement; for the same experimental data it is possible to get a larger or smaller apparent approach to complete displacement depending on the method used for estimating the degree of mixing. For example, Reboux [511] assumed that the flow of medium in a fluidized bed approached the regime of total displacement. On the basis of an analysis of experimental data for shallow fluidized beds, Ioffe and Pis'men [1050] also concluded that this was the case. On the other hand, Borechkov and Slin'ko [636] noted that gas mixing had an important effect on reactions carried out in fluidized beds. According to their estimate there is no approach to complete displacement. Gilliland and Mason [309] showed that, as regards the distribution of gas (tracer) residence times, a fluidized bed behaves very similarly to a system with complete mixing (Fig. 5-18). In their experiments a bed of glass beads (diameter 70 μ) was fluidized with air (w_f = 0.134–0.305 m/sec) in a 114-mm-diameter tube with L/D_T = 7.8. However, in this same work [309], in determining the closeness to complete mixing or total displacement from second-order reaction kinetics, it appeared that in many of the experiments the fluidized bed was operating in a regime close to total displacement. The contradiction here is obviously only apparent and as noted above is caused by failing to take into account the differences in the rates of passage of gas in the continuous phase and as bubbles. Obviously in most cases (when there is no pronounced degree of adsorption of the medium on the particles) the degree of mixing of the

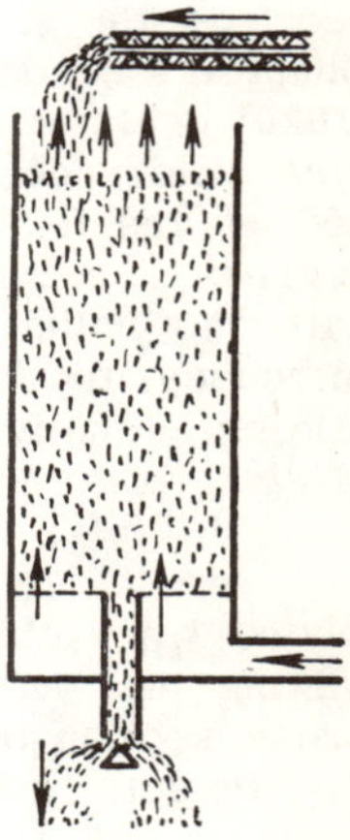

FIG. 5-20. Sketch of fluidized bed apparatus with material feed at top and discharge at bottom [1020].

medium is actually small, but this mixing is sufficient to retard significantly the course of some chemical reactions. Thus unconditional statements on the approach of the fluid mixing in a fluidized bed to one or the other of the limiting cases must be made with caution.

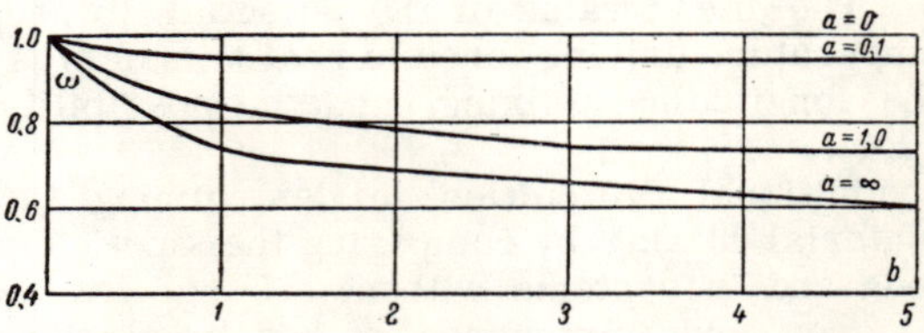

FIG. 5-21. Degree of countercurrent behavior (ω) of equipment as a function of the dimensionless group b. Numbers on the curves indicate the value of the quantity a.

Information on the effect of mixing in fluidized beds on the kinetics of chemical reactions of various orders can be found in many references [70, 107, 196, 430, 636, 755, and 893]. Due to lack of space, only brief mention will be made of the effect of mixing in fluidized beds on the process of heat transfer between a bed and walls (see Chapter 10) and between particles and medium. A paper by Baskakov [1020] deals with the latter problem. He analyzes the mean-temperature difference between the gas and the material fed into a fluidized bed at the top and discharged at the bottom (Fig. 5-21) for various degrees of mixing, and he compares the temperature driving force with the value for an ideal case of countercurrent flow. Baskakov introduces the following dimensionless parameter as a measure of the mixing intensity:

$$a = \lambda_{\text{eff}} / (\gamma_M c_M w_M H),$$

where λ_{eff} = effective thermal conductivity of the bed due to particle mixing and w_M = overall velocity of material (referred to the total cross section of the fluidized bed. Formally, a is the reciprocal of the Peclet mass transfer number for the material. The degree of countercurrent behavior of the apparatus

$$\omega = \frac{\Delta t_{\text{av}}}{(\Delta t_{\text{av}})_{\text{cc}}},$$

depends on a; there $(\Delta t_{\text{av}})_{\text{cc}}$ = mean-temperature driving force in countercurrent flow. Obviously the same expression can be set up using the quantities of heat transferred:

$$\omega = Q/Q_{\text{cc}}.$$

Baskakov also introduced a second group, on which ω also depends:

$$b = \frac{6\alpha H (1 - m)}{\gamma_m c_m w_f d}.$$

It is simple to show that

$$b = \alpha F / W_m,$$

where F = total surface area of the particles in the fluidized bed; W_m = water equivalent of the fluidizing medium.

Actually, $W_m = F_{sec}\, w_f c_m \gamma_m$, where F_{sec} = transverse cross-sectional area of the bed. The surface area of uniform spherical particles per cubic meter of bed is

$$S = \frac{6(1-m)}{d}.$$

The volume of the bed is

$$V_{bed} = F_{sec} H; \; F = SV_{bed} = \frac{6(1-m) F_{sec} H}{d}$$

and

$$\frac{\alpha F}{W_m} = \frac{6 \alpha H (1-m)}{\gamma_m c_m w_f d}.$$

Thus the group b is identical with the usual dimensionless group which appears in thermal calculations for ordinary surface heat exchangers.

Baskakov [1020] considers the case when the water equivalents of the heat transfer media are equal, when the advantages of countercurrent flow are most apparent, and hence when the effect of material mixing will be greatest (this statement is not exact, since at various values of $\alpha F / W_m$ the greatest deviation between the quantities of heat transferred in countercurrent and cocurrent flow is reached at different values of W_1 / W_2, and is only reached at $W_1 / W_2 = 1$ for $\alpha F / W \to \infty$; see, for instance, [98].

For the case $W_1 / W_2 = 1$, Baskakov obtained the following analytical expression for the mean-temperature driving force:

$$\Delta t_{av} = \frac{1 + ab + a^2 b (1 - e^r)}{1 + b + 2ab + a^2 b (2 - e^r) + ab^2}, \tag{5-18}$$

where

$$r = -\frac{ab + 1}{a}.$$

Figure 5-21 shows the relationship between W and the parameters a and b based on this equation. The graph shows that as the intensity of mixing of the material (parameter a) and $\alpha F / W_m$ (group b) increases, the degree of countercurrent behavior decreases and heat transfer is adversely affected. Thus an excess of αF (excessive bed height) has an important effect on mixing of the material. There is a similarity between the effect of $\alpha F / W_m$ and the surface areas of conventional heat exchangers, for which Fig. 5-22 (taken from [98]) shows that the advantage of using countercurrent flow increases as KF/W_1 increases. In the case of heat transfer between a gas and fine particles, K reverts to α, F is the surface area of all the particles in the bed, and W_1 or W_m is the water equivalent of one of the heat transfer media (the gas).

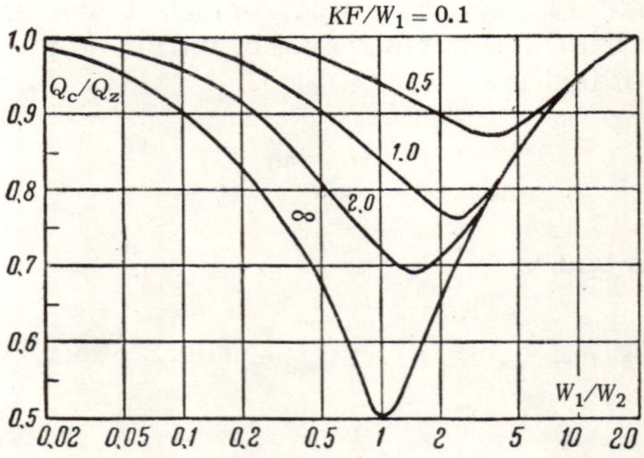

FIG. 5-22. Comparison of cocurrent and countercurrent flow.
Q_c = quantity of heat transferred in cocurrent flow;
Q_z = quantity of heat transferred in countercurrent flow, under otherwise equal conditions.

It can be seen from Fig. 5-22 that when $a < 0.1$, a heat exchanger with a countercurrent arrangement (material feed at the top and discharge at the bottom) is considerably more efficient than one with a top feed and discharge. Usually [1020] the weight flow rate of gas in a fluidized bed $w_f \gamma_m \leqslant$ 3-4 kg/m^2 · sec, $c_m \approx$ 0.25 kcal/kg · deg, and $H \leqslant 1$ m. An appreciable degree of countercurrent behavior ($a < 0.1$) then corresponds to an effective thermal conductivity $\lambda_{eff} \leqslant 300$ kcal/m · hr · deg.

When $a > 1$ the material mixing in a fluidized bed is so efficient that the relative positions of the feed and discharge points are unimportant. According to [1020], when $a \gg 1$, values of λ_{eff} as high as 26,000 kcal/m · hr · deg can be reached in industrial fluidized beds.

5.4 Means for Reducing Longitudinal Mixing

Apart from operating with small values of the fluidization index, various other methods for reducing longitudinal mixing in fluidized beds have been proposed. One consists in dividing the fluidized bed into a number of sections (Fig. 5-23) through which the gas and material pass in opposite directions. This gives rise to a so-called stagewise countercurrent operation. A second method involves placing in the fluidized bed horizontal screen baffles which are penetrable by the gas and solid material but which serve somewhat to restrain longitudinal mixing. A third method, in which rotating horizontal disks which do not cover the whole cross-sectional area [699] are placed in the fluidized bed has also been proposed for reducing longitudinal mixing. Even when such stirring devices are effective (nothing is said of this in the literature) they are inferior to the first two methods because of their complexity.

The idea of improving fluidized bed reactors by sectionalizing the reaction zone was proposed by Planovskiy in 1944 and developed by Orochko et al. [107, 652]. Details of the first method are given in these papers.

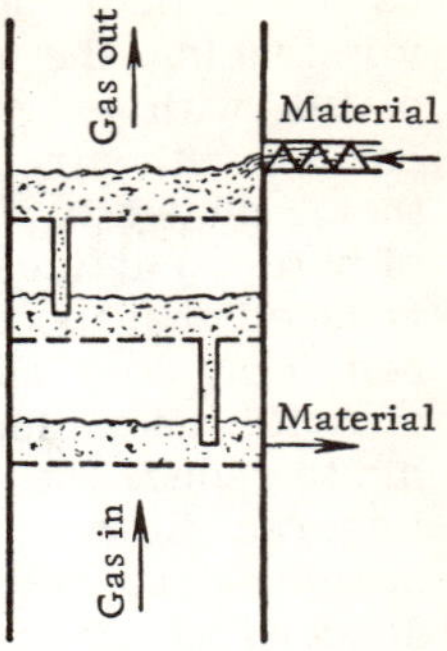

FIG. 5-23. Sketch of equipment with stagewise countercurrent flow.

Petrov [1065] studies the hydrodynamic aspects of fluidized beds with horizontal perforated and louvered baffles. He noted that such horizontal baffles divided the fluidized bed into zones with independent particle circulations and led to results close to those for stagewise sectionalizing in the reaction space. Petrov fluidized a narrow size fraction of sand (200-290 μ) with air in a glass tube (D_T = 70.5 mm); the bed weighed 1327 g. The perforated baffles had 1-, 2- or 3-mm-diameter openings in single or triple rows (10 mm between the rows); the louvered screens were of the form shown in Fig. 5-24. The experiments showed that the multirow perforated baffles reduced particle mixing more than the single-row baffles. The louvered screens had little effect on mixing, but, like the perforated baffles, they broke up the large bubbles.

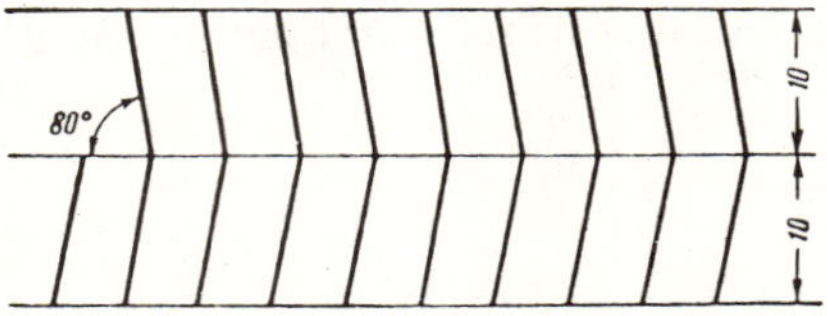

FIG. 5-24. Cross section of double-louvered screen [1065].

Among the foreign studies of fluidized beds "retarded" by screens and baffles, mention must be made of the work of Massimilla et al. [696, 808, 1121] and of Overcashier et al. [997]. In one of these papers, it is shown that horizontal screen baffles with openings of diameter 3.2-9.5 mm spaced 10 mm apart increased the fluidized bed density and lowered the particle velocity (5-10 times for glass spheres of diameter 710 μ, three times for alumina particles of 74-120 μ) [1121]. In [808] sand particles (170 μ) fluidized with air in a tube with D_T = 90 mm were retarded by horizontal wire screens placed 10 mm apart having openings 10 × 10 mm; the initial bed height was 860 mm and the weight flow rate of air was 760-3000 kg/m^2·hr. Concentration profiles of a tracer gas (CO_2) showed that the baffles significantly decreased longitudinal mixing.

Overcashier, Todd and Olin [997] studied the effect of horizontal baffles, with openings, on the mixing of air and material within a fluidized catalyst bed (γ_M = 1500 kg/m^3, d = 50-100 μ) in a column of relatively large diameter (D_T = 0.4 m). Helium and Freon were used as tracer materials for following the gas motion and radioactive Au^{198} for tracing the catalyst. The baffles studied were of two types: 1) plates with an open cross section of 15% and slitlike openings; 2) baffles with circular openings and short sections of tube inserted in them, projecting downward. The investigations showed that both types of baffle, particularly the one with the slitlike openings, considerably reduced the spectrum of residence times of gas and material in the bed, i.e., decreased their longitudinal mixing and the nonhomogeneity of fluidization. Increasing the superficial fluid velocity made it possible to reduce the spectrum of gas residence times in the retarded bed. Reverse flow of the gas ceased completely when the velocity in the openings reached 1.8-3 m/sec. In a countercurrent fluidized bed the descending stream of material passing through the baffle openings passes through a maximum as a function of the fluid velocity.

Immediately below the baffle a zone of low density is formed in the fluidized bed. In a countercurrent fluidized bed the most dilute zone occurs below the lowest baffle, and in cocurrent flow below the uppermost baffle.

An unusual case of a retarded fluidized bed occurs during fluidization of a highly dispersed material in a packing of relatively coarse spheres, which themselves may be stationary or moving.

Insertion of vertical elements (tubes, rods, screens) has the effect of retarding and hindering the formation of bubbles. According to statements in the literature, scaling up to full-size industrial equipment (with vertical inserts) results in the same behavior as in small-scale laboratory apparatus of the same hydraulic diameter.

Chapter 6

Carry-Over of Material From Fluidized Beds

6.1 Mechanism of the Phenomenon

Real fluidized beds always contain material having a range of particle sizes, and in most cases the superficial fluid velocity exceeds the free-fall velocity of the smallest particles in the bed charge or those that are formed in the bed by attrition of larger particles. There is therefore some carry-over of fines from all fluidized beds, and it is necessary to provide them with cyclones, electrostatic filters or other dust-removal equipment.

Carry-over from fluidized beds is usually but not always an undesirable phenomenon. Attempts are made to achieve maximum fines carry-over when a fluidized bed is used for removing undersize material from a granular mixture.

In any case, it is desirable to know the laws governing the carry-over of particles from fluidized beds. These are very complex, and have as yet been insufficiently studied. Much remains to be explained about the mechanism of the process. The correlations that have been proposed are suitable only for approximate calculations. Further study of the mechanism of particle carry-over from fluidized beds will make it possible to control these processes and optimize operating conditions. This is one of the most important problems connected with the introduction of fluidized bed techniques in various branches of industry.

The main factor causing and regulating the carry-over of material from a fluidized bed is the positive difference between the operating superficial fluid velocity and the free-fall velocity of the fine particles, $w_f - w_{ff}$. During steady-state motion of the fluid and material through the bed, fine particles reaching the upper surface are entrained with an absolute velocity $w_f - w_{ff}$. In this case the nominal concentration of material in the outlet gas (or the so-called carry-over value) would be equal to the concentration of a saturated stream determined from the choking condition of a pneumatic conveyor (see Chapter 3).

However, there are a number of facts that complicate the actual picture of carry-over, such as 1) ejection of particles from the fluidized bed by gas bubbles; 2) nonuniformity of the gas velocity field above the fluidized bed; and 3) resistance of the bed to "diffusion" of the fine particles to the surface of the fluidized bed.

The effect of particle ejection from the bed on the carry-over value has been considered in [857]. Gas bubbles that rise through the bed at large velocities, sometimes many times greater than the superficial fluid velocity, spray even coarse particles high above the fluidized bed. The free-fall velocity of these may be greater than the superficial fluid velocity. If the free height above the bed (to the gas outlet openings) H_f is less than the height to which the large particles are sprayed, these will also be entrained, falling into the zone of the outlet openings where the velocity is many times higher than the superficial fluid velocity. The height to which the coarse particles are sprayed and then separate from the stream and fall back into the fluidized bed is termed the disengaging height. Zenz and Weil [857] calculated the "theoretical" disengaging height from the condition that the particles are entrained from the surface of the fluidized bed with a velocity equal to the bubble velocity when it penetrates the surface, and fall back through a gas stream moving uniformly with the superficial gas velocity. It is assumed that each of the particles sprayed from the bed moves from then on in a free space under the action of inertia, gravity and drag forces.

The actual disengaging height appears to be greater than the theoretical value. This is mainly due [857] to the nonuniformity of the gas velocity field above the fluidized bed. Zenz and Weil assume that the gas bubbles penetrating the surface of the fluidized bed are analogous to intermittent jets, which are imposed on an otherwise irregular velocity profile over the apparatus cross section. A certain distance above the surface of the bed is required for dissipation of the velocities of these jets to the superficial fluid velocity. The actual disengaging height is therefore larger than the theoretical value.

For the same reasons, an increase should be expected in the entrainment for any increase in the nonuniformity of the gas stream above the fluidized bed, caused, for example, by a poor gas distributor (particularly if the bed is shallow and the initial gas distribution has an important effect on the upper parts of the bed). This is confirmed by the data of Fridland and Skoblo [1087] showing that an increase in the free cross section of the grid from 0.02 to 0.06, i.e., an improvement in the gas distribution uniformity, leads to a threefold decrease in the carry-over under otherwise identical conditions. Leva's remark [988] that entrainment increases on improving gas distribution is obviously in error, at least in unqualified form. For sufficiently deep fluidized beds, in which the nonuniformities in the upper parts increase as the initial height increases, it is reasonable to expect that entrainment will increase, i.e., that there will be a direct relationship between entrainment and bed height. On the other hand, the effect of bed height should be just the opposite for shallow fluidized beds: carry-over should increase with decreasing bed height, since defects in the gas distributor operation will then make themselves felt more strongly at the bed surface. Leva's remark [988] on the effect of bed height on entrainment corresponds to this second case. Fridland and Skoblo's experiments [1087], which covered a wide range of initial bed heights, have confirmed that the relationship between the rate of carry-over and the initial fluidized bed height is nonmonotonic.

Entrainment is significantly reduced if the bed provides a large resistance to the "diffusion" of fines toward the bed surface. There is even an equilibrium fines concentration which depends on the superficial fluid velocity for a given fluidized system, at least in the region of small relative bed expansions. The equilibrium residual fines concentration is related to the fact that, according to Trawinski, a certain quantity of fine material is locked up in the interstices between the coarse particles, depending on the degree of expansion, and is not transferred to the free surface of the fluidized bed, since at small fines concentrations the mobility and mixing of particles in the bed are reduced. The equilibrium fines concentration increases sharply as the superficial fluid velocity is reduced. For example, under the conditions of Leva's experiments [988], a decrease in the fluid velocity from 0.40 to 0.33 m/sec increased the equilibrium fines content from about 1.6 to 5%. Therefore, at small relative expansions a fluidized bed can act as a filter and retain any fine solid particles supplied in the feed stream of a fluidizing agent. The "dust capacity" of such a filter falls off sharply not only as the relative bed expansion R (or superficial fluid velocity) increases, but also as the ratio $d/D = n$ decreases (ratio of the diameter of the fines fed to the bed to the diameter of the coarse particles in the bed). Trawinski has derived the following approximate expression for calculating the equilibrium concentration of fine spherical particles in a bed of coarse particles:

$$x \approx \frac{0.565\,(3n + 3n^2 - R)}{1 + 0.565\,(3n + 3n^2 - R)}. \tag{6-1}$$

The rapid increase in carry-over with superficial fluid velocity in the region of small values of the fluidization index is explained not only by the increase in the difference $w_f - w_{ff}$ and in the nonuniformity of fluidization but also by the bed expansion, which directly facilitates travel of the fine particles from the depths of the bed to the surface. For this reason an increase in carry-over can be expected in feebly fluidized beds as the diameter of the coarse particles is increased, since the gaps between them, through which the fines can migrate, become larger; on the other hand, carry-over is reduced when the fine particles have irregular shapes and consequently stick between the coarse particles, hindering mixing in the bed.

The height of a fluidized bed can affect carry-over both by its effect on the nonuniformity of fluidization and in connection with the resistance to "diffusion" of the fines toward the bed surface in a feebly expanded bed. Because of this, there is a tendency for the rate of carry-over to increase as the bed height is decreased.

It must be remembered that some of these factors have only a limited effect on the magnitude of entrainment; they have an effect only when the height of the free space above the fluidized bed is less than the disengaging height. The effects of nonuniform gas distribution over the bed surface and of particle spraying at the surface are of this type. The carry-over intensity above the disengaging height can depend only on $w_f - w_{ff}$ and on factors that govern the travel of fines through the bed toward the surface. In allowing for the effects of various factors, within the disengaging zone or above it, account must be taken of whether the entrainment is being considered.

6.2 Experimental Data

Early investigations of carry-over from fluidized beds (up to and including 1955) were considered in detail in Leva's monograph [988]. Only some of these will be reviewed here.

Leva [406] studied the carry-over of an iron catalyst from air-fluidized beds of mixtures of two particle sizes in which the size ratios varied from 2.6 to 4.7. The initial concentration of fine particles in the bed was about 20% and the column diameter was only 33.5 mm. Some of the experiments were carried out with mixtures of catalyst and sand. Under Leva's experimental conditions the fines concentration (carry-over) in the air stream leaving the equipment increased rapidly with the superficial fluid velocity and the percentage fines concentration in the bed (Fig. 6-1). The equilibrium fines concentration (the quantity of powder stably retained in the bed) decreased sharply as the superficial fluid velocity increased.

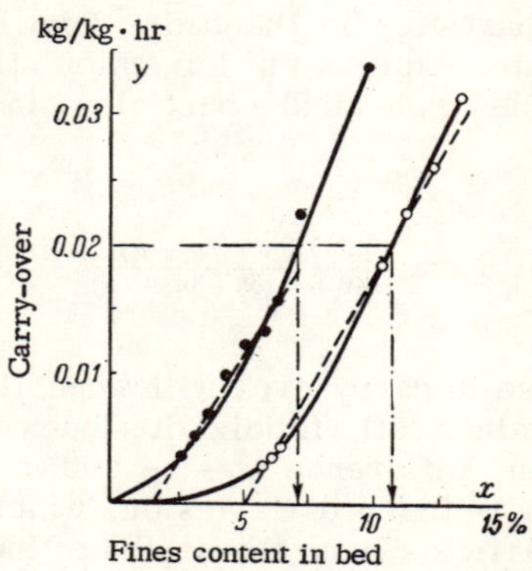

FIG. 6-1. Typical carry-over data, illustrating the equilibrium state of a bed of solid particles in a gas stream [406]: Two-component mixture of iron Fischer-Tropsch catalyst particles, fine particles 41 μ, coarse 191 μ. Superficial fluid velocities: ●—0.396 m/sec; ○—0.329 m/sec.

Similar investigations of Osberg and Charlesworth [489] (published at almost the same time) employed a fluid bed of 76-mm diameter. They fluidized coarse and fine glass beads with diameter ratios varying from 1.63 to 8.1. Air was the fluidizing medium, except for a few experiments with helium, CO_2 and Freon-12. The concentration of fine particles in the beds did not exceed 5%.

Hyman [355] studied carry-over from an air-fluidized bed in a 76-mm-diameter column using mixtures of fine (69 μ) and coarse (125 or 425 μ) glass spheres, i.e., with size ratios of 1.81 to 6.16. The initial fines concentrations ranged from 50 to 95%.

Based on the experimental data of these papers [355, 406, 489], Leva [988] gives a graph showing the timewise change of the fines concentration C/C_0 in fluidized beds (Fig. 6-2). When plotted on semilogarithmic coordinates, the data (after the first minute) from these three papers are approximated by straight lines, indicating that an exponential equation of the following type can be written:

$$C = C_0 e^{-k\tau}, \tag{6-2}$$

where C_0 and C = the initial value and the value of time τ of the fines concentration in the fluidized bed, %;
k = the carry-over constant, sec^{-1};
τ = the time, sec.

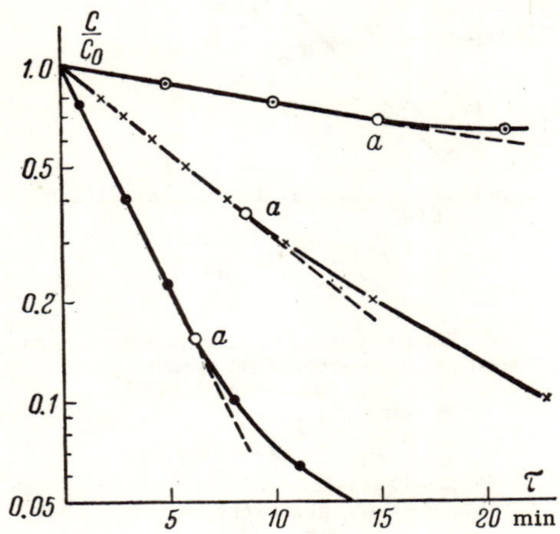

FIG. 6-2. Typical timewise changes of fines concentrations in beds during carry-over [988].
The points at which the carry-over mechanism changes are shown by O and the letter a.

Literature source	Solid particles	d of fine particles, μ	C_0, %	Superficial fluid velocity m/sec
● Osberg and Charlesworth [489]	glass spheres	18	1	0.122
× Hyman [355]	glass spheres	68.5	50	0.558
⊙ Leva [406]	Fischer-Tropsch catalyst	41	7.8	0.405

Using this expression, the constant k can be calculated from experimental data. However, after a certain length of time after the start of the experiment (points a, Fig. 6-2), the entrainment begins to decrease and the experimental lines deviate upward from the time axis. In some experiments, Osberg and Charlesworth [489] noted an anomalous deviation of the experimental lines toward the time axis, which Leva attributed to electrostatic effects during the fluidization of the glass spheres. In principle, such an anomalous effect could also be caused by the varying fluidized bed height in these unsteady-state experiments.

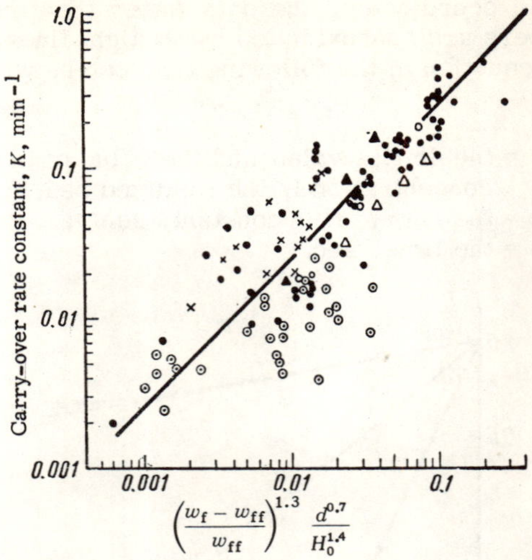

FIG. 6-3. Relationship for the carry-over rate constant (K) in two-component systems [988].
Osberg and Charlesworth [489]:
● —air
○ —CO_2
△ —Helium
▲ —Freon-12
} glass spheres;
× —air, glass spheres;
Leva [406]:
⊙ —air, sharp-angled particles of iron catalyst.

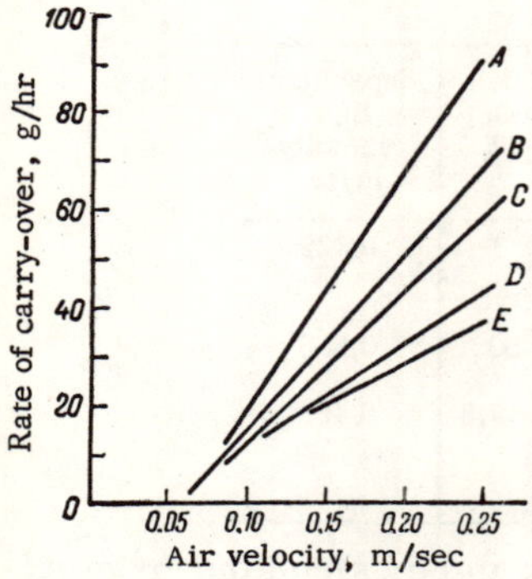

FIG. 6-4. Effect of height of the disengaging zone on carry-over of fines from a bed of coal particles with a narrow size range (75, μ) in a 51-mm-diameter tube, according to data of Jolley and Stanton [372, 988].
Superficial fluid velocities (m/sec): *A*) 0.92; *B*) 0.76; *C*) 0.61; *D*) 0.46; *E*) 0.30.

Osberg and Charlesworth [489] assumed that the entrainment rate constant was proportional to the construct:

$$\frac{(w_f - w_{ff})^{1.7} d^{0.7}}{d_p^2 H_0^{1.4}};$$

where d_p is the diameter of the fines.

Leva [988] took this relationship as a basis and, assuming further that the free-fall velocity above the fluidized bed was given by Stokes's law, he arrived at the correlation shown in Fig. 6-3, which was termed "encouraging." It is scarcely possible to agree with this. The scatter of the points is extremely large. The experimental points may deviate from the correlation by as much as a factor of 20. This can be partly explained as follows: in some of the experiments entrainment had to be determined at small free heights, smaller than the disengaging height. Leva makes no mention of this fact, though it is pointed out in the paper by Jolley and

Stanton [372], who showed that there is an important relationship between the rate of carry-over (g/hr) and the height of the free space above the fluidized bed (Fig. 6-4).

Zenz and Weil [857] dealt in detail with the role of the free height above the bed and derived an empirical correlation for estimating the disengaging height as a function of the bed diameter with the superficial fluid velocity as a parameter (Fig. 6-5). As shown in Fig. 6-5, the relative disengaging height (referred to the diameter of the vessel) increases with an increase in superficial fluid velocity and with a decrease in the vessel diameter. This graph includes data obtained for vessels of diameter 51, 76, 203, 305 and 610 mm and 4.88 m. The authors warn that the correlation is not very accurate, since the data were obtained for complete plants operating under various process conditions and with various inlet gas distributors and size analyses of the particle mixtures. As claimed by Zenz and Weil, laboratory experiments have shown that the disengaging height is not related to the size of the bubbles. They assume that the effect of the unstable velocity distribution caused by small bubbles is intensified by their high frequency to the same degree, and that the equilibrium disengaging height will be the same as for large bubbles moving more violently but less frequently.

Todd [1012] notes that carry-over from a fluidized bed can be reduced by using a few layers of coarse particles of some other lighter material "floating" on the main fluidized bed. The bulk density of the floating particles should be 0.2-0.95 times that of the main bed, γ_{bed}, at the operating superficial fluid velocity, and their diameter should be 10-700 times larger than the diameter of the particles in the fluidized bed. The reduction in carry-over here is obviously related to the equalization of the gas velocity profile at the exit from the bed and to the presence of obstructions tending to prevent the spraying of small particles from the main part of the bed.

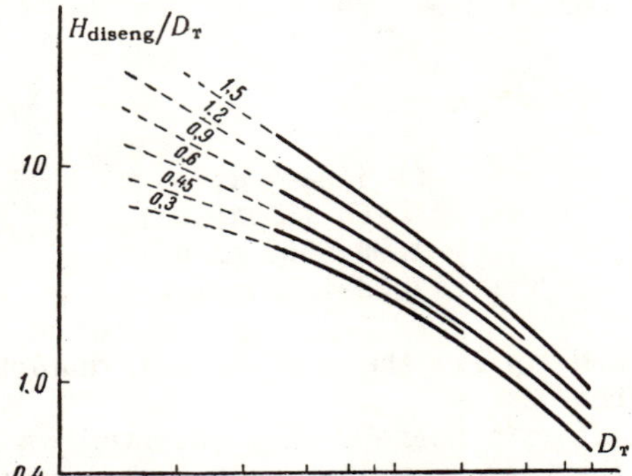

FIG. 6-5. Empirical correlation of Zenz and Weil [857] for calculating the disengaging height (H_{diseng}) above beds of particles with diameters up to 400 μ fluidized by gases at pressures up to 35 atm abs. The correlation becomes questionable at small values of D_T because of wall effects. The numbers on the lines refer to superficial fluid velocities, m/sec.

Wen and Hashinger [1135] investigated the entrainment of fines from beds of glass spheres (diameter d = 71-280 μ, bulk density γ_M = 2500 kg/m^3) and carbon particles (d = 97-7151 μ, γ_M = 1300 kg/m^3) fluidized with air or helium. The beds contained particles of two or more sizes and were fluidized in 50- and 100-mm-diameter tubes. The entrainment rate of each separate fraction from the bed did not depend on the presence of the others if the given fraction was considered together with the coarse grains as a two-component system. The difference between the linear gas velocity and the free-fall velocity of the particles is the main factor affecting the value of the specific carry-over constant K, kg/m^2·sec. The authors state that for large-diameter columns the dimensions of the bed and of the coarse particles have almost no effect on K as long as there is no channel formation or slugging break-through of the gas. This is not in agreement with the ideas presented earlier by Leva and Trawinski, according to whom entrainment depends directly on the diameter of the coarse particles in the bed. For two-component systems the authors found no effect of the initial fines concentration C_0 on the entrainment constant K as long as $C_0 < 25\%$, while for $C_0 > 25\%$ they found a quite significant relationship between K and C_0:

$$K_{>25\%} = K(C_0/25)^{-0.48}. \quad (6\text{-}3)$$

The independence of the entrainment constant on the fines concentration in the bed is again not in agreement with Leva's data, given earlier, which show that carry-over increases as the percentage of fines in the fluidized bed is increased.

In [1135] a "generalized correlation" is proposed for calculating the entrainment constant K for systems of wide particle size distribution:

$$\frac{K}{\gamma_g (w_f - w_{ff})} = 1.52 \cdot 10^{-5} \left[\frac{(w_f - w_{ff})^2}{gD}\right]^{0.5} \times \left(\frac{w_f d}{\nu}\right)^{0.725} \left(\frac{\gamma_M - \gamma_g}{\gamma_g}\right)^{1.15}, \quad (6\text{-}4)$$

which is applicable over the ranges:

$$d = 41\text{-}147\ \mu;$$
$$\gamma_g = 0.165\text{-}1.2\ \text{kg/m}^3;$$
$$\gamma_M = 1300\text{-}5000\ \text{kg/m}^3;$$
$$w_f = 0.22\text{-}1.32\ \text{m/sec}.$$

Here γ_M, d and w_{ff} are the bulk density, diameter and free-fall velocity of the fines.

Wen and Hashinger's correlation is essentially a special one, intended only for the case when the free height above the fluidized bed exceeds the disengaging height.

Fridland and Skoblo proposed a more generalized empirical correlation for entrainment [1086, 1087, 1088, 1183]. Entrainment studies were also made by Sechenov and Al'tshuler [780, 865, 936] and Razumov and Larionova [928, 1173].

Sechenov and Al'tshuler [780, 936] studied the fluidization of materials of wide particle size distribution at pressures of 0-15 atm gauge and showed that, as the pressure was increased, entrainment was reduced due to the decrease in the linear superficial gas velocity. Increased fluidization uniformity under pressure must also have had a favorable effect.

Razumov and Larionova [928, 1173] determined the entrainment from air-fluidized beds both with and without material circulation (continuous supply and removal). The material consisted of microspheres of natural clay catalyst (particle diameter d = 25-400 and 25-160 μ, bulk density of particles γ_M = 1600 kg/m^3) and a synthetic catalyst (d = 50-400 μ, bulk density γ_M = 1200 kg/m^3). The bed was 190 mm in diameter and had an initial height of 280 mm. There was a free height of 1900 mm above the bed. The free area of the grid was only 0.038 m^2. The superficial fluid velocity varied from 0.3 to 0.65 m/sec, and the rate of catalyst circulation from 70 to 450 kg/hr. Razumov and Larionova [1173] proposed an equation for calculating the entrainment from a fluidized bed when the height of the free space above the bed is sufficient for stabilization of the gas velocity:

$$\frac{w}{\gamma_g w_f} = 5.95\left[\mathrm{Fr}\cdot\frac{\gamma_g}{\gamma_M} f\right]^{1.365}, \qquad (6\text{-}5)$$

where w = entrainment, kg/sec·m^2;
$\mathrm{Fr} = w_f^2/gd_1$ = Froude number;
d_1 = maximum diameter of entrained particles, m;
f = fraction of material in bed that can be entrained at the given superficial fluid velocity.

The mean deviation between calculated and experimental data ranged from +30.2 to −38.2%.

6.3 Design Correlations

The work of Fridland and Skoblo [1087] is of great practical interest. They studied entrainment from fluidized beds of natural and synthetic microspherical aluminosilicate catalysts. Table 6-1 presents the characteristics of systems covered by their correlation that were also studied by other authors. The experiments of Fridland and Skoblo were carried out under unsteady-state conditions, i.e., without adding fines to the bed to replace the material carried over; however, the entrainment during an experiment did not exceed 2-3%, so that the experimental results can be taken as referring to the mean value of the fines concentration. Most of the particles were spheroidal in shape. The superficial fluid velocity varied from 0.35 to 0.58 m/sec. The fractional free area of the grids was 0.15-0.061.

An empirical relationship has been proposed [1087] by means of which the nominal concentration of material in the gas leaving the fluidized bed can be calculated when the free height H_f above the bed is larger or smaller than the disengaging height:

$$\frac{Q_M}{Q_g}\cdot 10^2 = A\frac{w_f^4 C^{0.5}}{d^{3.53} H_{ff}^n \varphi} H^k m_v^4, \qquad (6\text{-}6)$$

Table 6-1

Characteristics of Systems in Experiments on Particle Entrainment from Fluidized Beds [1087]

No.	Solid phase	Bulk density of particles, kg/m	Particle diameter, μ	Fluid-izing medium	Coeffi-cient m_v	Literature reference
1	Microspherical natural catalyst.	2350	200–400	Air	1	[1087]
2	Microspherical natural catalyst.	2350	30–630	Air	1	[1087]
3	Microspherical synthetic catalyst.	1880	20–500	Air	1.14–1.58	[1087]
4	Microspherical synthetic catalyst.	1880	30–630	Air	1.46–1.75	[1087]
5	Crushed synthetic catalyst . .	1880	20–630	Air	1.44–1.61	[1087]
6	Quartz sand	2640	40, 51, 67 135, 181, 226	Air	0.77–0.97	[406]
7	Iron catalyst.	4980	40, 51, 105, 135, 181, 191, 224	Air	0.42–0.71	[489]
8	Glass beads	2470	18, 26, 36, 57, 68, 145	Air	1	[489]
9	Glass beads	2470	19, 58	CO	0.81–1.3	[489]
10	Glass beads	2470	19, 58	He	0.89–1.4	[489]

where Q_M = carry-over, kg/hr;

Q_g = gas (vapor) throughput, kg/hr;

C = fines concentration in the fluidized bed (ratio of the weight of the particles whose free-fall velocity is less than the superficial fluid velocity to the total weight of the bed);

H = height of the fluidized bed, mm;

d = diameter of the fines in the bed (μ), defined as the weighted mean value given by the relationship

$$d = \sum_{1}^{n} x_i d_i;$$

x_i = weight fraction of the ith size fraction of fines in the bed (ratio of the weight of this fraction to the total weight of fines);

d_i = arithmetic mean diameter of a fine fraction by sieve analysis, μ;

H = free height above fluidized bed, mm;

φ = free area of gas-distributing grid (fraction of total);

m_v = correction factor, taking into account the physical properties of the gas-solid particle system.

The values of A, n and k are given in Table 6-2 for various values of H_f and H.

Fridland and Skoblo take the system air–natural microspherical aluminosilicate catalyst used in their experiments as the model, and

Table 6-2

Values of A, n and k for Various Values of H_f and H [1087]

Quantity	$H_f \leqslant H_{diseng}$		$H_f > H_{diseng}$	
	$H \leqslant 88$ mm	$H > 88$ mm	$H \leqslant 88$ mm	$H > 88$ mm
A	2.58· 10	10	5.29· 10	3.35· 10
n	1.01	0.01	0.273	0.273
k	-0.259	0.359	-0.259	0.359

on going to a sample (any fluidized gas–solid particle system) they determine a correction factor m_v for each system by means of an auxiliary equation derived from the condition of equality of rates of

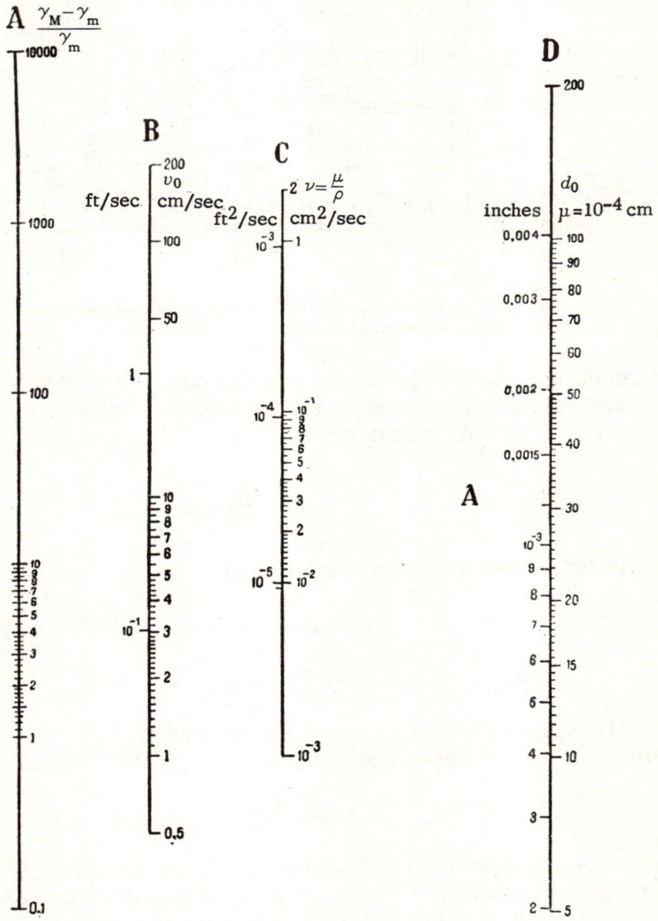

FIG. 6-6. Jottrand's nomogram [828] for determining the auxiliary quantities v_0 (scale B) and d_0 (scale D).

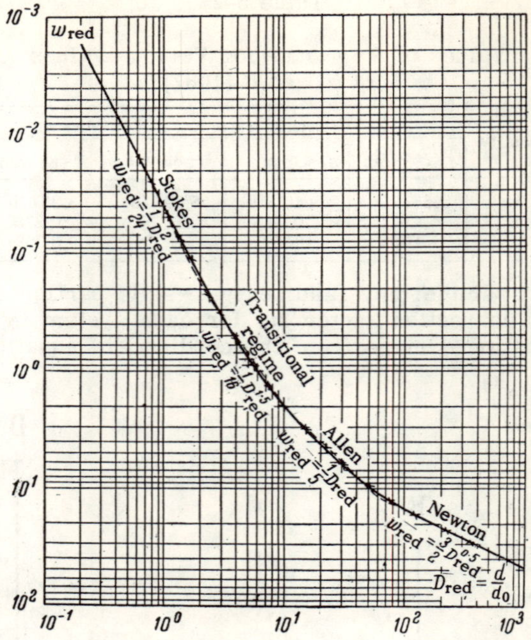

FIG. 6-7. Jottrand's graph [828] for determining the reduced velocity of free fall of particles, w_{red}.

entrainment Q_M/Q_g for the model and sample. As shown in [1087], if it is assumed that the relative velocity of the entrained particles is equal to their free-fall velocity,

$$m_v = \frac{w'_{ff}}{w_f - (w_f - w_{ff})MN}, \tag{6-7}$$

where the prime refers to the model, and

$$M = (1 - m_{sat})/(1 - m'_{sat});$$

$$N = \gamma_M \gamma'_g / (\gamma'_M \gamma_g);$$

m_{sat} is the voidage of an ascending gas stream saturated with solid particles. To a first approximation, $M = 1/N$ [1086]. Thus

$$m_v \approx w'_{ff}/w_{ff}. \tag{6-8}$$

The free-fall velocities in the model system and the sample can be determined conveniently from the graphs given by Jottrand [828], (Figs. 6-6 and 6-7). In using these, the auxiliary quantities v_0 and d_0 are determined first from the nomogram in Fig. 6-6 for the given properties of the material and medium [ν, $(\gamma_M - \gamma_m)/(\gamma_m)$]. The value of the reduced diameter $D_{red} = d/d_0$ is determined by dividing the actual particle

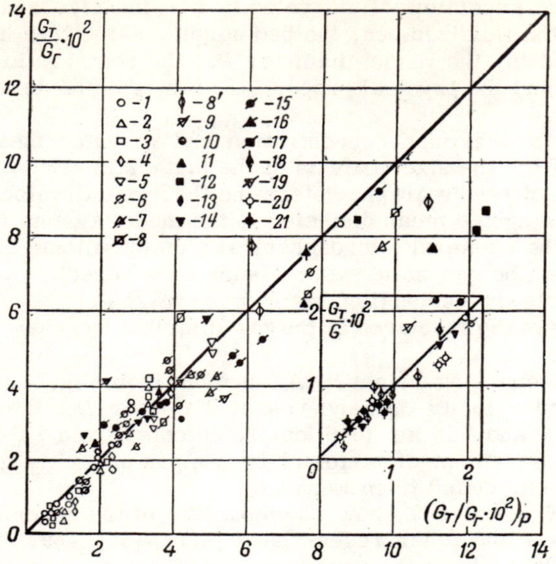

FIG. 6-8. Comparison of experimental data with Eq. (6-6) of Fridland and Skoblo [1087].
System air–natural catalyst [1087]:
H = 88 mm, 1) φ = 6%
2) φ = 3.4%;
H = 132 mm, 3) φ = 6%
4) φ = 3.4%
5) φ = 1.5%;
H = 176 mm, 6) φ = 6%;
H = 280 mm, 7) φ = 6.78%
8) φ = 4.84%
8') φ = 4.7%;
System air–microspheroidal catalyst [1087]:
H = 280 mm, 9) φ = 6.78%
10) φ = 4.7%;
System air–crushed synthetic catalyst [1087]:
H = 280 mm, 11) φ = 6.78%
12) φ = 4.7%;
System air–quartz sand [406]:
13) H = 172 mm
14) H = 129 mm;
System air–iron catalyst [406]:
15) H = 137 mm
16) H = 91 mm;
System air–glass beads [489]:
17) H = 214 mm
18) H = 139 mm
19) H = 100 mm;
20) System carbon dioxide–glass beads [489];
21) System helium–glass beads [489].
The ordinates of the points are the experimental values of carry-over; the abscissas are the calculated values.

diameter by the auxiliary quantity d_0, and the reduced free-fall velocity $w_{red} = w_{ff}/v_0$ is found from Fig. 6-7, whence w_{ff} can be calculated.

The method recommended by the author for calculating carry-over of material from fluidized bed equipment is as follows [1086].

Given: the superficial fluid velocity w_f, the size analysis of the material in the fluidized bed, the bed height and the free height above the bed H and H_f, the vessel diameter D_T, the construction of the gas distributor, and the physical properties of the gas and solid particles, γ_g, γ_M and ν.

1. Calculate the fines concentration and weighted mean diameter of the fines from the size analysis of the material.

2. From Jottrand's graphs determine the free-fall velocity of particles of the weighted mean diameter in the model system (air at 20°C and 1 atm abs., natural microspherical aluminosilicate catalyst) and in the sample (the gas-solid system being considered).

3. Calculate the correction factor $m_\nu \approx w'_{ff}/w_{ff}$.

4. Calculate the free area of the gas-distributing grid φ knowing its construction.

5. The disengaging height H_{diseng} is found from Fig. 6-5 for the superficial fluid velocity w_f and vessel diameter D_T, after which the values of A, n and k in Eq. (6-6) can be selected from Table 6-2.

6. The entrainment of material by the gas under the given conditions is then calculated from Eq. (6-6).

Figures 6-8 and 6-9 show a comparison of the experimental data of the authors and other researchers [406, 454, 489, 780] with the values calculated from the equations. The agreement is satisfactory.

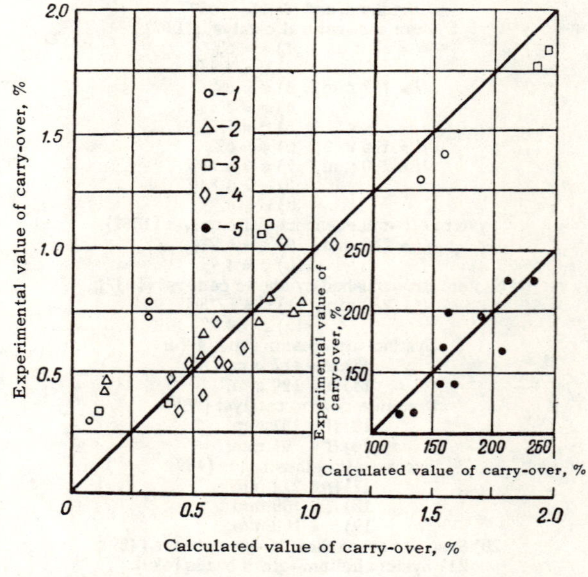

FIG. 6-9. Comparison between calculated and experimental data on particle carry-over from fluidized beds [1086].
1) system carbon dioxide–glass beads [489];
2) system helium–glass beads [489];
3) system Freon-12–glass beads [489];
4) system nitrogen–nickel catalyst [780];
5) system hydrocarbon vapor–coke [454].
The ordinates of the points are the actual values of carry-over; the abscissas correspond to the calculated values.

Table 6-3

Experimental Entrainment Data [1183]

Source	w_f, m/sec	Fines concentration in bed, C	H, mm	Diameter of fine particles d, mm	H_f mm	φ	m_v	Number of exponential points	Relative error, %
Gipronefmash....................	0.3–0.65	0.18–0.87	270	35–102	1,990	0.05	1.12–1.37	43	21.7
Molstadt and Moser [454]	0.92–1.31	0.15–0.65	3200–4700	200–260	4300–5200	0.05	1.39–1.5	10	3.6
Plant at Most (Czechoslovakia)........	2.5	0.16	4000	225	18,800	0.06	1.59	1	16.8
Gurvich [640]	0.4	0.8	4000	60	5,600	0.04	0.91	1	12.3
Othmer [490].....................	0.61–0.73	1	4000–4400	60	6000–6600	0.04	0.95	2	21.3
MINKh and GP [1087, 1088]..........	0.35–0.58	0.09–0.37	88–176	80–90	984–1116	0.04–0.06	1	4	14.7

Some of the initial data used in the comparisons are shown in Table 6-3, which also gives the calculated values of the correction coefficient m_v and the relative error of the correlation.

In Fridland and Skoblo's correlation there is some doubt about the inverse proportionality of carry-over with the free area of the grid φ, and also on the relatively high exponent $n = 0.273$ for the cases when $H_f > H_{diseng}$. Under these conditions it might be expected physically that carry-over would be independent of the gas distributor construction or of further increase of the free height above the bed. However, it is possible that the disengaging height H_{diseng} calculated according to Zenz and Weil is provisional and does not agree with the actual height at which complete stabilization of the velocity profile begins. The correlation by Fridland and Skoblo is the best of those proposed so far.

Chapter 7

Combined Heat and Mass Transfer Processes

The nature of heat and mass transfer in fluidized beds is determined by the phenomena of energy and material transport within and around the particles. There exist well-known analogies and interrelationships between the mechanisms of heat and mass transfer. In heat and mass transfer processes the transfer media for the heat and mass which are being exchanged are often the same. For instance, in the processes of molecular heat conduction and self-diffusion the very same molecules serve to transport both the heat and the mass. Analogously to the case of molecular transfer, organized molar motion or turbulent mixing in the presence of nonuniform heat or mass distribution in space involves transport of the substance contained in the moving liquid, gas, or bed, which may be heat, moisture, or various added materials. However, because of the nonuniform and nonisotropic nature of the mixing in fluidized beds the coefficients of turbulent transfer at various points and in various directions will not be the same. Analogies between heat and mass transfer processes are also somewhat restricted in nature; they cannot, for example, be extended to radiant heat transfer.

The interrelationships between heat and mass transfer and the effects of each on the other are widely known. A. V. Luikov [84]* has shown that, strictly speaking, heat transfer and mass transfer cannot be considered separately and independently of each other, but only as interdependent processes. For example, it is known that isothermal concentrational diffusion in a gas mixture leads to a transfer of heat and to the formation of a temperature gradient (Dufour effect). Conversely, the transfer of heat by conduction in a gas mixture leads to transfer of material (thermodiffusion, or Soret effect [910]).

As an example of a very simple interdependence suppose that a moist porous body is subjected to intensive heating. This will lead to intensive internal evaporation and formation of a stable overall pressure gradient in the body. Under the influence of this gradient there will be vigorous molar transfer of material (vapor), making the boundary layer on the body turbulent and affecting its convective transfer of heat. As a result of the boundary layer becoming turbulent and the

*Although Lykov is the correct transliteration, Academician Luikov prefers the above spelling.

ejection of submicroscopic particles of liquid into it (which then evaporate within the boundary layer) the heat transfer coefficient for a moist body is appreciably higher than that for a dry one. According to the data quoted in Luikov's monograph [84], the heat transfer coefficient for a moist gypsum plate of limited extent oriented along a stream was 42.6 kcal/m^2 · hr · deg, while the value calculated by the usual formula for "dry" heat transfer was only 17.9 kcal/m^2 · hr · deg.

Intensive heat and mass transfer processes are particularly complicated; there are major difficulties in understanding their mechanisms and in describing these mechanisms mathematically. It is sometimes helpful to use simplified model concepts (see, for example, Chapter 8); however, these usually involve a large number of simplifying assumptions, and the calculations are useful only to a first approximation. As a result, phenomenological (thermodynamic) means of describing heat and mass transfer are being developed intensively in addition to the usual molecular-kinetic methods. Based on very general principles, confirmed by experiments, the thermodynamic method makes it possible to arrive at very important conclusions without becoming involved in the mechanism of the phenomenon (which may be still unknown). However, in analyzing problems of dynamics (for instance, particle dynamics), we usually have to deal with velocities and forces. Classical thermodynamics is concerned only with equilibrium states and processes which occur infinitely slowly, i.e., it would be more correct to term it thermostatics. There are only limited possibilities in extending it to the study of real processes. It would be necessary to introduce time as an independent variable in thermodynamics—this is the fundamental and necessary parameter for describing the intensities of processes.

Quite recently the foundations have been laid for the so-called thermodynamics of irreversible processes as a result of the work of Onsager, Prigogine, de Groot, and others [910]; on the basis of this, it is possible to extend the thermodynamic method of investigation to real processes in which there are deviations from static equilibrium due to the presence in the system of steady and unsteady-state fluxes of heat, electricity, etc., having finite velocities. In addition to the natural laws, such as the laws of conservation and conversion of energy, two other principles can be formulated which yield the basis of the new thermodynamics: the principle of linearity and the principle of reciprocity.

The principle of linearity states that for a mass or energy flux, the rate of each process is proportional to the corresponding fundamental "force," and in addition, that the rate of a given process depends linearly on all the other driving "forces" present in the system. This is a more general statement than the assumption that each flux depends on only one of the driving forces. These "thermodynamic forces" are themselves concepts more general than forces in mechanics, and are taken to imply the causes of the motion or flux in the most general sense of the word, and not just in the narrow special sense of mechanical movement. The thermodynamic forces are proportional to the gradients of temperature, chemical potential, etc., and are therefore quantities characterizing the deviation of the system from static equilibrium.

The fluxes (of energy or mass) are related to the tendency of the system to go towards the equilibrium state and occur for quantities of different nature in various coordinate systems. In general, therefore, it is necessary to speak of generalized coordinates of "motion" rather than spatial coordinates.

Thus, according to the principle of linearity, the flux of any one form of energy or material is

$$I_1 = L_{11}X_1 + L_{12}X_2 + L_{13}X_3 + \ldots,$$

where X_1, X_2, etc., are the thermodynamic forces existing in the system; L_{11} is a coefficient (with the same subscripts), expressing the linear dependence of the given flux on the corresponding main thermodynamic force (for example, a quantity proportional to the temperature gradient for a heat flux); L_{12}, L_{13}, etc., are the cross coefficients (with different subscripts) expressing the linear dependence of the given flux (1) individually on each of the other types of thermodynamic force (2, 3, etc.) which exist in the system (for example, the dependence of the heat flux on quantities proportional to the electrical and chemical potentials).

The first subscript on a coefficient refers to the form of the flux, and the second to the type of thermodynamic force.

When there are i thermodynamic forces present in the system, i Onsager equations must be written with i terms on the right-hand side.

In Eckert's terminology, the thermodynamic equation of motion would be:

$$I_1 = L_{11}X_1 + L_{12}X_2 + \ldots + L_{1i}X_i;$$
$$\ldots\ldots\ldots\ldots\ldots\ldots$$
$$\ldots\ldots\ldots\ldots\ldots\ldots$$
$$I_i = L_{i1}X_1 + L_{i2}X_2 + \ldots + L_{ii}X_i.$$

The second principle in the thermodynamics of irreversible processes—the principle of reciprocity—states that the various processes occurring in a system mutually affect one another and are characterized by symmetry in the sense that the coupled cross coefficients (differing only in the order of the subscripts) in the Onsager equations are equal, that is, $L_{12}=L_{21}$; $L_{13}=L_{31}$, and, in general, $L_{ik}=L_{ki}$. Onsager showed that this reciprocity arises from the principle of so-called microscopic reversibility, which states that under equilibrium conditions any discrete process (and not only completely molecular processes) and its reverse process will both occur at the same rate in the medium. For instance, if a molecular process is complex and consists of two simple processes, i.e., migration of molecules and transfer of energy between these during collisions, it can be stated that when overall equilibrium of the system exists each of these processes will be individually in a state of dynamic equilibrium.

For any unrelated processes 1 and 2, the equality $L_{12}=L_{21}$ still exists, but degenerates to the trival result $L_{12}=L_{21}=0$. Thus, the principles of reciprocity and linearity, which indicate that processes usually depend on each other, and that this is always a mutual dependence, do not exclude those special cases where the processes are entirely independent, besides being reciprocal.

The principle of linearity is a generalization of many earlier experimentally established approximate relationships such as Fourier's law of heat conduction, Fick's laws of diffusion, etc.

Eckert terms the cross coefficients L_{12}, L_{ki}, etc., the entrainment coefficients (one flux acts on the other, as though entraining it).

According to Onsager, the selection of the thermodynamic forces is conveniently limited by the condition that multiplying the fluxes by the corresponding (fundamental) forces yields products whose sum is equal to the rate of formation of entropy (per unit volume of the system) multiplied by the temperature:

$$I_1X_1 + I_2X_2 + \ldots = \sum_i I_iX_i = T\frac{dS_v}{d\tau},$$

where $dS_v/d\tau$ is the rate of formation of entropy per unit volume of the system.

Here the term "formation of entropy" implies the increase of entropy in an irreversible process. It must be remembered that for non-isolated systems changes in entropy can occur due to the inflow of heat or mass from outside the system as well as by formation of entropy inside the system. Changes of entropy due to interactions of the system with the surrounding medium can be positive, zero, or even negative, depending on the nature of the interactions.

Taking into account Onsager's principle, A. V. Luikov and his co-workers have improved the laws of molecular energy and mass transfer in disperse media. The most recent publication of Luikov, Mikhaylov, Maksimov et al., makes it possible to develop a theory of molar-molecular transfer [84].

As Mikhaylov notes, differential equations for molecular and molar-molecular transfer in the presence of phase changes have been formulated in the light of the thermodynamics of irreversible processes and recent theoretical and experimental data. Unlike the earlier theories of heat conduction and diffusion, systems of equations are proposed on the basis of mathematical models of the processes rather than separate partial differential equations. Thus, molecular heat and mass transfer in a dispersed medium are described by a system of equations:

$$c_q\gamma\frac{\partial t}{\partial\tau} = \nabla(\lambda_q\nabla t) + E\rho c_m\gamma\frac{\partial\theta}{\partial\tau};$$

$$c_m\gamma\frac{\partial\Theta}{\partial\tau} = \nabla[\lambda_m(\nabla\Theta + \delta\nabla t)],$$

while for molar-molecular transfer, the system is

$$c_q\gamma\frac{\partial t}{\partial\tau} = \nabla(\lambda_q\nabla t) + E\rho c_m\gamma\frac{\partial\Theta}{\partial\tau} - \sum_i c_i q_{mi}\nabla t;$$

$$c_m\gamma\frac{\partial\Theta}{\partial\tau} = \nabla[\lambda_m(\nabla\Theta + \delta\nabla t) + \lambda_p\nabla p];$$

$$c_p\gamma\frac{\partial p}{\partial\tau} = \nabla(\lambda_p\nabla p) - Ec_m\gamma\frac{\partial\Theta}{\partial\tau}$$

instead of the earlier separate description of the analogous processes at small intensities by two independent equations

$$c_q\gamma\frac{\partial t}{\partial\tau}=\nabla(\lambda_q\nabla t)\text{ and }\frac{\partial c}{\partial\tau}=\nabla(c\nabla c).$$

In these equations, t, Θ, and p denote, respectively, the potentials for heat, mass and filtrational transfer; λ_q, λ_m, λ_p are the heat, mass, and molar (vapor) conductivities; c_q, c_m, c_p are the heat, mass, and vapor capacities; τ is time; δ is the thermogradient coefficient; γ is the medium density; E is the phase conversion criterion.

In the general case equations can be obtained for heat and mass transfer coefficients by using differential transfer equations derived from the fundamental laws of heat and mass transfer (the linear flux equations), with application of the laws of conservation of energy and quantity of material to some arbitrarily chosen volume of the body which is bounded by a closed surface.

Either of two methods can be used. The first method consists of establishing an analytical expression for the transfer potential distribution curves by means of an approximate solution of the differential transfer equations, for instance, by using integral transformations. The second method consists of using the principle of similarity. The differential transfer equations and conditions for uniqueness of the solution are used for establishing a system of similarity criteria. In some cases parametric criteria are also introduced when an important effect of these on the process can be expected from additional considerations concerning the mechanism or statement of the process. In investigations of heat transfer between particles and a gas stream in a fluidized bed this type of parametric criterion can include the fluidization index $N = w_f/w_{1.s}$ and the ratio of the actual pressure drop in the bed to the theoretical value, $\Delta P/\Delta P_{\text{theor}}$. The fluidization index characterizes the degree of development of fluidization, while $\Delta P/\Delta P_{\text{theor}}$ partly expresses the nonhomogeneity of the fluidized bed. A. V. Luikov showed that the application of integral transformations also shows great promise in using the theory of similarity, since the fundamental nature of the relationships which exist between the dimensionless groups in the description are also retained in the inverse transforms. Many nonlinear equations which cannot be solved in principle today, such as the equations of hydrodynamics, can be solved to the extent that the laws between the parameters of interest in the expressions can be obtained. These relationships are also retained in the inverse transforms. Without additional experiments, this makes it possible to solve which of the similarity criteria can be discarded without deleterious effects in a given real problem.

It is necessary to bear in mind the future possibility and desirability of applying these methods in order to obtain more complete solutions to heat transfer problems in dryers and similar fluidized bed equipment, but at present it seems more suitable to use a different, approximate approach. Because of the incompleteness of the existing experimental data even on "pure" heat transfer in fluidized beds, it must be assumed that there is insufficient information on the hydrodynamics of fluidized beds and on the physical properties of moist materials. Numerous research projects will be needed to accumulate this information.

The effect of the moisture content of granular materials on the forces of interaction between particles plus the poor accuracy in determining heat transfer coefficients for particles even in beds of perfectly dry material (see Chapter 8) are characteristic of fluidized beds, and are great disadvantages in the study of interactions between heat and mass transfer in such beds and the reduction of results to design formulas.

At the same time there is immediate practical interest in setting up even approximate methods for designing equipment utilizing fluidized beds for heat and mass transfer. These methods can be developed starting from a consideration of the special features which characterize transfer phenomena in fluidized beds and which permit a number of simplifying assumptions.

These features are determined primarily by the rapid particle mixing in fully developed fluidization, and their large bulk concentration, the comparatively small particle diameters, and the small relative velocity of the stream flowing around them. These will be considered in more detail in the following chapters dealing with various heat transfer problems in fluidized beds.

Chapter 8

Heat Transfer Between Particles and the Surrounding Medium

8.1 General Remarks

The main problems in fluidized bed heat transfer are: 1) heat transfer between the particles and medium; 2) the effective thermal conductivity of the bed; and 3) heat transfer between a fluidized bed and either a heating surface submerged in it or its surrounding walls.

The pertinent and related fields of heat transfer in suspended and fixed beds will also be considered briefly in addition to heat transfer between particles and medium in fluidized beds.

A large quantity of heat can be transferred per unit time in a heat exchange device if all of the three factors on the right-hand side of the Newton equation are very large; these factors are the heat transfer coefficient, the transfer area, and the mean-temperature difference. While the determination of the heat transfer coefficient represents the only difficulty in the design of simple surface heat exchangers, there are particular problems, in most bed-type exchangers, in determining the mean-temperature difference due to the nonuniform distribution of the medium streaming between the particles. In most bed-type heat exchangers the mean-temperature difference is extremely small, and it becomes necessary to find ways of overcoming this disadvantage, rather than of increasing the heat transfer coefficient between the particles and medium. When the particles are irregular in shape and have flat or concave sides it is also difficult to estimate accurately the heat transfer area.

1. Heat transfer coefficient between particles and medium

In the general case the heat transfer coefficient between the particles and medium must be regarded as the sum of three components: these are the coefficients of conductive, convective, and radiant heat transfer, so that, respectively,

$$\alpha = \alpha_{\text{cond}} + \alpha_{\text{conv}} + \alpha_{\text{rad}}. \tag{8-1}$$

Thus molecular and molar transfer are considered separately, i.e., α_{conv} completely excludes the transfer caused by the molecular thermal conductivity of the medium.

This separation of conductive-convective heat transfer coefficients into purely molecular and purely molar components is not exact, since conduction and convection are not completely parallel transfer processes. However, Eq. (8-1) is not to be taken as an exact design equation, and the separation of α_{cond} and α_{cond} is purely procedural, to show more clearly the effects of various factors on heat transfer to particles. These factors (for instance, particle diameter) affect the conductive and convective components of the heat transfer coefficient differently.

For fine particles the conductive component of α, caused by molecular heat conduction, is very large. The heat transfer coefficient to fine particles increases rapidly as their diameter is decreased. This follows directly from the so-called limiting Nusselt relationship, which gives the limiting minimum value of the heat transfer coefficient for spherical particles, assuming that in this case heat is transferred to the surface of the particles by molecular heat conduction alone through a gas sphere surrounding the particle. The outside diameter of this gas sphere is assumed to be infinitely large, i.e., the thickness of the layer to which the calculated temperature difference is applied is infinitely large. This minimum value of the heat transfer coefficient, $\alpha = 2\lambda_m/d$, can be obtained from the equation for steady-state heat conduction in a spherical wall. Actually,

$$Q = \frac{2\pi\lambda_m \Delta t}{\frac{1}{d_1} - \frac{1}{d_2}} = \alpha \Delta t \pi d_1^2,$$

whence,

$$\alpha = \frac{2\lambda_m}{d_1(1 - d_1/d_2)}, \tag{8-2}$$

where d_1 and d_2 are, respectively, the particle diameter and the diameter of the layer of medium around the particle. When $d_2 \to \infty$, (8-2) yields the minimum value of the heat transfer coefficient:

$$\alpha_{\min} = \frac{2\lambda_m}{d_1}. \tag{8-3}$$

The conductive component of the heat transfer coefficient is therefore inversely proportional to the particle diameter, i.e., values of the heat transfer coefficient as large as desired can be obtained simply by decreasing the particle size. Thus, intensification of heat transfer as the particle diameter is decreased is achieved in principle not only by the formation of a larger heat transfer surface area per unit weight of the material being heated or cooled, but also by an increase in the heat transfer coefficient itself. As shown elsewhere [45], the excellent conductive transfer of heat to spherical particles (or particles close to this shape) even through gas films of infinite thickness are related to the special features of this geometric shape—the increase in the cross section of the layer conducting heat is proportional to the square of the distance from the center. For infinite cylinders (for example, for wires or fibers) there is also an increase in the cross section of the conducting layer, but this is proportional to only the first power of the distance from the axis.

A further increase in α_{cond} is obtained by reducing the thickness of the film of medium to which the calculated temperature difference is applied. This is the case in a fixed bed or a gently fluidized bed; because of the high bulk concentration of the particles, the thickness of the layer of medium (the heat transfer agent) between the particles is very small. If a cubic particle array is assumed for the dense phase of a fluidized bed, the mean thickness of the "shell" of medium surrounding the particles is

$$\delta = \frac{d^3 - \pi d^3/6}{\pi (d+\delta)^2} \approx \frac{d^3 - \pi d^3/6}{\pi d^2} = 0.152d. \tag{8-4}$$

In this case, according to (8-2):

$$\alpha_{cond} = \frac{2\lambda_m}{d\,(1 - d/0.152d)} \approx \frac{10\lambda_m}{d}$$

or due to the numerous approximations n the model $Nu_{cond} \approx 10$.

During uniform expansion of the bed the mean thickness of the surrounding film of medium increases, i.e., α_{cond} falls.

Table 8-1 shows the values of α_{cond} calculated by Eq. (8-2) for various thicknesses of the medium envelope.

The known asymmetry of the gas "envelope" around the particles in the bed has not been taken into account, in order to avoid the problem of finding the exact expression for α_{min} (Nu_{min}) for a cubic particle packing. It is intended to demonstrate, on the basis of approximate calculations, that there is no sound basis for assuming, as Rowe [1252] has done, that a true film heat transfer coefficient corresponding to $Nu > 2$ is possible. As shown below, approximate calculations have also aided in demonstrating that the reason for the very small effective values of Nu in particle beds is the extreme sensitivity of the effective value of α on nonuniformities in gas distribution. There is no reason to expect that the actual (rather than effective) Nu would be smaller for any asymmetric gas "envelope" around a spherical particle in a bed, that the thickness of the symmetrical "envelope" (in its thickest positions imagined filled with medium) is obtained, and that the corresponding value of Nu is always greater than two.

Table 8-1

Conductive Component of Particle Heat Transfer Coefficient as a Function of δ

Relative thickness of medium, δ/d_1	d_1/d_2	$\frac{\alpha_{cond}}{\alpha_{min}}$
0.025	1.05	20.8
0.05	1.1	11.1
0.10	1.2	6
0.25	1.5	3
0.50	2	2
1	3	1.5
2	5	1.25
4.5	10	1.11
50	101	1.01

The convective component α_{conv} is taken to include all the part of α caused exclusively by molar transport. The characteristic velocity of the medium for the operation of fixed and fluidized bed heat exchangers is small, varying from a few meters/sec to a fraction of a meter/sec. It would therefore seem that α_{conv} for particles should be quite small, but this is not the case. This becomes obvious if the problem of steady-state heat transfer to a spherical wall is based on effective thermal conductivity rather than molecular conductivity. The local effective thermal conductivity

$$\lambda_{eff} = \lambda_m + \lambda_{conv} \qquad (8\text{-}5)$$

takes into account convective transfer (α_{conv}). If the convective flux passes through the same particle surface area and if α_{cond} is constant at all points:

$$\alpha_{cond} + \alpha_{conv} = \frac{2\lambda_{eff}}{d\,(1 - d_1/d_2)}.$$

In the presence of a laminar layer or sublayer the convective component of the particle heat transfer coefficient will be smaller than given by the preceding equation, since along the surface of the particle α_{conv} will approach zero, and the heat transfer coefficient will depend strongly on λ_{eff} for the medium in the immediate neighborhood of the particle surface. Even if molar transfer beyond the limits of this laminar "envelope" is so intense that λ_{eff} tends to infinity, the heat transfer coefficient for the particles will not always differ significantly from α_{min}. The assumption that $\lambda_{eff} \to \infty$ outside the laminar shell is equivalent to assuming that all of the temperature drop, Δt, occurs across this shell, i.e., this becomes the problem of steady-state thermal conduction in a spherical wall of thickness δ, having a thermal conductivity λ_m. From Table 8-1, which was based on this problem, it is found that when the thickness of the laminar sublayer $\delta = 2d$, the heat transfer coefficient is $\alpha = 1.25\,\alpha_{min}$, i.e., it does not exceed greatly the minimum value. This result can be formulated in a different way; turbulence in the medium beyond a layer of thickness equal to two particle diameters has little effect on heat transfer to the particles. The relative thickness of the laminar layer (sublayer) around the particles increases as particle size and hence free-fall velocities decrease and hence, α_{conv} becomes insignificant. However, in the dense phase of a fluidized bed, as well as in fixed beds, where the thicknesses of the layers of medium between the particles are small, flow turbulence can lead to a significant increase in α above the conductive value.

The radiant component of heat transfer between particles and gas is usually negligibly small (though this is not the case for heat transfer with the walls or with other particles) because of the considerable thermal transparency of gases and the small thicknesses of the gas layers around the particles. When a practically nonthermotransparent liquid is used,

$$\alpha_r \approx \frac{4.9\varepsilon_r\left[\left(\frac{T_s}{100}\right)^4 - \left(\frac{\Theta}{100}\right)^4\right]}{T - \Theta}, \qquad (8\text{-}6)$$

where ε_r is the reduced degree of blackness.

When the particles are wetted by a nontransparent liquid, the surface temperature of the latter takes part in radiant transfer $T_s \to \Theta$, here also $\alpha_r \to 0$, if the heat flux is referred as usual to the temperature difference $T - \Theta$, where T is the temperature in the middle of the stream.

There are as yet no reliable published data on the effects of mass transfer on heat transfer between particles and medium. It can be expected that the turbulence induced by passage of ordered streams of material through the boundary layers on fine particles will be smaller than in the case of large particles. The particle dimensions are quite small, and conditions do not arise for the formation of appreciable internal pressure gradients which could cause the spraying of even the very smallest liquid droplets from the particles. The ordered flux of material from the particle surface (for instance, water vapor during drying) in fluidized and fixed beds is usually quite small, since only materials of small initial moisture content are usually dried in such beds. Material with a high initial moisture content is frequently encountered in suspended beds. In experiments on the drying of various granular materials of high moisture content in a suspended state, Fedorov [170], obtained heat transfer coefficients which were very similar to those obtained earlier by Lyakhovskiy for dry materials. With some justification, Mikhaylov [170] assumes that during heat transfer to a material in a suspended state the boundary layers on the particles are always quite turbulent, even if there is no evaporation, and under these conditions the addition of evaporation does not introduce any appreciable hydrodynamic changes along the particle surface.

In summary, the heat transfer coefficients for dry particles not liberating or absorbing material should be used until new and more accurate data can be obtained for the case of heat transfer with simultaneous mass transfer.

2. Heat transfer to surfaces in beds

The active heat transfer area wetted by the stream of the medium in a fixed bed heat exchanger consists of some fraction of the total geometric surface area of the particles present in the heat exchanger. This fraction is theoretically equal to unity only if the particle surfaces are convex. The surfaces of particles of other shapes are completely wetted in suspended and uniformly fluidized beds. However, the active fraction of the surface area is usually quite large for dense beds of such particles as well, as indicated by the high values of the bed voidage. The value of the total particle surface area per unit volume of the bed is enormous for fine granular materials. Usually, heat transfer is not restricted by the value of the active particle surface area in fixed and fluidized beds.

The expression for the heat transfer surface area for a bed of uniform spherical particles is simply

$$F = SV_{\text{bed}} = \frac{6(1-m)V_{\text{bed}}}{d}. \tag{8-7}$$

For a suspended bed

$$1 - m = \mu_{\text{true}}\gamma_g/\gamma_M = \frac{\mu\gamma_g}{(1 - w_{\text{rel}}/w_f)\gamma_M}, \tag{8-8}$$

where μ = nominal material concentration (ratio of the weight flow rates of material and gas);

$\mu_{true} = \mu/(1 - w_{rel}/w_f)$ is the true weight concentration of material in the suspended bed;

w_{rel} = relative velocity of particles in gas stream.

Equation (8-8) was derived in [47].

3. Mean-temperature difference in bed heat exchangers

In contrast to the first two factors (the heat transfer coefficient and surface area for heat transfer) the temperature difference depends a great deal on the heat exchanger arrangement (whether there is co-current, countercurrent or cross-current flow), on the ratio of the water equivalents of the heat transfer medium and the material, and on a function of groups in which the first two factors and the water equivalent appear. The nature of this functional relationship is such that if there is a small temperature difference in the heat exchanger it is sometimes practically impossible to increase the quantity of heat being transferred by increasing the coefficient and surface area for heat transfer, since increasing these quantities leads to a further decrease in the mean-temperature difference. It is convenient to follow this change in the temperature difference in fixed-bed heat exchangers by using the formulas and graphs for ordinary surface heat exchangers [641]. Some of the quantities appearing in these formulas, such as the surface area for heat transfer and the water equivalent of the heat transfer medium, assume unusual forms. The expression for calculating the heat transfer surface area for a bed of uniform spherical particles is given above. The water equivalent of the medium is

$$W_m = w_f F_{sec} \gamma_m c_m = Q_m c_m, \quad (8\text{-}9)$$

where F_{sec} is the cross-sectional area of the bed perpendicular to the stream of medium, m².

The water equivalent of the material is

$$W_M = w_M F'_{sec} \gamma_M c_M = Q_M c_M, \quad (8\text{-}10)$$

where F'_{sec} is the cross-sectional area of the bed perpendicular to the stream of material; $w_M = w_{M,true}(1 - m)$ is the superficial velocity of the material (for motion as a single mass) at the given throughput (without interstices between the particles, i.e., at $m = 0$) spread over the total cross section of the apparatus; $w_{M,true}$ is the true velocity of ordered motion of the material particles.

Using the relationships given earlier for the surface area of the particles in the bed and the water equivalents, it is possible to analyze the operation of fixed-bed heat exchangers using the usual formulas and graphs. For example, if a countercurrent heat exchanger of the "fixed-bed" type is designed, the change in the gas temperature in the case of heating the particles will be

$$\delta t = (t_0 - \vartheta_0)\left[\frac{1 - e^{-(1 - W_g/W_M)\alpha F/W_g}}{1 - \frac{W_g}{W_M} e^{-(1 - W_g/W_M)\alpha F/W_g}}\right], \quad (8\text{-}11)$$

where t_0 and ϑ_0 are the gas and material temperatures ahead of the heat exchanger.

Graphs for Z (the quantity in the square brackets above) have been given, for instance, by Mikheev [98]:

$$\delta t = (t_0 - \vartheta_0)\, Z. \qquad (8\text{-}11a)$$

In Mikheev's graphs (Fig. 8-1), Z is given as a function of KF/W_1 and W_1/W_2. In the present case

$$\frac{KF}{W_1} = \frac{6\alpha\,(1-m)\,V_{bed}}{dQ_g c_g} \qquad (8\text{-}12)$$

and

$$\frac{W_1}{W_2} = \frac{Q_g c_g}{Q_M c_M}. \qquad (8\text{-}13)$$

The dimension of α and W must refer to the same time units.

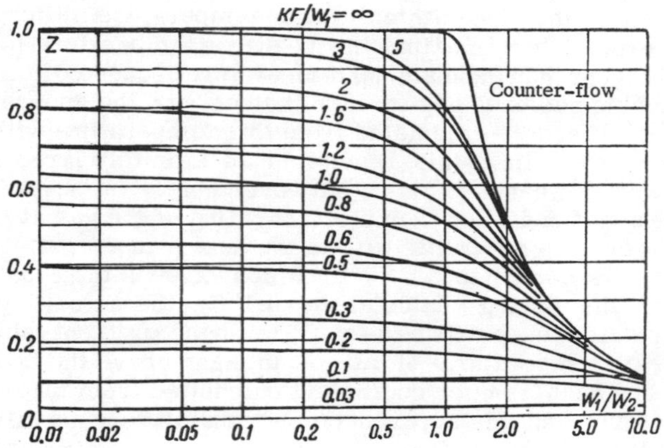

FIG. 8-1. Mikheev's graphs for calculating the auxiliary function Z [98].

The function Z (and the analogous function Π for co-current flow) indicates what fraction of the available temperature driving force (or more exactly, the difference between the initial temperature of the medium and material) is equal to the change in the medium temperature. Thus, if the numerical value of the function Z is increased to several times its original value, the quantity of heat transferred is increased by the same amount (at a given initial temperature of the stream and material and a given flow rate of the medium). In this sense and with these limitations, Z, or functions similar to it, such as Π, can be used as coefficients indicating the utilization of the available temperature driving force: equal values of Z indicate that the same quantities of heat are transferred in comparable heat exchangers.

It is easy to see from Mikheev's graphs [98] that if $W_1/W_2 \ll 1$ and KF/W_1 varies over wide limits, say, from 3 to 100 or more, the values of the functions Z and Π remain almost unchanged. Hence, if $KF/W_1 = \alpha F/W_g > 3$, increase of the active surface area of the bed or the heat transfer coefficient by a factor of ten will not lead to an appreciable increase in the quantity of heat transferred. Conversely, when estimating the value of the heat transfer coefficient between a gas and particles in a deep bed (with large $\alpha F/W_g$) from the overall quantity of heat being transferred, it is easy to be in error by a factor of ten or more due to inaccuracies in measuring the very small temperature differences or using formulas which include the difference between separately measured but very similar temperatures.

Even when the temperature of the material, the initial temperature of the gas, and its temperature after leaving the bed, are known accurately, appreciable errors can arise in calculating the mean-temperature driving force due to nonuniformity in the gas flow (with the possible exception of suspended beds). In the limit, when the heat exchanger is operating in the steady state and $\alpha F/W_g$ is infinitely large, the temperature of the gas contacting the particles, during intensive mixing within the fluidized bed, tends towards the temperature of the material, i.e., the logarithmic mean-temperature difference approaches zero. At the same time, the mean-gas temperature leaving the bed will be higher than the material temperature because of mixing with uncooled bubble gas breaking through the bed, and the apparent mean-temperature difference calculated from this temperature will be finite rather than zero, i.e., this results in an infinitely large error. In practical cases, when $\alpha F/W_g \neq \infty$, but nevertheless is large, the error can still be significant. For example, if only 10% of the gas at $t_1 = 100°C$ in a bed dryer does not participate in the heat transfer, while the remaining 90% is cooled to $t'_2 = 70°C$, which exceeds the constant temperature of the drying material by 0.1°C, the actual logarithmic temperature difference will be 5-25°C, while the apparent value, based on the mean temperature of all of the gas above the bed, will be 11.87°C. The heat transfer coefficient calculated from the quantity of heat transferred and this apparent temperature difference will be much lower than the actual effective value of α. It can be and is used, but its nominal nature must be borne in mind; the values of Nu based on it can turn out to be many times smaller than the minimum values for convection.

4. Temperature measurements in fluidized beds

Temperatures in fluidized beds are usually measured with hot junction thermocouples in direct or indirect contact with the particles. Protected thermocouples measure the temperature of the medium, but only inaccurately if the protected hot junction is large; the thermocouple also appreciably distorts the flow pattern. However, when there is a great deal of radiation from the particles (high-temperature fluidized beds), enclosing the hot junction by shielding the thermocouple does not avoid error in measuring the gas temperature.

There are quite diverse opinions regarding the temperature measured by unshielded thermocouples. One group of researchers

(Kettering et al., Richardson and Ayers) assume that they measure the temperature of the medium in a fluidized bed. Others (Ciborowski and Roszak, Shakhova, and others) assume that unshielded thermocouples measure some mean temperature lying between the temperatures of the medium and material, which practically coincides with the temperature of the medium. The proponents of this latter point of view indicate that if the gas flow through a bed is cut off abruptly, the temperature indicated by an unshielded thermocouple remains essentially unchanged. This seems plausible, since when the gas flow is cut off suddenly the thermocouple will assume the temperature of the material, due to the fact that the overall heat capacity of the solid material in the bed is thousands of times larger than that of the gas trapped in the interstices between the particles, and is even many times larger than the heat capacity of the thermocouple itself. It is necessary to stop the gas flow completely and suddenly, rather than slowly, in order to prevent the temperature of the settled particles at the point of measurement from being altered by the residual gas flow through the now fixed bed. The time of measurement should not be too long in order to avoid the effects of heat losses through the settled particles.

The second viewpoint is undoubtedly valid in the case when radiant transfer between the particles and thermocouple greatly predominates over convective and conductive transfer. However, if the fluidized bed has a low temperature, the transfer of heat to the thermocouple leads can be neglected, and the quantity of heat transferred from the particles to the thermocouple is purely by contact (passing the gas phase), then obviously an unshielded thermocouple will measure the temperature of the medium. Under the conditions prevailing in laboratory experiments, all the heat passing to the thermocouple is transferred to it by convection and conduction through a layer of the medium. In the quasi-steady state, when the operating regime of the fluidized bed is fully established and a thermocouple submerged in the bed indicates a constant temperature, the particles around it are constantly changing due to mixing in the bed and there is continuous heat transfer between the gas and the moving particles by unsteady-state heat conduction around the thermocouple position. In order to eliminate the effects of fluctuating nonuniformities in the fluidized bed, the thermocouple measuring system should possess sufficient inertia. In such quasi-steady-state regimes, there will be a constant mean value of the flux through the thermocouple junction, which means that there will be a constant temperature drop between the surface of the hot junction and the medium surrounding it. The size of the heat flux will be determined by the contact of a comparatively large hot junction with a zone of different gas temperature, and the junction will assume a temperature approximately equal to the arithmetic mean of the

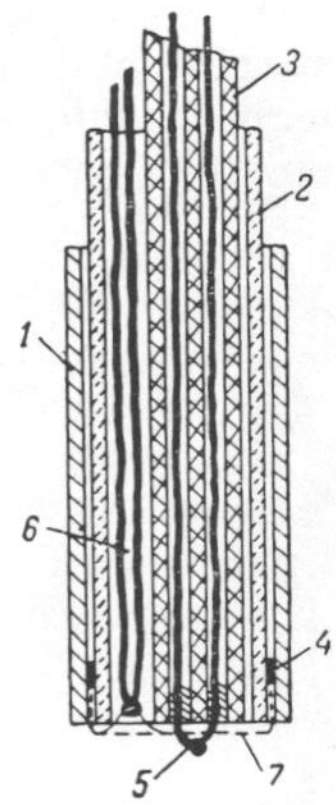

FIG. 8-2. Thermocouple probe used by Ciborowski and Roszak [789]. 1—steel tube (D = 6.5/4.5 mm) with polished and chromed surface; 2—glass tube, (D = 4 mm); 3—ceramic tube ("straw") with two channels; 4—layer of lacquer; 5—hot junction of bare thermocouple; 6—hot junction of thermocouple shielded from contact with particles; 7—silk screen.

temperatures of these zones. Thus, the particle temperature will affect the hot junction temperature only indirectly to the extent that the particles determine the longitudinal temperature gradient of the gas in the bed. It is not possible to detect the particle temperature directly by a hot junction, since the latter is separated from the particles by a film of gas.

The invariably lower temperature indicated by unshielded thermocouples (compared with shielded thermocouples) in the experiments of Ciborowski and Roszak is not convincing proof that bare thermocouples measure the particle temperature. In their probe (Fig. 8-2) the hot junction of the bare thermocouple 5 was positioned a little lower than that of the shielded thermocouple 6. The fluidized bed had a low temperature, of the order of 80°C, so that the radiant transfer component could be neglected. The fluidizing agent (air) was heated by the fluidized bed, which received heat from a steam jacket, i.e., the air was at a lower temperature than the particles in the bed. For fluidized beds of fine particles (but not coarse particles, for which Ciborowski and Roszak carried out calibration experiments), there was a characteristic sharp change in the gas temperature in the zone of active heat transfer (in the lower part of the bed around the screen). Hence, even if both thermocouples were measuring the gas temperature, under the experimental conditions of Ciborowski and Roszak, the lower (bare) thermocouple should have continuously indicated a temperature lower than that given by the shielded thermocouple. In addition, there could

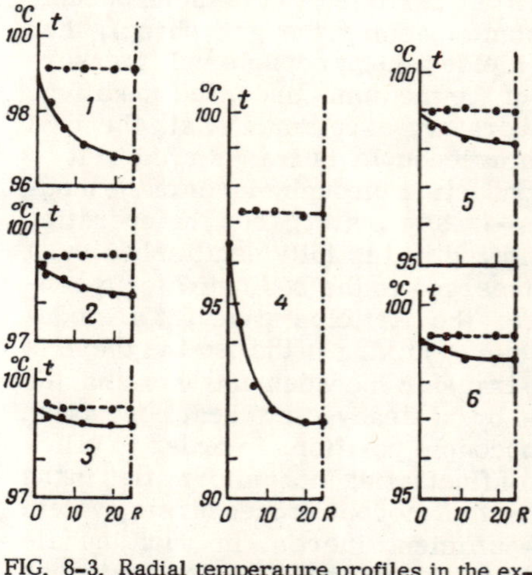

FIG. 8-3. Radial temperature profiles in the experiments of Ciborowski and Roszak [789]. ——— temperatures given by bare thermocouples; – – – temperatures given by shielded thermocouples; Particle diameter 210 microns; G = superficial fluid weight flow rate, kg/m^2 sec; h = distance from base of bed at which profile is measured, cm.

Curve	h	G	Curve	h	G
1	1	0.078	2	3	0.078
3	5	0.078	4	1	0.164
5	5	0.164	6	10	0.164

have been a heating effect of the surrounding steel wall of the thin channel in which the shielded thermocouple was placed, since the main part of the wall was situated in the region of higher bed temperature. The walls had a considerable thickness (1 mm). However, Ciborowski and Roszak obtained a radial change in $t_{\text{shielded}} - t_{\text{unshielded}}$ (Fig. 8-3) along the axis of the bed. This difference was considerable at the bottom of the bed, and gradually decreased on approaching the walls. This agrees with the impression that it is valid to regard $t_{\text{unshielded}}$ as the material temperature.

The problem is confusing in many respects and new and more careful measurements are required. It is also necessary to reexamine the assumption that the quantity of heat transferred to the wall or thermocouple purely through the surface area of contact between the thermocouple and particles is a small part of the total. This would be a surprising result since metallic powders have heat-insulating properties.

5. Entry effect

It is known that a so-called entrance effect is observed in fluidized bed heat exchangers [789]: the fluidizing agent passing through the gas-distributing grid changes considerably in temperature. Under the experimental conditions of Ciborowski and Roszak [789], the gas passing through the porous plate supporting the bed was heated 40-70°C. Leva [988] indicated a similar entrance effect, from transformation of energy due to the sharp constriction of the cross-sectional area available for gas flow. Coborowski and Roszak [789] gave a more probable explanation of this effect: the grid represents a type of additional heating surface, obtaining heat, for instance, from the warm main heating surface of the bed resting on it. The experimental data of Leva [988] and Heertjes et al. [341] confirm the views of Ciborowski and Roszak (Fig. 8-4). In Fig. 8-4 the letter E denotes the entrance effect, which is given by Heertjes as

$$(\Delta t_g)_{\text{grid}}/(\Delta t_g)_{\text{grid+bed}},$$

i.e., as the ratio of the gas-stream temperature drops in the screen and overall across the screen and bed.

It can be seen from Fig. 8-4 that the entrance effect increases with the superficial fluid velocity. This is to be expected, since as the superficial fluid velocity is decreased the heat transfer coefficient decreases more slowly than the water equivalent of the gas. The entrance effect also decreases with the tube diameter; this is obviously related to the decrease in heat transfer between the grid and walls. For thin grids, an increase in the thermal conductivity of the material of which they are made can play only a small part, substantially increasing the supply of heat to the grid from the walls only, but not from the fluidized bed. The information available in the literature on the types of grids used in experiments is unsatisfactory [988] and does not permit a more complete analysis of the data on entrance effects.

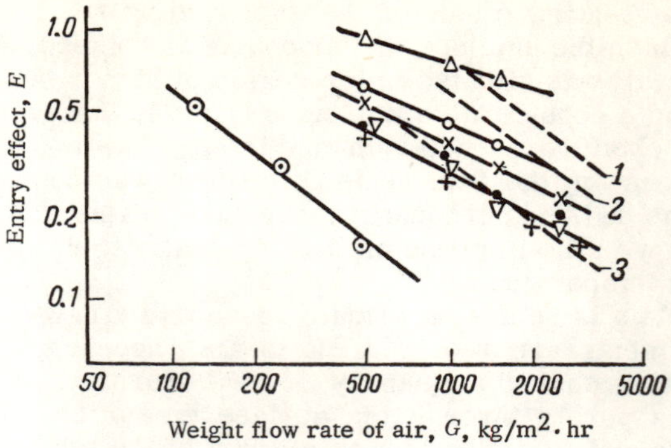

FIG. 8-4. Entrance effect in fluidized bed [988]. Data of Heertjes et al. [341]:
1—copper screen; 2—celluloid screen; 3—woven nickel screen. Data of Leva et al. [416]:

Symbol	System	Particle diameter, microns	Slope of line	Bed diameter, mm
●	Sand-air	77	-0.52	52.6
×	Sand-CO_2	77	-0.52	
○	Sand-air	115	-0.45	
△	Sand-air	50	-0.30	
+	Fe_3O_4-air	109	-0.52	
▽	Fe_3O_4-air	71	-0.52	
	Sand-air			
⊙	Sand-helium	97	-0.78	102.5

The entrance effect complicates studies of heat-transfer in beds with flowing media. It can be reduced by using thin screens and high superficial fluid velocity.

In industrial practice there is usually no need to be concerned with entrance effects.

8.2 Heat Transfer From Particles in Suspended, Fixed, and Spouting Beds

1. Suspended beds

Suspended beds have low volumetric concentrations of particles. The particles remain far apart from one another, and are uniformly wetted by the medium; the medium has a relatively uniform temperature over most of the cross section of the heat exchanger because of the intensive turbulent mixing. Hence in suspended beds the experimental determination of heat transfer coefficients from the particles is not complicated by nonuniform gas distribution.

For calculating heat transfer coefficients from particles in suspended beds the following empirical formulas, as recommended by Fedorov [170], are used:

Lyakhovskiy

$$Nu = 0.62Re^{0.5} \text{ for Re from 150 to 30,000} \qquad (8\text{-}14)$$

and Sokol'skiy

$$Nu = 2 + 0.16Re^{0.67} \text{ for Re} < 150 \qquad (8\text{-}15)$$

although they were developed from experiments with securely mounted spheres.

The tendency for Nu to fall to the minimum value for conduction ($Nu_{min} = 2$) at small Reynolds numbers is well confirmed by experiments on droplet evaporation in spray dryers [1109].

Equation (8-14) gives a safety factor in calculations compared with the formulas recommended by Kryukova [757]:

$$Nu = 8.2Re^{0.6} \text{ for Re from 200 to 1000} \qquad (8\text{-}16)$$

and

$$Nu = 1.06Re^{0.457} \text{ for Re from 1000 to 12,000.} \qquad (8\text{-}17)$$

In carrying out experiments with relatively large fixed particles (spheres, d = 20 - 50 mm, and cubes, a = 20 - 30 mm), Kryukova did not observe any effect of the particle shape and orientation up to $Re < 9 \times 10^3$. Rotation of these large particles at rates from 500 to 6000 r.p.m. led to an increase in the heat transfer coefficient compared with the value of α for a stationary sphere in the low Reynolds number region. This was not because of the creation of any special conditions (for example, formation of a turbulent boundary layer), but was obviously only due to the increased velocity of flow of the air stream relative to the particle, as shown by Kryukova [757].

The relative particle velocity varies in the acceleration zone from an initial value, which is usually close to the velocity of the medium stream, to the equilibrium velocity of free fall. This variation in addition to other problems imposes difficulties in calculating heat transfer in suspended beds of materials of wide particle size distribution. Various simplifications are therefore made in the practical methods of calculating heat and mass transfer in suspended beds. An account of such simplified design methods can be found, for instance, in the works of Fedorov [170], Lyuboshits [763] and Tagantseva [160].

Despite some advantages (simple construction, uniform wetting of particles), suspended beds also possess a number of disadvantages: low true volumetric particle concentration (otherwise choking begins, see p. 86), low-energy efficiency in pneumatic lifting (except for equipment with falling beds, where it is not required), difficulty in arranging countercurrent flow, and nonuniformity in heating the various fractions of materials with wide size distributions (temporary superheating of the fine particles (see [44]).

2. Fixed beds

Many equations are available for calculating heat transfer coefficients from particles in fixed beds. Several of these (with the conditions of the corresponding experiments) are given in the monographs of Chudnovskii [175] and Chechetkin [688]. By generalizing the experimental data of several workers, Chechetkin obtained the following empirical formula for nonmetallic packings:

$$Nu = 0.123 Re^{0.83} \text{ for Re from 50 to 2000.} \tag{8-18}$$

In calculating the value of $Re = w d_e/\nu$ the characteristic velocity is taken as $w = w/m$, and the characteristic dimension as the equivalent diameter, $d_e = 4m/S$, where S is the surface area of the particles per unit volume of the bed.

Timofeev proposed an empirical formula which describes (though not with very high accuracy) the experimental data on heat transfer coefficients for stationary packings (fixed beds) of any material:

$$Nu = 0.106 Re \text{ for Re from 20 to 200} \tag{8-19}$$

$$Nu = 0.61 Re^{0.67} \text{ for Re} > 2000 \tag{8-20}$$

where $Re = w\,d/\nu$, d = particle diameter.

The surface area for heat transfer is taken to be the total geometrical surface area of the spherical particles.

When it is necessary to consider the presence of temperature gradients within large particles Kitaev [60] proposed that the decrease in heat transfer can be allowed for by introducing a correction factor n in the heat transfer coefficient (for $Bi \leqslant 10$):

$$n = \frac{1}{1 + \frac{Bi}{\psi}}, \tag{8-21}$$

where ψ depends on the geometric particle shape. For spheres $\psi = 5$, for cylinders, 3.5, and for plates, 3.

The uniformity of gas distribution over the bed is not considered in any of the methods proposed for calculating heat transfer coefficients from the experimental data. Since in fact the distribution is not uniform, it can be shown that the mean-temperature drop is overestimated and the heat transfer coefficient is reduced by a larger or smaller degree depending on the degree of nonuniformity of the superficial fluid flow. The empirical formulas therefore do not give true and universal values of α, but an effective value, which is suitable for calculations only at a degree of nonuniformity in gas distribution close to that existing in the experiments on which the empirical formulas were based. The different particle packing uniformity in dense beds and differences in the experimental equipment of the various investigators are the main reasons for the known differences in the data on heat and mass transfer. Nevertheless, the differences are not very large, and according to all these empirical formulas the effective heat transfer coefficients in fixed beds are usually of the order of several dozen $kcal/m^2 \cdot hr \cdot degree$, i.e., quite large. Based on the experimental data of Chukhanov, Zabrodsky [47] has shown that "fixed-bed" heat exchangers with materials of narrow size fractions and with uniform hydraulic resistance and discharge are much more compact than suspended-bed heat exchangers.

In sharp contrast it appears that many industrial heat exchangers with moving dense beds operate poorly. For example, if an effective heat transfer coefficient is calculated from the initial temperature of the gaseous heat transfer medium, the particle temperature, the temperature of the outlet gas, and other characteristics under steady-state operation of a column dryer, the values obtained appear to be only a few kcal/m^2 • hr • deg. However, this is a fictitious (or effective) value of α, and indicates that the actual temperature difference is small and that the gas distribution is very nonuniform in the particular heat exchanger, with break-through and channels in the moving bed in an apparatus of complicated configuration. The low heat load at the particle surface in such equipment was shown in [641] to be related to the extremely high value of the group $\alpha F/W_g$. F is the surface area of the particles in the heat exchanger, and W_g is the water equivalent of the gas. This paper also mentioned a means of improving the performance of heat exchangers with moving dense beds of material by use of thin but simple configurations and steadily converging beds with high superficial fluid velocities.

It is obvious that the correction for the increase in the nonuniformity of gas distribution, which must be introduced in the effective heat transfer coefficient in going from a simple fixed bed to a moving bed will differ depending on the construction of the equipment. The accumulation and analysis of experimental data of heat transfer in moving beds and the construction of heat exchangers to correct for the observed poor operation would be desirable. This type of information would make it possible to establish values for the correction factors needed for converting data on the values of α for particles in stationary beds to the design of moving-bed heat exchangers.

3. Spouting beds

To a first approximation, spouting beds represent a combination of suspended beds (in the central fountain) and dense beds (in the peripheral annulus), flowing mainly hydrodynamically countercurrent to the gaseous heat transfer medium. Thermal countercurrent flow is not realized for the entering material due to the repeated and complicated circulation of each particle, which tends to equalize the material temperature in the apparatus.

No data from any systematic investigation of heat transfer between a gas and material in a spouting bed have yet been published. Approximately, it can be assumed that most of the heat transfer occurs in the peripheral annulus although the heat transfer coefficient and mean-temperature difference are much higher in the central fountain. It is important to note that at any given moment only a small part of the material is present in the central fountain, while the gas passing through the peripheral annulus in an efficiently spouting bed is a large part of the total.

It is assumed by Thorley et al. [1011] that:

1) the fraction of the gas passing through the peripheral annulus increases as the relative diameter d_0/D_T of the inlet openings increases, and as the cone angle of the equipment and the gas flow rate decrease;

2) as the flow rate is increased above the minimum value required to produce the spouting velocity, the "excess" gas tends towards the central spout (but improves the circulation of material).

Hence, knowing the fraction of the gas passing through the periphery at the initial level of the bed (for example, from experiments at the minimum spouting regime), it is possible to make an approximate calculation of the heat transfer between the gas and material by:

1) assuming that heat transfer from the gas passing through the periphery is complete, i.e., the gas exit temperature is equal to the material temperature; and

2) assuming that heat transfer in the fountain itself can be neglected.

The first assumption tends to increase the contacting efficiency, while the second tends to decrease it, i.e., to some extent the errors introduced tend to compensate each other.

This method of calculation is obviously lacking in accuracy.

8.3 Review of Investigations of Particle-to-Fluid Heat Transfer in Fluidized Beds

During 1945-1946, Fedorov carried out the extensive investigations of heat transfer between particles and gases in fluidized beds published in his doctoral thesis [169] in 1950. Fedorov determined heat transfer coefficients under constant rate drying conditions for moist granulated activated carbon (3 - 5 mm), Moscow coal (0.5 - 10 mm), and pieces of cardboard 5×6×3 mm and 10×5×3 mm. Air, preheated to 40-50°C, was used as the heat transfer and fluidizing medium; the weight flow rate of air varied from 0.9 to 4.0 kg/m^2 · sec. The fluidized bed rested on a wire screen in a cylindrical steel tube 100 mm in diameter; another screen was placed 300 mm above the bed support to prevent particle carry-over. The initial bed heights were 6 to 45 mm. Air temperature (by mercury thermometer) and air humidity (by psychrometer) were measured at the inlet and exit of the bed. The bed material was weighed before drying and after specified drying times. From the difference in weight (quantity of water evaporated) and the quantity of heat used in heating the material from its initial temperature to the wet-bulb temperature it was possible to determine the quantity of heat transferred to the particles. The surface area of the particles was calculated as

$$F = 6G_{\text{bed}}/\gamma_{\text{M}} d_{\text{e}},$$

where

$$d_{\text{e}} = \sqrt[3]{\frac{6P}{\pi n \gamma_{\text{M}}}};$$

and P and n are the weight of material and number of particles in the sample taken for determining the equivalent diameter.

In calculating the temperature driving force, Fedorov assumed the logarithmic mean of the psychrometric differences at inlet and exit of the bed.

Hence, as in the case of other investigators, Fedorov determined the mean-temperature driving force assuming uniform gas distribution in the fluidized bed, i.e., an increased driving force was obtained with a correspondingly lower value of the heat transfer coefficient compared to the true effective value.

From an evaluation of his experimental data, Fedorov proposed the dimensionless expressions:

$$Nu = 0.0151 Fe^{0.74} Re^{0.65} \left(\frac{H_0}{d_e}\right)^{-0.34} \quad \text{for } Fe = 30 \div 100; \tag{8-22}$$

$$Nu = 0.0283 Fe^{0.604} Re^{0.65} \left(\frac{H_0}{d_e}\right)^{-0.34} \quad \text{for } Fe = 100 \div 200, \tag{8-23}$$

where

$$Fe = d \sqrt[3]{\frac{4}{3\nu^2} g \left(\frac{\gamma_M}{\gamma_m} - 1\right)}.$$

However, he found that the data for the Moscow coal were better described by the equation:

$$Nu = 0.23 Re^{0.863} \quad \text{for Re from 20 to 500.} \tag{8-24}$$

The effective heat transfer coefficient increased as the size and specific weight of the particles increased. Fedorov noted that the intensity of mass transfer in a fluidized bed is so high that at a mean-particle size of 3 mm the gas was cooled practically to the material temperature even in a thin bed corresponding to a loading of 80 kg per square meter of grid area.

At about the same time (1949-1952), several investigations of heat and mass transfer in fluidized beds were published abroad [247, 389, 440, 516, and 584]. Of these the papers by Kettenring, Manderfield and Smith [389], Walton and co-workers [584], and Resnik and White [516] are of particular interest.

Kettenring et al. [389], carried out investigations of heat and mass transfer in fluidized beds of finer particles than those used by Fedorov. They studied simultaneous heat and mass transfer during the evaporation of water from moist particles (0.35-0.99 mm) of alumina and silica gel. The particles were of nonuniform shapes, but the heat transfer area was calculated as for a bed of spheres having a diameter equal to the mean of the screen openings between which the fraction being used was collected during screening. The bed diameter was 58 mm and the initial bed height 100-150 mm. A screen with 75 micron openings was used as the gas distributor. The superficial fluid velocity was controlled so that the Reynolds number fell between 9 and 55.

Kettenring et al., measured the gas temperature immediately below the screen, and also the temperatures at various heights in the bed using thermocouples with junctions which were not shielded from contact with the particles. They assumed that the thermocouples in the bed measured the gas temperature, and from the thermocouple e.m.f. they constructed a curve of the change in gas temperature over the height of the bed; Fig. 8-5 shows their data for a typical experiment. Starting at a height of 50 to 60 mm above the screen the measured

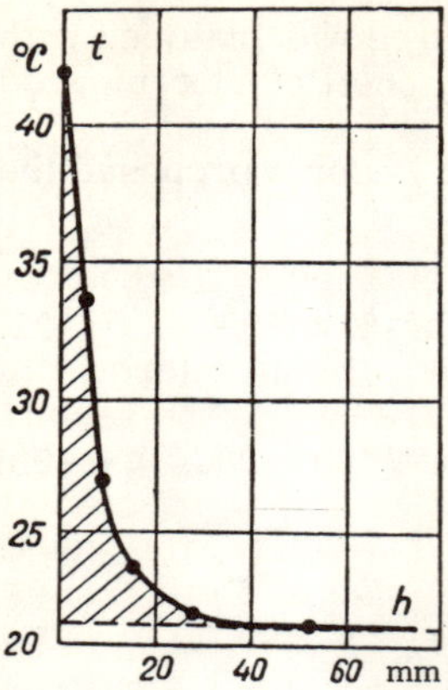

FIG. 8-5. Typical longitudinal temperature profile in a fluidized bed according to Kettenring, et al. [389]. h—distance from base of bed, cm; t—temperature, measured by exposed thermocouple, assumed as air temperature, °C; dashed line: equilibrium temperature, taken as the material temperature.

temperature ceased to change with increase in height and it was therefore assumed that thermal equilibrium between the gas and material had been attained. The measured equilibrium temperature t_e was taken as the particle temperature throughout the bed (as in ideal mixing of the solids) and the heat transfer coefficient calculated as

$$\alpha = \frac{Q_1}{F_1 \int\limits_0^{h_e} (t - t_e)dh},$$

where Q_1 is the overall quantity of heat transferred from the gas to the material per unit time, kcal/hr; F_1 is the surface area of the particles per unit height of the fluidized (expanded) bed; $\int\limits_0^{h_e}(t-t_e)\,dh$ is the area between the curve for t (Fig. 8-5) and the line for the equilibrium temperature t_e (for a linear variation of t this area would be expressed by the product of the mean-temperature difference and the bed height h_e at which thermal equilibrium is established).

The heat transfer coefficients calculated in this way fell in the range of 15-50 kcal/m^2·hr·degree C, while the Nusselt numbers appeared to be lower than 0.1 in a number of the experiments, i.e., many times smaller than the minimum conductive values.

Kettenring's data can be expressed in dimensionless form as:

$$\mathrm{Nu} = 0.0135\,\mathrm{Re}^{1.30}. \tag{8-25}$$

Leva [988] regarded the effective heat transfer coefficients obtained by Kettenring as being too low, but he appeared to base this opinion only on secondary considerations—errors arising from the assumption that the particle temperature throughout the bed did not differ from the equilibrium value. On the other hand Wamsley and Johanson [585], Chu [234] and Ciborowski and Roszak [951] assumed that the heat transfer coefficients obtained by Kettenring were rather too high on the basis that the measured equilibrium temperature t_e exceeded the adiabatic saturation temperature of the air entering the bed by 5-11°C. Assuming that the particle temperature could not exceed the adiabatic saturation temperature of the air, Wamsley and Johanson calculated that the "actual" heat transfer coefficients were 5 to 25 times smaller than those calculated by Kettenring. The validity of this evaluation is dubious, since it is improbable that the equilibrium temperatures measured in Kettenring's experiments by unshielded thermocouples could differ by 5-10°C from the actual particle temperature. Ciborowski, for instance, confirmed that the temperatures measured in this way are practically equal to the particle temperature. Hence the agreement of

Ciborowski and Roszak [951] with the view of Wamsley and Johanson is surprising if they admit the validity of measuring the particle temperature in a fluidized bed with unshielded thermocouples. It is more logical to explain the relatively high equilibrium temperature in these experiments [389] by the fact that the surface temperature of the particles was actually higher than the adiabatic saturation temperature, i.e., that drying was not occurring in the constant rate period.

If unshielded thermocouples read the particle temperature, the curve in Fig. 8-5 does not give the gas temperature beyond the second experimental point, and the area between this curve and the line for t_e does not correspond to the actual temperature difference. The actual line for the gas temperature should fall higher, and a larger area would occur in the calculation of α. However, the first point on Kettenring's curve should be displaced downwards, due to the entrance effect, which would compensate to some extent for the raising of the succeeding points. It is therefore difficult to say whether the effective heat transfer coefficients obtained by Kettenring et al., are on the high or the low side. The nonuniformity of the particle temperatures in the lower part of the bed can be explained by poor mixing of the particles in this zone under Kettenring's experimental conditions.

Thus, the heat transfer coefficients obtained in this work [389] are a form of effective coefficients, and are not true values because of nonuniform gas distribution and the approximations arising from considering the temperature of the material as constant throughout the bed, which is not valid under the experimental conditions.

Resnik and White [516] studied the rate of evaporation of naphthalene particles during the fluidization of fractions of wide size ranges (0.25-1.16 and 0.21-1.7 mm) with air, hydrogen and carbon dioxide. The bed diameters and initial heights were 22-24 mm and 13 to 25 mm, respectively. The evaporation process occurred under approximately isothermal conditions (25°C).

The particle surface area in the fluidized bed was calculated from the pressure drop in the stationary bed. This method was inaccurate (as mentioned by Chu [234]) because of the tendency of the naphthalene particles to form clusters and channels.

Resnik and White proposed the following equation for correlating their experimental data:

$$j_d/d^{1.5} = 0.19\,\mathrm{Re}^{\,0.27} \text{ for } \mathrm{Re} < 25; \tag{8-26}$$

where 0.19 is a dimensional coefficient, and d must be given in millimeters. Here j_d is the mass transfer factor.

Heat and mass transfer data are frequently represented in the form of relationships between the mass or heat transfer factors, j_d or j_h, and the Reynolds number [988, 788].

By definition,

$$j_d = \mathrm{St}_m \mathrm{Pr}_m^{2/3} = \frac{\alpha' P_{\mathrm{i.m}} M}{G} (\nu/D)^{2/3} = \frac{\alpha'_L C_{\mathrm{i.m.}} M}{G}; \tag{8-27}$$

$$j_h = \mathrm{St} \cdot \mathrm{Pr}^{2/3} = \frac{\alpha}{cG} \left(\frac{\nu}{a}\right)^{2/3}. \tag{8-28}$$

Here $St = Nu/Pe \cdot Pr$ is the Stanton number;

Nu, Pr, Pe are the Nusselt, Prandtl and Peclet numbers;

the subscript m denotes the mass-transfer (diffusion) analogs of the criteria for thermal similarity;

Pr_m is sometimes known as the Schmidt number and denoted Sc;

α' is the mass transfer coefficient, expressed in kg-moles and pressure units, kg-moles/m² • hr • atm.;

α'_L is the mass transfer coefficient expressed in kg-moles and concentration units, kg-moles/m² • hr;

$C_{i.m}$ is the mean concentration of inert (nondiffusing) gas, kg-mole/kg-mole;

$P_{i.m}$ is the mean pressure of the inert (nondiffusing) gas in the boundary layer, atm.;

M mean-molecular weight in the boundary layer;

G weight flow rate of stream, kg-mole/m² • hr.

According to Chu [234], it can be assumed, with sufficient accuracy for practical purposes, that for both stationary and fluidized beds,

$$j_d \approx j_h. \tag{8-29}$$

It is not difficult to convert j_d and j_h to Nu_m and Nu and vice versa. As an example, consider going from the mass transfer data of Resnik and White to a value of Nu for heat transfer:

$$j_h = St \cdot Pr^{2/3} \quad \text{or} \quad St = \frac{j_h}{Pr^{2/3}}. \tag{8-30}$$

From the definition of the Stanton number,

$$Nu = St \cdot Pr \cdot Re. \tag{8-31}$$

Substituting (8-30) into (8-31),

$$Nu = j_h Pr^{1/3} \cdot Re,$$

while using (8-29),

$$Nu = j_d Pr^{1/3} Re. \tag{8-32}$$

The equation of Resnik and White therefore gives

$$Nu = 0.19 d^{1.5} Re^{0.73} \cdot Pr^{1/3}, \tag{8-33}$$

where d is in mm.

Many papers have more recently appeared containing data on heat and mass transfer in fluidized beds [102, 178, 234, 235, 295, 324, 340, 480, 495, 585, 637, 638, 669, 792, 793, 807, 814, 840, 897, 906, 951, 952, 953, 975, 984, 1002, 1080, 1128, 1131, 1127, 354]. In 1953, Chu, Kalil and Wetteroth [235] published the results of an experimental investigation of mass transfer in fluidized and stationary beds. They improved on the method of Resnik and White, using (instead of flaky particles of pure naphthalene) particles of regular shape (such as glass spheres, turnip seed and lead shot) covered with a film of naphthalene. This made it possible to determine quite accurately the surface area of the particles and the bed voidage, which was varied over a wide range from 0.25 to 0.96. The range of particle sizes was 711 microns to 13.73 mm and the bed diameter 95.2 mm. The initial bed height was varied from 2.5 to 90 mm and air used as the fluidizing agent, with weight flow rates from 1300 to 21,000 kg/m²·hr, corresponding to particle Reynolds numbers from 15 to 1300.

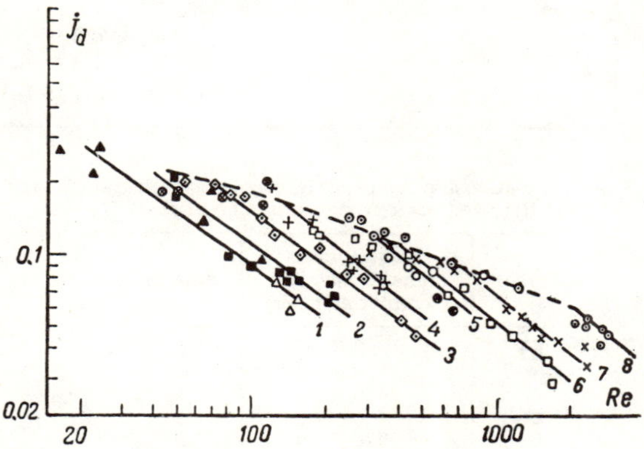

FIG. 8-6. Mass transfer factor j_d as a function of Reynolds number in stationary and fluidized beds, according to the data of Chu et al. [234].
1—glass spheres, mean diameter 0.71 mm; 2—lead shot, diameter 0.74 mm; 3—seed, diameter 1.95 mm; 4—lead shot, diameter 1.27 mm; 5—lead shot, diameter 1.95 mm; 6—small celite cylinders, diameter 5.5 mm; 7—medium celite cylinders, diameter 8.5 mm; 8—large celite cylinders, diameter 13.7 mm.
Particle diameter used as characteristic dimension. The upper dashed line corresponds to stationary beds.

The experimental results [234, 235] are shown in Fig. 8-6 in the form of a relationship between the mass transfer factor j_d and the Reynolds number. In terms of these coordinates, each fluidized system is characterized by its own linear relationship, while the stationary beds are characterized by a single line for all particle diameters. As the particle diameter is increased the lines for the fluidized beds are shifted to the right. However, by constructing a graph giving j_d as a function of a modified Reynolds number Re/(1—m), Chu et al., obtained a single line, as shown in Fig. 8-7. Chu [234] showed that this generalized relationship is applicable to stationary and fluidized beds regardless of the nature of the fluid medium (gas or liquid). The scatter of the points in Fig. 8-7 for mass transfer in the fluidized bed at high Reynolds numbers is quite large even for the small number of investigations which have been generalized. For example, at Re/(1—m) = 5 000, the generalized curve gives $j_d = 0.04$, while actual values are 0.022 for one experiment and 0.05 for another.

Chu et al., proposed the following analytical expressions for the functions shown in Fig. 8-7 as describing approximately heat and mass transfer between particles in fluidized and stationary beds:

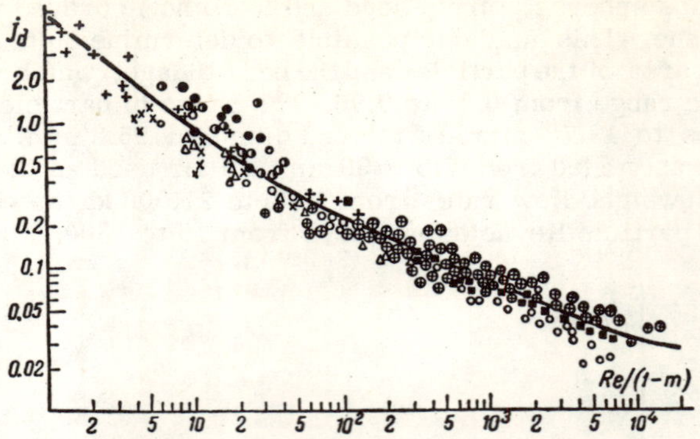

FIG. 8-7. Relationship for particle mass transfer in stationary and fluidized beds proposed by Chu et al. [234].

Symbol	Prandtl mass transfer number (Schmidt number)	State of bed
⊕	2.57	s, fb
▪	0.60	s
○	1460	s, fb
●	866	s
◐	776	s
+	368	s
×	690	s
△	164	s

s = stationary; fb = fluidized bed

$$j_d = j_h = 1.77\left(\frac{\text{Re}}{1-m}\right)^{-0.44} \text{ for } \frac{\text{Re}}{1-m} > 30; \qquad (8\text{-}34)$$

$$j_d = j_h = 5.7\left(\frac{\text{Re}}{1-m}\right)^{-0.78} \text{ for } \frac{\text{Re}}{1-m} < 30. \qquad (8\text{-}35)$$

Equations (8-34) and (8-35) are suitable only for beds of uniform size spherical particles.

Peters et al. [495] investigated heat transfer between cold air and sand particles in a fluidized bed. The particles were heated to 50-115°C by an electric heater submerged in the bed. The particle diameter was 0.23 mm and the heat transfer coefficient calculated from a heat balance according to the formula

$$\alpha = \frac{Gc}{hS} \cdot \frac{2(t' - t)}{2\vartheta - t' - t}, \qquad (8\text{-}36)$$

where t' and t were the gas temperatures at the point of measurement in the bed, and before entering the bed, respectively.

The particle temperature ϑ was taken to be the temperature measured by a bare thermocouple submerged in the fluidized bed. The gas temperature in the bed was assumed to be equal to the reading of another thermocouple also submerged in the bed, but with its junction shielded from contact with the particles by a small copper mesh cap.

The heat transfer coefficients calculated by Peters varied from 1 to 1.5 kcal/m^2 • hr • degree, yielding Nusselt numbers of the order of 0.02, i.e., smaller than those of Kettenring by a factor of about 10.

Peters did not comment on the provisional nature of Pris values of Nu and α, and attempted to justify the discrepancy between his experimental data and theoretical predictions only on the basis of the small relative air velocity.

It is obvious that the gas distribution was particularly poor in Peters' experiments. The superficial air velocity was 6.5-8.2 cm/sec with a sand particle diameter of d = 0.23 mm. The air entered the bed through a horizontal tube (D = 15 mm), placed close to the bottom of the vessel, having openings 1 mm in diameter spaced 10 mm apart along the upper side of the tube. With this type of gas distributor it is obvious that a large part of the bed above the horizontal sprayer tube was filled with channels due to the concentrated air jets.

Shakhova [178] and Wamsley and Johanson [585] carried out investigations of particle heat transfer in fluidized beds using a method based on the unsteady-state regime of heat transfer. The work of Shakhova is discussed in the monograph by A. V. Luikov [84]. Beds of quartz sand with equivalent particle diameters from 268 to 564 microns were fluidized with air above a porous glass screen in a cylindrical double-walled glass retort (i.d., 78.5 mm, height 900 mm).

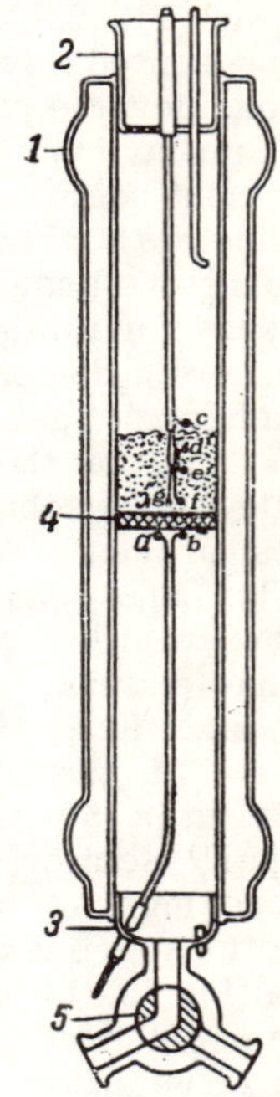

FIG. 8-8. Sketch of Shakhova's experimental apparatus [178] for investigating heat transfer in a fluidized bed.
1—body of vessel; 2—upper stopper; 3—lower stopper; 4—glass screen, 10 mm thick; 5—two-way cock; a, b—thermocouples for measuring gas temperature above bed; d, e, f, g—thermocouples for measuring temperatures in fluidized bed.

A vacuum was drawn (residual pressure 1-10 mm Hg) between the double walls to reduce heat losses, and the wall surfaces facing this space were silvered. At the top and bottom the retort was fitted with ground-glass fittings (Fig. 8-8), the upper one of which had an opening to allow the gas to escape. The lower one was attached to a tube with a two-way cock for feeding preheated gas and for measuring the pressure. Before an experiment the thermal regime was established by heating up the empty retort (without sand). Sand at a predetermined temperature was then poured in. The temperatures at a number of points along the height of the fluidized bed were all recorded simultaneously at half-minute intervals. After a specified time of heating of the bed the supply of gas was stopped, and measurements were made of the temperatures of the stationary bed particles. Experiments showed that the temperature of the fluidized bed measured by a thermocouple with a bare hot junction corresponds to the temperature of the solid particles.

The following relationship was obtained from an evaluation of the experimental data:

$$\mathrm{Nu} = 0.680 \cdot 10^{-2}\, \mathrm{Re}^{1.55} N^{-1} \tag{8-37}$$

for Re = 1138 - 9.51 and N = 2.67 - 12.82,

where N is the fluidization index.

The numerical values of the Nusselt number appeared to be very low ($\mathrm{Nu} \ll 2$). Sahkhova concluded that these low values of α and Nu represented only arbitrary effective values, due to the break-through of a considerable part of the gas without proper contact with the particles. Similar hypotheses, that part of the gas was passing through the bed without heat transfer contact with the particles, were made earlier by Wilhelm [603] and Toomey and Johnstone [567]. However, Toomey and Johnstone assumed that all of the excess quantity of gas (above that necessary for minimum fluidization) always passed through the bed in the form of a discontinuous phase—bubbles—and made no reference to the exchange of gas between the bubbles and the rest of the bed.

The fact that the initial two-phase theory of fluidization, which did not take this exchange of gas into account, is incompatible with the results of heat transfer experiments was indicated in the work of Wamsley and Johanson [585]. They fluidized beds of glass spheres (narrow fractions with mean diameters of 137, 216 and 455 microns), alumina (narrow fractions of particles with irregular shapes of mean size 913, 457 and 228 microns), and resin spheres with diameters of the order of 460-760 microns, using air and carbon dioxide in a 59-mm i.d. tube. Their gas distributor consisted of a "damper" section, filled with glass spheres (d = 6 mm) and a screen with openings of 75 micron diameter. Wamsley and Johanson carried out their experiments by the same unsteady-state method of Shakhova. In addition to the usual simplifying assumptions as to the absence of gradients in heating the particles, small heat losses, and the practically constant thermal properties of the gas and material in the bed, they made two others: 1) that the mixing of the material in the fluidized bed was sufficient to ensure that at any moment the temperature of the particles was uniform throughout the bed; and 2) that the gas passing through the bed was well mixed so that at any moment the gas temperature everywhere in the bed was the same and equal to the gas temperature leaving the bed. The final form of their equation for the instantaneous rate of heat transfer from the particles is then

$$Q_G c_G (t_{\mathrm{in}} - t) = \alpha F_{\mathrm{bed}} (t - \vartheta) \tag{8-38}$$

and the differential heat balance equation is

$$Q_G c_G (t_{\mathrm{in}} - t)\, d\tau = G_{\mathrm{bed}} c_M d\vartheta. \tag{8-39}$$

Here t_{in} and t are, respectively, the gas temperature at the inlet to the bed and at some point in the bed;

G_{bed} and F_{bed} are the weight of the bed and the surface area of the particles in it;

τ is the time duration of the experiment.

Equation (8-39) can be integrated graphically, since the change of $(t_{\mathrm{in}} - t)$ with time is known. The particle temperature ϑ given by (8-39) after some finite interval of time can be used for solving Eq. (8-3) and finding the desired heat transfer coefficient α, since the particular value of t corresponding to this interval of time is known from the experimental curve. However, it is very necessary to avoid inaccurate graphical integration. Wamsley and Johanson transformed Eqs. (8-38) and (8-39) to the expression

$$t_{\text{in}} - t = be^{-a\tau}, \tag{8-40}$$

where

$$a = \frac{\alpha F c_G Q_G}{G_{\text{bed}} c_M (\alpha F + Q_G c_G)};$$

$b = (t_{\text{in}} - t)_0$ is the gas temperature difference in the bed at the initial moment of time.

Hence, in terms of coordinates $\ln(t_{\text{in}} - t)$ and τ straight lines are obtained whose slopes ($-a$) determine the value of the heat transfer coefficients. It is therefore possible to obtain α from data for the "central part" of the experiment, where the experimental accuracy is the greatest. It is difficult to follow the temperature during the first few seconds of operation, due to the not quite instantaneous introduction of the material. At the end of the experiment $t_{\text{in}} - t$ becomes quite small as thermal equilibrium is approached, and the experimental error increases.

Wamsley and Johanson noted that the most remarkable feature of their data was the extremely low values of the heat transfer coefficient and Nusselt number (see the combined graph of $\text{Nu} = f(\text{Re})$ below), and as in the case of Shakhova, they correctly determined the primary cause. However, Wamsley and Johanson did not recognize the important relationship between the heat transfer coefficient and the weight flow rate of the fluid, as derived by Shakhova and other researchers [988]. This discrepancy was the basis for Leva's statement [988] that the data of Wamsley and Johanson were questionable. The discrepancy actually requires an explanation, though in principle it should be expected that in the presence of well-marked aggregates in two-phase fluidization there should be this lack of relationship between the heat transfer coefficient (the true value, at least) and the superficial fluid velocity, since the velocity of flow around the particles in the aggregates should be equal to the same minimum fluidization velocity at any superficial fluid velocity. Hence, the rate of flow around the particles in the aggregates is much lower than the rate of flow around particles in a stationary bed (held down to prevent fluidization) at the same high superficial fluid velocity.

Heertjes and McKibbins [342] determined fluidized bed heat transfer coefficients for the air drying of silica-gel particles in the constant-rate steady-state drying period. Narrow size fractions (0.3-0.42 to 1.0-1.2 mm diameter) were fluidized in a double-walled tube of 60 mm inside diameter; a vacuum was maintained between the walls. The weight rate of fluid flow varied from 1580 to 3070 kg/m$^2\cdot$hr, corresponding to particle Reynolds numbers of 8.8 to 52.3, Heertjes and McKibbins formed the opinion that the so-called "entrance effect" was negligible. As mentioned earlier, the entrance effect includes heating of the support screen by the gaseous heat transfer medium passing through it, and the transfer of this heat to the particles in the bed; the screen thereby acting to some extent as an additional heat transfer surface [789]. A thin nylon gauze was therefore used as the inlet gas distributor in this work [342]. The gas temperature in the fluidized bed was measured with a movable thermocouple whose hot junction was shielded from contact with the particles by a screen. The curve obtained for the change in gas temperature over the height of the bed was not used for graphical integration, as in the case of Kettenring et al., in evaluating α the authors employed another graphical procedure. The heat transfer coefficients were calculated on the basis of the expression

$$\ln \frac{(t - t_M)_0}{(t - t_M)_H} = \frac{A_1 \alpha}{c_G G F_{sec}}, \tag{8-41}$$

where t = variable value of the gas temperature over the bed height, °C;

t_M = constant temperature of the particles, equal to the wet-bulb temperature, °C;

A_1 = surface area of the particles per unit height of bed, m^2/m;

G = superficial fluid mass velocity, $kg/m^2 \cdot hr$;

F_{sec} = bed cross-sectional area, m^2.

The subscripts H and 0 indicate the height (H and zero) above the screen.

A graph of this relationship should give a straight line passing through the coordinate origin with a slope

$$n = \frac{A_1 \alpha}{c_G G F_{sec}}.$$

Straight lines were produced from the experimental points, from which the slopes n were determined, and from these, values of α. In the calculations the particles were assumed to be spherical, with a diameter equal to the arithmetic mean of the screen openings between which the fraction was collected. In the calculation of A_1 the bed voidage was assumed constant throughout the bed. The following empirical relationship was established by the method of least squares:

$$\alpha = 6.395 \, \mathrm{Re}^{0.76}, \tag{8-42}$$

where α is expressed in $kcal/m^2 \cdot hr \cdot degree$.

The authors also proposed another formula for expressing the experimental data:

$$\alpha = 126.5 G^{0.76} d^{0.67}; \tag{8-43}$$

where G is in $kg/m^2 \cdot hr$ and d in meters.

Interesting investigations of particle heat transfer coefficients in fluidized beds of dry material were carried out (at small values of the fluidization index) by Sharlovskaya [792, 793, 952, and 953].

This was the first time that the so-called quasi-steady-state regime was reported in such investigations. Essentially, this technique consists in setting up conditions such that the temperatures of the bed and medium vary in time according to the same linear relationship, and consequently that the temperature difference remains constant. The characteristics of the materials used by Sharlovskaya are given in Table 8-2.

Table 8-2

Characteristics of the Materials Used in Sharlovskaya's Experiments [952]

No.	Material	Particle shape	Calcd. diam., mm	Bulk weight of particles, kg/m^3	Heat capacity, kcal/kg · degree (C)
1	Porcelain spheres	Spherical	2.1	2475	0.213
2	Porcelain spheres	Spherical	4.63	1996	0.210
3	Porcelain spheres	Spherical	6.87	2072	0.211
4	Porcelain spheres	Spherical	8.9	1614	0.206
5	Quartz sand	Round	0.835	2626	0.1815
6	Marble fragments	Irregular	1.434	2412	0.199
7	Marble fragments	Irregular	2.22	2412	0.199
8	Marble fragments	Irregular	3.01	2412	0.199

The materials were fluidized in a cylindrical vessel (an asbestos-cement tube) of 100 mm internal diameter and 800 mm height. The beds were supported on a metal screen and the fluidizing medium (air) supplied from a compressor through a measuring orifice and electrical heater consisting of a series of spirals connected in parallel. The air-temperature drop in the bed was measured with a differential thermocouple. Initial thermal equilibrium of the system was attained by heating the vessel and bed at the required air temperature and velocity until the e.m.f. of the differential thermocouple became practically equal to zero. In the course of an experiment the temperature of the supply air was reduced and the bed was cooled according to a linear law following the previously selected regime by switching off the heating spirals. The characteristic form of the temperature curves obtained by this method is shown in Fig. 8-9. The temperature graphs are used to determine the rate of cooling b (the slope of the temperature change line), the temperature of the medium t_m (as the mean value of the air temperatures at the inlet and exit of the bed at the moment τ), and the mean-particle temperature $\bar{\vartheta}_\tau$:

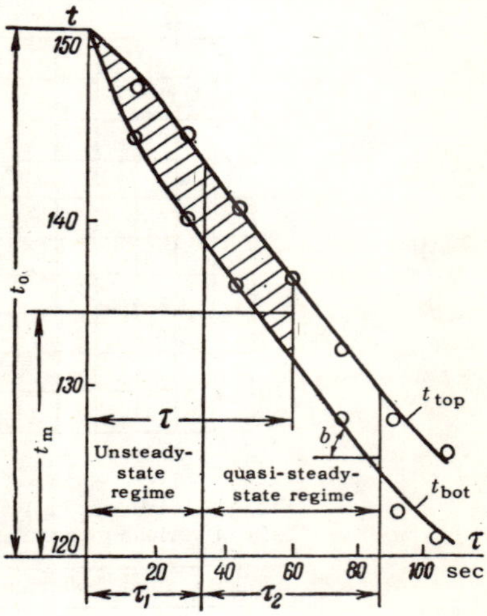

FIG. 8-9. Graphs of typical temperatures for Sharlovskaya's experiments [792, 952 and 953] on heat transfer between particles and gas in a fluidized bed.
t_{bot} = air temperature before entering fluidized bed; t_{top} = gas temperature at bed exit; t_m = mean temperature of medium in bed at time τ; t_0 = initial bed temperature.

$$\overline{\vartheta}_{\tau}=\vartheta_{0}\frac{cQ_{m}}{c_{M}G_{bed}}\int_{0}^{\tau}(t_{top}-t_{bot})\,d\tau+\frac{q_{wall}}{c_{M}G_{bed}},\tag{8-44}$$

where ϑ_0 = material temperature at start of experiment;
Q_c, t_{bot}, t_{top} = weight flow rate and temperature of air entering and leaving bed;
q_{wall} = quantity of heat obtained after time τ from wall (obtained from the change in temperature of the wall material, the heat capacity of which is known); q_{wall} is only 0.3-1% of the heat transferred to the air from the hot particles.

The heat transfer coefficient is calculated from the formula

$$\alpha=\frac{bRc_{M}\gamma_{M}}{3\,(\overline{\vartheta}_{\tau}-t_{m})}.\tag{8-45}$$

By evaluating the experimental data, Sharlovskaya obtained

$$\mathrm{Nu}=0.015\mathrm{Re}^{0.805}\mathrm{Fe}^{0.53}\left(\frac{H_0}{d}\right)^{-0.45}\tag{8-46}$$

for Re = 30-2500 and Fe = 27-294.

Here H_0 is the initial bed height and Fe is the Fedorov number.

The Reynolds number varied mainly as a result of changes in the particle sizes. For each particle diameter the range of superficial fluid velocities investigated was quite narrow. The approximate upper limit was the attainment of intensive "boiling." In comparing the data

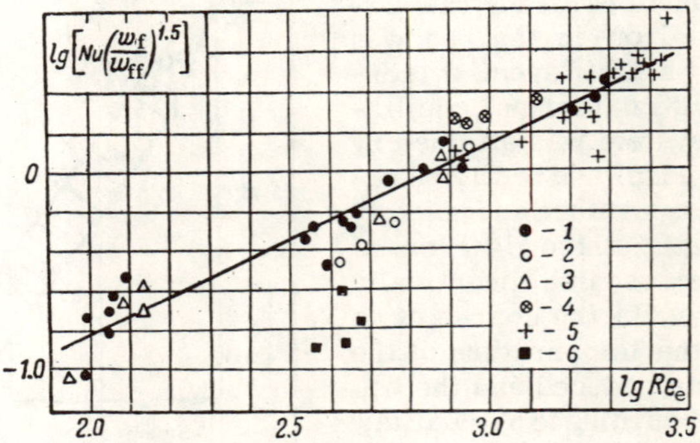

FIG. 8-10. $\lg[\mathrm{Nu}(w_f/w_{ff})^{1.5}]$ as a function of Re for fluidized beds of various materials, according to A. V. Luikov [911].

Symbol	Material	m	ξ	
1	Coal from Moscow area	0.5-0.6	0.75	
2	Granulated carbon....	0.44	0.8	
3	Charcoal..........	0.6-0.7	0.6	
4	Cardboard.........	0.46	0.7	Fedorov's data
5	Granulated superphosphate...........	0.35-0.50	0.98	
6	Grain (wheat).......	0.46	0.88	VTI data

of various researchers with her own, Sharlovskaya noted that there is not a single-valued relationship between Nu and Re under the conditions of a fluidized bed, but she did not indicate clearly that this is caused by the varying nonuniformity in gas distribution in such beds. Sharlovskaya made no distinction between the true and effective values of α and Nu, so that some of her conclusions are questionable. For example, her statement that the transition zone of a fluidized bed, where the true rate of flow around the particles decreases with increase in the superficial fluid velocity (see earlier, p. 55), is the least efficient from the point of view of heat transfer. It is true that in this zone the true heat transfer coefficient is actually lower than at much larger values of the fluidization index (not following the extreme point of view of the two-phase theory of fluidization). However, the effective heat transfer coefficients in the transition zone are higher than those just outside this zone, in the region of large Reynolds numbers, as can be seen, for example, on the graph in Fig. 8-11, where the experimental points corresponding to the transition zone are enclosed by a dashed line.

Ciborowski and Roszak [951] studied heat transfer with three sand fractions (d = 290, 590, and 1000 microns) fluidized in a copper tube of D_T = 49 mm inside diameter. Their investigation was carried out in the same apparatus (with minor modifications) used by these authors for determining the heat transfer coefficients between the wall and a fluidized bed (see below, p. 196). An unsteady-state thermal regime method was used, somewhat modified from that chosen by Wamsley and Johanson [585]. The same fundamental simplifying assumptions were made as in [585], only instead of eliminating heat losses, means were provided for calculating them. The walls were warmed with saturated steam. In addition, Ciborowski and Roszak, unlike Wamsley and Johanson, carried out their experiments without sudden introduction of the material into the apparatus; instead, it was loaded onto the screen in advance. The heat transfer coefficients were also calculated by a different method.

The material on the screen and affected by the steam jacket was initially fluidized with air at room temperature t' until steady conditions were established, and the temperature of the air leaving the bed became constant, $t_0 > t'$. During this time an equal quantity of air flowed along another pipe, which was heated to the required temperature $t_1 \approx 70°C$. This air was vented to the atmosphere through a valve by means of which the pressure in the pipeline was maintained the same as in the line feeding the fluidized bed. At the required moment the heated air could be directed into the fluidization chamber and the cold air vented to the atmosphere, by means of a multiway cock. However, since the hold-up of the section of the air pipeline between the cock and chamber was appreciable, there was a fall in temperature and then a slow rise after every such switchover. The experimental conditions required that the temperature t_1 be as constant as possible for the first several minutes after switching over. To achieve this [951] the air was initially heated to a temperature, t_2, higher than t_1, and after switching over the multiway valve, the voltage supplied to the terminals of the electric heater was reduced by means of an adjustable autotransformer. After a number of tests, the initial temperature t_2 and the relationship

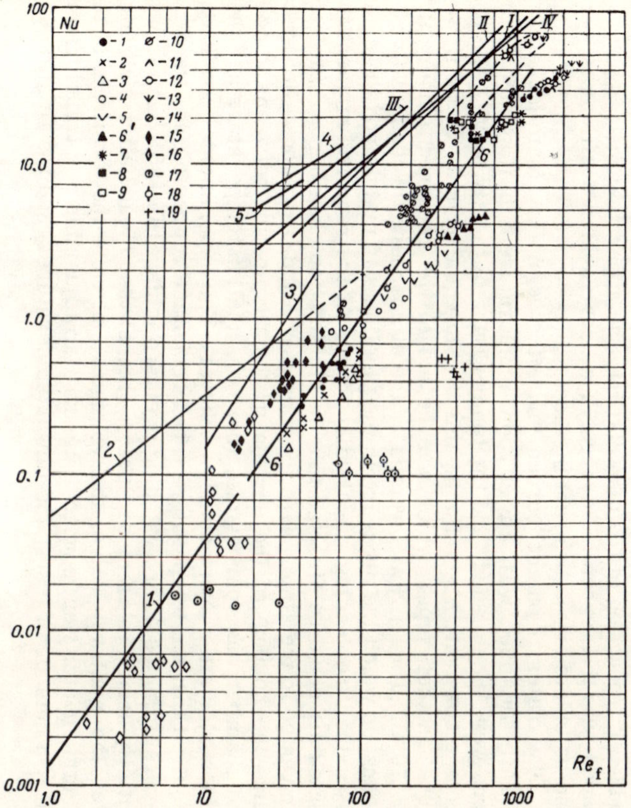

FIG. 8-11. Composite graph for heat transfer between particles and medium in fluidized and stationary beds.

Experiments of	Ref.	Symbol No.	Material	d, mm	h, mm
Sharlovskaya	952	1	Sand	0.832	40
		2	Sand	0.832	60
		3	Sand	0.832	180
		4	Marble particles	1.434	40
		5	Marble particles	2.2	62
		6	Marble particles	3.01	84
		6'	Porcelain spheres	2.1	45
		7	Porcelain spheres	4.63	40
		8	Porcelain spheres	4.63	60
		9	Porcelain spheres	4.63	80
		10	Porcelain spheres	6.87	84
		11	Porcelain spheres	6.87	110
		12	Porcelain spheres	8.9	96
		13	Porcelain spheres	8.9	148
Fedorov, acc. to	952	14	Carbon, cardboard	0.5–10	
Kettenring et al. acc. to	952	15	Silica gel	0.4–1	
Wamsley and Johanson, acc. to.	952	16	Glass spheres	0.1–1.0	
Shakhova, acc. to	952	line 7	Quartz sand	0.25–0.5	
Ciborowski and Roszak	951	17	Sand	0.21	
		18	Sand	0.59	
		19	Sand	1	

Line 2: correlation of Richardson and Ayers [1022]; Line 3: correlation of Sunkoori and Kaparthi [1131]; Lines 4 and 5: correlations of Chu [235] with $m = 4$ and $m = 5$ respectively; Line 6: $Nu = 1.25 \times 10^{-3} Re^{1.46}$ [cf. (8-59)].
The experimental points of Sharlovskaya lying in the "transition zone" are surrounded by a dotted line.
Stationary beds according to [952]: I. experiments of Furnas; II. experiments of Sanders; III. experiments of I. M. Fedorov; IV. experiments of Z. F. Chukhanov and E. A. Shapatina.

between the voltage and time were established so as to attain the temperature t_1 in the air flowing to the screen in the chamber. Graphs of these relationships were used in carrying out the experiments.

For finding the heat transfer coefficients of the particles, Ciborowski and Roszak used a relationship of the form

$$\log B = -n + \log B_0, \tag{8-47}$$

where

$$B = \alpha_{\text{wall}} F_{\text{wall}} (t_{\text{wall}} - t) - Q_{\text{G}} c (t - t_1);$$

$$B_0 = \alpha_{\text{wall}} F_{\text{wall}} (t_{\text{wall}} - t_0) - Q_{\text{G}} c (t_0 - t_1);$$

$$n = \frac{\alpha F (\alpha_{\text{wall}} F_{\text{wall}} + Q_{\text{G}} c)}{2.303 G_{\text{bed}} c_{\text{M}} (\alpha F + Q_{\text{G}} c + \alpha_{\text{wall}} F_{\text{wall}})} \tau.$$

Here $\alpha_{\text{wall}}, F_{\text{wall}}, t_{\text{wall}}$ are, respectively, the heat transfer coefficient between the fluidized bed and the wall, and the area and temperature of the inner surface of the wall wetted by the bed;

Q_{G} is the weight flow rate of gas, kg/hr;

G_{bed}, and F are the weight of the bed and the surface area of the particles in it.

The coefficient α_{wall} was known from previous parts of the experimental investigation. The time dependence of the temperature t of the gas leaving the bed was also known. Hence, for a given value of τ, B was calculated, and the coefficient n determined graphically; α was calculated from the expression for n.

As in the case of earlier investigators, Ciborowski and Roszak found a direct relationship between α and particle size (α = 2, 5.2 and 13 for d = 0.21, 0.59 and 1 mm). Like Wamsley and Johanson, they found no clear relationship between the effective heat transfer coefficient and the superficial fluid velocity under the conditions of their experiments. The reason for this is not apparent.

Komarovsky and Verteshev [897] studied particle mass transfer in stationary and fluidized beds in dissolution processes. They obtained the following relationship for mass transfer in a fluidized bed:

$$\text{Nu}_m = b \text{Re}_{\text{f}}^{0.252} \cdot \text{Ar}^{0.254} \text{Pr}_m^{0.63}, \tag{8-48}$$

where Re = Reynolds number;

Ar = Archimedes number;

Nu_m, Pr_m = Nusselt and Prandtl number for mass transfer (Sherwood and Schmidt numbers—translator);

b = coefficient depending on particle shape.

The particle size is used as the characteristic dimension, and the superficial fluid velocity as the characteristic velocity.

Equation (8-48) is valid over the ranges Re = 117-900 and Pr_m = 750-5400. Experiments were carried out with the systems NaCl, KNO_3, $CuSO_4 \cdot 5H_2O$-water and β-naphtol-water.

Richardson and Ayers [1002] determined particle heat transfer coefficients in shallow fluidized beds (H = 10 mm) of rectangular cross section (100 × 50 mm) under quasi-steady-state conditions: material was continuously supplied and removed with an unchanged initial temperature of the inlet gas and material. Glass and lead spheres were used as the fluidized materials, as well as petunia seeds. The fluidizing media were air and carbon dioxide. The characteristics of the particles and gases and the experimental conditions are shown in Table 8-3. The effective heat transfer coefficients were calculated from the change in the temperature difference over the height of the bed and the concept of uniform plug flow of gas in the bed:

$$\alpha = \frac{q}{S\int_0^H \Delta t \, dH},$$

where q = quantity of heat supplied by the gaseous heat transfer medium, referred to unit time and unit cross-sectional area of the bed, kcal/m^2 · hr;

Δt = variable value of the temperature driving force, °C.

Table 8-3

Phase Characteristics and Experimental Conditions of Richardson and Ayers [1002].

Phase characteristics and experimental conditions	Minimum value	Maximum value
Particle Characteristics		
Diameter, mm	0.114	0.55
Thermal conductivity, kcal/m · hr · degree (C)	0.74	29.8
Temperature conductivity, m²/hr	0.00186	0.088
Heat capacity, kcal/kg · degree (C)	0.03	0.44
Bulk weight, kg/m³	1120	11,200
Gas Characteristics		
Thermal conductivity, kcal/m · hr · degree (C)	0.015	0.025
Temperature conductivity, m²/hr	0.045	0.102
Heat capacity, kcal/kg · degree (C)	0.21	0.24
Specific weight, kg/m³	1.02	1.76
Viscosity, kg-mass/m · hr	0.058	0.074
Experimental Conditions		
Feed rate of solid particles, kg/hr	1.3	11.8
Weight flow rate of gas, kg/m² · hr	170	2360
Bed voidage	0.50	0.64
Reynolds number	0.53	34.4
Nusselt number	0.023	4.4

The quantity of heat transferred was determined from the gas flow rate and its change in temperature in passing through the bed.

$\int_0^n \Delta t dH$ was determined from the area between the temperature curve for the gas and the line of equilibrium temperature of the gas and material. The equilibrium temperature was reached at distances less than 2.5 mm from the cloth gas-distributing screen. The reading of a thermocouple with a bare hot junction submerged in the bed was taken as the gas temperature. The temperature driving force was possibly reduced in this way, with a corresponding overestimation of the effective heat transfer coefficient. For describing their data (to an accuracy of ± 25%), Richardson and Ayers proposed the equation

$$\mathrm{Nu} = 0.054\,\mathrm{Re}^{1.28}, \tag{8-49}$$

where the temperature driving force was determined in this manner. The heat transfer coefficients were independent of the feed rate and thermal properties of the material. The effective value of Nu calculated by Richardson and Ayers appeared to be much higher at a given Reynolds number than the values given by other investigators (see the combined graph, Fig. 8-11). Apart from the method of measuring the gas temperature, this might be explained by the low bed height; with a good distributor, the gas would penetrate a shallow bed more uniformly than a deep bed. It is known that the gas distribution becomes less uniform in the upper parts of a deep fluidized bed. However, according to the data of Richardson and Ayers, the effective value of the Nusselt number fell to 0.023, which is two orders of magnitude smaller than the minimum Nu for conduction. Richardson and Ayers proposed a new but questionable explanation for why $\mathrm{Nu} \ll 2$ in fluidized beds. They assume that particles arriving in the shallow zone of active heating (of the order of 2.5 mm from the screen) carry with them a gaseous envelope into which heat begins to penetrate by conduction. After the particles leave the heating zone, the heat in the gas envelope spreads both towards the particle and in the opposite direction—into the main mass of the gas. Overall, only a small amount of heat passes to the particles by penetrating the envelope. However, it is unrealistic to assume that particles can move about in a fluidized bed without renewing the gas in their "envelopes," particularly at the relatively small voidages (ϵ = 0.5-0.64) prevailing in the experiments of Richardson and Ayers; where then would the main gas stream pass in the bed? It is preferable to assume alternate heating and cooling of the particles or their surface layers. It is interesting to note that the low effective values of Nu, much smaller than two, are also observed in stationary dense beds, where there is no particle motion.

Recently, Syromyatnikov developed [1080] a new method for determining heat transfer coefficients between solid particles and the fluidizing agent (a liquid). He proposed carrying out measurements under steady heat transfer conditions, compensating continuously by positive internal heat sources (high-frequency heating) for the heat lost from the particles. The quantity of heat lost by the particles was determined from the increase in temperature of the liquid, while the particle surface temperature was found from the rate of cooling during calorimetric measurements (the current to the inductor was turned on and simultaneously the liquid flow was cut off, after which the temperature of the stirred liquid was recorded continuously). As long as the thermal resistance of the particles was not large, the determination of the temperature was reduced to a heat balance for the period of cooling. Obviously, a uniform temperature for all the particles can only occur in steady-state heat transfer for beds containing particles of uniform size.

Ruckenstein and Teoreanu [1127 and 1128] generalized the experimental data of several investigators [234 and 440], and obtained the following equations for heat transfer between particles and medium in uniform fluidized beds:

$$\mathrm{Nu} = 0.426\mathrm{Re}^{0.30} \cdot \mathrm{Pr}^{1/3}\mathrm{Ar}^{0.17} \quad \text{for } \mathrm{Re}\cdot\mathrm{Ar} < 2.15; \tag{8-50}$$

$$\mathrm{Nu} = 0.943\mathrm{Re}^{-1} \cdot \mathrm{Pr}^{1/3} \cdot \mathrm{Ar}^{0.69} \quad \text{for } \mathrm{Re}\cdot\mathrm{Ar} > 2.15, \tag{8-51}$$

where Nu, Re, Pr, Ar are, respectively, the Nusselt, Reynolds, Prandtl and Archimedes numbers; the characteristic dimension in Nu, Re and Ar is the particle diameter, and the characteristic velocity in Re is the superficial fluid velocity. These expressions are also valid for mass transfer, but Nu and Pr must be replaced by the mass transfer Nusselt and Prandtl numbers [Sherwood and Schmidt numbers].

In nonhomogeneous fluidized beds, Ruckenstein and Teoreanu assumed that the fraction of the gas passing in the form of bubbles, $(1-\xi)$ takes no part in heat and mass transfer with the material; thereby they obtained the following ratio between the effective, α_{eff}, and actual, α, heat transfer coefficients from the particles in a nonhomogeneous fluidized bed:

$$\frac{\alpha_{\mathrm{eff}}}{\alpha} = \frac{cQ/(\alpha F)}{cQ/(\alpha F) - (1-\xi)/\xi}, \tag{8-52}$$

where $cQ = W_G$ is the water equivalent of the gas;
F is the surface area of the material in the bed;
ξ is the fraction of gas passing through the discontinuous phase;

Frequently, $Qc/\alpha F \ll (1-\xi)/\xi$, and

$$\alpha_{\mathrm{eff}} \approx \frac{W_G}{F} \cdot \frac{\xi}{1-\xi}, \tag{8-53}$$

i.e., the effective heat transfer coefficient is quite independent of the true value.

Ruckenstein and Teoreanu's assumption that all of the gas passing in the form of bubbles takes no part in the heat transfer is essentially incorrect, and corresponds only to the initial oversimplified concepts of the two-phase theory of fluidization (Toomey). Hence Eqs. (8-52) and (8-53) cannot be regarded as completely valid; they apply only to the limiting case when there is no exchange of gas between the continuous and discontinuous phases. In this case the effective value,

α_{eff}, falls off particularly sharply with increase in the overall surface area, F, of the material in the bed, and is independent of the true heat transfer coefficient.

Sunkoori and Kaparthi [1131] determined particle heat transfer coefficients for granite and quartz in beds of closely sized particles fluidized by water in a 82.5 mm diameter tube with unsteady-state cooling of the particles, which were initially heated to 90-99 °C. The weight flow rates of water varied from 16.2 to 27.1 kg/m^2 · sec, the bed height from 21 to 46 cm, and the particle size from 0.5 to 1.68 mm. The particle heat transfer coefficients, α_{eff}, varied from 550 to 3020 kcal/m^2 · hr · degree, increasing as the particle diameter and superficial fluid velocity increased. The results, when compared with the data of [389, 584, and 585] on heat transfer coefficients during gas fluidization, illustrated the unfavorable effect of gas breakthrough. These authors' data are described by the dimensionless equation

$$\mathrm{Nu} = 0.00391\, \mathrm{Re}^{2.1}. \tag{8-54}$$

For very small particles the effective Nusselt number is reduced to Nu = 0.8, which obviously < 2 indicates some nonuniformity of fluidization, though much less than in the case of gas fluidization. The considerable scatter in the experimental points on Sunkoori and Kaparthi's graph of $\mathrm{Nu} = f(\mathrm{Re})$ is obviously not related to changes in the uniformity of fluidization in different experiments, but rather to experimental inaccuracies; for instance, to the initial disordered and poorly reproducible cooling of the particles during the charging period. It is also possible that there is an effect due to differences in the bed voidage at the same value of Re in the different experiments (different particle diameters and bulk weights). The smallest voidage should correspond to the largest true conductive heat transfer coefficient.

Luikov [911] evaluated the experimental data of Fedorov and additional data on the drying of superphosphate and wheat, and proposed the dimensionless relationships:

$$\mathrm{Nu} = 1.6 \cdot 10^{-3} \mathrm{Re}^{0.95} (w_f / w_{ff})^{-1.5} \tag{8-55}$$

and

$$\mathrm{Nu} \approx 0.024\, \mathrm{Re}^{0.84}, \tag{8-56}$$

where $\mathrm{Nu} = \alpha d_e / \lambda$ is the Nusselt number;

$\mathrm{Re} = (w_f / m_0)\, d_e / \nu$ is a modified Reynolds number;

d_e is the equivalent particle diameter $\sqrt[3]{6P/\pi n \gamma_M}$, as given by Fedorov, (p. 184).

Hence, Luikov uses in the Reynolds number the mean-true gas velocity in a slightly expanded bed instead of the superficial fluid velocity, since this represents more closely the conditions of flow around the particles. Equations (8-55) and (8-56) were not suitable for describing the results from drying wheat.

Luikov provided a table of calculated values of the voidages for a number of materials. However, the wide range given for m_0 (Table 8-4) for the same material leads to some uncertainty in calculations using the tabulated data.

Table 8-4

Design Values of the Bed Voidage [911].

Material and nature of particles	m_0	Material and nature of particles	m_0
Uniformly sized spheres	0.35-0.42	Cylinders, 12.7 × 12.7 mm . . .	0.36-0.41
Silica gel powder (2-80 microns)	0.37-0.49	Cubes	0.34-0.45
Sand (coarse)	0.36-0.44	Coal	0.41
Sand (fine)	0.40-0.48	Grain (wheat)	0.42-0.46
Cylinders, 9.5 × 19 mm	0.43-0.48	Granulated carbon	0.44

Donadieu [1110] investigated heat transfer between particles and air in a fluidized bed with unsteady-state heating or cooling of the bed. Assuming an exponential change of the gas temperature along the bed at each moment, he found

$$\frac{\mathrm{St}(1-m)}{V/F}=2.3\,\frac{\log(t_{\mathrm{in}}-\vartheta)-\log(t-\vartheta)}{h}, \tag{8-57}$$

where St = $\alpha/G_r c_r$;

V and F are, respectively, the volume and surface area of the particles;

t_{in} and t are, respectively, the gas temperatures at the bed inlet and at a distance h from the bottom of the bed (variable);

ϑ is the material temperature, which is assumed to be the same throughout the bed and equal to the gas temperature at the bed outlet.

A bed of glass spheres (d = 430 microns, containing only 10% of particles smaller than 350 microns and 10% larger than 570 microns) was fluidized in a double-walled tube of 200 mm internal diameter. A vacuum was drawn between the walls (residual pressure only a few mm Hg) to reduce heat losses. A porous steel plate 3 mm thick was used as the gas distributor; at the operating fluid velocities, it had a resistance well over ten $\mathrm{kg_f/m^2}$. The bed height was of the order of 430 mm.

In the course of the experiment the temperatures at a number of points in the bed (h = 0, 0.5, 1, 1.7, 2, 4.7 mm) were recorded simultaneously using thermocouples with bare hot junctions; it was assumed that these recorded the air temperatures.

Donnadieu obtained the dimensionless relationship

$$\mathrm{Nu}=0.0087\,\mathrm{Re}^{2.3} \tag{8-58}$$

for the narrow range Re = 6-10.

8.4 Comparison of Experimental Data

The data of several investigators of particle heat transfer in fluidized beds are plotted in terms of log (Nu) and log (Re) in Fig. 8-11; the experimental points by Ciborowski and Roszak [951] and the correlations by Richardson and Ayers [1002], Sunkoori and Kaparthi [1131] and Chu [235] have been added to the graph given by Sharlovskaya [952].

The graph is obviously inaccurate, and can be used only for general orientation, due to the poor agreement among the data. In particular, the data of Kettenring et al. [389] and Richardson and Ayers [1002] involve the calculation of the gas temperature from the readings of thermocouples with bare hot junctions, while Ciborowski and Roszak [951] and Shakhova [178] measured this temperature with thermocouples whose hot junctions were shielded by screens.

Nevertheless, Fig. 8-11 illustrates a number of the statements made earlier concerning heat transfer between particles and medium. In the case of many investigators (Shakhova, Wamsley and Johanson, Fedorov, Ciborowski and Roszak, Sharlovskaya, and others) when the experimental values of Nu were found to be lower than 2.0, then effective rather than true values of α and Nu were being determined. As shown earlier, this is explained principally by nonuniform gas distribution in the fluidized bed, since all of the methods used for determining the temperature difference depended on the concept of uniform gas distribution. In aggregates, where the weight of material is concentrated, and where only the fraction of gas necessary for minimum fluidization flows according to the two-phase theory of fluidization, there will be an excess of the quantity $\alpha F/W_G$ and an extremely low mean-temperature difference. Actually there will be some increase in this difference due to exchange of gas between the bubbles and aggregates, and, hence, a repeated flow of hot gas into the dense phase. To save space in the formulation only one case will be considered and referred to hereafter—the heating of a material by a gaseous heat transfer medium. The conclusions are unchanged if the material is cooled instead. The mean degree of nonuniformity of the bed and the intensity of gas transfer between the phases are quite different under the experimental conditions of various investigators. This necessarily means that there must be a large scatter in the experimental data on heat and mass transfer. Aggregation in a fluidized bed and the intensity of gas exchange depend on a number of factors, including the ratio of the densities of the material and medium, the fluidization index, the bed height, and the type of gas distributor, but up to now nothing quantitative is known of these relationships. As a result it is still not possible to give a fundamental generalized correlation for the mean-temperature difference or the effective heat transfer coefficient in a fluidized bed.

As can be seen from Fig. 8-11, the poorest heat transfer was obtained (with other conditions constant) in those experiments where there was well-marked aggregation. In Fig. 8-11 very poor heat transfer is demonstrated by the experiments of Wamsley and Johanson [585] and Ciborowski and Roszak [951]. At the same time, if it is conjectured that Nu is independent of Re, i.e., that the heat transfer coefficient is independent of the superficial fluid velocity for particles of a given diameter, there would be a more sharply marked aggregation in these experiments. Actually, the constancy of the particle heat transfer coefficients can be explained by flow around the particles in the aggregates being equal to the minimum fluidization velocity regardless of the bed superficial fluid velocity. This does not rule out the possibility of gas exchange between the phases.

The effective Nusselt numbers are very small for very fine particles; these are many times smaller than the minimum value for conduction, Nu=2. This indicates that these fine particles have a very much lower intensity of heat transfer even when gas breaks through on a microscale which cannot be noticed visually or measured by probes; examples are fine bubbles and channels above the gas distributor. Because of the very large specific surface area of the fine particles it might be expected that thermal equilibrium would be attained within the small aggregates even during operation with very small temperature differences within short distances, smaller than the "mixing length"—the distance over which a new stream of hot gas from the continuous phase can exist in the discontinuous phase. At the same Re, the difference in heat transfer for fine particles in a fluidized bed and for ideal conditions must be greater than for coarser particles, because of the larger values of the fluidization index, since the fraction of gas passing in the form of the discontinuous phase is increased. These very low values of the Nusselt number, far less than two, are characteristic of fluidized beds of fine particles.

The complicated overall picture of heat transfer between a material and gas in the presence of gas exchange between the aggregates and bubbles can be reduced to some equivalent model, in which the gas flow through the nonuniform fluidized bed is regarded as a combination of "microbreak-throughs" of excess gas over the height of one or a few rows of particles with subsequent complete transverse mixing of all the gas after these rows, then further "microbreak-throughs," and so forth. In this scheme a small bed nonuniformity corresponds to ideal mixing after each row of particles (Fig. 8-12a), while a large degree of nonuniformity corresponds to mixing after several rows (Fig. 8-12b). It is assumed that in the aggregates themselves the superficial fluid velocity is close to the minimum fluidization velocity.

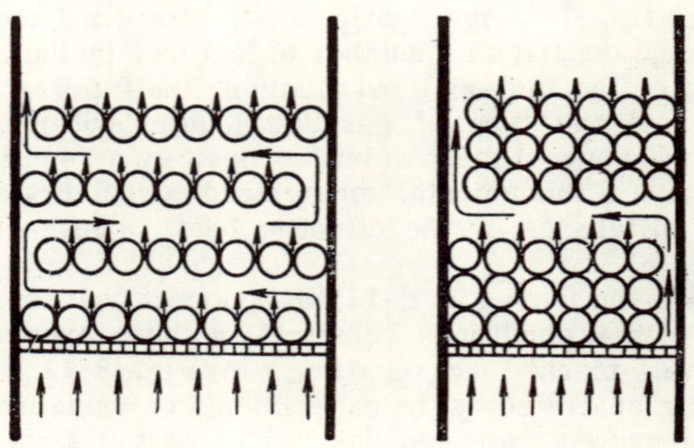

a b

FIG. 8-12. Mixing with particles for various degrees of nonuniformity.

Consider the transfer of heat to one particle in the first row of an aggregate. For a cubic particle packing within the aggregate the volume of gas wetting the particle per hour is 3600 $w_{\ell.s} d^2$. Equating the quantities of heat lost by the gas and taken up by the particle,

$$\pi d^2 \alpha_{1.s'} (\Delta t)_{av} = 3\,600\, w_{1.s} d^2 c_G \gamma_G (t_0 - t_1),$$

or,

$$(\Delta t)_{\mathrm{av}} = A(t_0 - t_1). \tag{8-59}$$

where

$$A \equiv \frac{3.6 \cdot 10^3 c_G \gamma_G w_{1.s}}{\pi \alpha_{1.s}}.$$

Table 8-5 gives numerical values of A calculated for particles of various diameters ($\gamma_M = 2500\ \mathrm{kg/m^3}$) fluidized by air ($t = 100°\mathrm{C}$). Only the component of α due to conduction is taken into account. The table also shows the temperature difference after the first row ($t_0 - \vartheta$) as a fraction of the initial temperature difference ($t_1 - \vartheta$) if the temperature of the particles in the first row, ϑ, is almost constant ($\vartheta \approx \mathrm{const}$). This would occur during drying in the constant rate period, and more generally whenever there is sufficiently good mixing of the material in the bed.

Table 8-5

Values of the Coefficient A for Particles of Various Diameters, and Fraction of Temperature Difference After First Row With ϑ = const.

d, m	A	$\frac{t_1 - \vartheta}{t_0 - \vartheta}$
10^{-4}	$3.31 \cdot 10^{-3}$	0
$3 \cdot 10^{-4}$	$3.8 \cdot 10^{-2}$	$3.54 \cdot 10^{-12}$
$5 \cdot 10^{-4}$	0.117	$2 \cdot 10^{-4}$
10^{-3}	0.533	0.155

In this case

$$(\Delta t)_{\mathrm{av}} = \frac{t_0 - t_1}{2.3 \log \dfrac{t_0 - \vartheta}{t_1 - \vartheta}},$$

whence, using Eq. (8-59),

$$\frac{t_1 - \vartheta}{t_0 - \vartheta} = (10^{1/2.3A})^{-1}.$$

It follows from the data of Table 8-5 that the fraction of the temperature difference retained after the first row of particles in an aggregate is frequently extremely small. For particles of the order of 500 microns or smaller there is practically complete cooling (heating) of the gas by the aggregate to the temperature of the material. If α_{conv} and the unsteady nature of the heat transfer are taken into account, this conclusion can also be extended to much larger particles.

Consider the rate of change of the mean-gas temperature over the height of a nonhomogeneous fluidized bed. Applying the concepts of the two-phase theory of fluidization to the microbreak-through conditions, the excess quantity of gas passing in the form of the discontinuous phase is close to

$$V_{excess} = w_f F_{sec} - w_{1.s} F_{sec}. \tag{8-60}$$

Equation (8-60) is in good agreement with the other concepts of this theory (that the fluid flows through the aggregates with a velocity $w_{1.s}$) only if the fraction of the cross section occupied by the jets breaking through is negligibly small, i.e., if the velocity of the discontinuous phase is large compared with $w_{1.s}$. This assumption is valid for approximate calculations. The fraction of gas passing through the aggregate is then

$$V_{ag} = V - V_{excess} = w_f F_{sec} - V_{excess} = w_{1.s} F_{sec}. \tag{8-61}$$

Because of the rapid change in gas temperature in passing through the aggregate of particles in the heat exchanger, the problem arises as to what gas temperature condition (8-61) refers. Since the gas temperature is close to ϑ for a large part of the aggregate, there is good reason to evaluate $w_{1.s}$ and w_f at the temperature ϑ. Thus, the weight flow rates of excess gas and of gas flowing through the aggregate will be respectively

$$M_{excess} = (w_f - w_{1.s})\gamma_\vartheta F_{sec} \text{ and } M_{ag} = w_{1.s}\gamma_\vartheta F_{sec}.$$

According to this scheme of break-through (Fig. 8-12), a quantity M_{ag} of gas cooled to ϑ is mixed after the first row of particles with a quantity M_{excess} having the initial temperature t_0. This case is the one most favoring rapid reduction of the gas temperature in a nonhomogeneous bed (minimum aggregate height, complete transverse mixing of the gas after each row of particles).

It is found from a mixing balance, neglecting changes in the mass heat capacity, that

$$M_{excess} t_0 + M_{ag}\vartheta = (M_{excess} + M_{ag}) t_{1\,mixt}.$$

Hence it is easily shown that the mean-gas temperature after the first mixing stage is

$$t_{1\,mixt} = \frac{(N-1)t_0 + \vartheta}{N},$$

where $N = w / w_{1.s}$ is the fluidization index.

Similarly, for the nth mixing after the nth microscale break-through,

$$t_{n\,mixt} = \frac{(N-1)t_{n-1\,mixt} + \vartheta}{N}. \tag{8-62}$$

The manner in which the mean-gas temperature (t_{mixt}) changes over the height of a nonhomogeneous fluidized bed can be calculated by

(8-62) when N, t_0 and ϑ are given. Table 8-6 gives the results of such calculations for the cases when $t_0 = 400\,°C$, $\vartheta = 100\,°C$, and $N = 2$ and 20 (a fluidized bed can exist at $N = 20$ only at sufficiently small values of Ar).

Table 8-6

Calculated Mean-Gas Temperatures, °C, Over the Height of a Nonhomogeneous Fluidized Bed for Two Specific Bed Conditions.

Temperature	Value of N		Temperature	Value of N	
	2	20		2	20
ϑ	100	100	$t_{6\,mixt}$	104.7	321
$t_{0\,mixt}$	400	400	$t_{7\,mixt}$	102.3	310
$t_{1\,mixt}$	250	385	$t_{8\,mixt}$	101.2	299
$t_{2\,mixt}$	175	371	$t_{9\,mixt}$	100.6	289
$t_{3\,mixt}$	137.5	358	$t_{10\,mixt}$	100.3	280
$t_{4\,mixt}$	118.8	345	$t_{11\,mixt}$	100.15	271
$t_{5\,mixt}$	109.4	333	$t_{12\,mixt}$	100.07	263

The table shows that the fall in the mean temperature over the cross section is quite slow, particularly at the large value of the fluidization index; this is due to microscale break-through of gas, in spite of the intensive exchange of gas between the continuous and discontinuous phases (complete radial gas mixing after each row of particles). Numerical calculations according to the data of Table 8-6 show that even with $N = 2$ and high values of the true heat transfer coefficients ($Nu \approx 10$), extremely low values of Nu_{eff} can be expected for fine particles in a nonhomogeneous bed. As usual, Nu_{eff} is evaluated over a bed height in which the mean gas temperature reaches within 0.1°C of the equilibrium value. According to the table, this occurs at a height equal to 12 rows of particles.

For $d = 10^{-4}$ it is found that

$$Nu_{eff} = 3.88 \cdot 10^{-2}.$$

Thus, in this case Nu_{eff} is $10:(3.88 \times 10^{-2}) \approx 260$ times smaller than the true value of Nu.

Obviously, still smaller values of Nu_{eff} are obtained for larger values of the fluidization index, as can be seen from the slow approximation of the mean-gas temperature to the equilibrium value at $N = 20$ (see Table 8-6).

Heat transfer may be even less in real fluidized beds of fine particles where the aggregates are not one but several rows of particles high and where there is incomplete radial mixing of the gas after each row.

This approximate analysis of the effect of microscale gas breakthrough on heat transfer in a fluidized bed leads to the conclusion that there should be marked changes in the effective particle heat transfer coefficients not only when there is a definite coherence of the particles (due, for instance, to molecular forces), but also for visually imperceptible changes of aggregation, (due, for instance, to feeble static electrification). In this case the sizes and times of existence of the unstable aggregates can vary, and hence also the intensity of gas transfer between the continuous and discontinuous phases, or, in terms of the schematic model, the number of complete mixing stages. This explains to a large degree, if not entirely, the differences between the effective heat transfer coefficient data of various investigators.

It is appropriate to stress the importance of accumulating experimental data on effective heat transfer coefficients α_{eff} for particles in fluidized beds. Even though α_{eff} and Nu_{eff} are only empirical and are not directly related to the true values α and Nu, it is important to know these effective values. The ratio Nu/Nu_{eff} characterizes the degree of imperfection in the gas distribution, and is a measure of the potential for improving heat transfer. Calculations based on α_{eff} can only be carried out with any degree of accuracy for conditions similar to those at which α_{eff} (Nu_{eff}) was obtained, including the conditions of initial gas distribution, static electrification, etc. Otherwise, if, for instance, only the value of Re is known, the possibility of errors of two- to threefold must be expected between the calculated values of α_{eff} and those obtained in use, and large safety factors have to be allowed in the calculations.

Any means of improving heat transfer between particles and gas in fluidized beds must include some method for increasing the uniformity of fluidization, for instance, retarding the bed with horizontal and vertical screens, the action of powerful sonic or supersonic vibrations, vibration of the gas distributor grid or elements placed in the bed, and improvement of the uniformity of the initial gas distribution. Screens break up bubbles, break up and slow down aggregates, and make it possible to increase the gas velocity in the aggregates.

The effect of various types of oscillations and vibrations on heat transfer between the gas and bed are thought to be due to increased contact between the highly mobile particles and break-up of aggregates, or in other words, to the prevention of bubble growth. The intensification of heat transfer between particles and gas due to the relative velocity of flow around the particles in the presence of pulsations or to the unsteady period of boundary-layer (sublayer) formation are of secondary importance only.* Since the particles in the dense phase of a fluidized bed are extremely close together, the boundary layer would be unchanged for completely steady flow and heat transfer around the particles.

The important problem is to achieve uniform initial gas distribution. The problem of increasing the effective particle heat transfer coefficient becomes really acute only in the development of equipment with shallow beds, which have a promising future due to their small hydraulic resistance. However a thin bed falls entirely within the sphere of influence of the gas distributing screen. "Classical" fluidization nonhomogeneity, with large bubbles and dense aggregates,

---

*It is probable that this is also valid for suspended beds and equipment with jets of gas suspensions with turbulent or imposed pulsations, as long as the relative velocity does not exceed 10 m/sec. In some systems imposed pulsations increase the true particle concentration.

cannot develop in shallow beds. On the other hand, there is considerable danger that either through channels or discontinuous channels (of the microscale break-through type) can form in the case of poor grid construction. In this case, the feeble motion of the particles about the grid sets up a zone of overheating of the material, the zone of gas cooling is extended, and α_{eff} is reduced still further.

Alternate heating and cooling of each individual particle can also contribute to the reduction in the effective values of the heat transfer coefficient and Nusselt number in a fluidized bed. Particles escaping from the thin hot zone at the bottom of the bed give up part of their heat to the main gas stream, and hence the zone of active heat transfer is extended due to the vertical mixing.

It is interesting to note that retarded particle motion, and particularly the absence of motion in the lower part of a poorly fluidized bed, leads to a reduction in the effective heat transfer coefficient compared with the actual value of α. The calculation of effective heat transfer coefficients is based on the assumption of a uniform material temperature throughout the fluidized bed equal to the equilibrium base temperature. If there is really an appreciable overheating of the material in the lower part of the bed due to poor particle mixing, the calculated temperature difference will be increased, and the effective value of the heat transfer coefficient will be reduced. Significant overheating of the material was observed in the lower parts of Ciborowski and Roszak's apparatus [789].

Both complications connected with particle movement obviously become unimportant in investigations in which it is ensured that the particle temperature is constant (drying experiments in the constant rate period).

The data of Sunkoori and Kaparthi [1131] on particle heat transfer in beds fluidized by liquids fall in the region of larger Nusselt numbers than for beds fluidized by gases; this is due not only to the larger Prandtl numbers, but also to the greater homogeneity of fluidization. For homogeneous fluidization it can be expected that there will be a scatter of the points in the relationship $\mathrm{Nu}=f(\mathrm{Re})$ for particles of various diameters or different bulk weights due to the absence of geometrical similarity of the beds at the same Reynolds numbers. For example, at a given Reynolds number there will be a small superficial fluid velocity for particles of large dimensions, which means that the bed voidage will be small and there will be increased conductive and convective components of the heat transfer coefficient. In the experiments of Sunkoori and Kaparthi a scatter of the points was observed in terms of the coordinates log Nu, log Re, but there was only a slight tendency for the points for fine particles of light materials to deviate downwards from the generalized correlation. The line plotted from the generalized correlation of Chu et al. [235] falls unexpectedly high in Fig. 8-11, even above the line for a stationary dense bed; it is possible that this represents a generalization of the data for very shallow beds only [1249], in which there was no time for nonuniformity to develop.

From these remarks it can be concluded that even in principle there cannot be a good correlation of the type $\mathrm{Nu}=f(\mathrm{Re})$ for nonhomogeneous fluidized beds for the following reasons.

1. For the same superficial fluid velocity and the same fluidized bed, the height of the zone of active heat transfer depends on the type of gas distributing device, which largely determines the gas flow in the lower parts of the bed.

2. Hydrodynamic similarity does not exist for such fluidized beds at a given Re, in particular, for fluidized beds of fine particles the fluidization index and gas flow may be many times larger than for beds of coarser particles.

The correlation can only be partly improved by introducing a group of the type N (the fluidization index) into the dimensionless equation. It is obvious that in principle relationships of the type $Nu = f(Re, Pr, Ar, N)$ also cannot be exact for nonhomogeneous fluidized beds.

Only approximate heat transfer calculations can be carried out with the correlations proposed to date. For the present it is convenient to use the simplest correlations $Nu = f(Re)$ from the data of Fedorov, Sharlovskaya and Shakhova, for example:

$$Nu = 1.25 \cdot 10^{-3} Re^{1.46}. \qquad (8\text{-}63)$$

Since there is a five-fold deviation in some of the experimental data of Fedorov and Sharlovskaya, this correlation gives a 2.5-fold deviation from the extreme points. This formula provides a safety factor in the calculations compared with most of Fedorov's data. When a good gas distributing device is used this safety factor will obviously be superfluous.

A very important conclusion to be drawn from the consideration of the experimental data on heat and mass transfer from particles in fluidized beds is that the possibilities have not been exhausted for achieving a considerable intensification (sometimes by ten- or even a hundred-fold) in heat transfer by reducing the nonuniformity of fluidization in the lower parts of beds, by improving gas exchange between the continuous and discontinuous phases, and by eliminating excess height from fluidized beds. It is possible that much will be achieved by improving gas distributors, thus making it possible to operate with shallow fluidized beds, taking measures against aggregation of the material, etc. This points to the urgent necessity of carrying out new and more detailed investigations of heat and mass transfer between the particles and medium in fluidized beds.

8.5 Heat Transfer Calculations

1. Determination of bed height

Thermal equilibrium (equalization of the mean gas and particle temperatures) is attained in fluidized beds even at small distances above the bottom of the bed in spite of the small effective heat transfer coefficients; this is due to the large bulk concentration of relatively small particles. According to Fedorov, the gas leaving a fluidized bed reaches the material temperature even for comparatively large particles (d_e = 3 mm) at bed thicknesses corresponding to a load of 80 kg/m^2 on the grid. In the laboratory experiments of Richardson and

Ayers [1002] using fine particles ($d = 114$-550 microns), in which there was a continuous supply and discharge of material from the bed, thermal equilibrium was attained at a height of 2.5 mm above the support screen. For fluidized beds higher than 20-30 particle diameters, kinetic calculations of heat transfer between the medium and material are therefore unnecessary, and only a static balance is required, assuming the gas temperature leaving the fluidized bed is equal to the temperature of the material in the bed and excluding the case of poor or incomplete fluidization. The considerable bed heights in the construction of existing fluidized dryers are sometimes chosen mainly because it is easier to avoid aggregation of the material and the disturbances in fluidization which arise if at any point in the bed only moist feed material (which tends to agglomerate) accumulates. In a shallow bed it is more difficult to avoid this agglomeration, particularly with a poorly constructed feed device, and it is usual to increase the bed height so that the moist material falls onto a "cushion" of partly dried material. In the future it will be necessary to seek more economic means than increasing the height and hydraulic resistance of the bed.

If approximate kinetic heat transfer calculations are to be carried out, the effective heat transfer coefficient must be determined from Eq. (8-59), the heat transfer area is chosen as the total geometric surface area of the particles given by Eq. (8-7), and the material temperature ϑ_{bed} is assumed to be uniform throughout the bed and 2-4°C lower than the gas temperature leaving the bed, as recommended by Luikov [911].

The temperature difference is then calculated as

$$\Delta t_{\text{av}} = \frac{t_1 - \vartheta_{\text{bed}} - 3}{2.3 \log \dfrac{t_1 - \vartheta_{\text{bed}}}{3}}, \tag{8-64}$$

where t_1 is the temperature of the gaseous heat transfer medium before entering the bed.

In the case of drying in the constant rate period, the material temperature can be assumed equal to the adiabatic saturation temperature of the gas entering the bed, $\vartheta_{\text{bed}} = t_M$. No internal particle temperature gradient is usually assumed. Only for coarse particles of low thermal conductivity ($\text{Bi} \geqslant 1$) is it necessary to take into account the reduction in heat transfer, due to the presence of temperature gradients within the particles, by correcting the effective heat transfer coefficient for spherical particles by the factor $1/(1 + \text{Bi}/5)$, assuming as usual that the particle surface temperature is equal to the mean-particle temperature. Obviously, in this case it is necessary to find initially an approximate value of α in order to estimate the value of the Biot number ($\text{Bi} = \alpha d/\lambda_M$). Kitaev et al. [60], recommend a similar correction factor for calculating the heating of lumps of material in beds with $\text{Bi} \leqslant 10$.

In the approximate calculations the physical properties of the gaseous heat transfer agent must be those corresponding to the equilibrium temperature. In continuously operating fluidized beds (i.e., with continuous feed and withdrawal of material from the bed) it seems that heat transfer calculations must be complicated by the differences in the residence times and temperatures of individual particles. However, Baskakov and Syromyatnikov [873] showed that for ideal mixing of the material in the fluidized bed these differences compensate one another, and in heat transfer calculations the residence time of the particles can be taken as equal to the mean time for which the particles would remain in the bed in the complete absence of mixing, and the particle temperature can be taken as the calorific mean value at the bed exit. It has also been shown [873] that for a material of nonuniform particle size, the weighted mean-particle diameter should be used in the calculations.

An example of an approximate calculation of heat transfer between a fluidized material and a gaseous heat transfer agent is given below.

Example. Calculate the initial height of a fluidized bed required for the case of simple heating of a dry material in order to utilize the heat from a gaseous heat transfer agent.

Given: Initial gas temperature $t_1 = 200°C$; acceptable temperature of discharge gas, 100°C; superficial gas velocity $w_f = 1.5$ m/sec (at $t = 100°C$); physical properties of the gas: $\nu_{100} = 23.1 \times 10^{-6}$ m²/sec; $\lambda_{100} = 2.76 \times 10^{-2}$ kcal/m · hr · degree (C); $\gamma_{100} = 0.95$ kg/m³; $c_{100} = 0.24$ kcal/kg · degree (C); particle diameter $d = 1$ mm, bed voidage $m_0 = 0.4$.

Solution. The calculation is based on 1 m² of bed cross section. The heating surface is

$$F = \frac{6(1-m_0)}{d} H_0 = \frac{6 \cdot 0.6 H_0}{10^{-3}} = 3\,600\, H_0 \text{ m}^2 .$$

The calculated temperature difference is

$$\Delta t_{av} = \frac{200 - 100 - 3}{2.3 \log \dfrac{200-100}{3}} = 27{,}7° \text{ C};$$

$$\text{Re} = \frac{1.5 \cdot 10^{-3}}{23.1 \cdot 10^{-6}} = 65.$$

According to Eq. (8-59)

$$\text{Nu} = 1.25 \cdot 10^{-3} \cdot 65^{1.46} = 0.551;$$

$$\alpha_{eff} = \frac{\text{Nu} \cdot \lambda_G}{d} = \frac{0.551 \cdot 2.76 \cdot 10^{-2}}{10^{-3}} = 15.2 \text{ kcal/m}^2 \cdot \text{hr} \cdot \text{degree}.$$

The quantity of heat transferred per hour is

$$Q = 3\,600 w_f \gamma c (t_1 - t_2) = 3\,600 \cdot 1.5 \cdot 0.95 \cdot 0.24 (200 - 100) = 123\,000 \text{ kcal/hr}.$$

However,

$$Q = \alpha F \Delta t_{av} = 15.2 \cdot 3\,600 H_0 \cdot 27.7,$$

whence,

$$H_0 = 0.081 \text{ m}.$$

Thus, the bed height is equal to 81 particle diameters, i.e., there is a large safety factor compared with the data of Fedorov, and even more compared with those of Richardson and Ayers. In order to carry out more fundamental calculations it would be necessary to do much experimental work on determining the relationship between α_{eff} and the gas distributing devices which are suitable for industrial scale equipment.

2. The number of stages

When high-temperature processes (for instance, certain chemical reactions) are carried out in fluidized beds, the problem arises of how to utilize the physical heat in the gas leaving the bed. Fluidized beds (one or several) can be used for this purpose by passing feed solids countercurrent to the efficient hot gases through one or more beds. The gas and material pass through the beds in series in opposite directions (stagewise countercurrent flow).

One bed in addition to the main fluidized bed is sufficient when the material being treated is able to take up all of the heat intended for recovery without being heated to a temperature exceeding the economically attainable exit gas temperature. This will be the case, for instance, if the material being treated has a high moisture content.

If the material is dry, these conditions are usually impossible to attain, and it is necessary to arrange for several additional fluid beds. Dement'yev (891] proposed a graphico-analytical method for determining the temperature of the ith bed when the total number of beds n is given; this method was used as an example to calculate a multibed furnace for calcining limestone. Dement'yev's calculations showed that in the problem which he considered, the utilization of 93% of the heat in the gas at 1000°C (for preheating the limestone) could be achieved in three additional beds. The increase in hydraulic resistance and overall height of the equipment argue against increasing the number of beds further.

Neglecting changes in the ratio of the weight heat capacities of the gas and material in the heat exchanging bed, and assuming that the temperatures of the gas and material are the same at the bed outlet (see Fig. 8-13), Tamarin [1253] found that the material temperature at the outlet of the nth stage of a multilayered multistage countercurrent heat exchanger can be approximated by:

$$t_{n-1} = t_n - \left[t_n + \theta_0 \frac{1+k^n}{(1-m)k^n}\right]\left[k^n + \frac{1+k^n}{1-k}\right] . \qquad (8\text{-}65)$$

The temperature of the exit gas is

$$t_0 = t_n + \theta_0\left[\frac{1-k^n}{(1-k)k^n}\right]\left[1 + \frac{1-k^n}{(1-k)k^n}\right] \qquad (8\text{-}66)$$

where $k = W_G/W_M$ is the ratio of the water equivalents (specific gravities) of the gas and material; the remaining notation will be clear from Fig. 8-13.

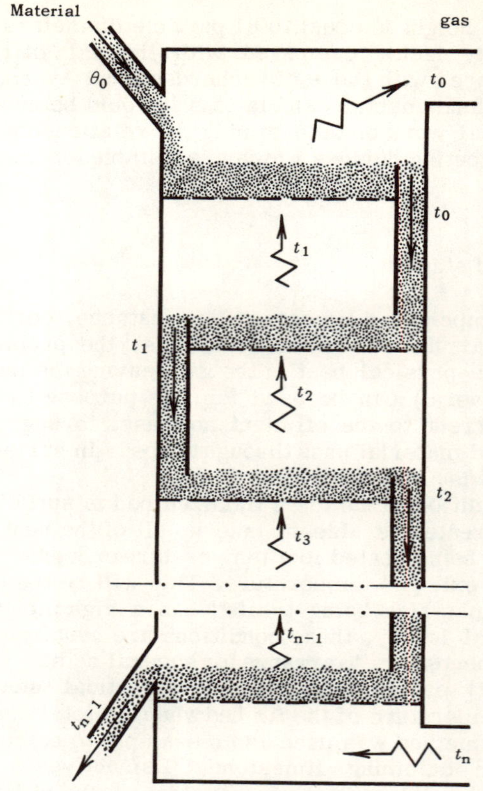

FIG. 8-13. Sketch of multistage stepwise countercurrent heat exchanger.

The ratio of the quantity of heat transferred in the equipment, using the given number of stages, to that which would be transferred in an apparatus with an infinite number of stages is

$$\eta = (t_{n-1} - \theta_0) / (t_n - \theta_0). \tag{8-67}$$

Using (8-65) and (8-66),

$$\eta = (k - k^{n+1}) / (1 - k^{n+1}). \tag{8-68}$$

The graph in Fig. 8-14 is constructed from Eq. (8-68).

Equation (8-67) and Fig. 8-14 can be used for calculating the number of stages required in a fluid-bed heat exchanger. For instance, if the material temperature (θ_0) and gas temperature (t_n) at the bed inlet are given as well as a desired material temperature at the outlet (t_{n-1}), then Eq. (8-67) can be used to determine the necessary value of η, and the corresponding number of stages, n, found from Fig. 8-14. This value of n is rounded to a whole number, and the corresponding exact value of η is found from the graph; the exact temperature of the material at the outlet is then found from Eq. (8-67).

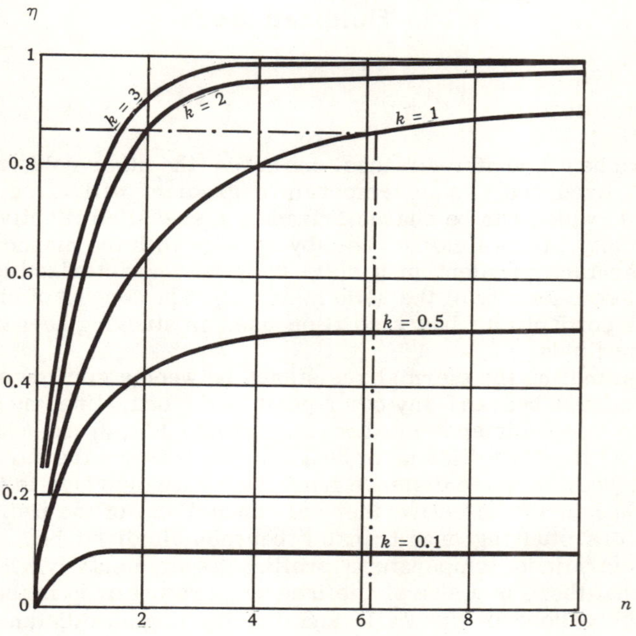

FIG. 8-14. η as a function of the number of stages in equipment, n, for ratios of the water equivalents of the gas and material k from 0.1 to 3.

Chapter 9

Effective Thermal Conductivities in Fluidized Beds

9.1 Introduction

If two heat transfer surfaces, one hot and the other cold, are placed in a fluidized bed, some temperature gradient will arise in the bed material, which can be characterized by a so-called effective thermal conductivity of the fluidized bed (by analogy with the manner in which a temperature gradient in a solid body is characterized by the true thermal conductivity of the solid material). The concept of an effective thermal conductivity has also been used in studying heat transfer in turbulent fluids.

In estimating the thermal resistance between a surface submerged in a fluidized bed and any other point in the bed, it seems obvious to separate this resistance into two components: $1/\alpha_{\text{wall}}$ and $\delta/\lambda_{\text{eff}}$, where α_{wall} is a "film" coefficient for heat transfer between the surface (wall) and the bed; δ is the distance from the heating surface to the chosen point; λ_{eff} is the effective thermal conductivity of the bed. Many investigators studying heat transfer between fluidized beds and walls have determined temperature profiles experimentally. It has been shown that there is a sharply defined temperature drop in the immediate neighborhood of the wall, and that starting at a distance of a few millimeters from the wall there is practically complete absence of a temperature gradient (Fig. 9-1) (Wicke and Fetting). This led to the view that the effective thermal conductivity of a fluidized bed is always very large, so that $\delta/\lambda_{\text{eff}}$ is negligibly small compared with $1/\alpha_{\text{wall}}$. In the large majority of investigations of heat transfer between walls and fluidized beds the film coefficient α_{wall} is therefore identified with the overall coefficient K from the wall to the bulk of the bed. This is permissible from approximate theoretical considerations of heat transfer between walls and a fluidized bed (cf. Chapter 10).

However, the experiments of Bondareva and Todes [1141] have shown that in some cases λ_{eff} for a fluidized bed falls to a quantity of the order of 10 kcal/m·hr·degree (°C). This illustrates the importance of studying the dependence of λ_{eff} on various bed properties. It is noted in [1141] that it is particularly convenient to use fluidized bed methods for carrying out chemical reactions which have large thermal effects and which are temperature sensitive. In such cases large temperature gradients within the bed are especially undesirable, and it is necessary to provide conditions under which the effective thermal conductivity of the bed is sufficiently large.

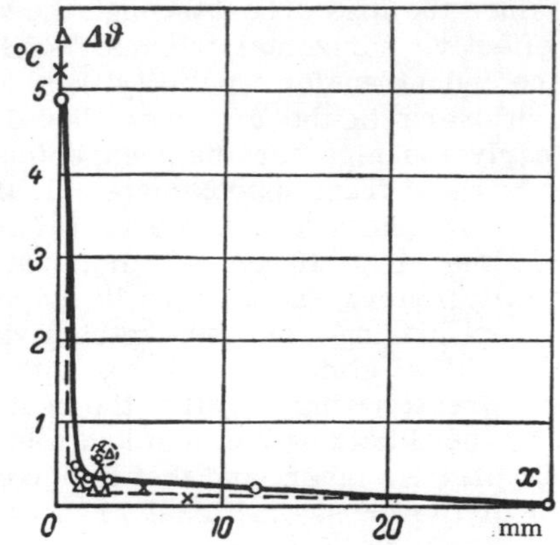

FIG. 9-1. Radial temperature profile in a fluidized bed according to Wicke and Fetting [600].
o—carborundum, $d = 0.065$ mm; △—sand, d = 0.315 mm; ×—glass spheres, $d = 0.9$ mm. The points joined by the dashed line correspond to the temperatures at a height 3-4 mm above the heater; $\Delta\vartheta$ is the temperature difference between the given point and the core of the bed, °C; x is the horizontal distance from the heater, mm.

9.2 Review of Investigations

There are relatively few published papers dealing with the effective thermal conductivities of fluidized beds [495, 662, 728, 892, 1024, 1044, 1111, 1141].

In 1953, Peters, Orlichek and Schmidt [495] published a paper in which they attempted to calculate the effective thermal conductivity of fluidized beds from experimentally determined temperature profiles. An appreciable temperature gradient acting in only one direction was established in their apparatus (one-dimensional problem). The equipment consisted of a rectangular glass cell 65 mm wide, 450 mm long, and 480 mm high, partly filled with sand (d = 0.23 mm). Fluidizing air was passed into the bottom of the vessel through a horizontal tube having 1-mm diameter holes drilled in its upper half. An electric heater with wire coils was used as the heat source; the coils were arranged so that heat was liberated quite uniformly over the bed cross section. The heater was placed at the center of the vessel length and the vessel thermally insulated on all sides, so that the heat loss through the walls was less than 7%.

The temperature distributions in the bed under steady-state conditions were used to calculate effected bed thermal conductivities in the horizontal direction. Steady state was reached 1.5-2.0 hours after switching on the heater and starting the blower.

Peters, Orlichek and Schmidt set up an equation taking into account the heat carried away by the fluidizing gas. In their apparatus this amounted to about 93% of the heat liberated by the heater. There were no special coolers, since the loss of heat through the walls was small. The values of the effective horizontal thermal conductivities (or in their terminology, the "heat transfer capability") calculated by Peters, Orlichek and Schmidt were of the order of 1000-1700 kcal/m·hr·degree, and are clearly too high for their experimental conditions. This is easily demonstrated from approximate calculations using the authors' raw data. The electric heater achieved 137 kcal/hr, the temperature gradient in the first 10 cm of distance from the heater amounted to 220 degrees/meter. The authors did not indicate the depth of the fluidized bed, but judging from the sketch given in their paper [495], the vessel was filled almost to the top with sand. Assuming (with a bias towards overestimating λ_{eff}) that the heat from the heater was not carried off by the fluidizing medium and was distributed only horizontally over the particle layer, and that the cross section of the latter was only 1/5 of the cross-sectional area of the vessel, then

$$F = \frac{1}{5} 0.065 \times 0.48 = 0.00624 \text{ m}^2$$

and the effective thermal conductivity is

$$\lambda_{eff} = \left| \frac{Q}{F \cdot \nabla t} \right| = \frac{137}{0.00624 \times 220} = 100 \text{ kcal/m} \cdot \text{hr} \cdot \text{degree}$$

which is an order of magnitude lower than the values calculated by Peters, Orlichek and Schmidt.

Bondareva [662 and 728] made a more successful attempt to determine numerical values by assuming that it was possible to neglect the quantity of heat carried off by the fluidizing agent. This is possible only in special cases. Bondareva's experimental apparatus consisted of a tube 82 mm i.d. and 500 mm long having a water-jacket cooler. An electric heater 200 mm high and 10 mm in diameter was placed axially inside the tube. The fluidized bed temperature was measured radially 100 mm above the screen supporting the bed, using four thermocouples which were positioned so that the first was 4 mm from the jacket of the heater and the last was 8 mm from the jacket.

Bondareva evaluated λ_{eff} and also the heat transfer coefficients α_1 from the heater surface to the bed and α_2 from the bed to the outer wall (cooler) from the formulas:

$$Q = \alpha_1 2\pi r h \Delta t_1 = -\lambda_{eff} 2\pi h \frac{\Delta t_2}{\Delta \ln r} = \alpha_2 2\pi R h \Delta t_3, \qquad (9\text{-}1)$$

where Q is the quantity of heat liberated by the heater;
r and R are the radii of the heater and cooler;
Δt_1 and Δt_3 are the temperature differences at the hot-wall – sand and sand – cold-wall boundaries;
Δt_2 is the temperature difference in the sand along a radius;
h is the height of the fluidized bed.

Bondareva therefore assumed that the radial heat flux passing from the heater to the cooler did not vary along the radius and was therefore not decreased by the fluid flowing in the bed. This assumption necessarily leads to some degree of overestimation in the calculated values of λ_{eff}. According to Bondareva's experimental data, the effective thermal conductivity of a fluidized bed depends on the pulsation velocity of the particles, and passes through a maximum as the superficial fluid velocity is varied. Bondareva related the decrease in λ_{eff} beyond the maximum to the decreasing number of particles per unit volume of the bed as it expands further. According to Bondareva, the numerical value of the maximum effective thermal conductivity for sand with a particle diameter of 100 microns was 170 kcal/m • hr • degree (°C), while beyond the maximum value, λ_{eff} fell to 10-20 kcal/m. hr • degree (°C), see Fig. 9-2.

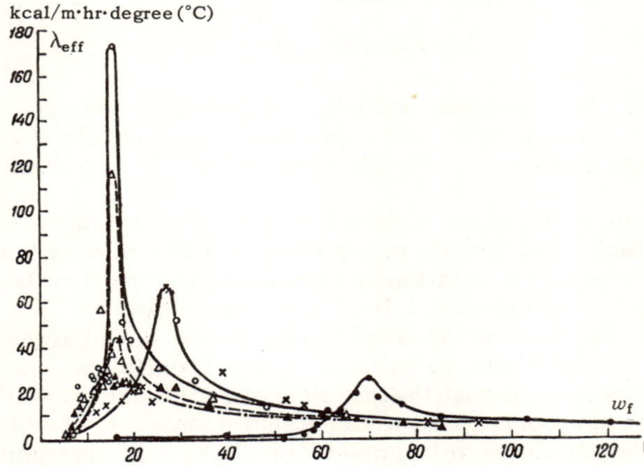

FIG. 9-2. Effective thermal conductivities in a fluidized bed of sand according to the data of A. K. Bondareva [635].
○—d = 100 microns; △—d = 153 microns; ▲—d = 233 microns; ×—d = 315 microns; ●—d = 477 microns.

The numerical values of λ_{eff} are obviously of the right order of magnitude at the maximum (10 to 100 kcal/m • hr • degree (°C)), remembering that these refer to the effective thermal conductivity of a bed in the horizontal direction. The effective bed thermal conductivity in the direction of the fluidizing stream must be several times larger due to the anisotropy of particle mixing; this was also recognized by Peters, Orlichek and Schmidt [495]. Hence the values of λ_{eff} for a fluidized bed are not the same in all directions. However, the value of λ_{eff} in a given direction and for a given bed and fluidizing agent at a given superficial fluid velocity can be regarded as a constant quantity, and is the same for the upper and lower parts of the bed. However, it can be expected that there will be differences in the local values of λ_{eff} and its dependence on the geometrical details of the apparatus, at least in the lower zones of the bed.

Unfortunately, there is some doubt regarding the validity of Bondareva's BSSR data for $\lambda_{(eff.max)}$. Measurements subsequently carried out at the Heat and Mass Transfer Institute by Borodula and Tamarin showed no maximum of λ_{eff} over the range of values of the fluidization

fluidization index occurring in Bondareva's experiments. In this range the effective thermal conductivity of the bed increased monotonically. The fall in λ_{eff} noted by Bondareva can be explained by defects in the experimental techniques; she measured the temperature profile at a height of 10 cm above the screen, while the total height of the expanded fluidized bed was held constant (20 cm) by overflow. As a result, so little material remained in the bed at large values of the fluidization index that the point of temperature measurement was above the level of the dense phase of the bed. Hence there appeared to be a reduction in the value of λ_{eff}.

By setting up an analogy between the effective thermal conductivity of a fluidized bed and the turbulent thermal conductivity of a fluid, Bondareva proposed the relationship

$$\lambda_{eff} = \psi c \rho_{bed} U R, \qquad (9\text{-}2)$$

where U is the pulsation velocity of the particles; ρ_{bed} is the bed density; c is the specific heat capacity; R is the scale of the motion, which is assumed equal to the tube radius; ψ is a proportionality coefficient.

The formula indicates a direct relationship between λ_{eff} and the particle velocity and density of the bed and substantiates the presence of a maximum in the relationship between the effective thermal conductivity and the superficial fluid velocity. However, the analogies made by Bondareva and also by Todes [1024] with the thermal conductivities of continuous media have important deficiencies. In the first place, the assumption that the turbulence of a fluidized bed is isotropic clearly does not correspond to the actual properties of fluidized beds (in which the particles mix anisotropically). It is also impossible to explain the small value and wide range of variation (10^{-2}–10^{-3}) of the coefficients ψ calculated from Bondareva's experimental data.

An alternate analogy between λ_{eff} for a fluidized bed and the molecular thermal conductivity of the gas can be reduced to a form [1044] similar to Eq. (9-2).

Both of these analogies are too primitive to be considered sound theoretical expressions for describing quantitatively the effective thermal conductivity of a nonhomogeneous fluidized bed. Recently, Bondareva and Todes [1141] published measured values of λ_{eff} for fluidized beds with a wide range of particle sizes in the bed. The measurements were again carried out in an 82-mm diameter tube with a rod-shaped electric heater 10 mm in diameter placed axially. A vanadium catalyst deposited on silica gel was used as the particulate material; Table 9-1 gives the screen analysis. The weight mean-particle diameter was approximately 100 microns. However, resistance measurements in the stationary bed gave an effective diameter of 37 microns, due to the large number of very fine particles. Under the experimental conditions the fine fractions were entrained from the main fluidized bed and formed an unusual "dilute phase" above it.

Table 9-1

Catalyst screen analysis [1141].

Particle size microns	Weight concentration, %
>250	0.5
250–150	22.0
150–100	27.3
100–75	13.7
<75	36.5

A metal cap with openings covered by flannel filters was placed over the upper end of the tube to prevent carry-over of the fine particles. A state of dynamic equilibrium was established between the quantity of powder attached to the inner surface of the filter and the amount carried back by the fountain-like discharge. Under these conditions λ_{eff} values up to 15 kcal/m·hr·degree were obtained. The values of λ_{eff} for a material of wide particle size distribution appeared to be much lower than for the sand having a uniform particle size approximately equal to the mean-particle diameter of the mixture.

However, from the data given in the paper it is difficult to determine the actual size analysis of the material in the zone where the effective thermal conductivity was measured. The experimental conditions were very far removed from the ordinary operating conditions of fluidized beds of wide particle size distributions.

Donnadieu [1111] obtained quantitative data which made it possible to estimate the value of the effective longitudinal thermal conductivity, $\lambda_{eff.long}$, of a fluidized bed. He determined the intensity of vertical particle mixing from the experimental values of the effective thermal conductivities of the bed. Donnadieu assumed that the equalization of temperature in a fluidized bed was caused by the transfer of heat by the solid particles; transfer by the gas was unimportant. Neglecting the temperature gradient within the particles he found

$$\frac{\partial^2\vartheta}{\partial h^2}=\frac{1-m}{A_M}\cdot\frac{\partial\vartheta}{\partial\tau}, \tag{9-3}$$

where ϑ = material temperature;
h = distance from bottom of bed;
τ = time;
$(1-m)/A_M$ = effective temperature conductivity or thermal diffusivity of the bed.

With the present assumptions,

$$A_M=\lambda_{eff.long}/c_M\gamma_M .$$

The experiments were carried out as follows: hot air was used for the preliminary heating of the stationary bed (without waiting for the temperatures of the material below the bed to equalize), and then the bed was rapidly fluidized with hot air at a lower temperature equal to the mean temperature of the preheated material ϑ_{av}, and the rate of temperature equalization over the height of the fluidized bed was recorded. At these air temperatures, Donnadieu neglected heat transfer between the air and material and made the apparently bold assumptions

$$\left(\frac{\partial\vartheta}{\partial h}\right)_{h=0}=0 \quad\text{and}\quad \left(\frac{\partial\vartheta}{\partial h}\right)_{h=H}=0.$$

Donnadieu found that

$$A_M\approx 0{,}23\,(1-m)\,H^2\,\frac{\log(\vartheta_1-\vartheta_{av})-\log(\vartheta_2-\vartheta_{av})}{\tau_2-\tau_1}, \tag{9-4}$$

where ϑ_1 and ϑ_2 are the temperatures of the material at a given point at times τ_1 and τ_2 after the beginning of fluidization.

The experiments were carried out in a 0.2-m diameter column with double walls between which a vacuum was produced to reduce heat exchange with the surrounding air. The particles were glass spheres of 130 microns diameter, heat capacity 0.19 kcal/kg · degree (°C) and a bulk weight of 2750 kg/m^2. The bed height was 0.31 m, and the temperature in the fluidized bed equalized within about 15 seconds. A value of A_M = 4.5 m^2/hr was obtained at a voidage m = 0.46.

The effects of particle diameter and air velocity on A_M were determined from special experiments with glass spheres of 130 and 800 microns at fluid flow rates from 180 to 3060 kg/m · hr. The following empirical relationship was obtained

$$\frac{A_M}{1-m}-\frac{A_{M.1.s}}{1-m_{1.s}}\approx 4.86\cdot 10^{-6}(G-G_{1.s})^2 \ \mathrm{m^2/hr}, \qquad (9\text{-}5)$$

where $m_{1.s.}$ and $G_{1.s.}$ are the bed voidage and fluid flow rate (kg/m^2 · hr) at the limit of stability; $A_{M.1.s}$ is the corresponding value or A_M obtained by extrapolation. The minimum fluidization rate was $G_{1.s}$ = 72 kg/m^2 · hr for the spheres of 130 microns diameter and 1008 kg/m^2 · hr for sheres of 800 microns diameter.

Thus the effective longitudinal temperature conductivity of a fluidized bed increases approximately proportionally to the square of the flow rate and hence decreases as the particle size is increased, since the minimum fluidization velocity increases.

The effective longitudinal thermal conductivity is

$$\lambda_{\mathrm{eff.long}} = A_M c_M \gamma_M \qquad (9\text{-}6)$$

or, from the empirical relationship (9-5)

$$\lambda_{\mathrm{eff.long}}=[4.86\cdot 10^{-6}(G-G_{1.s})^2+7.56](1-m)\,c_M\gamma_M. \qquad (9\text{-}7)$$

It is easy to calculate from (9-6) that the experimental value A_M = 4.5 m^2/hr given above for glass spheres corresponds to $\lambda_{(\mathrm{eff.long})}$ = 2350 kcal/m · hr · degree, which is several times larger than the thermal conductivity of silver.

More recently measurements of the vertical and horizontal effective temperature conductivities in free and restrained beds of sand fluidized with air at room temperature have been carried out at the Institute of Heat and Mass Transfer of the Academy of Sciences of the Belorussian SSR (Institut Teplo- i Massoobmena Akademiya Nauk BSSR) by Borodula and Tamarin. The experiments were carried out in two columns 82 and 1755 mm diameter. The temperature conductivity of the beds was studied by the method of instantaneous heat sources, which were set up by rapidly pouring a small portion of hot sand particles of the same size fraction into the bed. An approximately plane and instantaneous source of heat was thus set up at the upper boundary of the first bed in the column. The bed temperature at a fixed distance from the instantaneous heat source was recorded continuously during the experiment. The temperature conductivity of the fluidized bed was determined from the time required to reach a maximum temperature and the known distance from the heat source. Deviations of the actual heat source from an ideally planar and exactly instantaneous source made this method only approximate. However, it is more valid and convenient (rapid) than the method used by Donnadieu. Measurements have shown that the temperature conductivity of a fluidized bed in the vertical direction is often an order of magnitude larger than in a horizontal direction. The temperature conductivities in both directions increase with increasing air velocity. However, the vertical temperature conductivity increases more rapidly than that in the horizontal direction. Under the experimental conditions the vertical temperature conductivity of the bed increased appreciably with increase in the initial bed height H_0. H_0 did not exceed 175-200 mm, while in Donnadieu's it was 310 mm. This may explain the larger value obtained by Donnadieu for the longitudinal effective temperature conductivity.

The temperature conductivity of the bed in a horizontal plane varied only slightly as the initial bed height increased.

Figure 9-3 shows the experimental results for a sand size fraction of 0.794 mm diameter plotted as $A/(1-m)$ versus the fluidization index where A is the effective temperature conductivity of the fluidized bed. The experimental points fall satisfactorily along two solid curves; the upper curve corresponds to $H_0/D_T=2$ and the lower to $H_0/D_T=1$. The graph also shows the experimental points of Tamarin and Borodula for fluidized beds restrained by inserting horizontal wire screens. The baffles consisted of an assembly of screen disks 5×5 and 13×13 mm, with a wire thickness of 0.9-1.2 mm, placed 10-20 mm apart. The points for the experiments with the restraining baffles fell along a steeper curve (shown dashed). The experimental points for the baffled beds clearly tended towards the points for the same unbaffled bed as the fluidization index increased. Hence, in the region of large superficial fluid velocities (or more exactly, at large values of the fluidization index) the use of screens has little effect on the rate of vertical mixing of the solid phase in a fluidized bed.

The data given in Fig. 9-3, and particularly the temperature conductivities along the horizontal, A_h, for a free fluidized bed of small diameter undoubtedly need to be made more precise. The distance r between the level at which the "point" heat source was established and the location of the temperature measurement in the fluidized bed was small, and consequently so also was the time of the experiment, therefore the acceleration period for the charge of hot particles, which initially had no horizontal component of motion, could have an important effect. This obviously led to some lag in the propagation of the thermal wave and to values of the experimentally measured effective temperature conductivities which were smaller than would be obtained at distances further from the point of introducing the charge. This provides a basis for considering that the values of A_h obtained in this way provide a certain safety factor in calculating horizontal temperature gradients, since the actual gradients will be smaller. More precise values of A_h for freely fluidized beds will obviously be obtained from measurements by the method of instantaneous point sources with larger values of r which can be obtained in larger beds. However,

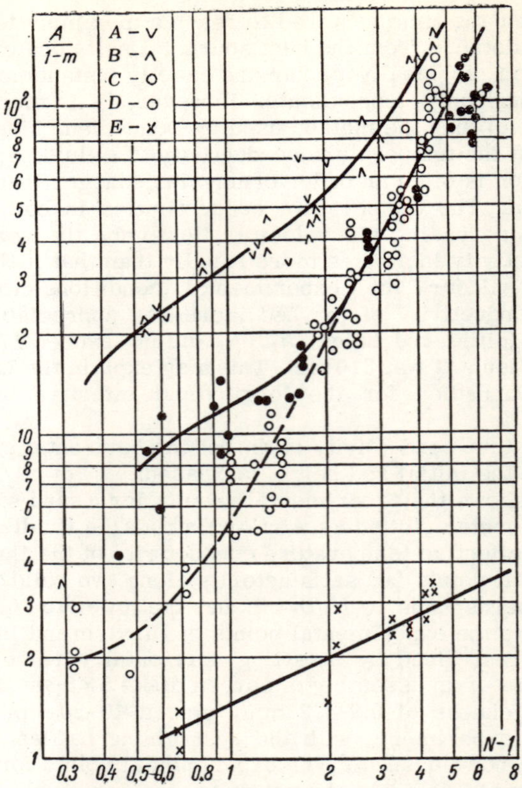

FIG. 9-3. Effective temperature conductivity of a fluidized bed as a function of the fluidization index. A, B—experiments of Borodula and Tamarin with sand beds (d = 0.794 mm) at H_0/D_T = 2 (temperature conductivity of bed along a vertical); C—the same, at H_0/D_T = 1; D—the same, for a bed restrained with screens at H_0/D_T = 1; E—horizontal temperature conductivity of fluidized sand bed, H_0/D_T = 1; Bed diameter D_T = 82 mm lines A, B, C) and D_T = 175 mm (line E).

excessively large values of r cannot be used. Under equal experimental conditions the accuracy of calculating A_h in a larger apparatus was greatly reduced at $r > 0.5$ m because of the poorly marked temperature maximum of the bed.

It can be expected that A_h will increase in going to fluidized beds of large hydraulic diameter. To some degree this has been confirmed by experiments carried out at the BSSR Academy of Sciences by Yuditskiy. He measured the effective temperature conductivity along the horizontal direction in shallow fluidized beds (H_0 = 30-35 mm) of large cross section (0.5 × 2.5 m) using a sand fraction of wide particle-size distribution (0.3-3.0 mm) fluidized with air.

A perforated plate of 0.8 mm thickness with openings of 1.3 mm diameter (free cross-sectional area = 9%) was used as the air distributor. Hot sand was admitted through a 65-mm diameter tube. The superficial fluid velocity varied from 0.6 to 3.2 m/sec. In various series of experiments the bed temperatures were recorded continuously at a height 12 mm above the screen and at distances 1.5, 0.5 or 0.25 m from the position at which the hot material was introduced. Values of A_h as large as 30-40 cm^2/sec were attained, which are many times larger than the values indicated in Fig. 9-3 based on laboratory size columns.

The hydrodynamic conditions in the experiments in thin beds of large cross section bore some unique features. There was marked eddy formation below the gas distributing screen, and with the small overall resistance of the screen and fluidized bed it is possible that this had some effect on the overall pattern of air flow through the bed and on the intensity of particle mixing.

There was considerable scatter in the experimental points for the effective temperature conductivities of fluidized beds measured by the method of modified instantaneous heat sources. This scatter can be explained by pulsations in the value of the effective temperature conductivity of the fluidized bed which actually exist and are not averaged out in a single experiment, rather than by errors of measurement or the noninstantaneous behavior of the heat pulse.

9.3 Comments on the Experimental Data

In correlating the effective thermal conductivities of fluidized beds one must consider the nature and cause of particle movement in terms of 1) the rate of mixing of the particles and the heat transfer medium; 2) the particle concentration in the fluidized bed (bed voidage); 3) the intensity of heat transfer between the particles passing through a given position in the bed and the medium. The relative importance of these factors is different for different degrees of development of the fluidized bed.

For real nonhomogeneous fluidized beds there has so far been little success in attempts to describe the vertical and horizontal effective thermal conductivities in terms of these three aspects of particle movement or by analogies with turbulent liquids and gases.

Apart from empirical relationships of limited applicability, the most promising approach to calculating λ_{eff} for nonhomogeneous beds with bubbles lies in establishing the laws for the formation, coalescence and break-up of fluid bubbles in fluidized beds, since the mixing of material in the bed is largely governed by the passage of bubbles. Rowe [1222] has dealt with the part played by bubbles in mixing beds.

In summary, the effective thermal conductivity of fluidized beds has barely been studied to date, and represents a gap in the state of knowledge which should be filled.

Chapter 10

Heat Transfer With Surfaces Wetted by Fluidized Beds

10.1 Theoretical Concepts

1. Mechanism of heat transfer

Even at low superficial gas velocities, of the order of a few tenths of a meter per second, there are extremely large heat transfer coefficients at surfaces (walls) submerged in beds of fine particles; the coefficients may reach 300-600 kcal/m^2 • hr • °C (60-120 BTU/w °F sq. ft.) or even higher.

Such high values of the heat transfer coefficient between walls and a fluidized bed of fine particles can be explained by the fact that here heat transfer is not limited by the thickness of a laminar layer or sublayer of gas, but by the thickness of the gas layer between the wall and the row of particles closest to it, which is many times smaller [741]. In contrast to heat transfer to a wall from a stationary dense bed, the particles in the rows closest to the wall in a fluidized bed have almost the same temperature as the core of the bed, since the residence time of each particle near the wall is very small. In a two-phase system there is therefore an unusual and extremely thin gas "thermal boundary layer," and a "turbulent core" of well-mixed particles in which there is little temperature gradient. Steady-state operation of a heat exchanger as a whole is advantageously coupled with unsteady-state heating of the particles themselves. The comparatively thick laminar boundary layer or sublayer which usually occurs in single-phase flow is eliminated, even in the absence of turbulence, by the moving particles in the region near the wall.

At any moment the thickness of the layer of medium between the wall and the row of particles closest to it varies from one position to another. The least calculated thickness of the layer would occur in the case of contact between the wall and all the particles of the first layer, and for spherical particles would be equal to one-sixth of the particle diameter. In the general case the calculated thickness of the gas wall layer δ_w will differ from $d/6$ by some distance b, which depends on the voidage of the fluidized bed and is not directly governed by the diameter.

2. Approximate equations for heat transfer by conduction

An average value of the distance (L) between the particles in a given row, and between rows, can be estimated from the bed voidage (m).

The volume of the bed is $V_{bed} = V_m/(1-m)$, where V_m is the volume of solid material.

The axial distance between particles of diameter d in a fluidized bed exceeds d, and as the bed expands, varies in proportion to the cube root of the ratio of volumes:

$$L''/L' = (V''_{bed}/V'_{bed})^{1/3} = (1-m'/1-m'')^{1/3}. \quad (10\text{-}1)$$

This distance is equal to d for a dense bed of spheres in cubical packing, for which $m = 0.476$.

For a fluidized bed of voidage m the mean distances between the particles become:

$$L = d\left(\frac{1-0.476}{1-m}\right)^{1/3},$$

i.e.,

$$L = \frac{0.807d}{(1-m)^{1/3}}. \quad (10\text{-}2)$$

Consider the derivation of a theoretical expression for the heat transfer coefficient between a fluidized bed of spherical particles and a wall, α_{wall}, with heat transfer through the gas layer occurring only by molecular heat conduction, and taking into account the convective and radiant components, for instance, by introducing an effective thermal conductivity λ_{eff} in the equation in place of the thermal conductivity of the medium, λ_m,

$$\lambda_{eff} = \lambda_m + \lambda_f + \lambda_r, \quad (10\text{-}3)$$

where λ_f and λ_r are the components due to convection (fluid flow) and radiation, respectively.

In the derivation, it is assumed that approximately

1. The temperature gradients in the particles are negligible compared with the temperature difference between the wall and the particle surface;

2. The particles are able to assume the temperature of the core of the fluidized bed in the time between two contacts with the wall;

3. The wall curvature is so small that it has an insignificant effect on heat transfer (the wall can be regarded as a plane);

4. Changes in the heat transfer coefficient over the height of the wall are negligibly small.

In a time $d\tau$ a single particle receives the following quantity of heat by conduction:

$$dQ = \frac{\lambda_m}{\delta_w} \cdot \frac{\pi d^2}{4} (t_{wall} - \vartheta)\, d\tau = c_M \gamma_M \frac{\pi d^3}{6} d\vartheta, \quad (10\text{-}4)$$

where δ_w = mean distance between the wall and the particle surface facing it;

λ_m = thermal conductivity of flowing medium;

t_{wall} = wall temperature;

ϑ = varying value of particle temperature.

Assuming $\delta_w = b + d/6$ the calculated area for heat transfer is equal to the projection of the particle on the wall. The heating of the particles is somewhat overestimated by neglecting the transfer of heat to the particles in the second row, i.e., heat transfer with the wall is underestimated.

From (10-4):

$$\frac{d\vartheta}{t_{wall} - \vartheta} = k_1 d\tau, \quad (10\text{-}5)$$

where

$$k_1 = 1.5\,\lambda_m / c_M \gamma_M \delta_w d.$$

Integrating equation (10-5) over the time τ during which a particle resides in the first row, it is found that

$$\ln \frac{t_{wall} - \vartheta}{t_{wall} - \vartheta_{f.b}}\, k_1 \tau, \quad (10\text{-}6)$$

where $\vartheta_{f.b}$ is the particle temperature in the main part (bulk) of the fluidized bed.

Equation (10-6) can be rewritten in the form

$$t_{wall} - \vartheta = (t_{wall} - \vartheta_{b.b}) \exp(-k_1 \tau). \quad (10\text{-}6a)$$

By subtracting the quantity $(t_{wall} - \vartheta_{b.b})$ from both sides of (10-6a) it is possible to isolate the heating of the particle after a residence time τ near the wall

$$\Delta\vartheta = \vartheta - \vartheta_{b.b} = (t_{wall} - \vartheta_{b.b})(1 - e^{-k_1 \tau}). \quad (10\text{-}7)$$

The time τ after which the particle passes from the first row to the second can be regarded as being equal to L/w_n, i.e., directly proportional to the distance L between rows of particles and inversely proportional to the velocity component of the particle motion normal to the wall (w_n). This means that the renewal time of the particles in the first row is

$$\tau = \frac{L}{w_n} = \frac{0.807d}{w_n(1-m)^{1/3}}.$$

The quantity of heat transferred in time τ per square meter of wall is

$$q_\tau = c_M \gamma_M \frac{\pi d^3}{6} \Delta\vartheta N,$$

where N is the number of particles in the first row resting on one square meter of wall.

Obviously,

$$N = \frac{1}{L^2} = \frac{(1-m)^{2/3}}{0.65d^2}, \quad (10\text{-}8)$$

i.e.,

$$q_{\tau} = \frac{\pi c_{\mathrm{M}} \gamma_{\mathrm{M}} d\,(1-m)^{2/3}}{3 \cdot 9}\,\Delta\vartheta.$$

In unit time, the quantity of heat transferred will be

$$Q = \frac{q_{\tau}}{\tau} = c_{\mathrm{M}} \gamma_{\mathrm{M}} w_{\mathrm{n}}\,(1-m)(t_{\mathrm{wall}} - \vartheta_{\mathrm{b.b.}})\,(1 - e^{-k_1 \tau}),$$

and so the heat transfer coefficient referred to the temperature difference $t_{\mathrm{wall}} - \vartheta_{\mathrm{b.b.}}$ between the wall and the bulk of the bed is

$$\alpha_{\mathrm{wall}} = c_{\mathrm{M}} \gamma_{\mathrm{M}} w_{\mathrm{n}} (1-m)\left(1 - \exp. {-\frac{1.5 \lambda_{\mathrm{m}} 0.807 d}{\delta_{\mathrm{w}} c_{\mathrm{M}} \gamma_{\mathrm{M}} w_{\mathrm{n}} (1-\mathrm{m})^{1/3}}}\right) = c_{\mathrm{M}} \gamma_{\mathrm{M}} w_{\mathrm{n}} (1-m)\left(1 - \exp. {-\frac{1.2 \lambda_{\mathrm{m}}}{\delta_{\mathrm{w}} c_{\mathrm{M}} \gamma_{\mathrm{M}} w_{\mathrm{n}} (1-\mathrm{m})^{1/3}}}\right). \qquad (10\text{-}9)$$

It is interesting to note that the more rapidly the particles move, i.e., the smaller the residence time of each particle on the heating surface, and hence the smaller the quantity of heat picked up by a particle in a single cycle, the larger will be the heat flux from the heating surface. As the particle mixing is accelerated, the number of particles actively taking part in heat transfer with the wall per unit time is increased many times more than the reduction in the quantity of heat removed by each particle.

The region of the fluidized bed filled with the first row of particles next to the wall (heating surface) is continually being renewed. It can be assumed that if other conditions are kept constant then from the change in the velocity (replacement) of the particles, their number and uniformity of distribution in the region being considered will average the same as in the fluidized bed as a whole. This means that the overall surface area of the first row of particles which is continuously turned towards the wall surface does not change, nor does the mean distance between the two heat transfer surfaces. The temperature difference between the wall and the particles, on the other hand, will vary, depending on the particle velocity, since the mean temperature of the first row of particles will vary. It will be higher for slower particle replacement, since the particles will become appreciably heated towards the end of their residence time in the first row. On the other hand, if the particles move rapidly, the mean temperature in the first row will fall, and will approach the temperature of the bulk of the fluidized bed asymptotically. This means that the temperature difference between the wall and the first row of particles increases, and heat transfer between the wall and the fluidized bed is facilitated. The limiting case when each particle is completely unheated, passing infinitely rapidly along the heating surface is physically impossible, since at any velocity the first row of particles will receive a finite quantity of heat per unit time which increases monotonically with the particle velocity, if the bed voidage is unchanged.

This descriptive mechanism requires modification if an appreciable temperature gradient can arise over the particles cross section; for example, under conditions of extremely brief heating processes.

Considering available information on the symmetrical problem of unsteady-state heating of a sphere having a uniform initial temperature ϑ_0 at all points, it can be shown for any value of the Biot number of the particles (which usually does not exceed one to ten) that as the Fourier number $\mathrm{Fo} = a\tau/R^2$ is reduced in the region of small Fourier numbers (of the order of 0.005 or lower) the relative surface temperature of the sphere $(1-\theta_s)$ approaches unity:

$$1-\theta_s = \frac{t_m - \vartheta(R,\tau)}{t_m - \vartheta_0} \rightarrow 1.$$

This means that in the initial period of heating the surface temperature of the sphere $\vartheta(R, \tau)$ is close to the initial temperature of the whole sphere, ϑ_0, and hence to the temperature at the center of the sphere, since the latter has as yet had no time to change. As Fo increases, the temperature gradient over the particle must pass through a maximum and then tend to zero again for complete heating, when all the points reach the equilibrium temperature t_m. The presence of this maximum in the temperature gradient across the particles will act initially to accelerate the fall in the temperature difference as Fo increases, and then act in the opposite direction. However, on the whole, there will be a monotonic decrease in the temperature difference as Fo increases—the effect of the steady increase in the particle temperature is greater than the effect of a short term peak in the value of the temperature gradient over the particle. For example, calculating the change in $(1-\theta_s)$ as a function of Fo for the simplest case of $\mathrm{Bi}=1$ confirms the validity of the conclusion that the effective temperature difference and heat transfer coefficient between the wall and fluidized bed increase continuously with increase in the particle velocity. The fact that the particle heating is nonsymmetrical, in spite of the particle rotation, does not change the substance of this conclusion.

As shown below, small absolute values of the exponent in equation (10-9) correspond to the maximum value of the heat transfer coefficient between the fluidized bed and wall, i.e., to the conditions of greatest practical interest.

The heat transfer coefficient between the wall and a given fluidized bed passes through a characteristic maximum as the superficial fluid velocity increases. Figure 10-1a shows examples of the curve $\alpha_{wall} = f(w_f)$ for a fluidized bed. The presence of a maximum is to be expected and is indicated in Eq. (10-9). As the superficial fluid velocity increases, so does the particle velocity w_n and the bed voidage m. Increase in particle velocity leads to an increase in α_{wall}, while increase in bed voidage has the opposite effect. Initially, when the voidage is small, and the normal particle velocity w_n is insufficient for the highest rates of heat transfer, an acceleration of the particles has a strong effect, i.e., α_{wall} increases with the superficial fluid velocity; later the situation is different. At high superficial fluid velocities the particle velocity w_n becomes so large that the particle temperature is unable to increase much more during the residence time of the particles in the first row. A further increase in the superficial fluid velocity, and hence in w_n, then has practically no effect on the value of the temperature difference being applied across the main thermal resistance, and will not facilitate further heat transfer. At the same time the negative effect of the voidage is retained or even increased as w_f increases. Therefore, after a more gentle increase, α_{wall} will begin to fall. A numerical example is considered by way of illustration of this effect in [744].

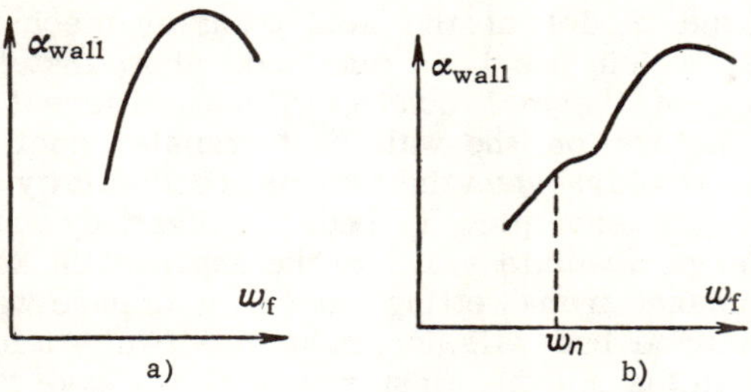

FIG. 10-1. Typical curves of $\alpha_{wall}=f(w_f)$ for fluidized beds.
a--presence of a maximum in the curve for the heat transfer coefficient α_{wall} as a function of the superficial fluid velocity; b—reduction in the rate of α_{wall} at the limit of stability for a bed fluidized with a medium having a high bulk heat capacity.

The structure of Eq. (10-9) indicates that at small absolute values of the exponent it degenerates into a simple expression. The quantity $(1-e^{-x})$ remains equal to x to an accuracy of no less than two figures when $x \leqslant 0.03$. Thus, if $c_M\gamma_M w_n$ is so large or λ_m is so small that the absolute value of the exponent remains less than 0.03, Eq. (10-9) can be replaced by

$$\alpha_{wall.max} = \frac{c_M\gamma_M w_n(1-m)\cdot 1.2\lambda_m}{\left(b+\frac{d}{6}\right) c_M\gamma_M w_n(1-m)^{1/3}} = \frac{1.2\lambda_m(1-m)^{2/3}}{b+\frac{d}{6}}. \qquad (10\text{-}10)$$

Under these conditions a further increase in the heat capacity of the material c_M, the bulk weight of the particles γ_M, or their normal velocity w_n will have no appreciable effect on heat transfer to the fluidized bed. Physically, this means that the bulk heat capacity of the material $c_M\gamma_M$ and the rate of replacement of the particles in the first row (which is directly proportional to w_n and depends inversely on the voidage m) are so large that the particle temperature increases very little during a single period of heating. As a result, making the heat capacity of the bed larger or using a more rapid replacement of the surface particles will not lead to an appreciable increase in the effective temperature difference applied to the main thermal resistance.

In the region of large absolute values of the exponent, the change from one material to another has an effect on the value of the heat transfer coefficient α_{wall}. It is important that in such a change the whole group $c_M\gamma_M w_n(1-m)$ changes, and not just individual factors. In changing from one material to another, some of these factors may change in opposite directions, and these changes compensate each other. For example, in going from glass to steel spheres, $c_M\gamma_M$

increases from $0.16 \times 2500 = 400$ to $0.11 \times 7900 = 870$, but under otherwise equal conditions (in particular, at the same superficial gas velocity) the particle velocity w_n decreases: the heavier steel particles move less rapidly. In such cases the heat transfer coefficient will hardly depend on the nature of the material even when far from the maximum.

This proposed model of the heat transfer mechanism, and the equation (10-9) which has been obtained, show that for media with different values of thermal conductivity λ_m different effects of the fluidized bed factors on the wall heat transfer coefficient must be expected. In particular, when the thermal conductivity of the medium is very large (for example, in beds fluidized by water) there will usually be a large absolute value of the exponent in Eq. (10-9), i.e., the effect of a change from wetting the wall with pure water to wetting it with the fluidized bed will not make itself fully felt in increasing α_{wall}. The particles in the first row will not have time to become appreciably heated in the time of contact, which means that there would be a large mean-temperature difference acting between the wall and the first row of particles dampened by the fluid's high conductivity. At large values of λ_m, heat transfer between the walls and a fluidized bed, where filtrational mixing of the medium near the walls is suppressed, will appear to be the less intensive heat transfer with a dense bed (although the absolute value of α_{wall} for a fluidized bed can be very large). For a fully developed fluidized bed, heat transfer to the walls by means of filtrational mixing of the stream, is reduced due to the high voidage of the bed and the absence of constant direct contact between the particles and wall. The absence of direct contact between the particles and the wall reduces filtrational mixing along the wall, which is very important for heat transfer.

For gas fluidized beds with small thermal conductivities λ_m and small heat capacities compared with liquids, there is a characteristically sharp increase in the value of α_{wall} at the transition from a dense bed to a fully developed fluidized bed. Filtrational mixing plays an unimportant part because of the low heat capacity of the gas.

In the initial stages of fluidization, close to the limit of stability, when there is still no motion of the particles normal to the wall, some reduction in the rate of increase of the heat transfer coefficient can be expected at the transition from a dense bed (Fig. 10-1b). The new effective heat transfer mechanism, related to mixing of the particles, has still not developed, and the old mechanism, related to the filtrational mixing of the medium, is somewhat reduced. For gas fluidized beds this "step" at the limit of stability on the line giving the heat transfer coefficient as a function of the stream velocity will obviously be small, since filtrational transfer plays a comparatively unimportant part, while even a feeble radial motion of the particles leads to an appreciable increase in heat transfer as a result of the small value of λ_m.

From Eqs. (10-9) and (10-10) it can be seen that as the particle diameter is reduced the heat transfer coefficient to the wall should increase. These equations also indicate that the absolute value of α_{wall} depends directly on the thermal conductivity of the medium (gas). Equation (10-10) also seems to indicate that $\alpha_{wall.max}$ is directly proportional to λ_m; however, this conclusion is incorrect. On going from one medium to another, for instance, from air to hydrogen, not only λ_m varies (increases) but also the bed voidage m corresponding to the maximum α_{wall} in Eq. (10-10). This is to be expected, since with a high λ_m and constant thickness of the gas layer separating the first row of particles from the wall, the particles will be heated rapidly, and in order to satisfy the conditions for approaching the maximum α_{wall}(the absence of an appreciable increase in the particle temperature during its residence time in the first row) more rapid replacement of the particles is necessary (larger value of w_n), which can be achieved at larger superficial fluid velocities w_f, and hence at larger bed voidages m. Because of the simultaneous increase in the thermal conductivity of the gas and the bed voidage corresponding to the maximum α_{wall}, Eq. (10-10) shows that the heat transfer coefficient increases more slowly than in proportion to λ_m.

It would therefore be incorrect to explain the absence of proportionality between λ_m and $\alpha_{wall.max}$ entirely by the restricting effects of differences in other gas properties such as viscosity. In the case of wall "wetted" by a fluidized bed it does not follow that there should be the same mechanism for the change of the boundary layer thickness with the kinematic viscosity of the gas as exists in streams of pure fluids. The thickness of the "boundary layer" δ_w will be determined mainly by the particle diameter and the mean distance between particles, which depends on the bed voidage. At higher values of λ_m, but otherwise equal conditions, $\alpha_{wall.max}$ in a given fluidized bed is reached at a larger voidage m. This leads to some increase in b in the expression for the thickness of the boundary layer $\delta_w = b + d/6$.

Thus, on going to a gas with a larger value of λ_m, for instance, from air to hydrogen, an increase must be expected in the component of δ_w, which does not depend on the particle diameter, i.e., the effect of the particle diameter on the value of $\alpha_{wall.max}$ is reduced.

The case considered above for attaining $\alpha_{wall.max}$ in the absence of appreciable cooling (heating) of the particles during their time of exposure at the wall, combined with a high particle concentration (or small effective bed voidage), is not the only case for attaining the maximum wall heat transfer coefficient. In general, $\alpha_{wall.max}$ will be reached whenever the positive effect of increasing the particle mixing velocity, as the superficial fluid velocity is increased, is compensated by the negative effect of the increasing effective bed voidage; this effect can arise even when there is an appreciable change in the particle temperature at the solid surface.

Unfortunately, it is still not possible to satisfactorily describe the rate of particle mixing in the immediate neighborhood of the wall as a function of w_f even for the very simple case of a bed of uniformly sized spherical particles. An elementary analytical determination of the position of $\alpha_{wall.max}$ is therefore impossible. Intensive motion of round particles is possible at a much lower bed voidage m than in the case of irregular, sharp-angled, needle-like or plate-like particles. For irregular particles which "adhere" to each other $\alpha_{wall.max}$ will be reached at a higher bed voidage, so that under otherwise similar conditions (size and density of bed, properties of gas) it will have a considerably smaller numerical value, i.e., for these cases it is necessary to introduce some correction factor or shape factor into the equations obtained for spherical particles.

The component of the particle velocity near the heat transfer surface is also important in determining α_{wall}, since it is this velocity which removes the particle as rapidly as possible from contact with the wall. In a uniform fluidized bed this particle velocity normal to the wall, w_n, has been considered in deriving the theoretical equation. Nonuniform fluidized beds are nonisotropic, and the vertical mixing of material may be more than an order of magnitude more intensive than in the horizontal direction; the vertical component of the particle velocity will determine α_{wall} not only for horizontal tubes, but also for comparatively large vertical surfaces.

3. Effect of convective heat transfer

Convective heat transfer through the gas wall layer δ_w at small superficial fluid velocities, which is characteristic of fluidized beds, is mainly due to the interruptions of the flow around the particles close to the wall, giving rise to eddy zones in the wakes of the particles. These zones expand and convective transfer becomes more important as the Reynolds number $Re = w_f . d/\nu$ increases, i.e., as the superficial fluid velocity and particle size increase. In a more formal manner, convective transfer in the bed can be regarded as the result of the simple passage of jets around the particles. It is interesting to note that this treatment leads to the transfer being proportional to the quantities w_f and d.

Up to now data have been lacking for the accurate calculation of convective transfer from walls to fluidized beds. Convective filtrational mixing within dense beds has been more carefully studied. For example, according to the data of Baum [107], the Peclet number $Pe_m \approx 10$ over the range of Reynolds numbers from 5 to 2000. Convective transfer from the walls to the first row of particles in a fluidized bed will be less important due to the large bed voidage and the absence of constant particle contact with the walls. Therefore the expression

$$\lambda_f \approx 360 w_f c_m \gamma_m d, \tag{10-11}$$

which corresponds to $Pe_m = 10$, will give too large a value for λ_f.

Equation (10-11) can be used to check if it is possible to neglect the convective component of λ_{eff} in actual cases, since the true λ_f will be even smaller than that given by the formula, and there will be a safety factor in the calculation. For example, when $d = 10^{-4}$, $w_f = 0.2$, $c_m = 0.24$, $\gamma_m = 1.2$, $\lambda_m = 2 \times 10^{-2}$, Eq. (10-11) gives $\lambda_f = 2.07 \times 10^{-3} \ll \lambda_m \approx 2 \times 10^{-2}$ and $Re \approx 1.27$, so that it is quite permissible to neglect the convective component of transfer.

Taking the experimental values of $\alpha_{wall.max}$ and w_{expt} by Sarkits for coarse particles of BAV catalyst and calculating the conductive component from $\alpha_{wall.max}$ using the empirical formula [742]

$$\alpha_{wall.max} = 33.7 \gamma_M^{0.2} \lambda_g^{0.6} d^{-0.36}, \tag{10-12}$$

the convective component of $\alpha_{wall.max}$ will correspond to

$$\lambda_f \approx 18 w_f c \gamma_m d. \tag{10-13}$$

In this instance filtrational mixing along the walls under the conditions of $\alpha_{wall.max}$ appears to be about 20 times less important than in the middle of a dense bed of the same particles at the same superficial fluid velocity.

4. Effect of radiation

Radiant heat transfer between the walls and a fluidized bed can be taken into account with a radiation component λ_r of the effective thermal conductivity of the layer. Since Eq. (10-9) expresses the conductive heat transfer between the wall and the first row of particles only, while radiant heat transfer occurs equally with all the particles of other rows which are "visible" to the wall, it is necessary to define λ_r for single substitution in λ_{eff} and Eq. (10-9) as:

$$-\frac{q}{\nabla t} = \frac{Q_{1.2}\delta_w}{\frac{\pi d^2}{4}(t_{wall} - \vartheta)},$$

where $Q_{1.2}$ is the heat taken up by radiation from the wall by one particle.

A correction factor must also be introduced to refer to the wall radiation to only the first row of particles. This correction factor must be the same as for the other components of the effective thermal conductivity, for instance, λ_m. It is found from Eq. (10-10) for the condition $\alpha_{wall.max}$ that

$$\lambda_m = \frac{\alpha_{wall.max}\delta_w}{1.2(1-m)^{2/3}} = \frac{\alpha_{wall.max}(t_{wall} - \vartheta)}{\frac{t_{wall} - \vartheta}{\delta_w} 1.2(1-m)^{2/3}} = \frac{-q}{\nabla t} \cdot \frac{1}{1.2(1-m)^{2/3}}. \tag{10-14}$$

Comparison of (10-14) with the usual expression $\lambda_m = -q/\nabla t$ for bodies of constant cross section shows that in this case the correction factor should be $1/[1.2(1-m)^{2/3}]$. However, it is much simpler and more convenient to take radiant heat transfer into account by defining the overall wall heat transfer coefficient as

$$\alpha_{wall} = \alpha'_{wall} + \alpha_r \tag{10-15}$$

where α'_{wall} is the conductive-convective component of the heat transfer coefficient calculated by Eq. (10-9), letting $\lambda = \lambda_m + \lambda_f$, and $\alpha_r = \frac{4.9\varepsilon_{red}\left[\left(\frac{T_{wall}}{100}\right)^4 - \left(\frac{\Theta}{100}\right)^4\right]}{T_{wall} - \Theta}$ is the radiant component of the heat transfer coefficient.

ε_{red} is the reduced degree of blackness of the wall-fluidized bed system; T_{wall} and Θ are the absolute temperatures of the wall and particles. It is necessary to estimate the values of ε_{red} and Θ.

The degree of blackness of a fluidized bed is determined in the same way as for a dusty gas stream. In the case of uniform distribution of dust particles of mean radius $\bar{r}$ in the stream, the degree of blackness caused by the dust is

$$\varepsilon_d = 1 - e^{-\frac{0.75cl}{\gamma_M \bar{r}}} \tag{10-16}$$

where l is the mean length of the radiant path (calculated layer thickness); and

c is the weight concentration of particles per unit volume of bed.

The mean length of the radiation path is

$$l = 4nV_{bed}/F, \tag{10-17}$$

where V_{bed} is the bed volume and F is the surface area surrounding this volume.

Here $n = 0.9$ for Bouguer numbers

$$B = 0.75\ cl/\gamma_M \bar{r} = 0.05 \div 50,$$

i.e.,

$$l = 3.6\ V_{bed}/F. \tag{10-17a}$$

If it is assumed that the layer thickness and hence the length of the radiation path are very small, the degree of blackness of the bed will be large for the concentrations characteristic of fluidized beds.

The weight concentration of the material per unit volume of the bed is

$$c = \gamma_M (1 - m),$$

which means that

$$\varepsilon_d = 1 - e^{-\frac{1.5l(1-m)}{d}}. \tag{10-18}$$

Usually m = 0.4–0.8 under fluidized bed conditions;

$$d = (0.1 - 3) \times 10^{-3}\,\text{m};\ l_{min} \approx 0.04\ \text{m}.$$

The minimum value of ε_d corresponding to l_{min}, m_{max} and d_{max} will then be

$$\varepsilon_d = 1 - e^{-\frac{1.5 \cdot 0.04 \cdot 0.2}{3 \cdot 10^{-3}}} = 0.98.$$

Usually the bed voidage does not exceed 0.65 and the particle diameter is less than 1.5 mm, in which case

$$\varepsilon_d = 1 - e^{-\frac{1.5 \cdot 0.04 \cdot 0.35}{1.5 \cdot 10^{-3}}} = 1 - 8.5 \cdot 10^{-5} \approx 1,$$

i.e., in practice, the degree of blackness of a fluidized bed is unity.

A fluidized bed may be considered an impervious body with a transfer surface equal to the heated surface (tube) surrounding it, it is therefore possible to use the usual methods of calculation assuming that radiation and absorption of energy occurs at the surface of the body.

The radiant heat transfer coefficient m is determined as in the case of one body (the fluidized bed) surrounded by another (the heated surface).

Since an element of the "tube" heating surface is convex in shape, the reduced degree of blackness in the radiant heat transfer formula will be

$$\varepsilon_{red} = \frac{1}{\frac{1}{\varepsilon_1} - \frac{F_1}{F_2}\left(\frac{1}{\varepsilon_d} - 1\right)}$$

with the simplification that in this instance $\varepsilon_d = 1$, giving

$$\varepsilon_{red} \approx \varepsilon_1, \tag{10-19}$$

i.e., the reduced degree of blackness is practically equal to the degree of blackness of the heated surface.

Thus, the radiant heat transfer coefficient referred to the heated surface of the tube and the calculated temperature difference $t_{wall} - \Theta_{f.b}$ will be

$$\alpha_r = \frac{4.9\varepsilon_1\left[\left(\frac{T_{wall}}{100}\right)^4 - \left(\frac{\Theta}{100}\right)^4\right]}{T_{wall} - \Theta_{f.b}}, \tag{10-20}$$

where Θ is the mean-absolute temperature of the surface of the fluidized bed which participates in radiant transfer at any moment with the wall (heated surface).

Here Θ will be lower than the mean temperature of the surface of the first row of particles, and is approximately equal to $\Theta_{f.b} - \Delta\vartheta/2$, since the particles further away also participate in radiant heat transfer with the heated surface, and these are practically at the temperature of the main part of the fluidized bed, $\Theta_{f.b}$. During intensive particle motion, when $\Delta\vartheta$ is small, it can be assumed that $\Theta = \Theta_{f.b}$, i.e.,

$$\alpha_{r_{max}} = \frac{4.9\varepsilon_1\left[\left(\frac{T_{wall}}{100}\right)^4 - \left(\frac{\Theta_{f.b}}{100}\right)^4\right]}{T_{wall} - \Theta_{f.b}}. \tag{10-21}$$

In some cases, for example, when fluidized beds are used in high-temperature heat exchangers, radiation may play the leading part in the transfer of heat. The use of Eq. (10-21) cannot be recommended in such cases without checking in advance whether the particle velocity w_m is sufficient to satisfy the assumption that $\Theta = \Theta_{f.b}$ made in deriving $\alpha_{r_{max}}$. To illustrate that $\alpha_{r_{max}}$ may sometimes reach high values, Table 10-1 gives calculated values of $\alpha_{r_{max}}$ with $\varepsilon_1 = 0.8$ for several pairs of values of $\Theta_{f.b}$ and T_{wall}. The table demonstrates that $\alpha_{r_{max}}$ increases sharply as T_{wall} is increased, particularly at small temperature differences between the wall and the fluidized bed. However, still further decrease of the difference has a smaller effect. Small values of $T_{wall} - \Theta_{f.b}$ are not economical because of the reduction of the radiant flux and of the total quantity of heat transferred.

Table 10-1

Relationship Between the Radiant Heat Transfer Coefficient and the Temperatures of the Fluidized Bed $\Theta_{f.b.}$ and Walls T_{wall}.

T_{wall}, °K	$\Theta_{f.b.}$ °K	$T_{wall}-\Theta_{f.b.}$	α_{rmax}, kcal/m²·hr·degree (C)
1000	300	700	55.5
2000	300	1700	369
1000	900	100	134
2000	1900	100	1190

Consider the derivation of a relationship between the normal velocity of particles and their heating during the time of exposure along the heated surface in the case of significant radiant transfer.

The approximate heat balance equation for the particles of the first row will be

$$\left\{4.9\varepsilon_1 \frac{\pi d^2}{4}\left[\left(\frac{T_{wall}}{100}\right)^4-\left(\frac{\Theta}{100}\right)^4\right]+\frac{\lambda_m+\lambda_f}{\delta_w}\frac{\pi d^2}{4}(T_{wall}-\Theta)\right\} d\tau = c_M\gamma_M\frac{\pi d^3}{6}\,d\Theta. \tag{10-22}$$

This general equation is easily integrated when numerical values of λ_m, λ_f, c_M, γ_M, ε_1 are known, and thus in any actual case it is possible to establish a relationship between the permissible value of the temperature increase and the required rate of particle replacement.

In the limiting case of a rarefied bed, when the conductive-convective component of the heat transfer coefficient can be neglected in comparison with the radiant component, the equation is simplified to

$$4.9\varepsilon_1\left[\left(\frac{T_{wall}}{100}\right)^4-\left(\frac{\Theta}{100}\right)^4\right] d\tau = \frac{2}{3}c_M\gamma_M d\,d\Theta. \tag{10-23}$$

Equation (10-23) must be integrated over the limits corresponding to the time τ for the replacement of particles receiving radiation from the wall. It is assumed as a first approximation that during the time of residence of the particles in the feebly irradiated zone they are able to assume the temperature of the bulk of the fluidized bed, $\Theta_{f.b}$,

$$\frac{1}{2T_{wall}^2}\left\{\int_{\Theta_{f.b}}^{\Theta}\frac{1}{2T_{wall}}\ln\frac{T_{wall}+\Theta}{T_{wall}-\Theta}+\right.$$

$$\left.+\int_{\Theta_{f.b}}^{\Theta}\frac{1}{T_{wall}}\arctan\frac{\Theta}{T_{wall}}\right\}=\frac{4.9\cdot 1.5\cdot 10^{-8}\varepsilon_1}{c_M\gamma_M d}\tau.$$

Substituting

$$\tau \approx \frac{0.807d}{w_n(1-m)},$$

it is found that

$$\left(\frac{1}{T_{wall}\cdot 10^{-3}}\right)^3\left[\frac{1}{2}\ln\frac{(T_{wall}+\theta)(T_{wall}-\theta_{f.b})}{(T_{wall}-\theta)(T_{wall}+\theta_{f.b})}+\right.$$
$$\left.+\arctan\frac{\theta}{T_{wall}}-\arctan\frac{\theta_{f.b}}{T_{wall}}\right]=\frac{118.5\varepsilon_1}{c_M\gamma_M w_n(1-m)^{1/3}}. \quad (10\text{-}24)$$

This equation can be used to determine the order of the quantity $\Delta\vartheta=\theta-\theta_{f.b}$ as a function of the attainable rate of replacement of the particles w_n or vice versa. It is not proposed here that the temperature θ should be measured, but that the temperature should be fixed from technical or other considerations and that then the normal particle velocity to ensure this temperature should be calculated.

This will be illustrated by a numerical example.

Example. Given: ε_1 = 0.8; c_M = 0.2; γ_M = 2300; m = 0.7; T_{wall}= 2000; $\theta_{f.b}$ = 1900.

It is necessary to find the velocity w_n to ensure that $\theta-\theta_{f.b}$ does not exceed 10°C. θ = 1910°K.

$$\frac{1}{2^3}\left[\frac{2.3}{2}\log\frac{(200+191)(20-19)}{(200-191)(20+19)}+\right.$$
$$\left.+\arctan\frac{191}{200}-\arctan\frac{19}{20}\right]=\frac{118.5\cdot 0,8}{0.2\cdot 3\,600 w_n\cdot 0.5^{1/3}}.$$

whence

$$w_n \approx 43 \text{ m/hr} = 1.2 \text{ cm/sec}.$$

The problem of determining $\alpha_{wall_{max}}$ for a high-temperature bed still remains. This cannot be taken as equal to the sum of the maximum conductive-convective term ($\alpha_{cc_{max}}$) and the maximum radiant term ($\alpha_{r_{max}}$) even though the equality $\alpha = \alpha_{cc} + \alpha_r$ is completely valid for any superficial fluid velocity. The reason for this can be interpreted as the noncoincidence of the superficial fluid velocities at the maxima of the convective-conductive and radiant components of the heat transfer coefficient. Of these two components, only α_{cc} is very sensitive to the bed voidage and therefore reaches its maximum $\alpha_{cc_{max}}$ at moderate values of m. In the case of α_r, its value in an ordinary bed of sufficient thickness will only increase with increase in the voidage, since then the surface will "see" through the intervals between the particles and bubbles not only the somewhat cool particles close by, but also the hot particles in new rows further away. It is also important that the particle motion is not retarded as the superficial fluid velocity w_f is increased. However, the regime of "normal fluidization with bubbles" only increases with w_f, and the maximum of d_r will therefore lie beyond the limits of this regime at high values of w_f. Thus as the fluidizing gas supply is increased there will be a monotonic increase in α_r, though the rate of increase will gradually decrease. This will lead to a wider zone $\alpha_{wall} \approx \alpha_{wall_{max}} = (\alpha_{cc} + \alpha_{r_{max}})$ in a high-temperature bed than in a low-temperature one. The extended zone of $\alpha_{wall} \approx \alpha_{wall_{max}}$ represents an important practical result of the presence of the radiant component of heat transfer between a body and a high-temperature fluidized bed, making it possible to operate with high α_{wall} at large voidages and large relative bed expansions, i.e., at small initial bed heights and small hydraulic resistances. This is useful in many cases. In the region of the gentle maximum the fraction of α_r in α_{wall} will vary along the curve $\alpha_{wall} = f(w_f)$; it will be higher for points with larger values of α_r.

For ordinary high-temperature fluidized beds at $\alpha_{cc_{max}}$, which is reached when the increase in m begins to affect α_{cc} more significantly than the corresponding increase in the particle velocity, the latter is insufficient to produce high values of α_r. The first row of particles is appreciably cooled, and the value of α_r at $\alpha_{cc_{max}}$ is very far from the maximum value mentioned above, equal to

$$\epsilon_1 \left[\left(\frac{\theta_{f.b}}{100}\right)^4 - \left(\frac{T_{wall}}{100}\right)^4\right] \Big/ (\theta_{f.b} - T_{wall})$$

In principle it is possible to obtain higher values of $\alpha_{wall_{max}}$ in high-temperature fluidized beds than in the very simple case considered here. An increase in $\alpha_{wall_{max}}$ is possible, for instance, by rapid mechanical mixing of the bed around an immersed stationary body or by an equivalent motion of the body. This equivalent motion need only be sufficient to ensure no prolonged contact between the surface of the body and the same particles.

In the case of very rapid replacement of the particles at the surface of the body, and a low bed voidage, it is possible to achieve a high $\alpha_{cc_{max}}$ and simultaneously (at the same w_f) to have α_r approach $\alpha_{r_{max}}$. In this "artificially" achievable case the following equality would be valid:

$$\alpha_{max} \approx \alpha_{cc_{max}} + \alpha_{r_{max}}. \tag{10-24'}$$

There are difficulties in developing a practical engineering solution to take advantage of this possibility of intensifying heat transfer in high temperature beds. This case can be approached in the fall of small bodies through an intensely fluidized high-temperature bed. On the other hand, this will be far from condition (10-24'), for example, in the relatively slow vertical motion of a long plate through a fluidized bed, or in the case of a bed moving under gravity along a vertical wall.

Under the conditions of laboratory experiments for determining mm, and under the conditions of fluidized bed heat exchangers using low potential heat (for instance, low-pressure steam), the size of the radiant component of the heat transfer coefficient between a wall and a fully developed fluidized bed is small compared with that of the conductive-convective component.

For example, under laboratory conditions, it is found that at $t_{wall} = 25°C$, $\Theta_{f.b} = 19.5°C$ and $\varepsilon_l = 0.8$,

$$\alpha_{r_{max}} = \frac{4.9 \cdot 0.8\,(2.68^4 - 2.925^4)}{25 - 19.5} \approx 3{,}9 \ll \alpha'_{wall}$$

As a result, radiant heat transfer can often be neglected in evaluating and analyzing experimental data on heat transfer in fluidized beds.

In an industrial dryer it was found that a $t_{wall} = 120°C$, $\Theta_{f.b} = 65°C$ and $\varepsilon_1 = 0.8$, $\alpha_{r_{max}} \approx 7.7$ kcal/m² · hr · degree, and again $\alpha_{r_{max}}$ is only a small fraction of the total heat transfer coefficient at the wall, which usually reaches 100-300 kcal/m² · hr · degree (C) under operating conditions.

5. Concluding remarks

Returning to the simplifying assumptions which were made in deriving the main heat transfer equation for a fluidized bed, (10-9), it must be noted that the errors introduced by ignoring the quantity of heat transferred from the first row of particles to the second during the time of residence of the particles in the first row is very small when the particles are moving rapidly. In this case the temperature difference between the rows of particles is much smaller than the temperature difference between the first row and the wall. Actually, the temperature difference between the rows of particles does not exceed $\vartheta - \vartheta_{f.b}$, where ϑ is the temperature of the particles in the first row.

It follows from Eq. (10-7) that the ratio of the temperature difference to the total calculated temperature difference $t_{wall} - \vartheta_{f.b}$ is

$$\frac{\vartheta - \vartheta_{f.b}}{t_{wall} - \vartheta_{f.b}} = 1 - e^{-\frac{1.21\lambda_M}{c_M \gamma_M w_n \delta_c (1-m)^{1/3}}}. \qquad (10\text{-}25)$$

For example, let $\lambda_m = 2.22 \times 10^{-2}$; $c_M = 0.2$; $\gamma_M = 2300$; $m = 0.5$; $\delta_c = 0.5 \times 10^{-4}$ and $w_n = 1$ cm/sec = 36 m/hr. In this case

$$\frac{\vartheta - \vartheta_{f.b}}{t_{wall} - \vartheta_{f.b}} \approx 0.04,$$

i.e., the temperature difference between the first and second rows of particles does not exceed 4% of the difference between the wall and the core of the bed.

The theoretical formula given above was derived on the basis of an extremely simple model for a uniform fluidized bed. However, it appears [744, 1152] to explain qualitatively the details of the complex relationship between α_{wall} and various factors even in nonhomogeneous fluidization.

The largest deviations between α_{wall} for homogeneous and nonhomogeneous fluidized beds must be expected at large reduced velocities, when the break-through of gas increases. It is interesting that theoretically α_{wall} should be larger for an aggregatively fluidized bed with slug flow than for a homogeneous bed, since b is large for the latter, which is determined by the mean voidage of the rarefied bed, and not by the voidage of the dense phase, which is close to $m_{1.s}$. On the other hand, in the slug-flow regime only the part of the wall in contact with the dense phase at any moment of time takes part in convective-conductive heat transfer; this fraction of the wall can be taken approximately as equal to the bulk fraction of the dense phase in the bed:

$$\frac{V_{d.f}}{V_{bed}} = \frac{1-m}{1-m_{1.s}}. \qquad (10\text{-}26)$$

Thus, to a first approximation, we have for a slug-flow fluidized bed

$$\alpha'_{wall} = \frac{1-m}{1-m_{eff}} \cdot \frac{1.2\lambda_{eff}(1-m_{1.s})^{2/3}}{b_{d.f}+d/6} = \frac{1.5\lambda_{eff}(1-m)}{b_{d.f}+d/6}, \qquad (10\text{-}27)$$

where $b_{d.f}$ is the component of δ_w by the particle aggregates, depending on the voidage; $m_{1.s} \approx 0.4$ (bed of spheres).

The mechanism for heat transfer to a fully developed ("boiling") fluidized bed must also be related to the case of heat transfer between walls and granular materials which are stirred by other means (mechanically, for instance, by means of agitators, movement over a rotating heat transfer body, vibration, etc.).

10.2 Review of Previous Research

During the period from 1949 to 1955 a number of papers were published on heat transfer between fluidized beds and heating surfaces [190, 202, 207, 225, 238, 239, 268, 283, 300, 377, 340, 370, 407, 410, 416, 419, 420, 450, 451, 452, 512, 566, 568, 574, 579, 593, 599, 600, 601].

In the 1949 and 1950 papers by Ciborowski [238 and 239] he reported on experimental studies of heat transfer between walls and deep fluidized beds (H/D_T of the order of ten or higher). Spherical alumina particles (fractions 44-53, 74-149-210 microns) were fluidized with air in glass tubes of 27/30 mm diameter. In the earlier work [238] the tube walls were electrically heated over a length of 1 meter. On the outside the heater was thermally insulated with a layer of asbomagnesite 100 mm thick. The tube wall temperatures were measured at eight points located at different heights above the bottom of the bed. Thermocouples were arranged along the tube axis; the hot junctions could be moved along a vertical line so as to measure the temperature in the bed at various heights. The heat transfer coefficients increased as the concentration of material in the tube increased. Even in the case of the material of narrow size range, the heat transfer between deep fluidized beds and the vessels walls was complicated by size-separation of the particles and the decrease of the material concentration along the height of the bed. As a result, the effects of the various factors on heat transfer to the wall were unclear to Ciborowski. For instance, at the same superficial fluid velocity he detected an apparent increase in the mean heat transfer coefficient at the wall as the particle diameter was increased. He noted that this unexpected effect was connected with the increase in the material concentration under these conditions when going from beds with fine particles to beds with coarser particles.

In the later work [239] Ciborowski heated only the upper part (540 mm) of the glass tube containing the fluidized bed, while the lower part (600 mm long) was cooled with a water jacket. The tube diameter was as before (27/30 mm), and the same alumina particles were fluidized, as well as coal particles and Fe_2O_3. Heat transfer coefficients to the wall were determined for both the heating and cooling zones. For the heated zone the temperature difference in the calculations was taken as $t_{wall} - t_{bed}$, as before. In the cooling zone the logarithmic mean temperature difference was used; this was calculated from the temperatures of the water at the jacket inlet and outlet and the bed temperatures at the tube axis. Large errors were possible in determining α_{wall} because of the gradual and poorly defined change in the wall temperature at the boundary between the heated and cooled zones. A considerable heat flow by conduction between the heated and cooled zones (which was not allowed for) could have bypassed the fluidized bed. In this work Ciborowski came to the same conclusion regarding the increase in the wall heat transfer coefficient as the material concentration and particle size were increased. Values of α_{wall} of several hundred kcal/m²·hr·degree (C) were attained. The nonuniformities of the temperatures along the bed axis, measured by the bare thermocouple junctions, did not exceed 6-10°C and decreased as the material concentration and superficial fluid velocity increased.

In 1949 Mickley and Trilling [451] published heat transfer coefficients between air fluidized beds of glass spheres (450-40 microns) and heated walls. In one group of experiments a heater in the form of a tube 12.5 mm in diameter and 875 mm long was placed along the axis of a 73 mm diameter tube which contained the fluidized bed. In another group of experiments the fluidized bed was heated peripherally.

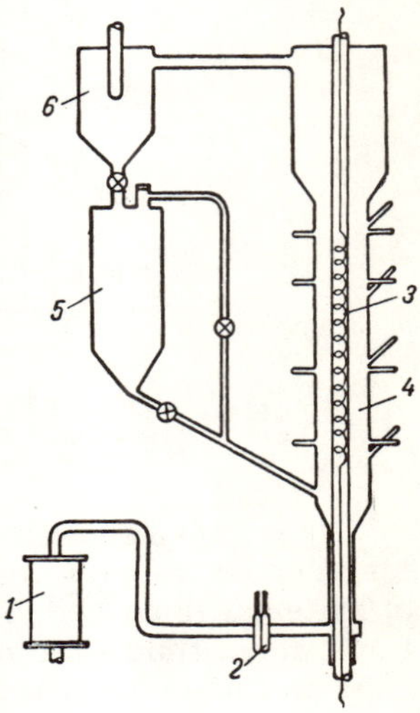

FIG. 10-2. Experimental equipment of Mickley and Trilling [451]:
1) filter; 2) orifice meter; 3) heater; 4) tube; 5) solids feed hopper; 6) cyclone.

Figure 10-2 illustrates the experimental equipment of Mickley and Trilling with an internally heated fluidized bed. Air at the prescribed velocity passed through an orifice meter 2 into the annular gap between the rod heater 3 and the lower conical contraction of the tube which holds the fluidized bed. Material was supplied from the hopper 5 to the tube 4 until the required mean particle concentration ν_{bed} was reached, as determined approximately from the measured pressure drop in the air stream through the fluidized bed. It was assumed that this pressure drop was equal to the weight of the column of particles divided by its cross section. In Mickley and Trilling's experiments the air velocity was so high that a considerable portion of the bed was lifted out of the cylindrical part of tube 4 into the expanded section above, which served as a separator for returning particles to the bed. The spheres carried over from the separator were removed by the cyclone 6 and returned to the system through the hopper 5. The heater wall temperatures, the bed temperature, and the pressure were measured at several cross sections over the height of the bed.

The special feature of Mickley and Trilling's experiments was that they were carried out at low particle concentrations, so that sometimes the bed voidage reached 98%. Carrying out the experiments in this zone, close to a suspended bed, was made possible by returning the carry-over. Particle concentration is nonuniform over the height of such a bed, decreasing rapidly with height.

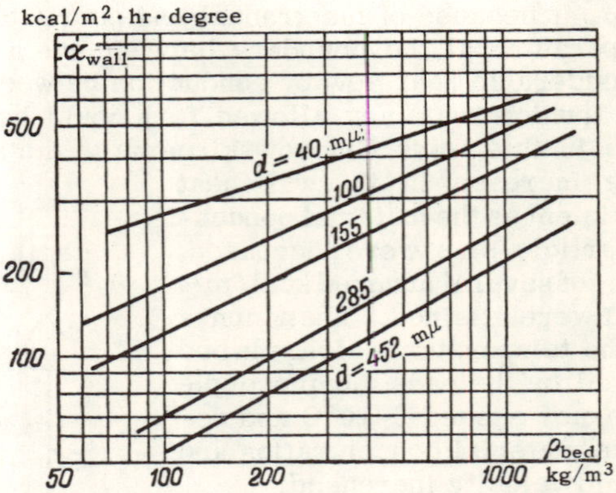

FIG. 10-3. Heat transfer coefficient as a function of bed concentration for particles of various diameters based on the data of Mickley and Trilling [451].

Figure 10-3 shows the relationship derived by Mickley and Trilling for the heat transfer coefficient α_{wall} as a function of the concentration of material in the bed; Fig. 10-4 shows α_{wall} as a function of the weight flow rate of fluid. The nature of the two relationships indicate that the experiments corresponded to higher velocity and lower concentrations than those which would lead to $\alpha_{wall.max}$. Mickley and Trilling [451] offered no explanations for these relationships, but indicated them in the empirical relationship

$$\alpha_{wall} = 0.0242\left(\frac{\gamma_{bed}^2}{d^3}\right)^{0.238}, \tag{10-28}$$

The applicability of this equation was checked for γ_{bed} from 12.2 to 415 kg/m and particle diameters (102÷452) × 10^{-6} m.

Failure of some of the fine particle (d = 40 millimicrons) data to satisfy Eq. (10-28), was attributed to particle agglomeration.

The authors proposed Eq. (10-28) only as a representation of the experimental data which were obtained, without suggesting that it indicated the mechanism of the process, since the effect of a number of important factors on α_{bed} was not studied.

In considering the radial temperature profiles in the fluidized bed, Mickley and Trilling stated that the main thermal resistance was concentrated in a boundary layer at the walls of the heater, while further in the bed there was intensive heat transfer due to particle mixing. They assumed that the effective thickness of this "boundary" layer was reduced by eddies produced by the particle motion.

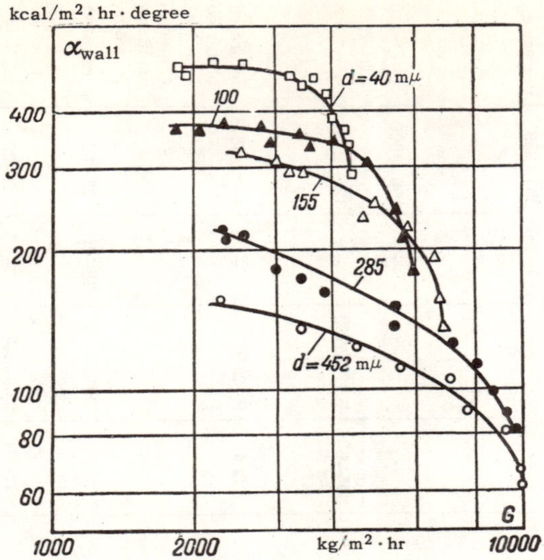

FIG. 10-4. Heat transfer coefficient α_{wall} as a function of the gas mass flow rate G, data of Mickley and Trilling [451].

The results of experiments with externally heated beds in 25 and 102 mm diameter tubes were represented by another empirical equation

$$\alpha_{wall} = 7.17 \cdot 10^{-3} \left(\frac{\gamma_M G}{d^3}\right)^{0.263}. \qquad (10\text{-}29)$$

The ranges over which this equation was checked and the units in which the various quantities were measured are:

d,	diameter of spheres:	0.070–0.452 mm
γ_M,	specific weight of material:	2240–2835 kg/m^3
G,	weight flow rate of air:	659–12,940 kg/m^2·hr
	bed voidage:	0.56–0.98.

Levenspiel and Walton [419] fluidized coal particles (0.132–1.177 mm) with air in a 102 mm diameter tube, and found a characteristic maximum in the relationship between the wall heat transfer coefficient and the superficial fluid velocity.

Jolley [370] determined α_{wall} for high-temperature (800–1000°C) and low-temperature (100–120°C) beds of coke fluidized with air. He placed a massive metallic cylinder in the fluidized bed for a measured time, and then transferred it rapidly to a calorimeter in order to determine the quantity of heat taken up from the bed and therefrom to calculate the heat transfer coefficient. Jolley mentioned the important effect of radiant transfer on α_{wall} for the bed of hot coke.

Leva et al. [407 and 417], published data on measurements of α_{wall} and of temperature profiles in a fluidized bed heated externally with a steam jacket. Particles of sand, iron catalyst and silica gel were fluidized with various gases (Table 10-2) in a 64-mm diameter tube.

Table 10-2. Investigations of Heat Transfer Between Fluidized Beds and Surfaces

Author	System	m	γ_M, kg/m^3	d, microns	Type of heater	D_T, mm.	H_p, mm.	H, mm.	G, kg/m^2.hr	Temperatures, °C		Type of gridplate
										Surface	Bed	
1. Mickley, Trilling [451]	air-glass spheres (narrow fraction), circulation	0.48–0.98	2415–2835	41–452	coaxial heater, $\phi = 12.5$ mm	73	875	–	182–9650	104–162	48–111	no screen
The same	the same	0.56–0.98	2415–2835	70–452	external electric heater	25; 102	–	–	637–12750	195–267	151–239	the same
2. Levenspiel, Walton [419]	air-coal	dense phase	–	132–1177	air cooled wall	102	610	–	244–5370	–	–	–
3. Jolley [370,988]	air-coke (hot and cool)	–	–	1590	massive cylinder heated by bed	178	–	–	–	–	100–1000	–
4. Leva et al. [416,407]	gas (air, CO_2, H_2)-particles (sand, iron catalyst, silica gel)	0.35–0.75	1282–8000	39–454	steam-heated walls	51; 102	635; 661	305–635	7.17–5350	–	126–212	–
5. Baerg et al. [202]	air-particles (see Table 10-3)	0.388–0.75	1905–6950	60–878	coaxial electric heater	140; 318	102	254	9–2960	–	4.7 to 18.3	perforated plate, covered with screen of openings $\varnothing = 40\mu$
6. Campbell, Rumford [225]	air-sand, aluminum, graphite, copper catalyst	dense phase	–	241–3840	central cooling coil	59	–	–	186.5–488	15–56	31–63	perforated
7. Vreedenberg [579, 580]	air-sand, iron ore	the same	2640–5280	234–601	water-cooled horizontal and vertical tubes, $\varnothing = 34$ mm	565	–	1700	318–1466	–	40–220	–
8. Van Heerden et al. [337]	gas (air, CH_4, CO_2, town gas, mixture of H_2 with N_2)-particles (carborundum, iron oxide, coke, lead, Deward alloy)	the same	600–11100	50–800	water-cooled wall	86	102	407	215–3800	–	10–30	–

9. Dow, Jakob [268, 988]	air-particles (coke, iron powder, aluminosilicate catalyst)	0.52–0.69	1935–7450	111–171	steam heated wall	51–76	584;673	51–330	244–1465	–	93.3–104.4	–
10. Miller, Logwinuk [452]	gas (air, He, CO_2)-particles (carborundum, Al_2O_3, silica gel)	dense phase	1120–3890	88–249	central cooled tube, $\varnothing = 9.5$ mm	51	559	–	31–977	–	49–212	–
11. Wicke, Fetting [283, 600]	gas (air, H_2, CO_2)-particles (carborundum, sand, aluminum, lead)	dense phase	2560–11200	65–3000	internal heating by small cylindrical element $\varnothing = 13.3$	100	–	100 (lead powder -50)	–	–	–	–
12. Ciborowski [238]	air-alumina	dense phase	1940	44–210	electrically heated walls	27	1000	1100–2380	–	140	–	–
13. Toomey, Johnstone [568, 988]	air-glass spheres	dense phase	2670–2865	55–848	water-cooled walls	120	series of sections 127 mm each	335–625	116–7530	–	–	–
14. Bartholomew, Katz [207]	air-particles (sand, aluminum, marble)	0.54–0.95	2560–2670	84–251	electrically heated tube wall	102	762	762	469–4565	–	125–315	–
15. Levenspiel, Walton [420, 988]	air-particles (glass spheres, coal)	0.417–0.862	1018–2880	76–4330	as last	102	three sections 51, 128, 51 mm		470–19000	–	–	–
16. Trense [574]	air-glass spheres	dilute phase	–	150–1100	central electric heater	100	500	76–455	470–19000	–	–	–
17. Mickley, Fairbanks [450]	gas (air, He, CH_4)-glass spheres	dense phase	2210–2450	68–82	small electric heating probe, $\varnothing = 6.3$ mm	102	150	about 400–500	49–732	–	–	–
18. Massimilla, Bracale [697]	air-glass spheres	–	–	700	water cooled walls	90	10 sections of 100	200–800	1630–6300	–	–	–

Author	System	m	γ_M, kg/m³	d, microns	Type of heater	D_T, mm.	H_p, mm.	H, mm.	G, kg/m².hr	Temperatures, °C		Type of gridplate
										Surface	Bed	
19. A.V. Chechetin [668]	air-particles (activated carbon, silica gel, iron-aluminum catalyst)	dilute phase	–	250–840	central cooler, ø = 12.7 mm	–	1000	–	about 2900–13500	–	115–163	–
20. Wesser, Mardus [713]	water-glass spheres	–	~2500	1000–5900	(see No. 11)	–	–	–	–	–	–	–
21. Lemlich, Caldas [831]	as above	0.44–0.95	2430–2530	48–680	steam heated wall	51	610	19500–310000	–	–	–	screen with opening of 44 microns on screen with openings of 420 microns
22. Richardson, Mitson [842]	liquid (water, aqueous glycerine solution)-particles (glass)	0.627–1.0	–	460–4260	heated wall	50	810	–	–	–	–	–
23. A.K. Bondareva [635]	air-river sand	dense and dilute phases (to 0.95)	2300	100–470	central heater (ø = 10 mm) and water cooled walls	82	heater height 200 mm	200	230–5180	100	–	fabric grid (canvas)
24. J. Ciborowski, J. Roszak [789, 813, 965]	air-particles (river sand, coke)	–	1610–2620	150–2500	steam heated walls	49	1000	1000	120–8000	100	99–99.5	porous glass plate, thickness 3 mm
25. V.G. Ainshtein [860]	air-sand	dense phase	2651–2659	163–285	cylindrical electric heater ø = 22, length = 110 and 210 mm; ø = 30, length 210 mm	275 and 152	22, 30, 110 and 210	350	280–1450 (horizontal tube); 280–1730 (vertical tube)	–	22–60 and 116–133	grid of porous fireclay

26. Vreedenberg [850]	air-particles (cracking catalyst, round and irregular sand)	as above	1600–2660	63.5–316	horizontal cooled tube, ϕ = 17, 33.6, and 51 mm, 850 mm above screen	565	17; 33, 6 51	1200 init.	37–580	–	20–340	–
27. N.N. Varygin, I.G. Martyushin [877]	air-particles (quartz sand, ferrosilicon, hematite, glass spheres, carborundum)	–	2500–6800	82.5–1160	unsteady-state cooling of silver sphere submerged in bed, ϕ = 20 mm	82.5 and 157	sphere ϕ = 21 mm	–	200–5600	–	20–30	–
28. Ernst [972]	air-quartz particles	dense phase	2650	100-200 300-500 500-700	submerged heater 150 × 150 × 10 mm, or horizontal tube, ϕ = 6 mm, also heated hoop of height 5 mm	150	150; 6; 5	1000	7–4000	–	–	porous plate
29. V.B. Sarkits [931], Sarkits et al.	air-particles (vanadium catalysts BAV and VKhZ-spherical, alumino-nickel, iron-chromium, iron-magnesium catalysts, silica gel, marble, quartz, fireclay, coke)	dense and dilute phases	bulk density 497–1330	127–4500	cooling coil (coil diameter 20 mm, tube diameter 4 mm)	49, 73, 100	80 and 160	120	130–17000	–	130–150	perforated steel grid, screen and bed of crushed quartz of height 30 mm

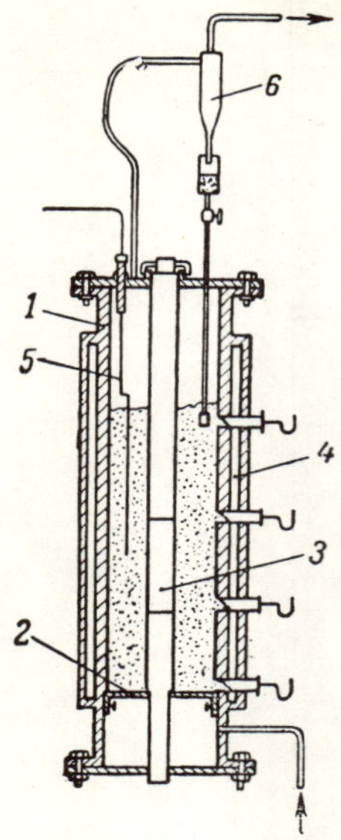

FIG 10-5. Experimental equipment of Baerg, Klassen and Gishler [202]. 1) thick-walled brass cylinder; 2) air distributor; 3) heater; 4) water jacket; 5) movable thermocouple; 6) cyclone.

The following empirical equation was proposed for beds of particles of 161 to 452 microns in diameter:

$$Nu_{wall} = 0.5Re, \qquad (10\text{-}30)$$

where $Nu_{wall} = \alpha_{wall} d / \lambda_g$ and $Re = dG / \mu$.

For beds of finer particles (d = 39-109 millimicrons):

$$\alpha_{wall} = 0.276 c_g G^{1.15}. \qquad (10\text{-}31)$$

As mentioned later by Leva [988], the values of α_{wall} [407 and 416] at small weight flow rates appear to be smaller than in more recently published investigations of such systems.

In the very detailed work of Baerg, Klassen and Gishler [202] published in 1950, they reported heat transfer coefficients α_{wall} between a heated wall and fluidized beds of iron filings, round sand, molding sand, crushed silica gel, glass beads, cracking catalyst, and alumina over a wide range of fluidizing air velocities. The particle diameter was varied from 60 to 878 microns. A diagram of the experimental equipment is shown in Fig. 10-5. The granular material was

Table 10-3

Maximum Heat Transfer Coefficients in Experiments of Baerg, Klassen and Gishler [202]

No.	Material	d, mm	α_{max}, kcal/m^2 · hr · deg
1.	Iron filings	0.375	360
2.	Iron filings	0.300	403
3.	Iron filings	0.195	429
4.	Spherical sand . .	0.878	224
5.	Spherical sand . .	0.323	343
6.	Molding sand . . .	0.276	351
7.	Molding sand . . .	0.218	384
8.	Molding sand . . .	0.160	419
9.	Molding sand . . .	0.128	438
10.	Crushed silica gel	0.738	248
11.	Crushed silica gel	0.503	294
12.	Crushed silica gel	0.314	348
13.	Glass spheres. . .	0.460	379
14.	Glass spheres. . .	0.158	512
15.	Glass spheres. . .	0.060	637
16.	Cracking catalyst	0.139	333
17.	Cracking catalyst	0.097	374
18.	Silica	0.117	423
19.	Silica	0.095	476

poured into the thick-walled brass cylinder 1 of 140 mm inside diameter and 560 mm height on top of an air-distributing grid 2, which consisted of a perforated plate covered with a screen having 40 micron openings. The height of the bed at rest was 254 mm. Heat was supplied to the fluidized bed from the electrical element 3, which had an active height of 102 mm and a diameter of 32 mm. A water jacket 4 was fitted around the cylinder 1 to maintain a high temperature difference between the walls and the fluidized bed. The wall temperatures of the heater were measured with stationary thermocouples, while the bed temperature was measured with a movable thermocouple 5. The air flow rate was measured with a rotameter. Particles entrained from the cylinder 1 by the air stream were removed by the cyclone 6 and returned to the fluidized bed. The particle diameter was determined microscopically as the arithmetic mean for 100 particles of the given fraction. The bulk density of the stationary bed was determined approximately by means of a measuring cylinder. The heat transfer coefficient α_{wall} was determined from the electrical power supplied to the heater, the temperatures of the wall and the bed, t_{wall} and ϑ, and the surface area of the heater. The bed temperature was held close to room temperature.

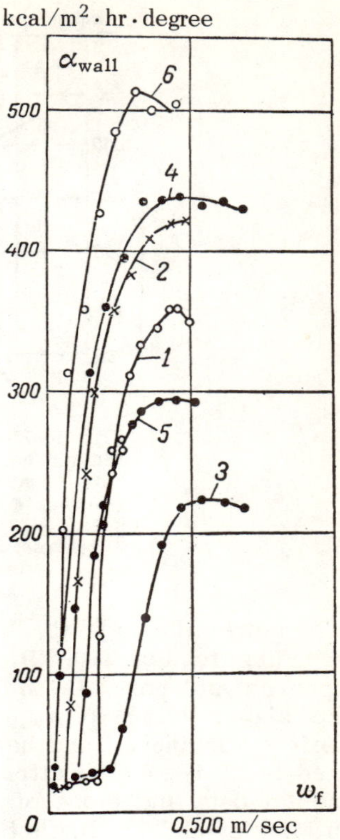

FIG. 10-6. Typical curves of heat transfer coefficient as a function of superficial fluid velocity taken from the experimental data of Baerg, Klassen and Gishler [202].
1) sponge iron, d = 375 microns; 2) sponge iron, d = 195 microns; 3) sand (round), d = 878 microns; 4) molding sand, d = 128 microns; 5) silica gel, d = 503 microns; 6) smooth glass spheres, d = 128 microns.

For each bed Baerg et al., determined α_{wall} over a wide range of superficial fluid velocities from the region below the limit of stability until $\alpha_{wall.max}$ was reached. They showed that the region of superificial fluid velocities for which the value of α_{wall} was close to $\alpha_{wall.max}$ was quite wide, and that the maximum was a "plateau." The values of $\alpha_{wall.max}$ obtained by Baerg, et al. [202], are given in Table 10-3. Figure 10-6 shows some typical curves for the relationship between α_{wall} and the superficial fluid velocity w_f.

After correlating their data on air fluidization, Baerg et al. proposed the following empirical formula for $\alpha_{wall.max}$, which gain is not indicative of the heat transfer mechanism,

$$\alpha_{wall.max} = 239.5 \log (7.05 \cdot 10^{-6} \gamma / d), \qquad (10\text{-}32)$$

where d is the particle diameter, mm, and γ is the bulk density of the stationary bed (before fluidization), kg/m^3.

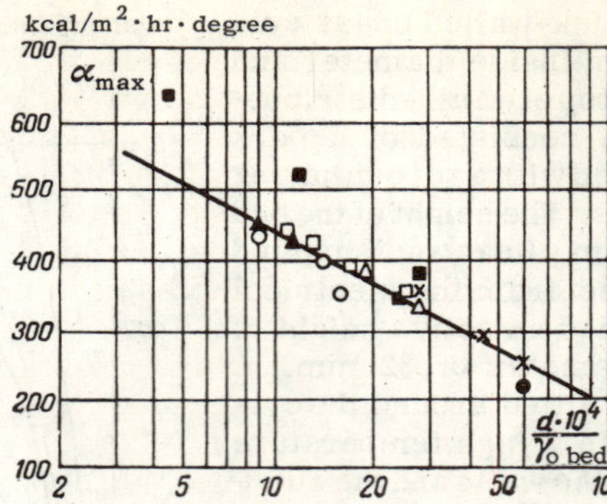

FIG. 10-7. Relationship proposed by Baerg, Klassen and Gishler [202] for calculating the maximum heat transfer coefficient $\alpha_{wall.max}$. ○—iron; ●—sand (round grains); ▪—molding sand; ×—pulverized silica gel; □—smooth glass spheres; △—cracking catalyst; ▲—Al_2O_3. The line corresponds to Eq. (10-32).

Figure 10-7 shows the experimental points and a line corresponding to Eq. (10-32). The experimental data for smooth glass spheres do not fall on the general line. Baerg et al., attributed the higher values of $\alpha_{wall.max}$ for the glass spheres as being due to their smoothness, and hence to the increased mobility of the fluidized bed. They also attributed the difference between their experimental data and those of Mickley and Trilling to the higher voidage in Mickley and Trilling's bed.

An empirical formula for α_{wall} was prepared which approximately described their experimental data:

$$\alpha_{wall} = \alpha_{wall.max} - 269e^{-0.00246\,(G-0.216\gamma)}. \qquad (10\text{-}33)$$

Campbell and Rumford [225] investigated the effect of the properties of the material on heat transfer between a fluidized bed and a submerged cooling coil. The thermal conductivities of the particles varied over a wide range: from λ_M = 1.19 (sand) to λ_M = 327 kcal/m·hr·degree C (copper). Over this range of thermal conductivities of the particles, the heat transfer coefficient of the coil varied about 3-fold. However, the change in α_{wall} was not caused so much by λ_M as by the accompanying high bulk density of the materials having the high thermal conductivities. Actually, the role of λ_M must be quite small (see the analysis of experimental data below).

Vreedenberg [579, 580] carried out an experimental investigation of heat transfer in fluidized beds of large dimensions. The bed diameter was 0.565 m, and the height 1.2-1.7 m; sand was the fluidized material. Heat transfer was investigated between the fluidized bed and water-cooled horizontal and vertical (1-1.18 m) tubes of 34 mm diameter at atmospheric and elevated pressures and at various temperatures of the fluidizing agent (air).

Vreedenberg found that the heat transfer coefficient was independent of the air pressure if the linear superifical air velocity was kept constant. Because of the considerable nonuniformity of the beds which he investigated, Vreedenberg did not observe any regular relationship between the changes in α_{wall} and the position and height of the cooled surface, though he used vertical tubes of considerable length, for which some degree of deterioration of heat transfer could be expected because of overheating the layer of particles near the surface moving along the length of the tubes.

To describe his experimental data, Vreedenberg gave the formula:

$$\frac{\alpha_{wall}d}{\lambda_g} = 1.25\left[\frac{G\nu}{(G\nu)_{1.s}}\right]^{0.35}. \quad (10-34)$$

This is empirical, even though it consists of dimensionless groups. It is not applicable to the data of other investigators. Vreedenberg also proposed a second, more general, empirical formula which encompasses the experimental data of Mickley and Trilling and of Baerg, et al., it is more complicated, and gives very large discrepancies with the experimental data of others. The earlier work of Vreedenberg is described in more detail in Leva's monograph [988].

Van Heerden, Nobel and van Krevelen [337] measured the transfer of heat from a heated fluidized bed to a surrounding cooled wall. They fluidized carborundum, coke, iron oxide, and lead particles from 50 to 800 microns in diameter. Air, argon, carbon dioxide, methane, town gas, and mixtures of hydrogen and nitrogen were used as the fluidizing agents. Since they confined their investigations to incipiently fluidized beds (the so-called dense phase of a fluidized bed), as the upper limit of which they assumed a fluidized index of $N = 20$, they reached the maximum heat transfer coefficient $\alpha_{wall.max}$ in only a few cases.

They proposed two empirical formulas for the dense phase of a fluidized bed in the form of relationships between dimensionless groups.

The first was

$$\frac{\mathrm{Nu}}{\mathrm{Pr}^{0.5}\Phi^{0.45}}\left(\frac{\gamma_{1.s}}{\gamma_g}\right)^{0.18}\left(\frac{\gamma_g c_g}{\gamma_{1.s}c_M}\right)^{0.36} = 0.035\left(\frac{G}{G_{1.s}}-1\right)^{0.36}, \quad (10-35)$$

where $\Phi = \gamma_g \gamma_{1.s} d^3 / g\mu$;

Nu = Nusselt number;

Pr = Prandtl number;

$\gamma_{1.s}$ = bulk density of dense bed obtained from the fluidized bed by gradual reduction of the stream velocity to zero;

The other symbols are as used previously.

This relationship is valid only over the range $1 < G/G_{1.s} < 20$ and $\mathrm{Re} < 5$.

In the second formula they included the Reynolds number and a shape factor B (equal to unity for spheres) in place of $G/G_{1.s}$ and Φ, respectively. It also is only valid for $\mathrm{Re} < 5$ and $G/G_{1.s} < 20$.

It should be noted that the maximum heat transfer coefficients for beds of very fine particles are reached only beyond the limits of $G/G_{1.s}$ (at $G/G_{1.s} < 20$), and for coarse particles the maxima are reached beyond the limits of Re (Re > 5 in this case).

In a more recent paper [340], van Heerden, Nobel and van Krevelen made some general assumptions about the mechanism of heat transfer in fluidized beds. They paid special attention to the intensive mixing of particles in a fully developed fluidized bed, the high heat capacity of the particles compared with the gas, and the good conduction of heat in the small particles, and assumed that the rapid transfer of heat between the various parts of a fluidized bed occurred as a result of particle mixing. They particularly emphasized the role of the high particle heat capacity in intensifying heat transfer, and using a method based on analogy, confirmed these assumptions by special mass transfer experiments.

However, their statement that in a fluidized bed "the gas stream must be regarded as being everywhere essentially laminar, and that the interstitial gas serves only as a means for transferring heat from particle to particle and from particle to wall by normal conduction exclusively" may not be rigorous. In a fluidized bed the value of the Reynolds number, referred to the particle diameter and the superficial fluid velocity, may fall in the region of ten to a hundred. For instance, Re $\approx$ 20 even for quite small particles (d = 0.31 mm) at a superficial fluid velocity w_f = 0.5 m/sec in an air stream at room temperature. Flow around the particles will lead to breakaway of the streamlines. Filtrational mixing will be observed, i.e., to some extent there will be a convective mechanism of heat transfer. However, such convective transfer will be insignificant in the case of fine particles, while the conductive component of the heat transfer coefficient will be large. The gas boundary layer on the wall will experience local disturbances or even turbulence on contact of the particles if one accepts the view of Z. F. Chukhanov [177], who states that the points of contact of the particles are centers of artificial turbulence even at small Reynolds numbers. However, the intensity of such disturbances of the boundary layer has not been investigated directly and its dependence is not certain.

Based on the poor agreement between the experimental data and the empirical formulas of different workers, M. Leva [407] proposed that a real generalized fundamental relationship for heat transfer in fluidized beds could be obtained only by including the particle velocity, and that in order to solve the problems of heat transfer, further studies of the dynamics of fluidized beds and their relationship to the constructural details of the equipment were required.

Leva proposed a mechanism of heat transfer between fluidized beds and walls in which the most important part was played by the motion of particles along the wall. Leva suggested that when the particles moved in this way, they had a scouring action on the gas boundary layer, and so decreased its thickness. This scouring action of the particles moving along the boundary layer caused some doubt as to this concept of defining a boundary layer. Also, it is incorrect to neglect (as Leva did) the transverse component of the particles velocity (normal to the wall). Although the transverse particle velocity in a fluidized bed is small compared with the longitudinal velocity (for a vertical wall), it nevertheless plays a very important part (see previous reactions).

Leva gave the following empirical formula for the heat transfer coefficient:

$$\alpha_{\text{wall}} = 32.3 \cdot 10^{6} \cdot \lambda_{g} d \left(\frac{dG\eta}{\mu R}\right)^{0.6}, \tag{10-36}$$

where η = the so-called efficiency of fluidization; it is a dimensionless quantity taking into account the vertical velocity component of the particle; Leva, et al. [595, 693], gave graphs for determining the values of η to use in calculations;

$R = H/H_0$ is the ratio of expansion of the fluidized bed.

The numerical coefficient in (10-36) is dimensional.

Equations (10-36) and (10-37) apply only in the "dense phase" of a fluidized bed, starting at a fluidization index of $N \approx 2$.

In later papers [410, 595, 693], Leva, et al., developed this concept further. Wen and Leva [595] proposed a modified empirical formula for "dense phase" heat transfer between the fluidized bed and a surrounding wall which was improved by introducing the physical constants of the material, and which finally assumed the form of a relationship among three dimensionless groups:

$$\frac{\alpha_{\text{wall}} d}{\lambda_g} = 0.16 \left(\frac{c_M \rho_M d^{1.5} g^{0.5}}{\lambda_g}\right)^{0.4} \left(\frac{Gd\eta}{\mu R}\right)^{0.36}, \tag{10-37}$$

where c_M and ρ_M are the specific heat and density of the material; other symbols are as used previously; $Gd\eta/\mu R$ is a modified Reynolds number.

Wen and Fan [712] prepared a nomogram for calculating the wall heat transfer coefficient from this formula.

Agarwal and Storrow [190] investigated heat transfer between fluidized beds of various materials and surrounding steam heated walls. They paid special attention to the entry effects and the jump in gas temperature on entering the bed. As the temperature driving force Agarwal and Storrow chose the logarithmic mean of the initial (below the grid) and final gas temperatures. This gives values which are much too high compared to the actual temperature driving force, i.e., the heat transfer coefficients were reduced. Insufficient data were given to permit recalculation of their results.

Dow and Jakob [268] also investigated heat transfer between a fluidized bed and a steam-jacketed wall (see main table). They noted the presence of a core in the fluidized bed in which the radial temperature drop was negligible, and a boundary film of gas and particles along the walls; this film represented the main thermal resistance to heat transfer between the wall and the bed. Finally, there was a "thermal mixing zone" or "zone of active heat transfer" between entering gas and particles near the bottom of the bed, where a radial temperature drop was noted over the whole width of the bed. They evaluated their experimental data in the form of a relationship between dimensionless groups, and in the original paper made a rather unusual choice of the tube diameter as the characteristic dimension in Re and Nu. N. I. Gel'perin [988] rearranged the Dow and Jakob equation using the particle diameter as the characteristic dimension, and obtained

$$\mathrm{Nu} = 0.55\left(\frac{D_{\mathrm{bed}}}{H}\right)^{0.65}\left(\frac{D_{\mathrm{bed}}}{d}\right)^{-0.03}\left[\frac{(1-m)\gamma_M c_M}{m\gamma c}\right]^{-0.25}\mathrm{Re}^{0.8}, \qquad (10\text{-}38)$$

where H is the height of the fluidized bed.

Gamson [300] evaluated the experimental data of Mickley and Trilling [451] and proposed empirical correlations which were different for externally and internally heated fluidized beds, but were more complicated than the original correlation of the authors.

Miller and Logwinuk [452] investigated heat transfer between the dense phase of a fluidized bed and a central vertical cooled tube. Unfortunately the original experimental data are not given in the paper, so they cannot readily be analyzed. The correlation given by the authors is not regarded as very satisfactory. Some of the experimental values for α_{wall} differ as much as 50% from the values predicted by their correlation.

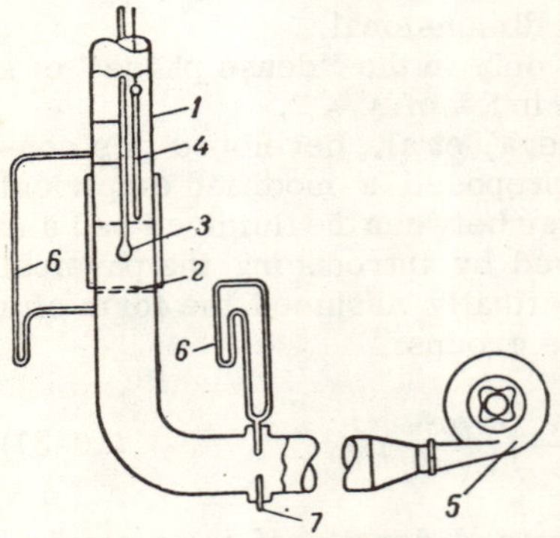

FIG. 10-8. Experimental equipment of Wicke and Hedden [601].
1) tube; 2) gas distributor; 3) heater; 4) thermometer; 5) blower; 6) micromanometer; 7) orifice meter.

In 1952 Wicke and Hedden [601] published their data on transfer coefficients for surfaces submerged in fluidized beds. They concluded (as did Leva) that the intensification of heat transfer is caused by disruption of the gas boundary layer on the heating surface by "bouncing" particles. Wicke and Hedden proposed that a decrease in the mean particle concentration is the cause of the decreases in α_{wall} after passing through the maximum.

Figure 10-8 is a diagram of Wicke and Hedden's experimental equipment. The granular material was poured in to form a bed 10 cm deep in the vertical tube 1 above the perforated gas distribution plate 2, and was fluidized by a supply of gas from the blower 5. A miniature electric heater 3 which was 13.3 mm in diameter and about 30 mm long was submerged in the fluidized bed at the end of a rod (tube). The heater surface temperature was obtained from the resistance of a thermometric nickel coil placed in the immediate neighborhood of the heater body surface. The temperature of the fluidized bed was measured with a mercury thermometer with 1/10°C divisions. Much experimental information on heat transfer coefficients between walls and fluidized beds, obtained in this apparatus, was published by Wicke and Fetting [283, 600]. The fluidized beds consisted of carborundum, sand, aluminum and lead particles and glass spheres. Particle sizes varied from 0.065 to 3 mm. Table 10-4 presents data on the materials used, and Table 10-5 summarizes the experimental results.

Table 10-4

Data on the Materials Used in the Experiments of Wicke and Fetting [283]

Expt. No.	Type of material	Particle size, mm	Mean particle diam., mm	Particle shape
1, 2, 3	Carborundum polishing sand	0.045-0.085 0.1-0.2	0.065 0.15	Triangular tetrahedral and prismatic (smooth)
4, 5, 6	River sand	0.22-0.42	0.315	Rounded polyhedra (rough)
7	River sand	0.4-0.5	0.45	Rounded polyhedra (rough)
8	River sand	0.5-1	0.75	Rounded polyhedra (rough)
9, 10, 11	Aluminum grit	0.22-0.4	0.31	Circular, polyhedral smooth grains
12	Aluminum grit	0.4-0.5	0.45	
13	Aluminum grit	0.5-1	0.75	The same
14	Lead powder	0.10-0.15	0.125	Spheres, hemispheres (smooth)

Figures 10-9 and 10-10 give experimental curves demonstrating the relationships which were found between α_{wall} and the superficial fluid velocity, particle diameter, and type of gas.

Wicke and Fetting investigated in detail the decrease in the heat transfer coefficient with increase in height of the heating surface, due to the heating of the stream of particles moving along its surface. They noted that feasible means of practical heat exchanger construction (for

Table 10-5

Experimental Results of Wicke and Fetting [283]

No.	Particle material and type of gas	Particle diam., mm	$w_{1.s}$, cm/sec	$\alpha_{1.s}$, $\frac{kcal}{m^2 \cdot hr \cdot deg}$	w_{opt}, cm/sec	α_{max}, $\frac{kcal}{m^2 \cdot hr \cdot deg}$
1.	carborundum, air	0.065	0.7	30	17.5	610
2.	carborundum, CO_2	0.065	1*	30*	**	440
3.	carborundum, air	0.15	3	50	19	490
4.	sand, air	0.315	8	42	28.5	310
5.	sand, H_2	0.315	16	135	62	930
6.	sand, CO_2	0.315	12	40	32	240
7.	sand, air	0.45	11	46	30	250
8.	sand, air	0.75	26	69	48	220
9.	aluminum, air	0.31	10	44	22	245
10.	aluminum, H_2..........	0.31	21	177	58	780
11.	aluminum, CO_2	0.31	17	50	31	160
12.	aluminum, air	0.45	16	51	31	215
13.	aluminum, air	0.75	24	61	41	190
14.	lead, air	0.125	18*	60*	45	340
15.	glass, air	0.9	44	72	70	195

*Approximate values.
**New decrease in α_{wall} not observed.

The following symbols are used in the table:
$w_{1.s}$ superficial gas velocity through dense bed close to limit of stability, cm/sec;
$\alpha_{1.s}$ heat transfer coefficient from dense bed to heating surface close to limit of stability, kcal/m^2 · hr · degree C;
α_{max} maximum heat transfer coefficient from wall to given fluidized bed, kcal/m^2 · hr · degree C;
w_{opt} superficial gas velocity through fluidized bed on reaching α_{max} cm/sec.

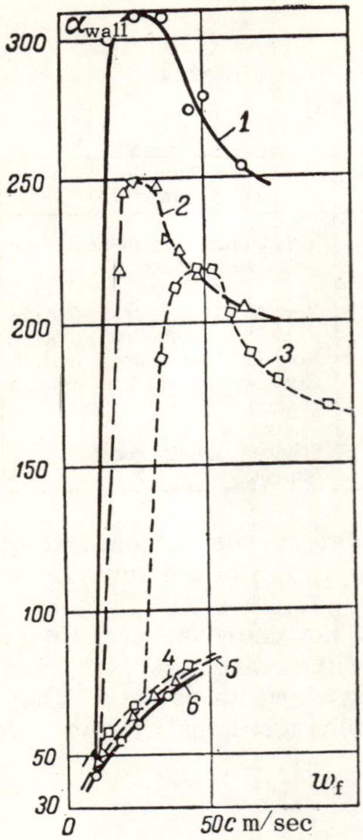

FIG. 10-9. Heat transfer coefficients between heater walls and beds of sand of various particle diameter according to Wicke and Fetting [600]. ○—d_{avg} = 315 microns; △—d_{avg} = 450 microns; □—d_{avg} = 750 microns; 1, 2, 3) fluidized beds; 4, 5, 6) stationary beds (held between screens).

instance, the use of short heating surfaces) would keep this undesirable effect small.

Wicke and Fetting proposed a model for heat exchange between a wall and a fluidized bed, assuming, like Dow and Jakob, that there is not only a boundary layer of gas but also a "boundary layer" of particles moving along the wall hindering heat transfer between it and the bulk fluidized bed. This circulating stream of particles is not isolated; particles move into it from the interior parts of the fluidized bed. As a result, there is a more or less intensive radial exchange of particle positions within the circulating stream. After some time, or some distance of travel, all the particles in the circulating stream will have been exchanged with particles from the core of the fluidized bed. Correspondingly, the heat flux density can be broken down into radial and axial components which vary over the height of the bed.

On the basis of their model, Wicke and Fetting proposed the formula

$$\alpha_{wall} = \frac{K_z}{2H_{wall}} \cdot \frac{H_0}{H}\left(1 - e^{-\frac{\alpha_d \cdot 2H_{wall}}{K_z}}\right), \tag{10-39}$$

where $K_z = \gamma_M (1 - m_{1.s}) c_M v_M \delta_g$;
H_{wall} = section of wall height active in heat transfer;
$H\ H_0$ = bed expansion;
δ_g = thickness of gas thermal boundary layer;
$m_{1.s}$ = bed voidage at limit of stability;
v_M = velocity of circulating stream of particles;
δ = thickness of "boundary layer" of particles moving along wall;
$\alpha_d = \lambda_g/\delta_g$.

The authors did not propose this equation as a design formula, and thus far it has not been possible to compare it with other data on the thickness of the gas boundary layer or the velocity of the stream of particles, which are required for determining α_d and K_z. The formula is intended only for comparing the experimental data of various investigators; values of m and the other quantities necessary for using the formula are calculated from a known value of $\alpha_{wall.max}$ and other experimental data depending on the details of the experiment. Wicke and Fetting used it to evaluate the results of their experiments and also the experimental data of Baerg [202] and van Heerden, et al. [337].

The agreement was not very good. The main deficiency of the formula and model proposed by Wicke and Fetting is the absence of analysis of the nature of many separate factors (particle diameter, thermal conductivity of the material, etc.) on the intensity of heat transfer in fluidized beds.

Toomey and Johnstone [568] took cinephotos through a transparent section of the tube wall and determined the particle velocities near the walls in a fluidized bed, and also the changes in the heat transfer coefficient between the bed and wall as a function of height in the bed. For this purpose the walls had water jackets which were sectioned over the height of the bed. Unfortunately it was not possible to find a clear relationship because of changes in factors other than those being investigated which also had an effect on α_{wall}.

Bartholomew and Katz [207] carried out experiments on heat transfer between the dilute phase of a fluidized bed (m = 0.54–0.95) and a surrounding heated wall. The bed was very deep (762 mm) with a diameter of 102 mm. For finding the average heat transfer coefficient over the bed height, they proposed the correlation:

$$\mathrm{St} = \frac{1.56 + \ln(\mathrm{Re}\cdot c_m^{-2/3} - 0.012)}{-0.227\mathrm{Pr}^{2/3}\cdot c_m^{0.42}}, \qquad (10\text{-}40)$$

where $\mathrm{St} = \alpha_{wall}/c_g G_g$; $c_m = d^3 g \gamma_g(\gamma_M - \gamma_g)/\mu^2$;
$g = 9.81$ m/sec²;
μ = gas velocity, kg/m·hr.

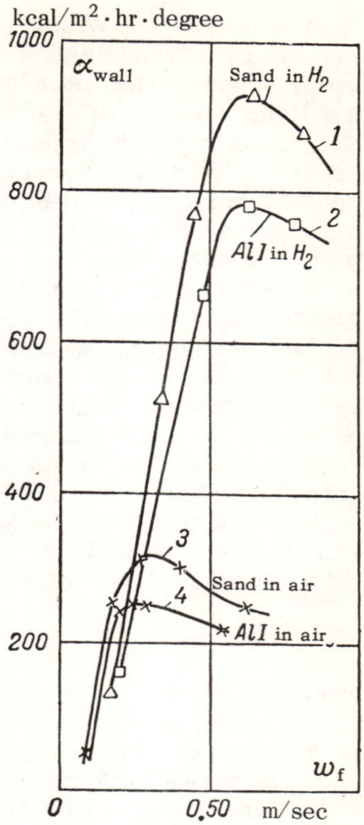

FIG. 10-10. Heat transfer coefficients between heater walls and beds of sand and aluminum particles fluidized with air and hydrogen, according to the data of Wicke and Fetting [600]. 1) sand (d_{avg} = 315 microns) in hydrogen; 2) aluminum (d_{avg} = 310 microns) in hydrogen; 3) sand (d_{avg} 315 microns) in air; 4) Al (d_{avg} = 310 microns) in air.

Their experimental conditions are shown in the prior summary Table 10-2. Leva [988] came to the conclusion that the correlations of Bartholomew and Katz and of Mickley and Trilling were quite different in form for similar operating conditions and for producing similar numerical values of α_{bed}.

Levenspiel and Walton [420] determined α_{wall} over a wide range of bed voidages for fluidized beds of glass catalyst spheres and coal and a heated tube wall (D_{bed} = 102 mm). Their values of α_{wall} were much lower than those of most other investigators, and the reason must lie in defects in the experimental methods. They proposed a dimensionless empirical relationship

$$\mathrm{St} = 0.6\,\mathrm{Re}^{-0.7}, \qquad (10\text{-}41)$$

where $\mathrm{St} = \alpha_{wall}/c_g G_g$.

Levenspiel and Walton also attempted to give a theoretical equation for the heat transfer coefficient for vertical walls wetted by fluidized beds, based on the boundary-layer theory for the medium and a very artificial concept of the structure of the fluidized bed. In addition to the usual simplifying assumptions that the bed consists of uniform spherical particles and that there is no temperature gradient in the central part, they assumed the particles in the fluidized bed to be arranged in horizontal layers. The distance between layers was constant and determined by the bed voidage; it increased with bed expansion, but in each horizontal layer the particles remained close to one another and to the wall. The boundary layer was disturbed in the plane of each horizontal layer because of contact between the particles and the wall, but it was formed anew in the intervals between the layers, and up to the next point of disturbance it achieved some thickness which depended on the distance between the rows, i.e., on the bed voidage. Assuming that the growth of the boundary layer along the vertical heat transfer surface wetted by the fluidized bed proceeded as in flow along a plate of length equal to the distance D_r between the horizontal layers of particles, and that the resistance to heat transfer was due entirely to this boundary layer of medium, Levenspiel and Walton [419] found

$$\alpha D_r/\lambda_g = 4.78/\delta\,(D_r G/\mu)^{1/5}, \tag{10-42}$$

where $\delta = (1 + \beta^{5/4})^{9/5} - \beta^{9/4}$; $\beta = 0.544\,(1 - m)\,(D_r G/\mu)^{1/5}$.

Chu [234] noted a disagreement between the theoretical and experimental values obtained by Levenspiel and Walton. Later, Ruckenstein and Schorr [843] attempted to improve the model by assuming that the wall boundary layer was disturbed by the particle motion.

A short note [574] was made in the 1955 doctoral thesis of R. Trense, which dealt with heat transfer in air-fluidized beds. Trense experimentally determined heat transfer coefficients between fluidized beds of glass spheres (d = 0.15–1.10 mm) and tall axially submerged heating surfaces (l = 500 mm); the conditions studied extended beyond the "dense phase" of fluidization. The spheres were charged onto a gas-distributing screen through a vertical tube (D_T = 100 mm) to form beds having thicknesses, before fluidization, of 76–455 mm. The weight flow rate of air varied from 470 to 19,000 kg/m ·hr, and the heat transfer coefficients varied from 34 to 107 kcal/m ·hr·degree C. Trense proposed an empirical formula for correlating his data; the actual data were not published.

Mickley and Fairbanks [450] measured heat transfer between a small electric heating probe and a bed of glass spheres fluidized by various gases. They assumed that heat transfer from the heater wall was accomplished by aggregates—"packets" of suspension. The packets received heat from the wall by means of an unsteady-state heat transfer. Based on this mechanism, Mickley and Fairbanks obtained an equation for the heat transfer coefficient of the following type:

$$\alpha_{wall} = \sqrt{\lambda_p \gamma_p c_p S}, \tag{10-43}$$

where λ_p, γ_p and c_p are, respectively, the thermal conductivity, bulk density and heat capacity of the packets; S is a mixing factor reflecting the motion and geometry of the bed. By Mickley and Fairbanks' assumptions the particle diameter has no direct effect on α_{wall}.

The work of Reed and Fenske [512] published in 1955 clearly indicated the extremely important role of particle motion in intensifying heat transfer between a bed and a submerged surface. In these experiments the heating surface was subjected to mechanical vibrations of various amplitudes and frequencies.

In recent years there have been more intensive studies of the problems of heat transfer between heating surfaces and fluidized beds [668, 686, 693, 697, 712, 713, 728, 734, 741, 742, 744, 782, 789, 830, 831, 832, 842, 843, 580, 584, 860, 877, 920, 921, 931, 932, 933, 950, 965, 967, 968, 972, 979, 988, 989, 1064, 1082, 1152, 1134, 1216, 1233].

Massimilla and Bracale [697] carried out an interesting comparative investigation of heat transfer in free beds and beds subdivided by horizontal screens. Beds of glass spheres (d = 0.7 mm) were fluidized with air in a 90-mm diameter column containing ten 100 mm high elements. The ten sections were thermally insulated from one another by gaskets, and had separate water cooling jackets, so that the heat transfer coefficient could be determined separately for each section. The free bed had an initial height of 200–800 mm, and the restricted bed from 400 to 800 mm. On the basis of a graph giving the mean wall heat transfer coefficient for the whole column α_{wall} as a function of the superificial fluid mass velocity (Fig. 10-11), the authors concluded that the values of α_{wall} were practically the same for the free and the restricted beds. However, it can be seen quite clearly from Fig. 10-11 that in the region of small mass flow rates of fluid α_{wall} for the free fluidized bed is higher than the value for the restricted bed, while at high flow rates the reverse is true. Hence, in the region of low mass flow rates G, where the particles are moving too slowly, the screens appear to have an undesirable effect on heat transfer at the walls, while in the region of high G, where the particle velocity is quite high even with the restrictions, the screens facilitate intensification of heat transfer, by keeping in check the increase in the bed voidage.

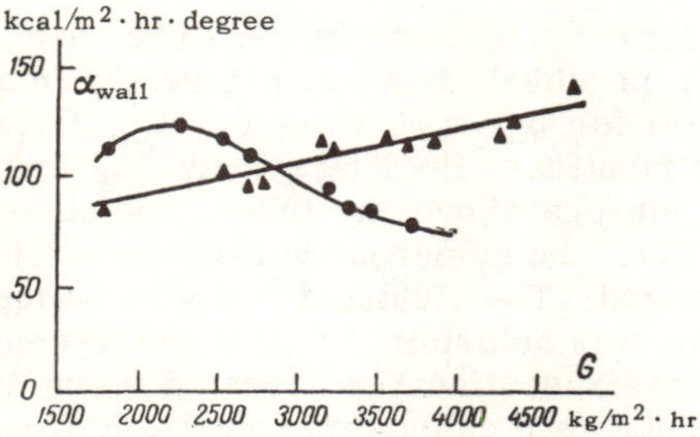

FIG. 10-11. Average heat transfer coefficient between fluidized bed and wall as a function of the air mass flow rate, according to Massimilla, et al. [697]. Bed height 400 mm, particle diameter 0.7 mm.
●—free fluidized bed; ▲—fluidized bed restrained with screens.

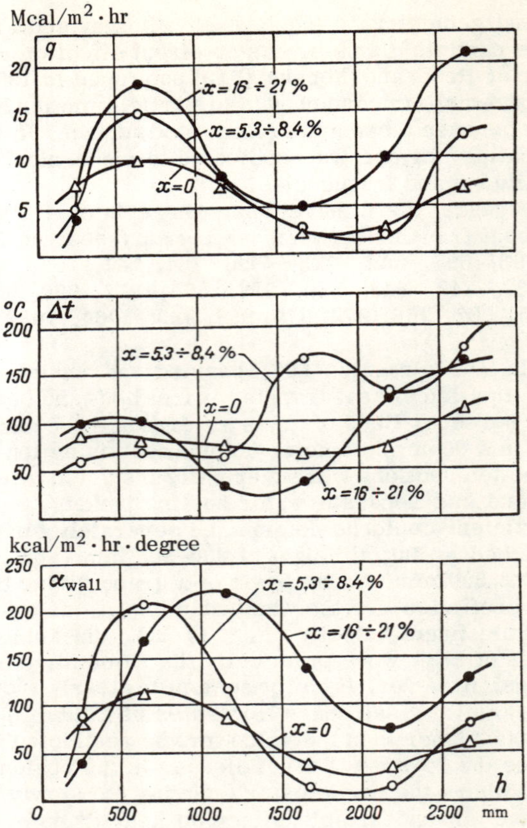

FIG. 10-12. Variation of heat flux (q), temperature driving force (Δt), and heat transfer coefficient (α_{wall}) with height (h) of the fluidized beds of A. V. Chechetkin [942]. Parameter: steam content of fluidizing air (x).

In his monograph on high-temperature heat transfer media, Chechetkin [668] provided data from investigations of complicated cases of heat transfer between walls and deep fluidized beds of wide particle size distributions. He determined α_{wall} for a number of sections of the column one above the other, and obtained different heat transfer coefficients and numerous variations as the superficial fluid velocity was changed. The fluidized bed was heated with an external jacket, while α_{wall} was determined with a water-cooled tube calorimeter divided into six sections and located along the bed axis. Heat transfer coefficients were calculated separately for each section. The bed was fluidized with air of varying steam content. Figure 10-12 shows the change in α_{wall} over the height of a fluidized bed of activated carbon particles (d = 1-4 mm) at a superficial fluid velocity w_f = 1.44-1.62 m/sec and an initial bed height H_0 = 234-262 mm. As the steam content x increased, so did α_{wall}, due to the increase in the thermal conductivity of the medium and the condensation of steam on the surface of the water-cooled tube. Figure 10-12 also shows that in the deep fluidized beds the largest values of α_{wall} were reached in each case at different heights in the lower and denser half of the bed. Such curves for a given operating regime of a fluidized bed apparatus permit determining the most suitable point for positioning a heat exchange surface within the height of the bed.

Table 10-6

Compositions and Properties of Gases Used in the Experiments of Jakob and Osberg [686]

Gas	λ_g at 0°C, kcal/m · hr · degree C	Temperature coefficient 1/degree C	Pr
H_2	0.1503	0.00279	0.718
He	0.1264	0.00210	0.638
81% H_2	0.1070	0.00279	0.444
19% N_2	0.1070	0.00279	0.444
78.9% H_2	0.0943	0.00279	0.477
21.1% N_2	0.0943	0.00279	0.477
37.6% H_2	0.0438	0.00284	0.519
62.4% N_2	0.0438	0.00284	0.519
33.3% H_2	0.0405	0.00284	0.526
66.7% N_2	0.0405	0.00284	0.526
Air	0.0210	0.00290	0.705
CO_2	0.01264	0.00470	0.809
Freon-12	0.007113	0.00670	0.732

The paper by Jakob and Osberg [686] contains experimental data on the effect of gas thermal conductivity on the heat transfer coefficient. These authors investigated heat transfer between a fluidized bed of glass spheres (d = 30-290 microns) and a submerged horizontal wire heated by an electric current. The wire diameter (0.132 mm) was of the same order of magnitude as the particle diameter. Eight gases with thermal conductivities from 0.00715 to 0.15 kcal/m · hr · degree were investigated. Table 10-6 gives the compositions and properties of the fluidizing gases. The experimental results are given in Table 10-7. Very high transfer coefficients α_{wall} were obtained under the conditions of these experiments. The data confirm that there is a direct proportionality between α_{wall} and the gas thermal conductivity and an inverse proportionality to the particle diameter.

Jakob and Osberg considered an unsteady-state mechanism of heat transfer between the wire and the fluidized bed, and proposed a formula containing two empirical constants dependent on particle diameter:

$$\alpha_{wall} = \alpha_0 (1 - m)(1 - e^{-p\lambda_r}), \qquad (10\text{-}44)$$

where α_0 and p are experimental constants given in Table 10-8; and α_{wall} is the heat transfer coefficient between the wire and fluidized bed.

In 1957, Wesser and Mardus [713] published data on heat transfer to surfaces submerged in beds of glass spheres fluidized with water.

Table 10-7

Experimental Results of Jakob and Osberg [686]

Particle diameter, mm	Gas	Wire temp., °C	Gas temp., °C	Fluidization index	$1-m$	α_{wall}, $\frac{kcal}{m^2 \cdot hr \cdot deg}$
0.031	H_2	65	16	118*	0.481	6150
	He	75	20.5	248*	0.406	5200
	81% H_2	75	19.4	230*	0.437	5445
	19% N_2	75	19.4	230*	0.437	5445
	78.9% H_2	75	18.9	216*	0.440	5120
	21.1% N_2	75	18.9	216*	0.440	5120
	33.3% H_2	75	15.5	258*	0.422	3610
	66.7% N_2	75	15.5	258*	0.422	3610
	Air	100	22.2	188*	0.415	2500
	CO_2	100	20.5	80*	0.425	1830
	Freon-12	75	11.6	145*	0.429	1340
0.061	H_2	65	15	5	0.517	5120
	He	75	19.4	81*	0.469	4830
	81% H_2	75	18.9	8	0.496	4680
	19% N_2	75	18.9	8	0.496	4680
	78.9% H_2	75	18.3	5	0.502	4630
	21.1% N_2	75	18.3	5	0.502	4630
	33.3% H_2	75	14.4	5	0.512	2900
	66.7% N_2	75	14.4	5	0.512	2900
	Air	100	21	4.5	0.513	1925
	CO_2	100	20	7	0.474	1487
	Freon-12	100	18.9	4	0.510	1060
0.153	H_2	65	14.4	1.9	0.534	4215
	He	75	19.4	1.9	0.549	3950
	81% H_2	75	17.8	1.9	0.537	4100
	19% N_2	75	17.8	1.9	0.537	4100
	78.9% H_2	75	17.2	1.9	0.549	3780
	21.1% N_2	75	17.2	1.9	0.549	3780
	33.3% H_2	75	13.9	—	0.545	2340
	66.7% N_2	75	13.9	—	0.545	2340
	Air	100	20.5	1.5	0.559	1413
	CO_2	100	21	1.7	0.558	1120
	Freon-12	100	18.9	1.7	0.546	731
0.292	H_2	65	13.3	1.95	0.525	3440
	He	75	17.8	1.6	0.518	3140
	81% H_2	75	17.2	1.6	0.542	3170
	19% N_2	75	17.2	1.6	0.542	3170
	78.9% H_2	75	16.1	1.7	0.525	3070
	21.1% N_2	75	16.1	1.7	0.525	3070
	33.3% H_2	75	13.9	—	0.517	1830
	66.7% N_2	75	13.9	—	0.517	1830
	Air	100	20	1.25	0.537	1023
	CO_2	100	20	1.3	0.535	717
	Freon-12	100	18.3	1.15	0.543	498

*Calculated from E. Graf, Thesis, Zurich, 1955.

Their experiments were carried out in an apparatus, almost identical to that used by Wicke (described above), which makes it convenient in comparing the results. They also determined the heat transfer coefficients of the surfaces when in contact with stationary beds of material and with pure water (with no change of material in the tube). The particles had diameters of 1.0, 3.3, and 5.9 mm.

Table 10-8

Experimental Constants α_0 and p in Experiments of Jakob and Osberg [686]

Particle diameter, mm	α_0, kcal/m^2 · hr · degree C	p, m · hr · degree/kcal
0.031	12,970	24.2
0.061	10,300	20.2
0.153	8,100	16.8
0.292	7,560	12.1

The data of Wesser and Mardus confirmed the presence of a maximum in the curve of α_{wall} versus the velocity of the stream fluidizing the bed, as well as an increase in the heat transfer coefficient in going from a surface wetted by the liquid in an empty tube to a surface in contact with a liquid passing through a stationary or fluidized bed (Fig. 10-13). However, in contrast to the data for beds permeated by gases, in the experiments with water it appeared that the heat transfer coefficient in the fluidized bed was smaller than in the dense bed (Fig. 10-13). Over a certain range of superficial fluid velocities immediately after fluidization (Fig. 10-14) there is no appreciable increase in α_{wall} as the water velocity is increased.

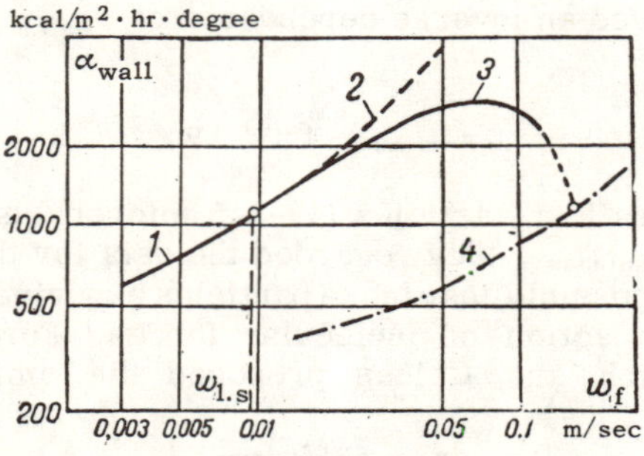

FIG. 10-13. Heat transfer coefficient at the heater wall in the experiments of Wesser and Mardus [713].
1) bed up to limit of stability ($w_{1.s}$); 2) stationary dense bed at $w_f > w_{1.s}$ (held between screens to prevent fluidization); 3) fluidized bed; 4) tube without bed.
Particle diameter d = 1 mm. The section of $\alpha \approx$ const. near the limit of stability is omitted for convenience.

Wesser and Mardus gave no conclusive explanation for the experimental results which they obtained, but mentioned that Wicke's model for heat transfer and fluidization did not correspond to their observations.

In attempting to explain the results of their experiments, Wesser and Mardus came to the conclusion that the internal motion of the spheres in the fluidized bed could have no effect on heat transfer at the walls. An explanation of the "anomaly" was given by Zabrodsky [744] on the basis of theoretical arguments given earlier.

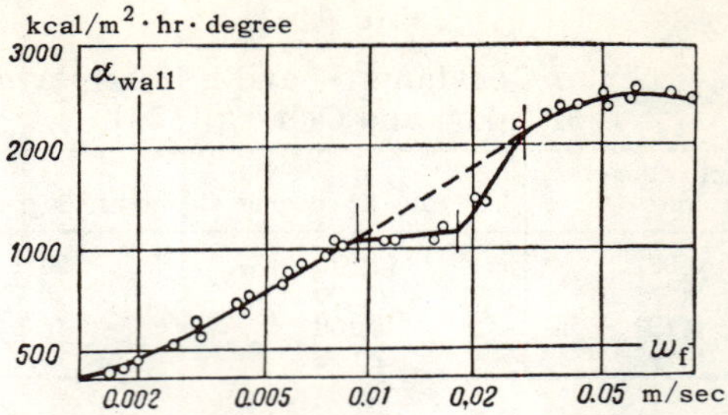

FIG. 10-14. Heat transfer between heater and bed of spheres (d = 1 mm) in a water stream according to Wesser and Mardus [713].

In contrast to Wesser and Mardus, Lemlich and Caldas [831] fluidized a bed of very fine glass spheres (40-680 microns) with water. The column had a diameter of 51 mm and its walls were steam heated. In these experiments the presence of a bed increased the heat transfer coefficients appreciably, and a maximum was observed in the relationship between α_{wall} for the fluidized bed and the superficial fluid velocity. For particles as small as those used by Lemlich and Caldas filtrational mixing was obviously insignificant, and as in the case of gas fluidized beds, they observed an inverse dependence of $\alpha_{wall.max}$ on the particle diameter:

$$\alpha_{wall.max} = 4.03 d^{-0.77}. \qquad (10\text{-}45)$$

Only their smallest particles (49-145 microns) gave unexpectedly low values of $\alpha_{wall.max}$; they regarded the data for these particles as unreliable. It is possible that these particles experienced some aggregation under the action of molecular forces. For the region up to $\alpha_{wall.max}$, Lemlich and Caldas proposed the empirical equation:

$$\alpha_{wall} = 0.055 \frac{G\lambda_m}{\mu} \qquad (10\text{-}46)$$

where G, μ, λ_c are, respectively, the weight flow rate (kg/m²·hr), the viscosity (kg/m·hr) and the thermal conductivity (kcal/m·hr·degree C) of the liquid.

For the region beyond $\alpha_{wall.max}$:

$$\mathrm{St} \cdot \mathrm{Pr}^{2/3} = 1.4 \left(\frac{D_{bed}}{d}\right)^{0.79} \mathrm{Re}_{bed} \cdot m, \qquad (10\text{-}47)$$

where $\mathrm{Re}_{bed} = w_f D_{bed}/\nu$.

Richardson and Mitson [841, 842] fluidized beds of spherical particles (d = 460-4260 microns) of various materials (glass, Pb, Fe, Cu) in a 50-mm diameter tube using water and aqueous glycerine solutions. Under the conditions of these experiments, $\alpha_{wall.max}$ was reached at bulk concentrations of particles in the fluidized bed of the order of 7%. It is possible that as the concentration $(1-m)$ is increased above 7%, the negative effect of laminarization of the core of the stream of particles was more pronounced than the positive effect of increasing the number of particles transferring heat from the wall. The Reynolds number, referred to the tube diameter, varied from 3000 to 23,900. The authors explained the intensification of heat transfer when a bed was present in the tube by the turbulent effect of the particles on the liquid boundary layer.

For the region up to $\alpha_{wall.max}$, i.e., for $(1-m) > 7\%$, the authors proposed a quite complicated formula:

$$Nu = 119 Pr^{0.4} \left(\frac{c_M}{c_L}\right)^{0.28} \left(\frac{\lambda_M}{\lambda_L}\right)^{0.02} Re_{ff}^{N} \cdot Re_{bed}^{0.075}, \qquad (10-48)$$

where $N = 0.02\,(3.45 + \gamma_M/\gamma_L)$;
$Re_{ff} = w_{ff} d/\nu$;
$Re_{bed} = w_f D_T/\nu$.

According to (10-48) α_{wall} is practically independent of the liquid velocity and of the ratio of the thermal conductivities of the particles and liquid.

Bondareva [728] determined α_{wall} for a centrally placed electric heater (rod of 10 mm diameter) submerged in a air-fluidized bed of river sand in an 82-mm diameter tube; simultaneously she measured the effective thermal conductivity of the bed. The numerical values of α_{wall} which she obtained were much larger than those of other investigators, since the latter identified α_{wall} with the heat transfer coefficient from the wall to the core of the bed, while Bondareva separated the total thermal resistance to heat transfer into $1/\alpha_{wall.film}$ and δ/λ_{eff}. Here $\alpha_{wall.film}$ denotes the "film" heat transfer coefficient at the wall when the resistance of the effective thermal conduction is taken into account separately; δ denotes the distance from the wall to the position at which the bed temperature is measured. The numerical values of $\alpha_{wall.film}$ are lacking in accuracy, since they require exact knowledge of the temperature profiles. The coefficients $\alpha_{wall.film}$ obtained by Bondareva are shown in Fig. 10-15. The maximum $\alpha_{wall.film}$ lies in the region of low relative bed expansions (of the order of 1.2). No other investigators have given data for the determination of such coefficients. Part of the discrepancy between the numerical values of α_{wall} obtained by various investigators may be due to neglecting the thermal resistance δ/λ_{eff}.

In 1958, Zabrodsky published two papers [741 and 744] in which an approximate theoretical equation was derived for heat transfer between a fluidized bed and a boundary surface (see earlier). On the basis of the concepts developed in these papers, an analysis was given of the experimental data of various investigators. Later, 1959-1960, Ernst [972] and I. P. Mukhlenov, D. T. Traber, and V. V. Sarkits [931, 932, 1082] came to very similar conclusions. In experiments to determine α_{wall} for a horizontal 21-mm diameter tube submerged in a fluidized bed of milled peat, Zabrodsky showed [744] that λ_g of the gas had an important effect on α_{wall} only in the immediate neighborhood of the heating surface but not in the core of the bed. Thus the increase in the heat transfer coefficient between the wall and fluidized bed resulting from increasing the wall temperature, is not due solely to the radiant component.

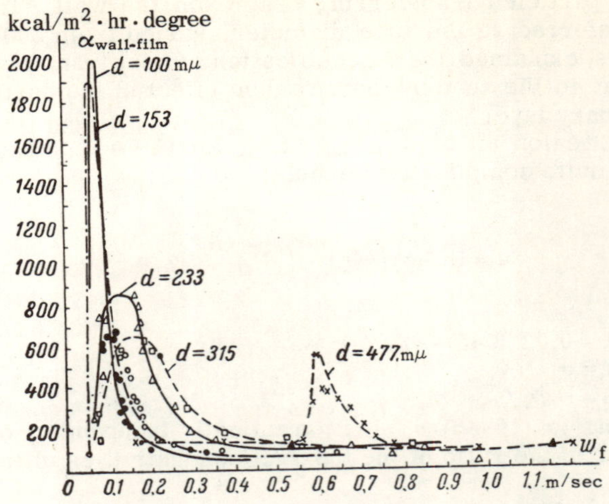

FIG. 10-15. "Film" heat transfer coefficient wall.film between heat wall and a fluidized bed of sand according to A. K. Bondareva [728]. Parameter: particle diameter.

In 1958-1959, J. Ciborowski and J. Roszak published a series of papers [789, 813, 950, 965] dealing with investigations of high-temperature heat transfer between fluidized beds and walls ($H/D_{\text{bed}} = 20$). They fluidized eight narrow fractions of river sand and four fractions of coal coke in a copper tube with a porous steel grid plate, using air as the fluidizing medium. The particle diameters varied from 0.15 to 2.5 mm.

The copper tube was surrounded by two steam jackets, of which the outer served mainly as a heat insulator for the inner. Condensate from the inner jacket, for heating the fluidized bed, flowed into a copper vessel. A feed tube was fitted to the upper part of the copper tube by means of which it was possible to maintain a constant bed level in all the experiments from the lowest to the highest velocities. The bed temperature was measured by a probe with two thermocouples (shielded and unshielded hot junctions) which could be traversed horizontally and vertically. Measuring temperatures by such probes were discussed in preceding sections. The heat transfer coefficients at the tube walls were determined as

$$\alpha_{\text{wall}} = Q / F \Delta t \tau,$$

where Q is the quantity of heat transferred to the bed, as determined from the weight of condensate drawn from the inner steam jacket. Bearing in mind the possibility of cooling of the condensate, the inaccuracy in determining the quantity, and possible variations in the wetness of the steam, this method cannot be regarded as very precise.

By this method Ciborowski and Roszak determined α_{wall} for relatively long heating surfaces ($H = 100$ mm). The temperatures measured by the unshielded thermocouple were taken as the particle temperatures, and appeared to be practically constant over the cross section at distances more than 3 mm from the walls. The gas temperatures (measured by the shielded thermocouples) decreased gradually from the walls to the tube axis, particularly in the lower parts of the bed.

In addition to α_{wall}, the heat transfer coefficient evaluated from the integral mean temperature measured by the unshielded thermocouple, $(t_{bare})_{avg} = \int_0^H \frac{t_{bare} d_H}{H}$, Ciborowski and Roszak also introduced another heat transfer coefficient, denoted by $\alpha_{wall.shielded}$, evaluated on the basis of the gas temperature as measured by the shielded thermocouple at the tube axis. These temperatures are simpler to measure (they appear to be more stable during bed oscillations). Since $t_{wall} - t_{shielded} > t_{wall} - t_{bare}$, it is found that $\alpha_{wall.shielded} < \alpha_{wall}$. However, physically $\alpha_{wall.shielded}$ is more arbitrary than α_{wall}, since according to this scheme $\alpha_{wall.shielded}$ is also a film coefficient, and for its calculation temperatures are taken not at the boundaries of some thin "film," but at points separated by the length of the tube radius in the presence of an appreciable radial temperature gradient, as indicated by the bare thermocouples.

Ciborowski and Roszak found a nonmonotonic dependence of $\alpha_{wall.max}$ on the particle diameter (under their experimental conditions, the dependence changed in the region between d = 0.59 and d = 1 mm) as shown in Fig. 10-16. They offered no explanation and considered

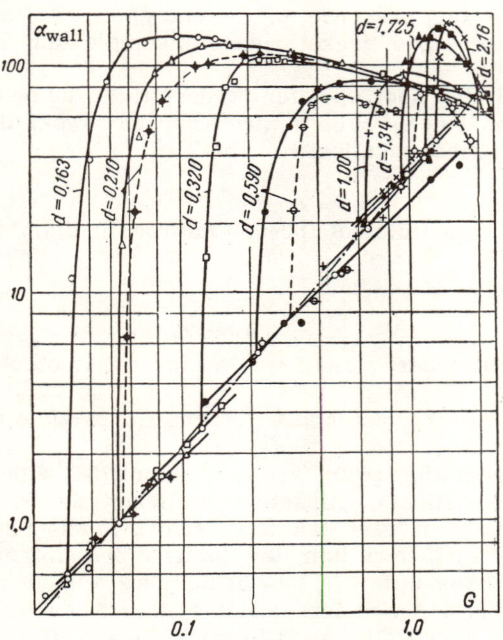

FIG. 10-16. Heat transfer coefficient at wall (α_{wall}, kcal/m^2·hr·degree) as a function of fluid mass flow rate (g, kg/m^2·hr) for beds of quartz sand and coke according to Ciborowski and Roszak [950].
Solid lines are for sand, dashed lines for coke. Parameter: particle diameter, mm.

the results not completely reliable. However, the nonmonotonic variation is in agreement with theoretical considerations [744], and was more recently confirmed in the experiments of V. B. Sarkits [931].

The authors stated that by visual observation it was possible to distinguish a number of zones of variable height in the deep fluidized bed, where many of the particles move downward in intermittent motion, but the greater part are in continuous motion within the bed.

Ciborowski and Roszak dealt in some detail with the complications in calculating α_{wall} caused by entry effects (the gas temperature jump at the grid). Usually the temperature difference is taken to be the integral mean value above the grid, while the quantity of heat is taken as the total amount, including that transferred by the gas at the grid; this gives an overestimate of α_{wall}. The authors therefore proposed to calculate α_{wall} from the logarithmic mean-temperature difference, basing this on the initial gas temperature below the grid, i.e., far from the calculated heat transfer surface. Usually such values of α_{lg} are lower than α_{wall}, but, as indicated by the authors themselves, the use of α_{lg} is theoretically incorrect. In practice, it seems best to use the usual α_{wall} rather than α_{lg}, particularly since the heating action of the grid is mainly if not entirely of a secondary nature in the sense that it is caused by heat conducted to the grid by the particles of material, which obtain the heat from the walls not by thermal conduction through the constructional elements of the equipment itself. Thus, it can be assumed that in the presence of an entry effect almost all the heat liberated by the steam jacket passes to some extent through the wall surfaces.

For calculating heat transfer in deep fluidized beds, at values of the fluidization index $G/G_{1.s} = 1.05 \div 1.8$, Ciborowski and Roszak proposed two empirical equations:

$$\alpha_{lg} = 55d^{0.468}(G - G_{1.s})^{0.52}; \quad (10\text{-}49)$$

(for particles of 163-590 microns in diameter)

and

$$\alpha_{lg} = 50\varphi^{-0.54}(G - G_{1.s})^{0.27} \quad (10\text{-}50)$$

(for particles of 1-2.16 mm in diameter).

Here φ is an experimentally determined particle shape factor as defined by Leva.

In his dissertation [860] (and in a paper [734] with N. I. Gel'perin and V. Ya. Kruglikov), Ainshtein published the results of an experimental study of heat transfer from a single tube (vertical and horizontal) to a 275-mm diameter fluidized bed (152 mm in a few experiments). Three different tube heaters were used; one was 30 mm in diameter and 210 mm long and the other two were both 22 mm in diameter and 110 and 210 mm in length. The air distribution grid was made of foamed fire clay. The fluidized solids consisted of narrow fractions of sand with equivalent diameters of 163, 224 and 285 microns; in most experiments the initial bed height was 350 mm.

About 100 experiments were carried out with the horizontal tube placed at five different distances, h, (54 to 221 mm) above the grid.

The experimental data fell along the following line with a scatter not exceeding 10%:

$$Nu_e = 0{,}96 Re_e^{0.34} \cdot Pr^{0.33} \left(\frac{h}{D_{bed}} \right)^{0.16}, \tag{10-51}$$

or, specifically for air fluidization,

$$Nu = 0{,}86 Re_e^{0.34} \left(\frac{h}{D_{bed}} \right)^{0.16}, \tag{10-51a}$$

where the mean velocity in the interstices between the particles, $w_{avg} = w_f / m$, is used as the characteristic velocity, and the hydraulic radius $\delta = dm/(6)(1-m)$ is taken as the characteristic length.

Ainshtein checked this relationship over the ranges of Re_e from 0.215 to 2.35 and h/D_{bed} from 0.2 to 0.8. The vertical tube was placed at 15 different distances r from the bed axis (from 29 to 106 mm). The more than 500 experimental points fell on the following line with an error not exceeding 15%:

$$Nu_e = 1{,}31 Re_e^{0,285} \cdot Pr^{0.33} W^{-0.2} \left(1 - \frac{r}{R} \right)^{0.36}, \tag{10-52}$$

or, specifically for air fluidization,

$$Nu_e = 1.18 Re_e^{0,285} W^{-0.2} \left(1 - \frac{r}{R} \right)^{0.36}, \tag{10-52a}$$

where $W = w_{avg}/w_{1.s} = N/m_{1.s}$, and R is the radius of the fluidized bed.

Table 10-9

Experimental Results of N. N. Varygin and I. G. Martyushin [877]

Material	Properties of granular material		Temp. in bath, °C	Experimental results	
	Equiv. particle diameter, microns	Spec. weight of particles kg/m3		Optimum velocity, m/sec	Maximum heat transfer coefficient, kcal/(m²·hr·degree C)
Ferrosilicon . .	82.5	6800	30	0.3	620
Hematite	173	4630	30	0.35	420
Carborundum . .	137	3200	25	0.22	430
Quartz sand . . .	140	2590	30	0.2	426
Quartz sand . . .	198	2610	30	0.25	367
Quartz sand . . .	215	2620	30	0.27	358
Quartz sand . . .	428	2600	30	0.36	310
Quartz sand . . .	515	2600	30	0.42	265
Quartz sand . . .	650	2600	20	0.5	238
Quartz sand . . .	1110	2600	20	0.68	180
Glass spheres .	1160	2500	20	0.76	175

Note: The equivalent particle diameter is determined from the formula

$$d_e = \frac{1}{\sum \frac{a_i}{d_i}},$$

where a_i is the weight fraction of the size fraction on the sieve;
d_i is half the sum of the passing sieve sizes, microns.

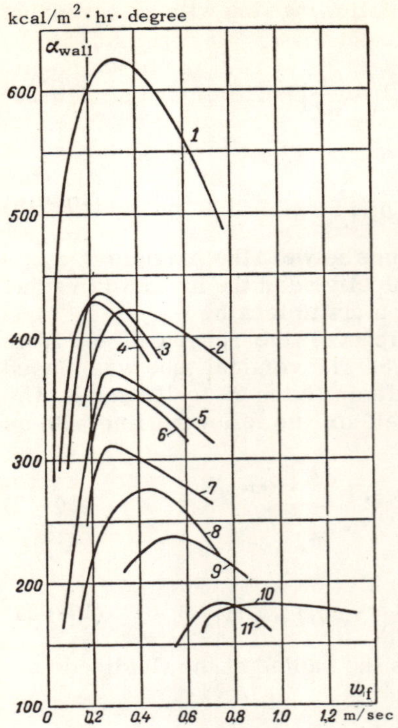

FIG. 10-17. Wall heat transfer coefficients as functions of the superficial fluid velocity, data of Varygin and Martyushin [877].

Particle sizes in microns: 1) ferrosilicon, d = 82.5; 2) hematite, d = 173; 3) carborundum, d = 137; 4) quartz sand, d = 140; 5) quartz sand, d = 198; 6) quartz sand, d = 216; 7) quartz sand, d = 428, 8) quartz sand, d = 515; 9) quartz sand, d = 650; 10) quartz sand, d = 1110; 11) glass spheres, d = 1160.

The experiments with the vertical tube were carried out over the ranges r/R from 0.21 to 0.77, $\mathrm{Re_e}$ from 0.215 to 2.9, and W from 1.55 to 9.1. Under the conditions of Ainshtein's experiments the bed height had no effect on heat transfer. Increasing the bed diameter without changing h or r led to poorer heat transfer to the horizontal tube and better transfer to the vertical tube. Heat transfer improved as the distance h of the horizontal tube above the gas distributor grid was increased, and as the vertical tube approached the tube axis. Changing the geometrical dimensions of the heaters over narrow limits had no noticeable effect on heat transfer.

Varygin and Martyushin [877] determined α_{wall} in a fluidized bed of 20-mm diameter silver spheres by the regular thermal regime method under practically gradientless cooling. The spheres were preheated to 800°C in an electric furnace, and then dropped into the fluidized bed, always at the same spot. The hot junction of a thermocouple was built into a sphere, and its cooling continuously and automatically recorded on a potentiometer with a chart speed of 2.67 mm/sec. The fluidized bed was either 82.5 or 157 mm in diameter. There was no observable effect of bed diameter on the cooling process. Table 10-9 gives the characteristics of the granular materials and some of the experimental results.

Figure 10-17 gives curves showing α_{wall} as a function of w_f. Using dimensional analysis, Varygin and Martyushin found the following empirical formula for determining $\alpha_{wall.max}$ from the data:

$$\mathrm{Nu}_{max} = 0.86\,\mathrm{Ar}^{0.2}; \text{ for Ar from 30 to 135,000,} \qquad (10\text{-}53)$$

where $\mathrm{Ar} = d^3\gamma_M\gamma_g/\mu^2 g$.

For the preliminary design of industrial equipment with horizontal tubes at large pitch, the authors suggested an operating heat transfer coefficient equal to $(0.8\div0.9)\ \alpha_{wall.max}$, assuming no difficulties in selecting the optimum w_f from a heat transfer point of view.

Ernst [972] carried out studies of heat transfer between wall surfaces and both moving beds and fluidized beds to determine the effects of various factors as well as the mechanism of heat transfer.

Narrow size fractions of quartz particles 100-200, 300-500 and 500-700 microns (in most of the experiments) were fluidized in 150-mm diameter tubes. Vertical particle velocities were determined through a transparent window in the tube walls using a ruler and stop-watch. A short section of the tube wall was used as the heating surface, and in most experiments with the fluidized bed a plate heater (height 150, width 50, and thickness 10 mm) was submerged in the bed. In some cases the heating surface consisted of a horizontal 6-mm diameter tube. In one of the series of experiments the submerged heater was vibrated by a crank mechanism moving at up to 200 r.p.m. The flow rate of circulating material varied from 0 to 4 tons/hr. The maximum flow rate corresponded to a descending velocity of the material in the dense bed of about 4 cm/sec.

Ernst assumed that heat is transferred from the walls by rapidly moving particles or packets of particles which have the temperature of the core of the bed on arriving at the heating surface, and that the main thermal resistance is concentrated near the wall itself and corresponds to the smallest distance between the particles and the heating surface. This is an even greater simplification than was made in [744]. Ernst lists three models for the mechanism of heat transfer through the gas sublayer:

1. Heat is transferred through the gas layer by steady-state molecular thermal conduction.

2. Passage of heat through a moving gas layer is caused by unsteady-state take-up of heat by the gas. In this case there is a complicated relationship between heat transfer and the velocities of the gas and the moving particles.

3. Heat transfer occurs in an unsteady-state fashion to packets of particles coming into contact with the heat transfer surface.

Ernst noted that none of these models gave a clearly defined picture of the complicated process of heat transfer in fluidized beds. For long contact times of packets of particles with the heating surface the mechanism approximates to model 3, while at small contact times it approximates to model 1, since then the temperature of the packets contacting the surface is not able to increase appreciably above the temperature in the core of the bed during their total residence at the wall; optimum heat transfer is achieved under these conditions.

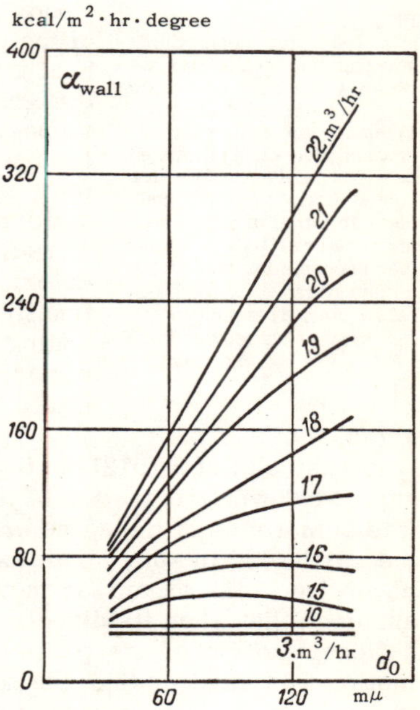

FIG. 10-18. Brötz' data [942] on the heat transfer coefficient (α_{wall}) between a fluidized bed and a wall as a function of the diameter of the openings in the grid (d_0). Parameter: flow rate of fluidizing air.

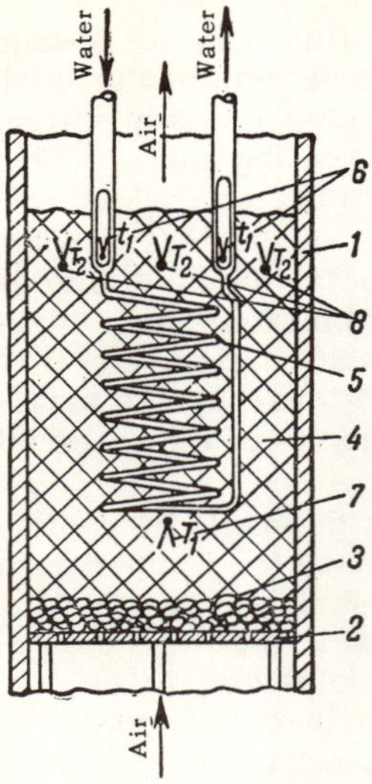

FIG. 10-19. Sketch of Sarkits [931] for determining heat transfer coefficients between a fluidized bed and a submerged coil.
1) quartz tube; 2) grid; 3) crushed quartz; 4) fluidized bed of catalyst; 5) water cooled coil; 6) thermocouples for measuring water temperatures (t_1 and t_2); 7) thermocouple for measuring temperature (t_{bare}) of bed below coil; 8) thermocouple for measuring temperature (t_{2bare}) of bed above coil.

Ernst's experiments confirmed an inverse relationship between α_{wall} and both the height of the heating surface and the height at which the heater is placed in a given fluidized bed (due to the increase of bed voidage with height in the bed); they also confirmed that α_{wall} is independent of the bed depth under otherwise similar conditions. At a given linear superficial fluid velocity, Ernst did not observe an effect of gas pressure on α_{wall} over a range of pressures from 1 to 100 atm. No effect of electrostatic charge on heat transfer was observed under the experimental conditions for the quartz particles (larger than 100 microns). Electrostatic effects were not investigated with the finer particles. Ernst mentioned that the type of gas distributing device had a considerable effect on α_{wall} if the heat transfer surface was submerged in the lower part of the fluidized bed. He gave a graph indicating a sharp decrease with "poor gas distribution," but gave no information on the differences between the "poor" and "good" gas distributors. The form of the optimum gas distributor for purposes of heat transfer with the wall is not clear. From the graph of Brötz (Fig. 10-18) given in the monograph of Syromyatnikov and Volkov [942], it appears that α_{wall} increases sharply as the pore size of the grid increases, i.e., as the uniformity of the initial gas distribution is made worse. Obviously some (but not an extreme) deterioration of the gas distribution will intensify the particle mixing. There is some analogy here to the improved mixing of the particles at the transition from a uniform liquid fluidized bed to a bed fluidized by a gas.

Ernst claimed [972] that α_{wall} was independent of the rate of organized downward motion of the fluidized beds, although this velocity was sometimes of the same order as the vertical pulsation velocity of the particles. In some of his experiments there was even a decrease in α_{wall} due to the organized motion of the material. However, it can be assumed that when its direction coincided with the direction of the pulsational motion of the particles, the positive effect of the organized motion was smaller than the negative effect in the other half-period of the pulsation. It is also possible that the heater was submerged in a zone of the fluidized bed where there is usually a "natural circulation" ascending current of particles, and the organized downward motion of the whole bed only retarded the transfer of particles around the heating surface. Vertical mechanical oscillations of a 150 mm high heating surface submerged in a fluidized bed led to a marked increase in heat transfer at low gas velocities, but at high gas velocities there was even some reduction in heat transfer because of an increase in bed voidage around the vibrating heater.

Ernst proposed the following formula for calculating the heat transfer coefficient under optimum conditions:

$$\alpha_{wall.max}=\frac{6(1-m)\lambda_g}{d}\left[\left(1+\frac{2\delta_{w.min}}{d}\right)\times\ln\left(1+\frac{d}{2\delta_{w.min}}\right)-1\right]. \quad (10\text{-}54)$$

He calculated [972], $\delta_{w.min}\approx$ 10 microns from the experimental data for all the particle fractions investigated. This is difficult to check, since the corresponding data on the voidage of the fluidized bed and its method of measurement were not published.

Sarkits [931] published the results of a very extensive study of heat transfer between a fluidized bed and a submerged coil cooler in a series of papers written with I. P. Mukhlenov, D. G. Traber, T. P. Bondarchuk and E. A. Protopopova [920, 921, 932, 933, 1064, 1082]. The experiments were carried out in 49, 73 and 100 mm diameter quartz tubes wrapped with external heater windings. The water-cooled 20 mm diameter coil was made of 4 mm outside diameter tubing, and had a height of 80 or 60 mm. The temperature difference used in the calculations was taken as

$$\Delta t=\frac{(t_{2bare}-t_1)-(t_{1bare}-t_2)}{2.3\log\frac{t_{2bare}-t_1}{t_{1bare}-t_2}},$$

Table 10-10

Characteristics of Materials Used in the Experiments of V. B. Sarkits [931]

No.	Material	d, mm	γ_{bulk}*, kg/m^3	c_M, kcal/kg · degree C
1.	Vanadium catalyst BAV	0.127	497	0.233
		0.257	503	0.233
		0.38	507	0.233
		0.75	518	0.233
		1.5	527	0.233
		2.5	535	0.233
		3.5	540	0.233
2.	Alumina-nickel, methane conversion catalyst	1.5	1295	0.224
		2.5	1300	0.224
		3.5	1330	0.224
3.	Iron-magnesium, carbon monoxide conversion catalyst	0.38	842	0.315
		0.75	895	0.315
		1.5	948	0.315
		2.5	954	0.315
		3.5	960	0.315
4.	Iron-chromium, carbon monoxide conversion catalyst	0.38	1140	0.207
		1.5	1172	0.207
5.	Vanadium catalyst VKhZ (spherical)	2.5	728	0.249
6.	Silica gel	0.38	895	0.285
		1.5	915	0.285
7.	Marble	0.38	1243	0.250
		1.5	1290	0.250
8.	Quartz	0.38	1265	0.223
9.	Chamotte	1.5	648	0.258
10.	Coke	1.5	620	0.270

*γ_{bulk} = bulk density of material "as poured."

where t_1 and t_2 are the water temperatures at the inlet and outlet of the coil, and t_{1bare} and t_{2bare} are the bed temperatures below and above measured by bare thermocouples (see Fig. 10-19).

The characteristics of the granular materials used in the experiments are given in Table 10-10. The calculated diameter is determined as the arithmetic mean of the corresponding sieve openings. Fractions 0.25-0.5, 0.5-1, 1-2, 2-3, 3-4 and 4-5 mm were used. Other than the spherical vanadium catalyst, the particles had irregular shapes and rough surfaces. Figures 10-20, 10-21 and 10-22 give the experimental curves derived from more than 1000 experiments for the relationships between α_{wall} and w_f with particle size and material as parameters. The relationships shown in Figs. 10-20 and 10-21 are of the usual nature; Figs. 10-22 and 10-23, however, show that Sarkits and his co-authors investigated the nonmonotonic nature of the relationship between $\alpha_{wall.max}$ and the particle diameter more systematically than Ciborowski and Roszak [789]. As already mentioned, this type of relationship is to be expected because of the increasing importance of filtrational mixing with large particle diameters. The experimental values of $\alpha_{wall.max}$ and some additional data are given in Table 10-11. Sarkits also showed by experiment that the fluidized index N does not characterize in a single-valued manner the relative expansion of a bed of particles in going from one particle diameter to another. Beds of large particles expand more at the same value of N [931].

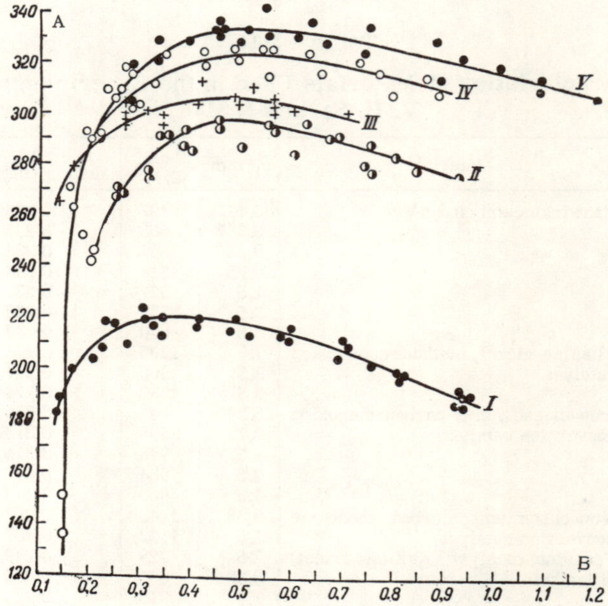

FIG. 10-20. Effect of air velocity and particle physical properties on heat transfer between a fluidized bed and a submerged coil [1082]. Mean particle diameter 0.38 mm: I—BAV catalyst; II—iron-chromium catalyst; III—silica gel; IV—quartz; V—marble. A—heat transfer coefficient, kcal/m^2·hr·degree; B—air velocity, m/sec.

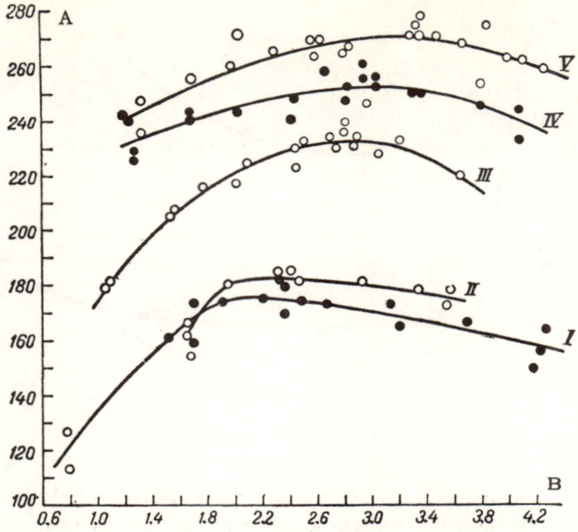

FIG. 10-21. Effect of air velocity and particle physical properties on heat transfer between a fluidized bed and a submerged coil [1082].
Mean particle diameter 1.5 mm. I—coke; II—chamotte; III—silica gel; IV—iron-chromium catalyst; V—marble. A, B as in Fig. 10-20.

Sarkits arrived at the following concept of the mechanism of heat transfer between walls and fluidized beds [931]: the solid particles are "main heat sources" maintaining a high temperature gradient at the heat transfer surface because of the frequent interchange of cool and hot particles. In the laminar gas flow regime a reduction in the size of the solid particles and an increase in their concentration reduces the length of the path along which the heat must flow in the gas phase at the heat transfer surface, and increases the temperature

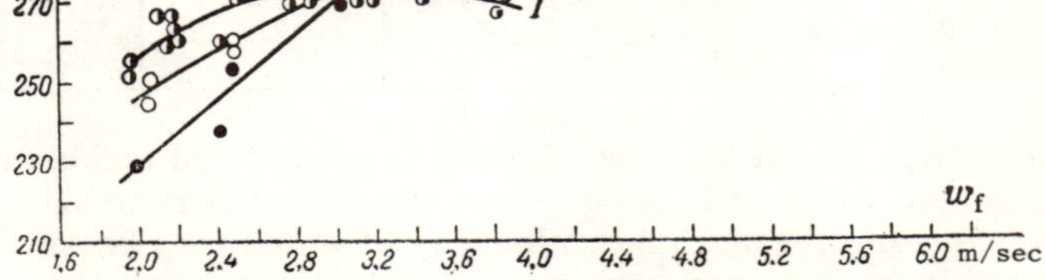

FIG. 10-22. Effect of air velocity and size of coarse particles on heat transfer between a fluidized bed of alumina-nickel catalyst and a submerged coil [1082].
Mean particle size (mm): I—1.5; II—2.5; III—3.5.

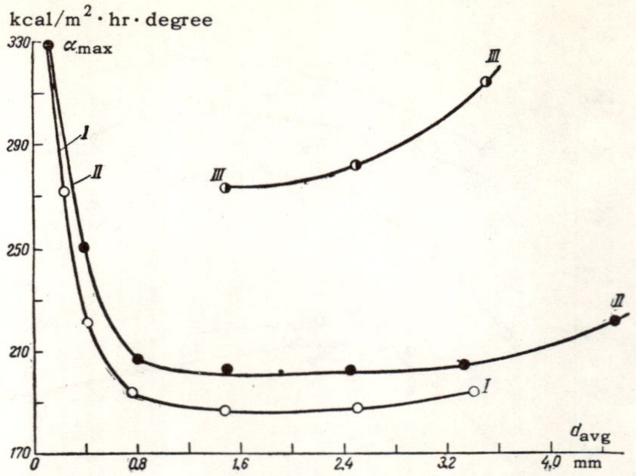

FIG. 10-23. α_{max} as a function of mean-particle size d_{avg} [1082].
I—BAV catalyst, bed diameter D_T = 49 mm; II—BAV, D_T = 73 mm; III—alumina-nickel catalyst, D_T = 49 mm.

Table 10-11

Experimental Results of V. B. Sarkits for α_{max} [931]

No.	Material	d, mm	w_{opt}, m/sec	α_{max} $\frac{kcal}{m^2 \cdot hr \cdot deg (C)}$	Re_{opt}	Ar_n
1.	Vanadium catalyst BAV	0.127	0.17	325	0.777	15.2
		0.257	0.15	175	0.31	127
		0.38	0.3	220	4.10	414
		1.5	1.9	190	102	26,500
		2.5	2.7	185	243	124,000
		3.5	3.1	195	391	334,000
		4.5	3.3	218	534	739,000
2.	Alumina-nickel catalyst	1.5	3.1	272	167	65,000
		2.5	4.0	282	360	302,000
		3.5	4.8	313	604	848,000
3.	Iron-chromium catalyst	0.38	0.45	295	6.15	930
		1.5	2.9	260	156	56,000
4.	Silica gel	0.38	0.4	307	5.47	729
		1.5	2.6	240	140	46,000
5.	Marble	0.38	0.47	333	6.42	1,013
		1.5	3.1	277	167	64,800
6.	Quartz	0.38	0.45	320	6.15	1,032
		1.5	2.2	198	119	32,600
7.	Coke	1.5	2.16	192	117	31,100
8.	Vanadium catalyst (spherical)	2.5	2.5	218	225	169,000

gradient and hence the heat transfer coefficient. In the turbulent regime of the gas at the heat transfer surface there still remains a laminar boundary layer of gas, "the thickness of which may be smaller than the particle radius and depends on the linear velocity of the gas stream." Sarkits and his co-authors made no attempt to give a theoretical heat transfer equation, but proposed a number of empirical relationships in dimensionless form.

The equations

$$Nu=0.0133Re^{0.41}Ar_N^{0.27} \cdot Pr^{0.33} \cdot \left(\frac{c_M}{c_g}\right)^{0.45} \left(\frac{D_{bed}}{d}\right)^{0.16} \left(\frac{H_0}{d}\right)^{0.45} \quad (10\text{-}55)$$

and

$$Nu = 0.00705Re^{-0.14} \cdot Ar_N^{0.49} \cdot Pr^{0.33} \left(\frac{c_M}{c_g}\right)^{0.45} \left(\frac{D_{bed}}{d}\right)^{0.16} \left(\frac{H_0}{d}\right)^{0.45} \quad (10\text{-}56)$$

are valid for the region of laminar flow with $Ar_N = 15 \div 1000$; $c_M / c_g = 0.8 \div 1.3$; $D_{bed}/ d = 129 \div 575$; $H_0/ d = 263 \div 945$; the equations

$$Nu = 0.0528Re^{0.85} \cdot Ar_N^{0.25} \cdot Pr^{0.33} \left(\frac{c_M}{c_g}\right)^{0.1} \left(\frac{D_{bed}}{d}\right)^{0.13} \left(\frac{H_0}{d}\right)^{0.16} \quad (10\text{-}57)$$

and

$$Nu = 0.018Re^{-0.12} \cdot Ar_N^{0.56} \cdot Pr^{0.33} \left(\frac{c_M}{c_g}\right)^{0.1} \left(\frac{D_{bed}}{d}\right)^{0.13} \left(\frac{H_0}{d}\right)^{0.16} \quad (10\text{-}58)$$

are valid for the turbulent region, with $Ar_N = 2.6 \times 10 \div 8.5 \times 10^5$; $c_M / c_g = 0.8 \div 1.2$; $D_{bed}/ d = 14 \div 67$; $H_0/ d = 17 \div 120$.

In these equations,

$$Ar_N = d^3 g \gamma_{bulk} / \nu^2 \gamma_g ;$$
$$Nu = \alpha_{wall} d / \lambda_g ;$$
$$Pr = c_g \mu g / \lambda_g ;$$
$$Re = w d / \nu .$$

Equations (10-55) and (10-56) refer to the regime extending from the limit of stability to the optimum superficial fluid velocity w_{opt}, and Eqs. (10-57) and (10-58) to the regime from w_{opt} to the onset of particle entrainment.

Equation (10-55) was examined over the range $Re = 0.34 \div 6.4$ and gives a maximum error $\varepsilon = \pm 25\%$; Eq. (10-56) gives $\varepsilon = \pm 15\%$ (for $Re = 0.78 \div 16.5$); Eq. (10-57) gives $\varepsilon = \pm 15\%$ (for $Re = 40 \div 600$); and Eq. (10-58) gives $\varepsilon = \pm 10\%$ (for $Re = 100 \div 900$).

The maximum heat transfer coefficient $\alpha_{wall.max}$ is reached at

$$Re_{opt} = 0.2Ar_N^{0.5}, \quad (10\text{-}59)$$

in the laminar region, and at

$$Re_{opt} = 0.66Ar_N^{0.5} \quad (10\text{-}60)$$

in the turbulent region.

The following expressions are proposed for calculating $\alpha_{wall.max}$:

$$Nu_{max} = 0.087 Ar^{0.42} \cdot Pr^{0.33} \left(\frac{c_M}{c_g}\right)^{0.45} \left(\frac{D_{bed}}{d}\right)^{0.16} \left(\frac{H_0}{d}\right)^{0.45} \quad (10\text{-}61)$$

for the laminar region, and

$$Nu_{max} = 0.019 Ar_N^{0.5} \cdot Pr^{0.33} \left(\frac{c_M}{c_g}\right)^{0.1} \left(\frac{D_{bed}}{d}\right)^{0.13} \left(\frac{H_0}{d}\right)^{0.16} \quad (10\text{-}62)$$

for the turbulent region.

The maximum error of these equation is $\pm 10\%$.

Equation (10-61) has been examined over the ranges

$$Ar_N = 15 \div 1000; \quad c_M/c_g = 0.801.3;$$
$$D_{bed}/d = 129 \div 575; \quad H_0/d = 263 \div 945.$$

Equation (10-62) has been examined for

$$Ar_N = 2.6 \times 10^4 \div 8.5 \times 10^5; \quad c_M/c_g = 0.8 \div 1.2;$$
$$D_{bed}/d = 14 \div 67; \quad H_0/d = 17 \div 120.$$

Here

$$Nu_{max} = \frac{\alpha_{wall.max} d}{\lambda_g} \quad Re_{opt} = \frac{w_{opt} d}{\nu}.$$

Despite the large number of experiments represented by these dimensionless equations, they are only special empirical relationships whose applicability to the design of various types of industrial equipment is questionable. The authors recognized, for instance, that existing contradictions in the literature on the effects of equipment diameter and bed height are not accidental, but related to differences in some experimental condition. Thus, formulas containing relationships between α_{wall}, H_0 and D_{bed} are not universal. Unfortunately the authors made no attempt to determine the magnitude of these effects by comparing their experimental data with those of other investigators. Such a comparison is difficult because in their formulas Sarkits, et al., used the bulk density of the particles "as poured" instead of the particle density. Thus, the Archimedes number appearing in Sarkits's formula (which we have denoted by Ar_N) is not comparable with the Ar in the equations of other investigators. The "as poured" bulk density can vary considerably.*

Kharchenko and Makhorin [1254] carried out an experimental investigation of heat transfer between a spherical copper alpha-calorimeter (D = 60 mm) and high-temperature (up to 1050°C) beds of sand and chamotte fluidized by flue gases. The bed diameter was 220 mm. Typical experimental results are shown in Fig. 10-23a. At the higher bed temperatures the region of $\alpha_{wall.max}$ is extended

*The use of γ_{bulk} can be justified by the fact that it expresses both the value of γ_M and the surface state of the particles (under otherwise equal conditions, smooth particles are more mobile and give a higher γ_{bulk} and also a higher α_{max}).

appreciably; $\alpha_{wall.max}$ depends approximately linearly on the bed temperature. The sharp-angled chamotte particles gave values of $\alpha_{wall.max}$ approximately 0.725 times the values for the sand beds under similar conditions.

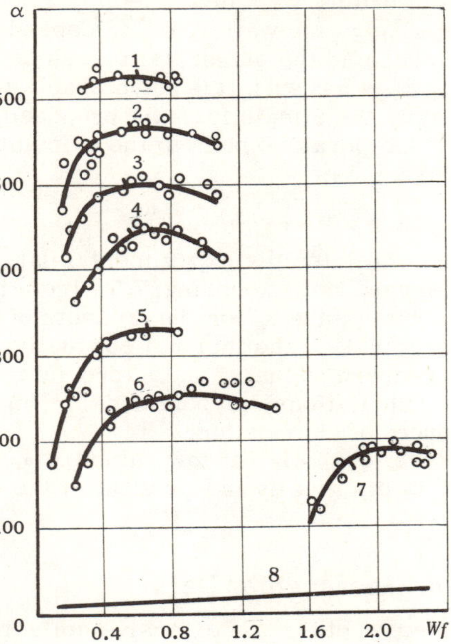

FIG. 10-23a. Heat transfer coefficients α, (kcal/$m^2 \cdot$hr$\cdot$degree), as a function of superficial fluid velocity w_f (m/sec) in the experiments of Kharchenko and Makhorin [1254].
Experiments with particles of quartz sand $d = 0.34$ mm at various bed temperatures: 1) 900°C, 2) 700°C; 3) 500°C; 4) 300°C; with particles of chamotte at bed temperature of 500°C: 5) $d = 0.42$ mm; 6) $d = 0.71$ mm; 7) $d = 1.66$ mm, and calculated data for empty tube (8).

In analyzing their experimental data, the authors [1254] made the assumption that the radiant component was unimportant in heat transfer between high temperature beds and solid bodies. They gave the following arguments in support of this point of view [1254]:

1. There can be no radiant transfer between a fluidized bed and a body as between a diathermous fluid and a wall.

2. The maximum heat transfer coefficient $\alpha_{wall.max}$ varies linearly with the bed temperature.

However, their analogy between a fluidized bed and a "nontransparent fluid" did not take into account that in contrast to Newtonian fluids and even to beds moving under gravity, in the case of a fully developed fluidized bed in which there is intensive mixing of the particles there is an appreciable temperature difference between the surface of a wetted body (or wall) and the rows of particles contiguous to it. Even in the worst case, when a layer of particles is formed along the wall due to adhesion heat transfer by radiation will not be eliminated, but only transferred to the interface between the stationary and moving particles. In this instance the degree of blackness of the surface of the heated body would be somewhat reduced.

Since $\alpha_{cc\,max}$ in ordinary fluidized beds is reached when α_r is still low, the value of $\alpha_{wall.max}$ as well as $\alpha_{cc\,max}$ depends almost linearly on the bed temperature, and the numerical values of $\alpha_{wall.max}$ will essentially equal $\alpha_{cc\,max}$. As a result, it is permissable to assume $\alpha_{wall.max} \approx \alpha_{cc\,max}$, and to apply the simple formula proposed earlier by Zabrodsky [742] for low temperature beds to the calculation of heat transfer in high temperature beds:

$$\alpha_{max} = 33.7\,\gamma_M^{0.2}\lambda_g^{0.6}d^{-0.36}.$$

This equation correlates the experimental data [1254] for spherical particles, with errors not exceeding 7.5% from the correlation proposed by the authors, (using λ_g for the products of combustion of fuels at the bed temperature). If the thermal conductivity of air is used instead of λ_g, a reduced value of α_{max} (providing effectively a safety factor in design calculations) is obtained for high temperature beds fluidized by products of combustion.

In regimes close to $\alpha_{cc\,max}$, at low values of α_r, the time of heating a body in a bed is practically independent of the degree of blackness of its surface.

10.3 Analysis of Experimental Data

Let us consider whether the extensive collection of various types of experimental data agrees with the theoretical relationships proposed earlier. It will be most convenient to consider in turn the relationships between the wall heat transfer coefficient and the various properties of the fluidizing agent, the granular material and the operating conditions.

In analyzing the effects of the individual factors it must be borne in mind that the heat transfer coefficient is a function of many variables, so that to evaluate the effect of one all others must have been held constant.

Also, the optimum heat transfer is not related in any way to any one hydrodynamic state of the fluidized bed, or to a single degree of bed expansion, regardless, for example, of the particle diameter, friction of one particle against another, or the thermal conductivity of the fluidizing medium. The fluidization index corresponding to the maximum heat transfer coefficient, N_{opt}, as well as w_{opt}, depends significantly on these factors.

The theoretical formulas given earlier can be used for the qualitative analysis of the complex phenomena of heat transfer in fluidized beds.

1. Maximum heat transfer coefficient

By analogy with other cases of heat transfer (in particular, in dense beds), investigators initially expected a monotonic increase of the heat transfer coefficient α_{wall} as the superficial stream velocity increased.

However, as already mentioned, the relationship appears to be much more complex, and contains a maximum.

The presence of a maximum in the heat transfer coefficient between a fluidized bed and a wall could be anticipated because of the opposing effects of particle velocity and bed voidage, which both increase with the superficial fluid velocity. The presence of a maximum is confirmed by the experimental data of numerous authors who have investigated a sufficiently wide range of superficial fluid velocities (for example, Wicke and Fetting, Baerg, Klassen and Gishler, Jakob and Osberg, Wesser and Mardus, Sarkits, et al., Ciborowski and Roszak, Varygin and Martyushin, and others) (see Fig. 10-9).

The experiments of Mickley and Trilling [451] were carried out in the high-voidage zone of a fluidized bed, corresponding to the descending right-hand branch of the relationship between α_{wall} and the superficial fluid velocity. It is therefore not surprising that they found a monotonic increase of α_{wall} as the superficial fluid velocity was decreased (Fig. 10-4). On the other extreme, V. G. Ainshtein and Leva, et al., confined themselves to the "dense phase" region of fluidized beds, and never reached the maximum value of the heat transfer coefficient.

It is interesting to note that the maximum in the experimental curves is quite flat, i.e., the heat transfer coefficient is close to the maximum over a wide range of superficial fluid velocities. This is important in practical heat exchanger design.

The characteristic case of the onset of the maximum was discussed earlier—the exponent in Eq. (10-9) is small, less than 0.03, and the dependence of α_{wall} on the properties of the material and the particle velocities vanishes. The physical significance of this condition has already been discussed. According to the theoretical formula for $\alpha_{wall.max}$, the effect of the particle density γ_M is contained in the bed voidage, since the voidage at the optimum conditions for heat transfer, m_{opt}, depends on γ_M. In order to reach $\alpha_{wall.max}$ the particles must move rapidly. This requires a high superficial fluid velocity, at which a bed of light particles will have a larger voidage than a bed of heavy particles of the same size. $\alpha_{wall.max}$ for the light particles will be correspondingly smaller. This argument is confirmed by the fact that it is possible to describe the experimental data on $\alpha_{wall.max}$ very well with an empirical formula in which γ_M appears as the only physical property of the material, and the relationship between $\alpha_{wall.max}$ and γ_M is exponential ($\alpha_{wall.max} \sim \gamma_M^{0,2}$).

Wicke and Fetting, without commenting on the nature and limits of the region in which the heat transfer coefficient $\alpha_{wall.max}$ was independent of the physical properties of the particle material, nearly exemplified this behavior in comparing the experimental curves for particles of sand and aluminum of the same mean diameter. The sand and aluminum were similar in specific heat (0.19 and 0.22 kcal/kg · degree, respectively) and in density (2320 and 2670 kg/m), but differed appreciably in their thermal and electrical conductivities.

According to Wicke, the heat transfer coefficients for fluidized beds of aluminum were considerably lower than those for beds of sand. This cannot be explained by the high thermal conductivity of the aluminum, but is probably due to some indirect factor, such as dissimilar granulometric compositions of the beds of aluminum and sand at the same nominal mean-particle diameter. There are also other possible causes: Wicke's experiments gave sharply reduced values of $\alpha_{wall.max}$ only for the metals which he used (aluminum, lead). This anomaly was not observed in the experiments of Baerg, Klassen and Gishler for iron. This forces the conclusion that there were effects in Wicke's experiments which artifically retarded the motions of the metallic particles. In all cases Wicke's experimental data on metals should be treated with caution.

The superficial fluid velocity has only an indirect effect on the value of the maximum of α_{wall} through changing the bed voidage and particle velocity.

Recent experimental data [1254] confirm the theoretical discussion given at the beginning of the chapter concerning the elongation of the zone of $\alpha_{wall.max}$ in high-temperature beds, and the decrease of $\alpha_{wall.max}$ when the particle shape differs appreciably from a sphere.

2. Effect of particle thermal conductivity

The experiments of various investigators have not indicated a marked dependence of heat transfer in a fluidized bed on the thermal conductivity of the particle material. This is to be expected cn the basis of the proposed theoretical formulas, in the derivation of which, the assumption was made that there was practically no temperature gradient within the particles because the thermal conductivity of the particles was sufficiently large.

It has been shown [742 and 744] that this assumption is valid in the symmetrical heating of particles with an intensity corresponding to the mean heat transfer coefficient of the particles. However, under real fluidized bed conditions, the particles near the heating surface do not have an intensive rotation, and are subjected to particularly unsymmetrical heating. Because of this, the local conductive heat transfer in the particles may reach very large values in the neighborhood of the points of contact between the particles and the heating surface. Recently, Botterill, Redish, et al., calculated that appreciable temperature gradients can arise under such conditions even in glass spheres as small as 0.2 mm diameter. This is reasonable, since the large local heat fluxes within the particles occur at positions of very high local values of α for the particles.

The practical independence of α_{wall} on λ_M, demonstrated by the experimental data, cannot therefore be derived simply from theoretical arguments, as proposed earlier [742].

3. Effect of particle diameter

Sarkits [931] carried out experiments over a very wide range of particle diameters (0.127 to 4.5 mm) for the same material; these confirmed conclusively the earlier data of Ciborowski and Roszak [789] concerning the nonmonotonic relationship between $\alpha_{wall.max}$ and the particle diameter; there is a strong inverse dependence in the region of fine particles, and a weaker direct dependence for coarse particles. Such a dependence is in complete agreement with the theoretical discussion given at the beginning of the chapter; the conductive component of $\alpha_{wall.max}$ varies inversely and monotonically with the diameter, while the convectional (diffusional) component depends directly on the particle diameter and the optimum superficial fluid velocity. The latter also increases with the particle diameter. Hence, for small particles, when α_{cond} is large and α_{conv} is negligible, there is a resultant inverse dependence of $\alpha_{wall.max}$ on the particle diameter, while for large particles, when the convective component of heat transfer predominates, $\alpha_{wall.max}$ increases with increasing diameter. It is to be expected that in the case of heavy particles, for which the superficial fluid velocity and diffusional component of wall heat transfer are higher under otherwise equal conditions, the direct dependence of $\alpha_{wall.max}$ on diameter will begin earlier (with respect to d) and be more pronounced (see Fig. 10-23). Figure 10-22 indicates that the direct relationship which is observed between $\alpha_{wall.max}$ (and, in general, α_{wall}) and the particle diameter is completely explained by the convective component of heat transfer. At low superficial fluid velocities (below w_{opt}), where diffusional mixing is suppressed, the inverse dependence of α_{wall} on particle diameter is retained even for large particles (2.5-3.5 mm), while in the region of high w_f a direct relationship develops between α_{wall} and particle diameter.

Only the inverse relationship between α_{wall} and the particle diameter is indicated in the overwhelming majority of experimental investigations [202, 451, 686, 877, 972, and others], since these studies were not carried out with coarse particles varying in diameter from one experiment to another.

It can be expected on theoretical grounds that the dependence of $\alpha_{wall.max}$ on the particle diameter will be less pronounced with a gas having a thermal conductivity. The experimental data of Jakob and Osberg [686] confirm this conclusion. For example, in going from particles of 0.29 mm diameter to particles of 0.061 mm diameter in air, there is a 1.88-fold increase in the maximum heat transfer coefficient, while in hydrogen the increase is only 1.49-fold.

4. Effect of particle bulk density

There may be an appreciable decrease in the maximum heat transfer coefficient of a fluidized bed of material having a low bulk density (due to the low particle density). In this case, the limit of stability is reached at small superficial fluid velocities, and the bed has a high voidage even at small superficial fluid velocities so that the particle velocities cannot be large; as a result, the particles are overheated and the maximum in α_{wall} moves further into the region of high bed expansion. Under otherwise equal conditions, the shift in the maximum of α_{wall} into the region of high bed voidage is accompanied by a reduction in its absolute value because of the inverse dependence of α_{wall} on the bed voidage (see p. 328).

The experimental data of Sarkits, et al. [1082], clearly demonstrate the marked dependence of $\alpha_{wall.max}$ on γ_M. The experimental data of a number of other investigators, which have been expressed in an empirical formula by Zabrodsky [741], also indicate a minor but direct relationship between $\alpha_{wall.max}$ and γ_M.

5. Effect of the surface state of spherical particles

In the experiments of Baerg, Klassen and Gishler, the maximum heat transfer coefficients for smooth glass spheres appeared to be considerably larger than those for other particles of the same diameter (see Fig. 10-7). Subsequently these authors carried out experiments with the same glass spheres with etched surfaces and demonstrated a decrease in $\alpha_{wall.max}$. Baerg, Klassen and Gishler established the effect of surface characteristics but made no attempt to explain the increase in $\alpha_{wall.max}$ as probably due to decreased friction between the particles. Theoretical considerations made it possible for Zabrodsky to interpret this phenomenon [744].

It is to be expected that smooth spheres can acquire an intensive motion at very small voidages of the fluidized bed, which is impossible for rough or irregular particles. In the case of smooth spheres, therefore, $\alpha_{wall.max}$ is reached at low bed voidages, i.e., under very favorable conditions for achieving a high numerical value of the heat transfer coefficient. The fact that the maximum value of α_{wall} for smooth spheres is actually reached at considerably smaller values of the fluidization index (which means at smaller values of the voidage of the fluidized bed) than in the case of molding sand, for instance, is directly confirmed by the experimental data of Baerg, Klassen and Gishler. For smooth glass spheres (diameter 158 microns) the maximum is reached at a superficial fluid velocity of 0.304 m/sec, while for sand particles (160 microns) it occurs at 0.396 m/sec, even though incipient fluidization velocity for both spheres and sand particles is nearly identical due to the similarity in particle densities of the two materials.

6. Effect of particle heat capacity

Theoretical considerations indicate a direct relationship between α_{wall} and particle heat capacity, c_M, until the optimum heat transfer is reached at $\alpha_{wall.max}$, since at large c_M, smaller particle velocities and bed voidages are required to ensure that overheating of the particles is small during their exposure time along the wall. Further relationships are lacking, since under the conditions of α_{max} and beyond this value, the particles are replaced so rapidly that even when c_M is reduced they are not appreciably overheated. The attainable numerical values of $\alpha_{wall.max}$ are related to this indirect effect of c_M. This is confirmed by the presence of a factor $c_M^{0.45}$ in the empirical formula of Sarkits, et al., for $\alpha_{wall.max}$ in the laminar regime. However, in their formula for $\alpha_{wall.max}$ in the turbulent regime, c_M appears only to the 0.1 power, while in the empirical formulas of Zabrodsky [741] and Varygin and Martyushin [877], it is altogether absent. This is due to the nearly single-valued inverse dependence of c_M on the specific weight (or bulk density) of the material γ_M. Consequently, c_M can be approximately replaced by γ_M in the formulas for $\alpha_{wall.max}$.

7. Effect of medium thermal conductivity

From the experimental results of Varygin and Martyushin [877], Sarkits [931], Wicke and Fetting [283, 600], Jakob and Osberg [686], Wesser and Mardus [713] and Miller and Logwinuk [452], it is possible to determine that the maximum heat transfer coefficient for a fluidized bed is almost directly proportional to the thermal conductivity of the fluidizing medium.

For example, according to the data of Wesser and Mardus, $\alpha_{wall.max} \approx 300$ kcal/m²·degree for a bed of glass spheres ($d \approx 1$ mm) fluidized by water ($\lambda_0 = 0.515$ kcal/m·hr·degree), while according to the data of Wicke, $\alpha_{wall.max} \approx 180$ kcal/m²·hr·degree when a similar bed is fluidized by air ($\lambda_m = 0.023$ kcal/m·hr·degree). Thus $\alpha_{wall.max}$ increased by a factor of 16.7 when λ_m was increased by a factor of 22.4. This comparison is not entirely valid, and the effect of λ_m is considerably overestimated because of the sharp increase in the diffusional component of $\alpha_{wall.max}$ as well as in the conductive component when replacing a gas with a liquid.

Wesser and Mardus observed some unexpected features of heat transfer in beds fluidized by water, which behaved anomalously compared with air-fluidized beds. The initial transition to fluidization from the dense bed state (artificially held down to prevent fluidization at high superficial fluid velocities) was accompanied by a fall in heat transfer, which was not observed in beds fluidized by gases. On the other hand, according to the experiments of Baerg, Klassen and Gishler and of Wicke and Fetting, the transition to a fluidized bed is always accompanied by a sharp increase in the heat transfer coefficient.

For beds fluidized by water, it was also observed that immediately beyond the limit of stability, the increase in α_{wall} ceased as superficial fluid velocity increased (Fig. 10-14), which the investigators did not observe in gas-fluidized beds. As indicated earlier in the chapter, both "anomalies" are mainly due to the sharp changes in the thermal conductivity and heat capacity of the medium when gas is replaced by water, and can be predicted on the basis of the arguments developed earlier.

8. The effect of nonuniformity of fluidization

Theoretical concepts of the mechanism of wall heat transfer derived for homogeneous fluidized beds can explain qualitatively many of the facts observed in experiments with nonhomogeneous beds. However, a number of additional considerations have to be taken into account for a more detailed analysis of the data for nonhomogeneous fluidized beds. Following Mickley and Fairbanks [450], it is convenient to use as a first approximation for nonhomogeneous gas fluidization the concept that the heating surface is not wetted by individual particles, but by whole aggregates of particles. However, it may be assumed that the contact density of the aggregates with the surface (the effective thickness of the gas layer between them) depends on the particle diameter just as does the contact density of the first row of particles in a homogeneous bed, with the simplification that the voidage of the aggregates is assumed to be constant and equal to $m_{1.s}$. In correlations for nonhomogeneous beds, an increase in the relative time of surface wetting by the discontinuous phase (the pure gas) is equivalent to an increase in voidage as the superficial fluid velocity increases.

In this sense it is possible to speak of a time-averaged voidage, at a given position in a nonhomogeneous bed, which is larger than the voidage of the aggregates, and of a dependence of this effective voidage m_{eff} on the superficial fluid velocity. To a first approximation it can be assumed that m_{eff} is numerically equal to the integral mean of the instantaneous voidages in the lower part of the nonhomogeneous fluidized bed being considered (up to height H_0). It can also be assumed that there are no preferred paths for bubble motion in the bed, i.e., that the probability of a bubble passing is the same at different points over the cross section of the bed, and that m_{eff} is the same over the points of a given cross section. In beds of uniformly sized particles m_{eff} should be smaller in the upper parts of the bed than in the lower parts, due to the acceleration of the rising (merging) bubbles. However, the break-up of some of the bubbles leads to the opposite effect, as do the facts that real beds do not consist of particles of exactly the same size, and that size segregation of the material is possible when a range of particle sizes is present. In the presence of segregation, the smaller particles accumulate at the top, and at a given superficial fluid velocity their value of the fluidization index and their quantity of "excess gas" is larger.

If differences in m_{eff} are neglected at the initial bed height, i.e., if it is assumed that m_{eff} is uniform over the whole volume, then for the main dense part of a nonhomogeneous fluidized bed of constant cross section m_{eff}, the value of F_{sec} will be

$$m_{\text{eff}} = \frac{\Sigma V_{\text{bubble}} + m_{1.s}(H_0 F_{\text{sec}} - \Sigma V_{\text{bubble}})}{H_0 F_{\text{sec}}} = m_{1.s} + \frac{\Sigma V_{\text{bubble}}(1 - m_{1.s})}{H_0 F_{\text{sec}}}, \tag{10-63}$$

where ΣV_{bubble} is the volume of bubbles in the volume $H_0 F_{\text{sec}}$ in the lower part of the bed, and H_0 is the bed height at the limit of stability.

Bubbles can only be formed in the volume $H_0 F_{\text{sec}}$ by displacing upwards (above the level H_0) a quantity of material

$$G_{\text{upl}} = \Sigma V_{\text{bubble}}(1 - m_{1.s})\gamma_M.$$

Hence,

$$\Sigma V_{\text{bubble}} = \frac{G_{\text{up}}}{(1 - m_{1.s})\gamma_M} = \frac{\Delta p_{\text{up}} \; F_{\text{sec}}}{(1 - m_{1.s})\gamma_M}. \tag{10-64}$$

From (10-63) and (10-64),

$$m_{\text{eff}} = m_{1.s} + \frac{\Delta p_{\text{up}}}{\gamma_M H_0}$$

A similiar problem of the density distribution in a nonhomogeneous bed was solved by Fan, et al. [1242].

It is possible to find m_{eff} by measuring the pressure drop in the bed above the level H_0. As the fluidization index N increases, larger bubbles with higher velocity pass through the bed; because of this a nonhomogeneous fluidized bed expands more slowly than a homogeneous bed. The experiments of Bakker and Heertjes [1097] on the distribution of voidage in beds as well as experimental data obtained in our laboratory confirmed that Δp_{up}, and hence m_{eff}, increase very slowly for nonhomogeneous beds as the superficial fluid velocity is increased above the optimum value.

This leads to the expectation that in nonhomogeneous beds there will be a less rapid decrease in the heat transfer coefficient beyond the maximum value if the heat transfer surface being considered does not project into the dilute phase of the nonhomogeneous bed. This was confirmed by experiments which covered a wide range of the fluidization index.

Many details of heat transfer between heating surfaces and fluidized beds, connected with the flow of the particles and aggregates along the heating surface, have been explored in experiments recently carried out in our laboratory by Antonishin, including variations in the local heat transfer coefficients, differences in the mean heat transfer coefficients for horizontal and vertical tubes, the effect of tube diameter, etc.

In the downflow of particle aggregates around a horizontal tube in a fluidized bed, it is characteristic to have dense "caps" of particles on the top of the tube and "gaps" more or less free of particles below the tube. This situation, which can be observed visually, appears to be close to the well known picture of downflow of a dense moving bed around a horizontal tube. However, a more careful investigation indicates that there are important differences. In the fluidized bed the "cap" of particles is mobile, though of considerable density, and the surface of the tube below it becomes a zone of stable high local heat transfer coefficients. As the fluidization index is increased, more particles are thrown into the "gap" below the tube, and the local heat transfer coefficients for the lower parts of the horizontal tube increase considerably. In closely staggered bundles of tubes (with small vertical pitch), one tube may fall as it were into the wake of another (or several other) tubes, and the manner in which the local heat transfer coefficients vary with the fluidization index may become quite complex.

However, in Antonishin's experiments in the regime of intensive bed mixing, it was observed that the mean values of α over the surfaces of horizontal tubes are very similar regardless of the position of the tube within a bundle (Fig. 10-24). They are also close to α for a single horizontal tube, although the magnitudes and manner of variation of each of the local heat transfer coefficients are quite different in these cases. This can be partially explained by the fact that in all cases some value close to the effective bed voidage m_{eff} is obtained for each of the tubes.

At the same time the accessibility of various points on the tube surface for "wetting" by passing bubbles can vary widely depending on the pitch of the tubes within the bundle, i.e., the local heat transfer coefficient may vary at such points. However, such a change in the local heat transfer coefficient is compensated by opposite changes at

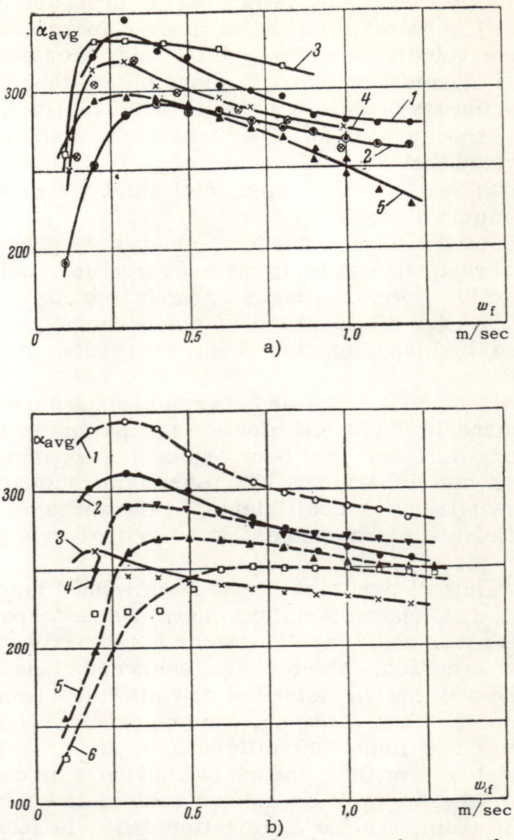

FIG. 10-24. Relationships for the mean heat transfer coefficient.
a—mean heat transfer coefficient as a function of superficial fluid velocity for a 14 mm tube, submerged in the dense phase of a fluidized bed of 268 micron sand: 1—single vertical tube on bed axis; 2—single horizontal tube placed in the middle of the bed diameter; 3—same, displaced to the wall by a distance equal to 1/4 of the bed diameter; 4—horizontal tubes in unstaggered bundle of 28 mm pitch; 5—vertical tubes in bundle with 20 mm pitch.
Error of measurements 5-10%.
b—mean heat transfer coefficient as a function of superficial fluid velocity for centrally placed single tubes submerged in the dense phase of a fluidized bed of 268 micron sand:
1, 2, 3—vertical tubes of diameters 14, 21, 28 mm, respectively; 4, 5, 6—horizontal tubes of diameters 14, 21, 28 mm, respectively.

other points. For instance, if a large bubble of pure gas begins to pass some point, i.e., the time-averaged local voidage increases and the local heat transfer decreases, then at the same time at other points the local voidage decreases and the local heat transfer coefficient increases. There can also be compensation for local changes of heat transfer coefficients at two points on the tube surface if they are disposed successively on the path along which the particles flow around the tube, i.e., one obtains particles already heated around the second point. In this case, an increase in heat transfer at the point first reached by the particles must lead to a reduction at the following point.

In the case of vertical tubes submerged in fluidized beds there also exist breakaways in the flow of particles due to the horizontal motion of aggregates. Aggregates which encounter the vertical tube in their path are deformed and split by the tube, as by a wedge, so that a cavity (gap) is formed in the downstream side of the tube, depending on the direction in which the aggregate was moving. In our experiments the mean heat transfer coefficients for vertical tubes submerged in the dense phase of a bed were close to the values of α_{avg} for horizontal tubes. Usually α_{avg} for vertical tubes (short heaters on their surfaces) was 5-20% higher than α_{avg} for horizontal tubes.

For closely staggered horizontal tube bundles, α_{avg} differed little (5-15%) from α_{avg} for a single tube. The distance between the tubes was of the order of the tube diameter, i.e., 14-28 mm, while the particle diameters in the bed were 0.2-0.49 mm. In the vertical bundles the distances were approximately the same, but α_{avg} had values higher than for horizontal tubes. There is therefore a very promising future for the construction of fluidized bed heat exchangers with very compact surfaces.

Antonishin's experiments showed a pronounced dependence of α_{avg} on the tube diameter in the zone of the maximum heat transfer coefficient. This was the case for both horizontal and vertical tubes; the tube diameters were 14-28 mm. This dependence was obviously related to the appreciable increase in the temperature of the solid particles as they flowed around the tubes. Under otherwise equal conditions, the heating of the particles will be greater (i.e., α_{avg} will be smaller) in flowing past tubes of large diameter than in the case of small diameter tubes.

The effect of tube diameter on α_{avg} for a single vertical tube submerged in a fluidized bed suggests that the flow transverse to the tube, rather than along it, is important in determining the effectiveness of heat transfer, though it appears that longitudinal mixing should ensure a short contact time between the particles and short heating surfaces. However, the longitudinal effective thermal conductivity of a bed is much greater than the transverse (horizontal) conductivity. It is probable that dense contact between a vertical tube and the particles is obtained only when moving aggregates flow perpendicularly against the tube. In such a case, the particles are so retarded around the walls that their residence time is not at all determined by the high coefficient of longitudinal effective thermal conductivity, measured for the core of a bed, but is equal to the time for transverse flow of the aggregate around the tube. The appreciable heating of the retarded particles around a tube submerged in a bed plays a role analogous to poor thermal insulation in ordinary convective heat transfer between tubes and single-phase fluids.

As the intensity of bed mixing increases, the "poor insulation" becomes worse, and finally its effect can be neglected, i.e., as the fluidization index is increased still further, the dependence of α_{avg} on the tube diameter should become less marked. This has been confirmed by experiment. The values of α_{avg} for horizontal tubes of various diameters converge very rapidly.

From this analogy with the critical insulation diameter it can also be concluded that for small-diameter tubes submerged in fluidized beds, the maximum heat transfer will occur at low velocities and appreciable heating of the particles, i.e., sooner than in the case of tubes of larger diameter.

10.4 Generalized Correlations

The design formulas proposed by various authors should only be used in calculations for cases close to the experimental conditions. Attempts to find a generalized correlation for calculating heat transfer coefficients between fluidized beds and heating surfaces have been made by Leva (for the case of heat transfer to surrounding walls [988]; by Vreedenberg [850, 1134]; and by Wender and Cooper [854] (see also [1138] pp. 432 and 441).

Table 10-12

Applicability of the Correlations (10-65) and (10-66) to the Experimental Data of Vreedenberg [850]

No.	Material	Mean equiv. particle diameter, microns	Weighed mean diameter, microns	γ_M, kg/mi	Particle shape	Diameter of heating surface mm	G, kg/m · hr
1.	Regenerated cracking catalyst	63.5	68	1600	Angular	16.9	37.1
						16.9	54.7
						16.9	231
						16.9	349
						33.6	37.1
						33.6	55
						33.6	229
						33.6	349
						33.6	536
						51	55.8
						51	346
2.	Sand for sand-jet apparatus	136	187	2660	Round	16.9	164.5
						16.9	278
						33.6	165
						33.6	278.5
						51	165.5
						51	271
3.	Sand from dunes	213	230	2660	Round	16.9	200
						16.9	342
						33.6	202
						33.6	346
4.	River sand	316	393	2660	Round	16.9	357
						16.9	580
						33.1	367
						33.1	587
						33.6	360
						33.6	572
						51	357
						51	576

Vreedenberg gave separate correlations for single vertical and horizontal tube heaters. In both cases he further subdivided the data into two groups depending on the value of the modified Reynolds number for flow around the particles. Values of particle $Re_m < 2050$ were placed in the category of fine and light, while $Re_m > 2550$ was referred to the category of coarse and heavy. Each of these categories had its own correlation. In the transitional region of Re_m between 2050 and 2550, Vreedenberg recommended the use of the average of the values given by the two correlations.

Vreedenberg [850] also gave some new experimental data on heat transfer between horizontal tubes and fine light particles (Table 10-12). Bed temperature, weight fluid flow rate, mean-particle diameter, particle shape, particle specific weight, and diameter of the water-cooled tube were varied in his experiments (see Tables 10-12 and 10-13).

Table 10-13

Experiments With Horizontal Tubes
(h = 0.85 m; H = 1.2 m) [850]

No.	Material	Equiv. mean-particle diameter, microns	γ_M, kg/m³	G, kg/m²·hr	t_{bed}, °C	Range of $Re_M = \frac{Gd\gamma_M}{\gamma\mu}$	Correlation
1.	Regenerated cracking catalyst	64	1600	36-538	40-260	49- 740	(10-65)
2.	Sand for sand-jet apparatus	136	2660	166-277	40-200	670- 1,290	(10-65)
3.	Sand from dunes	213	2660	202-346	40-240	1,270- 2,550	(10-65)
4.	Sand from Maas.	316	2660	357-587	20-260	3,360- 6,610	(10-66)
5.	Round and "silver sand". . .	215	2660	321-543	40-260	2,050- 4,060	(10-66)
6.	Old sand from Maas	353	2660	532-861	40-320	5,620-10,900	(10-66)
7.	Mixture of angular "silver" sand and old sand from Maas	260	2660	425-712	40-340	3,270- 6,690	(10-66)
8.	Round "silver" sand	259	2660	328-540	40-280	2,520- 5,500	(10-66)
9.	Iron ore (wide range of particle sizes).	130	5150	842	40-220	6,160- 7,250	(10-66)
10.	Iron ore (narrow range of particle sizes).	247	5150	860	40-240	12,900-15,100	(10-66)

The experiments were carried out with the tube placed high above the gas distributor (h = 0.85 m) so as not to fall into the zone of channel formation and to ensure that it would be possible to use the pressure drop in the zone in which the heating surface was placed for determining the bed voidage which appears in the correlation.

For calculating heat transfer between horizontal tubes and fluidized beds of fine light particles, Vreedenberg proposed the correlation [850]:

$$\frac{\alpha_{wall} d_T/\lambda_g}{(c_M \mu/\lambda_g)^{0.3}} = 0.66 \left[\frac{G d_T \gamma_M (1-m)}{\gamma_c \mu m} \right]^{0.44}, \qquad (10\text{-}65)$$

and for coarse and heavy particles

$$\frac{\alpha_{wall} d_T/\lambda_g}{(c_M \mu/\lambda_g)^{0.3}} = 420 \left(\frac{G d_T \gamma_M}{\gamma_g \mu} \cdot \frac{\mu^2}{12.96 \cdot 10^6 d^3 \gamma_M^2 g} \right)^{0.3}, \qquad (10\text{-}66)$$

where μ is the gas viscosity, kg/m·hr;
g is the acceleration of gravity, m/sec ;
d_T is the tube diameter, m.

The coefficient 12.96×10^6 is introduced since μ and g have different units of time.

The maximum deviations of the experimental data from the correlations (10-65) and (10-66) were 42% and 29%, respectively.

In his correlation of α_{wall} for vertical tubes [1134] Vreedenberg included the experimental data of ten other investigators [202, 222, 283, 450, 451, 452, 482, 568, 600, 601] for voidages $m < 0.7$ as well as his own [580]. Vreedenberg stated that depending on the conditions of fluidization, the heat transfer coefficient at the bed periphery can be higher or lower than the value at the bed axis, but both are lower than at intermediate points. This does not agree with the data of Ainshtein. Furthermore, according to Vreedenberg, in all cases the effect of eccentricity in the position of the submerged vertical tube is smaller than the deviations between the experimental curves and the correlation line. Vreedenberg therefore recommended the use of his correlation for tube bundles as well.

For heat transfer between vertical tubes and fluidized beds of fine light particles ($Re_m < 2050$), the correlation lines are broken into

$$\frac{\alpha_{wall}(D_{bed} - d_T)}{\lambda_g}\left(\frac{d_T}{D_{bed}}\right)^{1/3}\left(\overline{c_M \mu}\right)^{1/2} = A\left[\frac{G(D_{bed} - d_T)\gamma_M}{\mu\gamma_g}\right]^B, \qquad (10\text{-}67)$$

where $A = 0.27 \times 10^{-15}$, $B = 3.4$ for $G(D_{bed} - d_T)\gamma_M/\mu\gamma_g \leqslant 2.37 \times 10^5$;
$A = 2.2$; $B = 0.44$ for $G(D_{bed} - d_T)\gamma_M/\mu\gamma_g > 2.37 \times 10^5$.

Here D_{bed}, d_T are, respectively, the diameters of the fluidized bed and the outside of the tube.

For coarse heavy particles ($Re_m > 2550$) the correlation lines correspond to the equations

$$\frac{\alpha_{wall}(D_{bed} - d_T)}{\lambda_g}\left(\frac{d_T}{D_{bed}} \cdot \frac{d}{D_{bed} - d_T} \cdot \frac{\lambda_g}{c_M \mu}\right)^{1/3} = E\left[\frac{G(D_{bed} - d_T)}{3600\gamma_g d^{3/2} g^{1/2}}\right]^F, \qquad (10\text{-}68)$$

where $E = 0.105 \times 10^{-3}$; $F = 2.0$ for $G(D_{bed} - d_T)/3600\gamma_g d^{3/2} g^{1/2} < 1070$;
$E = 240$; $F = -0.1$ for $G(D_{bed} - d_T)/3600\gamma_g d^{3/2} g^{1/2} \geqslant 1070$.

The coefficient 3600 is introduced here since G is in kg/m²·hr and g in m/sec . For both of the vertical tube correlations the experimental values of α_{wall} differ from those calculated by the correlation by no more than a factor of 2.2. This is rather poor agreement.

The empirical correlation proposed earlier by Wender and Cooper [854] for the data of various investigators appear more satisfactory. The correlation for heat transfer between a fluidized bed and the surrounding walls (429 experiments) is given in Fig. 10-25 in the form of a relationship between the group

$$\frac{\alpha_{wall} D/[\lambda_g(1 - m)(c_M\gamma_M/c_g\gamma_g)]}{1 + 7{,}5e^{-0{,}44}(H_s/D_T)(c_g/c_M)} \qquad \text{(the ordinate } y\text{)}$$

and $Re \equiv Gd/\mu$. The mean deviation from the experimental data is 22.1%.

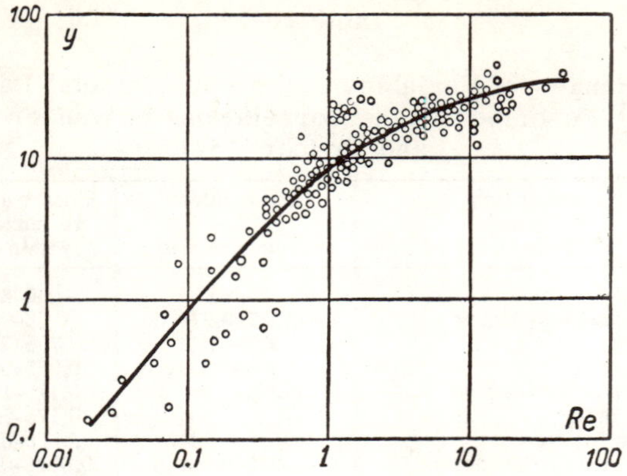

FIG. 10-25. Wender and Cooper's [854] correction of experimental data for heat transfer between fluidized beds and surrounding walls (see Table 10-14).

The correlation for 323 experiments on heat transfer between fluidized beds and submerged surfaces at a radial distance r from the bed axis was

$$\frac{\alpha_{wall} D}{\lambda_g (1-m)} \left(\frac{\lambda_g}{c_g \gamma_g}\right)^{0.43} =$$

$$= 0.0119 C_r \left(\frac{GD}{\mu}\right)^{0.23} \left(\frac{c_M}{c_g}\right)^{0.8} \left(\frac{\gamma_M}{\gamma_g}\right)^{0.66} \quad (10\text{-}69)$$

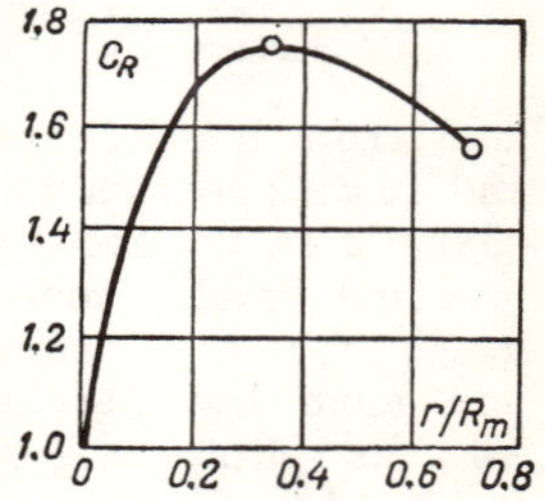

FIG. 10-26. Correction factor (C_R) based on the data of Vreedenberg for use in Wender and Cooper's [854] Eq. (10-69). R_T = bed radius; r = eccentricity of position of heating surface in bed.

Here $\lambda_r/c_g\gamma_g$ is a dimensional group. The value of the correction factor C_r is given in Fig. 10-26. The mean deviation of the correlation from the experimental data is 19.5%. The ranges of the variables over which the correlations are applicable are indicated in Table 10-14.

Vreedenberg [1134] noted that if attempts were made to include other than the selected experimental data in the correlation of Wender and Cooper, the deviation of the experimental data from the calculated values was even larger than in the case of his own correlation. Wen and Leva [989] criticized the correlation of Wender and Cooper for heat transfer to surrounding walls, and demonstrated its deviation from the relatively reliable data of van Heerden, et al.; they mentioned that Wender and Cooper's correlation gave values of the heat transfer coefficient which were generally too high.

Thus, neither of the correlations proposed to date can be recommended with confidence for practical calculations of the heat transfer coefficient α_{wall} between heating surfaces and fluidized beds.

To obtain a reliable correlation for α_{wall} (perhaps in the form of several relationships for fluidized systems of various geometry) it is necessary to carry out very broad and systematic investigations. In order to estimate the relative "ability" of a fluidized bed to transfer

Table 10-14

Ranges of Variables in the Experimental Data Described by the Correlations of Wender and Cooper [854]

Quantities and dimensions	Heat transfer to surrounding walls	Heat transfer to surfaces within bed
d, 10^6 m	50–850	40–880
c_M, kcal/kg · degree	0.117–0.276	0.20–0.316
γ_M, kg/m³	833–5250	795–2870
$1-m$	0.050–0.564	0.055–0.603
c_g, kcal/kg · degree	0.235–1.24	0.125–1.24
γ_g, kg/m³	0.11–2.96	0.176–7.77
μ, kg/m · hr	0.0695–0.1042	0.0378–0.1374
λ_g, kcal/m · hr · degree	0.0246–0.161	0.0091–0.131
D_T, m	0.025–0.12	0.073–1.93
H_s, m	0.043–1.015	0.101–5.34
H, m	0.043–2.54	0.253–8.11
α_{wall}, kcal/m² · hr · degree	16–850	49–883

Here H and H_s are the heights of the bed and the heat transfer surface.

heat from heating surfaces in equipment of various types and sizes and in different parts of the bed, it is most suitable to investigate the changes in the local heat transfer coefficients over the bed by means of some small probe, and to construct "isoalphs" (lines of constant α_{wall}) for the bed, as suggested by Antonishin. It is prescribed that such investigations be carried out on laboratory equipment as well as pilot-scale and industrial scale equipment, to investigate fluidized beds with gas distributors of the industrial types. This obviously facilitates bridging the gap between laboratory and industrial data.

If for technical or other special reasons it is possible to operate with the dense phase of the fluidized beds close to $\alpha_{wall.max}$, it is better to use an empirical formula for $\alpha_{wall.max}$ to calculate the surface area for heating, rather than an unreliable though generalized correlation for α_{wall}; an example of such an empirical formula [741] is given by:

$$\alpha_{wall.max} = 33.7\,\gamma_M^{0.2}\,\lambda_g^{0.6}\,d^{-0.36} \qquad (10\text{-}70)$$

with a correction for diffusional mixing (in gases, for particles larger than 0.7-1 mm). For this λ_f is approximated by $18 c_g \gamma_g w_{opt} d$, while w_{opt} is calculated from the empirical formula by Sarkits (10-60):

$$\mathrm{Re}_{opt} = 0.66\,\mathrm{Ar}_n^{0.5} \qquad \mathrm{Ar}_n = \frac{d^3 g}{\nu^2} \cdot \frac{\gamma_{bulk}}{\gamma_g}.$$

According to the suggestion of Varygin and Martyushin [877] more confidence in the calculated value of α_{wall} is obtained by taking

$$\alpha_{wall} \approx 0.8\,\alpha_{wall.max} \qquad (10\text{-}71)$$

10.5 Comparative Heat Transfer Between Walls and Dense, Suspended, and Spouting Beds

1. Stationary dense beds

It is to be expected that heat transfer will be poorer in flow through fixed beds than in fluidized beds, since the particles close to the wall are not replaced and become overheated compared with the core of the bed (in the case of heating the bed). Apart from radiation, the removal of heat from the wall occurs by means of the diffusional mixing of the fluid medium and as a result of the molecular thermal conductivity λ_0 of the two-phase fluid-particle system. Since λ_0 for the two-phase system is of the order of 0.15-0.2 kcal/m·hr·degree, i.e., many times larger than the thermal conductivity of the gas alone, heat transfer from the walls of a tube is improved when the cross section is filled with a fixed bed. With a liquid stream, on the other hand, filling the cross section with a fixed bed of solid particles usually reduces the rate of heat transfer between the walls and the stream, since the thermal conductivity of the two-phase system differs little from the thermal conductivity of the fluid medium alone, and diffusional mixing does not compensate for the absence of a turbulent core, which would be present in a single-phase stream of pure liquid. Information on heat transfer in stationary beds can be found in many references [175, 830, 950, 744, 1216, 1239 and 965].

2. Moving dense beds

As a result of the high particle concentrations (small bed voidages) moving dense beds present the possibility of obtaining even higher wall heat transfer coefficients than fluidized beds as long as contact times between the particles and heating surface are short, such as when the surface is wetted by a fluidized bed. To this end it is desirable to provide high particle velocities and heating surfaces in the form of horizontal tubes of small diameter placed in staggered array [972, 739]. In Ernst's work [972] very high values of α_{wall} (of the order of 600-1000 kcal/m^2·hr·degree) were obtained for heating surfaces consisting of short sections of tubes containing moving beds. In flow around a horizontal tube the mean heat transfer coefficient would be considerably lower, due to the presence of a zone of retarded particle motion and a zone free of particles. In the case of heat transfer to long vertical tubes, where there is no zone of stagnant particles, other circumstances prevent the achievement of high heat transfer coefficients, which would otherwise appear to be easily obtained by giving the bed a high velocity. When the bed moves under gravity with high velocity it undergoes breakup and separations, i.e., at any given moment, not all of the heating surface is in contact with the bed actively taking part in heat transfer. From Kalender'yan and Gorbis' [1155] observations of such decrease in the bed density and the appearance of separations they explained the presence of a maximum in the relationship for the heat transfer coefficient between vertical wall and a moving bed. Under the conditions of their experiments with particles of 0.4 mm diameter in a channel of 97.5 × 54 mm cross section, a decrease in heat transfer began even at a bed slip velocity of $w_M \approx$ 28 cm/sec; the maximum value of the heat transfer coefficient between the wall and such a bed did not exceed 100 kcal/m^2·hr·degree. With mixed size particles, heat transfer coefficients of the order of 200 kcal/m^2·hr·degree were obtained. The dependence of α_{wall} on the particle diameter found by Ernst [972] agrees with the concepts given earlier concerning conductive transfer between the walls and a fluidized bed.

Some increase in α_{wall} for a moving bed can be expected in the presence of a permeating stream of fluid medium, since there will be additional heat transfer by diffusional mixing of the medium [841]. This component of heat transfer is most important for large particles and high superficial fluid velocities, particularly for media having large bulk heat capacities (liquids).

The wall heat transfer coefficient for an incipiently fluidized bed near the limit of stability, but without hydraulic mixing of the particles, should be lower than the heat transfer coefficient with equal fluid flowing through a dense bed (held down to prevent fluidization), due to the lower particle concentration and the reduction in diffusional mixing. In principle it is possible to obtain high values of α_{wall} for such incipient fluidized beds by operating in a regime of ascending, descending, or horizontal bulk displacement of the material.

3. Suspended beds

If suspended beds are regarded as very highly developed fluidized beds, it can be expected that the values of α_{wall} will be relatively low due to the small particle concentration. This is confirmed by experiments. The radiant portion of such heat transfer can be taken into account as suggested by G. L. Polyak [see Eq. (10-16)]. For describing convective heat transfer between the walls of a vertical tube and a suspended bed, Gorbis [1139] proposed the relationship

$$\frac{\alpha_p}{\alpha_g} = 1 + 6.7\mathrm{Re}_g^{-0.3}\,\mathrm{Re}_T^{-0.33}\,\frac{c_M}{c_g}\,\mu,$$

where α_p, α_g are, respectively, the wall heat transfer coefficients with the suspended bed and with a stream of pure gas (at the same velocity); $\mathrm{Re}_g = w_{stream} D_T/\nu$ is the Reynolds number for the gas stream; $\mathrm{Re}_T = w_{rel} d/\nu$ is the Reynolds number for the particles of the suspended bed, around which flow occurs with a relative velocity w_{rel}; μ is the weight concentration of particles in the suspension, based on the throughputs.

This relationship was confirmed by Gorbis over the ranges:

$$\mathrm{Re}_g = 700 \div 65{,}000; \quad \mathrm{Re}_T = 5 \div 800; \quad \mu < 32; \quad D_T/d = 12\text{-}133.$$

4. Spouting beds

It is well known that particle concentrations and velocities vary widely in the different zones of a spouting bed. The conditions for heat transfer to the surrounding walls of a spouting bed are somewhat better than those for a dense bed with permeating fluid moving along a tube due to the presence of radial particle movement. Data on values of the heat transfer coefficients between walls and spouting beds are given in [979, 830]. The values of α_{wall} for a spouting bed [830] are 25–75% lower than for a fluidized bed at the same superficial fluid velocity.

Chapter 11

Means for Overcoming the Deficiencies of Fluidization as a Processing Technique

The inherent disadvantages in fluidization as a technological process include: 1) the difficulty in arranging for thermal countercurrent behavior or cross-current operation; 2) the moderate superficial fluid velocity, and hence limited gas throughput per unit area of apparatus cross section; 3) the nonuniform residence times of individual particles of material in continuously operating fluidized beds; 4) the entrainment of fine particles of material from fluidized beds with wide ranges of particle sizes; 5) the substantial pressure drop across deep fluidized beds and their grids; 6) the nonhomogeneity of fluidized beds and the difficulties in fluidizing at low superficial velocities close to the limit of stability; 7) the erosion of the body of the apparatus and the attrition of the granular material.

Many of these deficiencies are of a relative nature, i.e., they are disadvantages in carrying out one process but might be advantages in another process. For instance, good mixing of the material in a fluidized bed is a disadvantage from the point of view of carrying out countercurrent operations, but an advantage when it is necessary to maintain a uniform material temperature over the whole bed or for carrying out intensive heat transfer to a surface submerged in the bed. Entrainment of fine particles is a drawback if it is necessary to treat equally all the fractions of a material of wide particle size range in a fluidized bed, while it is an advantage if it is necessary to remove the dusty fractions from a granular material. The moderate gas loading per unit cross-sectional area of a fluidized bed is a deficiency when it is necessary to pass a large flow rate of a gaseous heat transfer medium or reagent through the bed, but it becomes a real advantage if the gas is only a fluidizing agent imparting physical fluidity to the material (horizontal transport of material on air slides).

These deficiencies are not all characteristic only of fluidized beds. It is no less difficult to organize a countercurrent process in a suspended bed of material than in a fluidized bed. The pressure drop in a fixed bed is many times higher than in a fluidized bed of the same particle size analysis operating at the same superficial fluid velocity. The limiting fluid velocity (the gas loading per unit cross section) is usually lower for a fixed bed than for a fluidized bed.

Let us consider some of the means available for eliminating these deficiencies (or, more specifically, means for altering various characteristics of a fluidized bed in a desired direction).

Creation of thermal counterflow and cross-flow

The following methods have been suggested:

1) multistage fluidized bed equipment;

2) retardation of the fluidized bed by screens;

3) operation at low values of the fluidization index;

4) operation at low values of the fluidization index in a fluidized bed with a conically expanding cross section.

In multistage fluidized bed equipment the material and gas pass in succession through a series of beds, with the gas moving from the lowest to the uppermost fluidized bed and the material moving in the opposite direction. This produces a so-called stagewise countercurrent operation (see [82, 107, 652 and 8911] and Chapter 5). The stages can be placed in rows, retaining, however, the counterflow of the gas and material from stage to stage.

Placing in a fluidized bed horizontal screens (see Chapter 5) which are permeable not only to the gas but to the material retards the mixing of the particles and creates a certain "degree of countercurrent behavior" within a single fluidized bed [1065, 808, 696]. It is known that packings (stationary dense beds) of coarse spheres can be used as well as screens for retarding the motion of particles in well-dispersed fluidized materials. The highly dispersed material is fluidized in the interstices of the coarse spherical packing.

In operation at low values of the fluidization index it is also possible to provide for counterflow or cross-flow, since the bed has a low voidage and almost no particle mixing can occur. When there is a large pressure drop across the fluidized bed the gas passing through the bed expands considerably. Hence, in a bed of constant cross section the linear gas velocity and the fluidization index will be much larger at the top of the bed than at the bottom. At the top there is intensive mixing of the material, while the lower parts are almost at the limit of stability and there is no mixing. To entirely avoid mixing in such cases a bed with a cross section gradually expanding towards the top should be used (conical tubes or conical inserts in cylindrical tubes) [1228].

At small values of the fluidization index there is also no difficulty in arranging cross-flow of the material and gas. Material fluidized in a trough with almost no mixing will slide from the feed point to the discharge point perpendicularly to an ascending gas stream. In order to establish cross-flow at high values of the fluidization index it is possible to use the device proposed by Grek [1038]. A fluidized bed of annular cross section (in plane) is forced to move from the feed point to the discharge point by a surrounding continuous moving vertical barrier, moved by a continuous pull chain.

Reducing the equipment cross section

For a given fluidized bed gas throughput it is possible to reduce the plan area of the equipment by: 1) using multistage fluidized bed devices; 2) operating with coarse-grained or pelletized materials; 3) operating at increased pressures; 4) using devices for centrifugally compressing the fluidized bed.

Under conditions where it is unnecessary to provide countercurrent flow, staged fluidized beds can be placed one above the other and operated in parallel with respect to both the gas and the material flows. The total transverse cross-sectional area of the bed is thereby not reduced, but it is placed on several levels, and the floor area of the apparatus is reduced.

Operating with coarse-grained materials also makes it possible to reduce the overall cross-sectional area of the bed. The linear fluidization velocity, and hence the throughput of gas per unit area of the bed cross section, are increased on changing from a fine to a coarse material. If the initial material is very fine or consists of particles covering a wide range of sizes, it is sometimes useful to make it into granules or pellets before fluidization [960]. However, there are limits to the extent to which the particle dimensions of the fluidized material can be reasonably increased due to the attendant decrease in the active surface area per unit weight of the bed and to the appearance of temperature gradients or other inhomogeneities within the particles.

Operation at elevated pressure also leads to a greater weight flow rate of gas per unit cross-sectional area of the bed, since the linear velocity for minimum fluidization in the laminar regime of fluid flow between the particles is almost unchanged as the pressure is increased [865, 780], and in the "turbulent" regime it decreases quite slowly as the pressure increases.

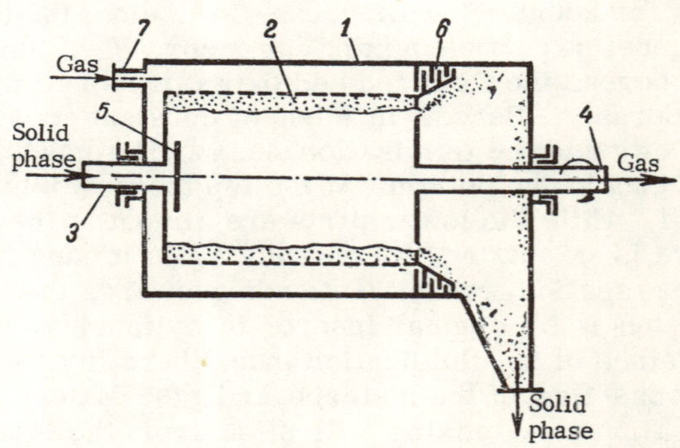

FIG. 11-1. Apparatus with centrifugally compressed fluidized bed [1034].
1—shell of apparatus; 2—rotor with screen side walls; 3—hollow shaft for introducing material; 4—hollow shaft for removing gas; 5—baffle; 6—labyrinth seal; 7—tube for supply of fluidizing gas.

Gel'perin, Ainshtein and co-workers [1034, 1035] proposed fluidizing a granular material in a centrifugal force field on the inner surface of a rotating cylindrical screen. Figure 11-1 shows a sketch of the simplest (single-layer) apparatus with a "centrifugally compressed" fluidized bed. The fluidizing gas is fed into the shell 1, and passes from the periphery to the axis of the apparatus through the screen-like side walls of the cylindrical rotor 2, fluidizing the material which is thrown against the inner surface of the screen by the centrifugal force. Material is fed into the apparatus through the open shaft 3. The efficient gas leaves through the second hollow shaft 4. The baffle plate 5 prevents the material being fed in through 3 from passing directly into 4 and facilitates the spreading of the particles on the screen. In this type of equipment it is easy to achieve a marked increase in the gas throughput per unit area of bed compared with ordinary fluidized beds at the same gas pressure and material particle size. As already mentioned, the limit to the throughput is set by the critical velocity of entrainment of the particles from the bed. For an ordinary fluidized bed, this cannot exceed the velocity of free fall of the particles in a gravitational force field. However, in the present type of apparatus, entrainment is opposed by a centrifugal acceleration, which may be several times larger than that due to gravity (when the rotor is spinning rapidly). A rotating bed (even one consisting of uniform-sized particles) passes from the dense to the fluidized state gradually, since the peripheral particles experience the greatest centrifugal acceleration and hence are fluidized later than the others.

The pressure drop in a rotating fluidized bed (neglecting the gravitational acceleration in comparison with the centrifugal acceleration) is given as [1034]:

$$\Delta P_{\text{rot}} = \frac{G\omega^2}{2\pi g l}, \qquad (11\text{-}1)$$

where G, l are the weight of the bed and its length (along the radius of rotation);
g is the acceleration of gravity;
ω is the angular velocity at which the bed rotates.

The value of the pressure drop ΔP_{rot} does not depend on the degree of bed expansion.

Experimental values of ΔP_{rot} appear to be lower than those given by Eq. (11-1), which the authors [1034] explain as due to the fact that the effects of the gravitational acceleration are not taken into account. The discrepancy decreases as the speed of rotation increases. No break-through of bubbles has been observed in rotating fluidized beds.

References [1034 and 1035] give sketches of single-layer and multilayer continuously operating equipment. The drawbacks of the method are the relative complexity of the equipment and the presence of massive rotating parts.

Equalizing particle residence times in fluidized beds

A simple method for equalizing particle residence times in fluidized beds is to operate with periodic charging and discharging of the material. This method is generally not attractive in view of the advantages of continuous over batchwise processes. As alternate solutions there remain multistage equipment, retarding screens, operation at low values of the fluidization index, and the use of moving vertical baffles.

The efficiency of equalizing particle residence times in a fluidized bed by using equipment with two or more stages has been considered in detail by Fedorov [170]. If it is unnecessary to arrange for stagewise countercurrent flow of both material and gas, the material can be made to flow from one stage to the next in succession, while the gas is fed in parallel to all the stages (Fig. 11-2). Figure 11-2 illustrates a five-stage fluidized bed dryer arrangement. With respect to uniformity of the residence times of the individual particles in the apparatus, the worst possible situation which can arise is ideal mixing of the material in each of the stages. In this case, Fedorov [170] has shown that for an apparatus with n stages,

$$1 - y = 1 - e^{-\tau/\tau_{av}}\left[1 + \frac{\tau}{\tau_{av}} + \frac{(\tau/\tau_{av})^2}{2!} + \ldots + \frac{(\tau/\tau_{av})^{n-1}}{(n-1)!}\right] \quad (11\text{-}2)$$

where τ_{av} is the residence time of the material in the complete absence of mixing;
$1 - y$ is the fraction of particles which reside in the apparatus for a period of time less than τ.

As a result, we find for particles residing in the apparatus for less than τ_{av}:

$$1 - y_{av} = 1 - e^{-1}\left[1 + 1 + \frac{1}{2!} + \ldots + \frac{1}{(n-1)!}\right]. \quad (11\text{-}3)$$

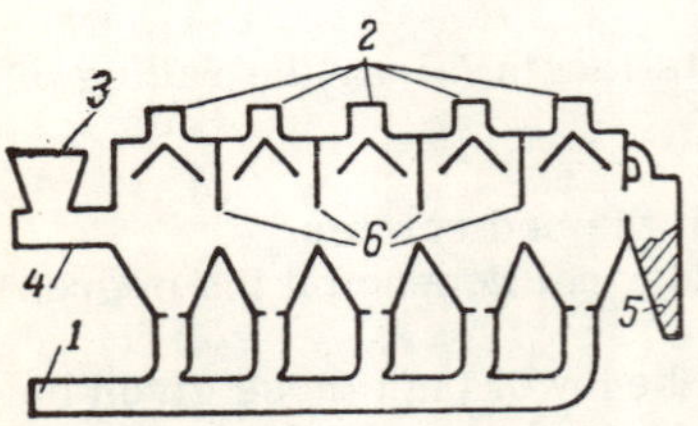

FIG. 11-2. Multistage fluidized bed apparatus [170].
1—tube for supplying fluidizing gas-heat transfer medium; 2—tubes for removing gas; 3—hopper; 4—material feeder; 5—tube for removing material; 6—openings for cross flow of material from stage to stage.

For a single-stage apparatus, this fraction is 0.632, becoming 0.264 for two stages, 0.081 for three, 0.020 for four, and 0.004 for five stages; the fraction decreases rapidly as the number of stages increases.

In an editorial note in the book by Fedorov [170], Mikhaylov pointed out that the arrangement shown in Fig. 11-2 requires the supply of gas to each stage by a separate blower. Otherwise, if the quantity of material in one stage increased for any reason, this would decrease the supply of gas to this stage. Because of this redistribution of gas between the stages, the quantity of fluidized material in the given stage would continue to increase. Material would begin to enter this stage from the stages preceding and following it. Because of this system maldistribution of fluidization in the flooded stage would cease rapidly. On the other hand, with independent blowers for each stage it is possible to vary the gas supply to any one of the stages over quite a wide range, and at the same time to increase or decrease the quantity of material fluidized in it without disturbing the operation of the remaining stages [170].

Reducing bed pressure drop

A direct means of reducing pressure drop is to operate with shallow fluidized beds, in principle this is possible in many cases because of the well-distributed high surface area of fine granular materials and the high values of transfer coefficients. Remarks have been made earlier (Chapter 1) concerning the constructional requirements and pressure drops of the gas-distributing grids. In order to make thin fluidized beds feasible in continuously operating equipment, the feed device must distribute the material uniformly over the bed. Hanging of the material in the bunker above the feed device must be eliminated. When a moist material has a tendency to stick together, it is sometimes convenient to add free-flowing, pretreated material.

Increasing bed uniformity

It is possible to increase the uniformity of fluidization by 1) improving the inlet gas distributor; 2) retarding the fluidized bed; 3) operating at increased pressure; 4) imposing vibrations on the system; 5) mixing the bed mechanically.

The uniformity of initial gas distribution is very important to activate the lower parts of deep beds and to permit operating with shallow fluidized beds (see Chapter 1).

"Retardation" of a fluidized bed by horizontal screens and vertical inserts breaks up large aggregates of particles and bubbles and produces more uniform gas distribution [1065 and 433]. Sechenov and Al'tshuler [865 and 780] showed that uniformity of fluidization improves with increased pressure.

The imposition of vibration or mechanical mixing on the bed is useful for stimulating fluidization, especially at low values of the fluidization index and with materials which have a tendency to channel. Syromyatnikov [149] suggested the use of vibrating air-distributing grids to eliminate stagnant zones in fluidized bed combustion chambers. In 1955, Reed and Fenske [512] observed that vibrators submerged in a bed improved fluidization over the whole volume of the bed. Simultaneously, Morse [460] showed that low-frequency (50-500 c/s) sonic vibrations (above 110 db) could appreciably improve the fluidization of nonfree-flowing highly dispersed powders such as alabaster and cement. This method however has little practical application. Gorling, et al. [826], suggested the use of a pulsating feed of the fluidizing agent to prevent channel formation in the reduction of finely ground iron ore in a fluidized bed. In 1958, Beber, et al. [804], obtained a U.S. patent on a method of mechanically stimulating fluidization by using vibrations of the gas-distributing grid similar to the work of Syromyatnikov. Beber, et al., carried out experiments to determine the effects of grid vibrations as well as mixing of the bed with a paddle rotated at 57 r.p.m. near the grid. They gave approximate ranges for the amplitude (from a few ten-thousandths to several cm) and the frequency (from 60 to 10,000 c/s) "depending on the rate of fluid flow and the properties of the material." Recently, Leva [1117] also carried out experiments using mechanical mixing in gently aerated beds, varying the number and inclinations of the mixers and their rates of rotation. This method of stimulating fluidization did not appear advantageous from an energy point of view [1117]: rotating the stirrers required about 50 times more power than that for simply fluidizing the bed.

Decreasing particle carry-over

Entrainment of material from fluidized beds can be reduced or eliminated by using granulated or pelleted materials, leaving a sufficient free height above the fluidized bed, increasing the uniformity of fluidization (decreasing bed fluctuations), or using "covering beds" of coarse but light particles.

The effects of granulation or pelletization are obvious and need no explanation.

The effect of free height above the fluidized bed and the covering layer of coarse light particles [1012] has been considered earlier (pages 146 and 151).

Increasing the uniformity of fluidization reduces carry-over by decreasing the amount of material ejected by local spouting.

Problems of material attrition and equipment erosion

Considerable particle attrition can occur in fluidized bed equipment, particularly if the material is continuously circulated through the system or fluidized for tens or hundreds of hours in periodically operating beds. However, in such instances, it appears that attrition occurs mainly in the pneumatic transport lines and not within the fluidized beds. In pneumatic transport lines the velocities of the gas and particles relative to the walls are many times larger than in a fluidized bed. Lines with high-velocity gas streams should always be avoided in the handling of friable particles; hex-metal grids should not be placed in the bends of pipelines to protect the walls from erosion, for this will also increase particle attrition. The presence of fines in the material reduces particle attrition within the fluidized bed.

Reboux [511] noted that the nature of any chemical reactions occurring in the system has a very important effect on the attrition of the material in the fluidized bed and in other elements of the equipment. In some cases the reaction products sharply reduce the mechanical strength of the particles and lead to rapid size reduction. For example, under the conditions of an industrial process, the concentration of particles larger than 90 microns, in a bed of Fischer-Tropsch catalyst, fell from 98.3% to 58% after 170 hours of fluidization. However, this was accompanied by degeneration of the material. Its density fell from 2.00 to 1.09, and the carbon content increased from zero to 24.7%. The process of particle enlargement is also encountered in fluidized beds.

Erosion of walls surrounding fluidized beds is usually insignificant [279, 327, and 511]. On the other hand, there may be very great wear in the bends of pneumatic transport lines. To reduce pipe wear it is advisable to operate at the minimum possible gas velocities, since erosion is proportional to approximately the third power of velocity [511]. It is also advisable to use elbows with a large radius of curvature [717]. Square elbows are good for reducing tube erosion if the particle material is sufficiently wear-resistant; the accumulation of deposited material protects the elbows from wear [717, 1138].

Bibliography

Some notes on the Russian items in the Bibliography

The titles of all the Russian papers and books have been translated into English to assist the reader in determining the subject of a given reference. The names of authors, journals, etc., have been transliterated in accordance with the scheme mentioned in the Translator's Note (p. xiii). The following notes may also prove useful.

I. Book Publishers

(Izd., Izdat., abbreviations for Izdatel'stvo = press, publishing house).
Aznefteizdat: Azerbaijan Petroleum Press.
GITTL (Gosudarstvennoe Izdatel'stvo Tekhnicheskoi i Teoreticheskoi Literatury): State Technical and Theoretical Press.
GONTI (Gosudarstvennoe Obedinennoe Nauchno-Tekhnicheskoe Izdatel'stvo): United Scientific and Technical State Press.
Gosenergoizdat: State Power Engineering Press.
Goskhimizdat: State Press for Chemical Literature.
Gostoptekhizdat: State Scientific and Technical Press of the Petroleum and Mineral Fuel Industry.
Izd. Akad. Nauk BSSR: Press of the BSSR Academy of Sciences.
Izd. Akad. Nauk SSSR: Press of the USSR Academy of Sciences.
Izd. GOSINTI (Izdatel'stvo Gosudarstvenn. Nauchno-Issledovatelsk. Instituta Nauchnoi i Tekhnicheskoi Informatsii): Press of the State Scientific Research Institute of Scientific and Technical Information.
Izd. Kirgiz. Fil. Akad. Nauk SSSR: Press of the Kirgiz Affiliate of the USSR Academy of Sciences.
Izd. Mosk. Energet. In-ta: Press of the Moscow Power Institute.
Izd. "Sov. Nauka": "Soviet Science" Press.
Mashgiz: Mechanical Engineering Press.
Metallurgizdat: Metallurgy Press.
ONTI (Obedinennoe Nauchno-Tekhnicheskoe Izdatel'stvo): United Scientific Technical Press.
Pishchepromizdat: Food Industry Press.
Promstroiizdat: Building Industry Press.
Ugletekhizdat: Coal Technology Press.

II. Journals referred to most frequently

Az. N. Kh., Az. Neft. Khoz. (Azerbaidzhanskoe Neftyanoe Khozyaistvo): Azerbaijan Petroleum Industry.
Byulleten' Izobretenii: Bulletin of Inventions.

Byull(eten') Tsvet(noi) Met(allurgii): Non-Ferrous Metallurgy Bulletin.
Doklad. Akad. Nauk SSSR (Doklady Akademii Nauk SSSR): Reports of the USSR Academy of Sciences.
Gaz(ovaya) Prom(yshlennost'): Gas Industry.
Inzh.-Fiz. Zhurn. (Inzhernerno-Fizicheskii Zhurnal): Engineering Physics Journal.
Izmer(itel'naya) Tekhnika: Measurement Techniques.
Izv. Akad. Nauk SSSR (Izvestiya Akademii Nauk SSSR): News of the USSR Academy of Sciences (various divisions).
Izv. Sib. Otd. Akad. Nauk SSSR: News of the Siberian Section of the USSR Academy of Sciences.
Izv. Vost. Fil. Akad. Nauk SSSR: News of the Eastern Affiliate of the USSR Academy of Sciences.
Izv(estiya) VUZov: VUZ News (various divisions) (VUZ = Vysshee Uchebnoe Zavedenie = Higher Institutes of Learning).
Khim. Mashinostr. (Khimicheskoe Mashinstroenie): Chemical Equipment Construction.
Khim. Nauka i Prom. (Khimicheskaya Nauka i Promyshelennost'): Chemical Science and Industry.
Khim. Prom. (Khimicheskaya Promyshlennost'): Chemical Industry.
Khimiya i Khim(icheskaya) Tekhnol(ogiya) Topliv: Chemistry and Chemical Technology of Fuels and Oils.
Kislorod: Oxygen
Koks i Khimiya: Coke and Chemistry.
Nauchn(ye) Doklady Vyssh(ei) Shkoly: Scientific Reports from the Higher Schools.
Neft. Khoz. (Neftyanoe Khozyaistvo): Petroleum Economy.
Teploenergetika: Thermal Power Engineering.
Tsvet. Met. (Tsvetnye Metally): Non-Ferrous Metals.
Za Ekonom. Topliva: Fuel Economics.
Zav. Lab. (Zavodskaya Laboratoriya): Plant Laboratory.
Zhurn(al) Fiz(icheskoi) Khim(ii): Journal of Physical Chemistry.
Zhurn(al) Prikl(adnoi) Khim(ii): Journal of Applied Chemistry.
Zhurn(al) Tekhn(icheskoi) Fiziki: Journal of Technical Physics.

III. Other Abbreviations

Akad. Nauk (Akademiya Nauk): Academy of Sciences.
ArmSSSR: Armenian SSR.
Avt. Svid. (Avtorskoe Svidetel'stvo) SSSR: USSR Inventor's Certificate.
AzSSR: Azerbaijan SSR.
BSSR: Belorussian SSR.
Byull., Byulleten': Bulletin.
Dokl., Doklad., Doklady: Reports
Dokt. Dissertat.: Doctoral Thesis.
In-t, In-ta (Institut): Institute.
Izd., Izdat. (Izdatel'stvo): Press, publishing house.
Izv. (Izvestiya): News.
Kand. Dissertat.: Thesis for Candidate's Degree.
Politekhn(icheskii): Polytechnical.
Sb., Sborn. (Sbornik): Collection of papers, symposium.
SSSR: USSR.

Tr. (Trudy): Proceedings.
U.P.I. (Ural'skii Politekhnicheskii Institut): Urals Polytechnic Inst.
USSR: Ukrainian SSR.
Vestnik: Bulletin, herald.
V.N.I.G.I. (probably Vsesoyuznyi Nauchnyi Institut Goryuchikh Iskopaemykh): All-Union Scientific Institute of Mined Fuels.
Vses(oyuznyi): All-Union.
Zhurn(al): Journal.

References published up to 1957

1. L. A. Akopyan and A. G. Kasatkin, Hydrodynamics of fluidized beds of granular material (in Russian), Khim. Prom., 1955, No. 2.
2. K. V. Alferov, Bunkers, Valves and Feeders (in Russian). Mashgiz, 1946.
3. K. V. Alferov and R. L. Zenkov, Bunker Plant (in Russian). Mashgiz, 1955.
4. S. P. Artyushin, Drying and Dust Collection in Coal Beneficiation Plants (in Russian). Ugletekhizdat, 1956.
5. M. E. Aerov, Catalytic and and adsorption processes with moving beds (in Russian), Khim. Prom., 1951, No. 1.
6. M. E. Aerov and N. N. Umnik, Measurement of gas velocities in granular beds, (in Russian), Zhurn. Prikl. Khim., 1950, 23, No. 10, 1009-1017.
7. M. E. Aerov and N. N. Umnik, Thermal conductivities of granular beds (in Russian), Zhurn. Tekhn. Fiz., 1951, 21, No. 11, 1351-1363.
8. M. E. Aerov and N. N. Umnik, Heat transfer in tubes filled with granular packings (in Russian), Zhurn. Tekhn. Fiz., 1951, 21, No. 11, 1364-1371.
9. M. E. Aerov and N. N. Umnik, Diffusivities in granular beds (in Russian, Zhurn. Prikl. Khim., 1954, 27, No. 3, 265-272.
10. V. A. Baum, Investigation of mixing processes in fluid streams flowing through tubes filled with granular material (in Russian), Izv. Akad. Nauk SSSR, Otd. Tekhn. Nauk, 1953, No. 9.
11. V. G. Bakhurov and G. K. Boreskov, Effective thermal conductivities of contacting masses (in Russian), Zhurn. Prikl. Khim., 1947, 20, 721-738.
12. M. S. Bernshtein, Type of efflux and grain pressures in silos (in Russian), Stroit. Prom., 1945, Nos. 10-11.
13. G. K. Boreskov and L. G. Rikhter, Pressure drops in contact masses (in Russian), Khim. Prom., 1946, No. 6, 5-10.
14. V. M. Borishanskii, A study of the aerodynamics of fuel beds (in Russian), Kotloturbostroenie, 1948, No. 1.
15. B. N. Brounshtein and O. M. Todes, The hydraulics of beds of solid particles suspended in an ascending gas stream (in Russian), Neft. Khoz., 1951, No. 4.
16. B. N. Brounshtein and O. M. Todes, Fundamentals of the theory of pneumatic transport (in Russian), Zhurn. Tekhn. Fiz., 1953, No. 1.

17. I. A. Burovoi, Automatic control of fluidized bed calciners (in Russian), Tsvet. Met., 1955, No. 3.
18. I. A. Burovoi, I. M. Bernshtein and G. M. Shteingard, Automation of the fluidized bed roasting of zinc concentrates (in Russian), Tsvet. Met., 1956, No. 7, 47-53.
19. M. L. Varlamov, Oxidation of sulfur dioxide on vanadium catalysts in a fluidized state (in Russian), Zhurn. Prikl. Khim., 1954, 27, 360-367.
20. A. A. Voitekhovskii, A. P. Zinov'eva and D. I. Orochko, Some problems in the suspension and fluidization of powder-like materials (in Russian), Tr. V.N.I.G.I., 1949, No. 3.
21. A. A. Voitekhovskii, A. P. Zinov'eva and D. I. Orochko, Tr. V.N.I.G.I., 1951, No. 3, 190-212.
22. V. F. Volkov, An investigation of gas formation in a suspended bed of fine-grained fuel (in Russian), Trudy U.P.I., 1956, Sb. 61, 15-24.
23. V. F. Volkov, Gasification of heavy coal-tar products in a suspended state (in Russian), Trudy U.P.I., 1956, Sb. 61, 75-80.
24. I. Gasterstadt, Pneumatic Transport (in Russian, transl. from German), Izd. Prombyuro V.S.N.Kh., 1927.
25. A. Garbonenko, The V.T.I. furnace for the suspended combustion of fuel (in Russian), Novosti Tekhniki, 1939, No. 30.
26. K. M. Grinev, M. N. Krasheninnikov and A. P. Krotkov, Pneumatic Transport in the Cement Industry (in Russian), Promstroiizdat, 1951.
27. S. F. Gudkov and A. B. Chernyshev, A study of the methane synthesis process in an apparatus with a fluidized catalyst bed (in Russian), Izv. Akad. Nauk SSSR, Otd. Tekhn. Nauk, 1955, No. 5.
28. R. B. Gun, Reconstruction of the Winkler-Koch cracking plant (in Russian), Neft. Khoz., 1947, 25, No. 6, 41.
29. A. A. Gukhman and S. I. Ilyukhin, The effect of temperature on the state of heat transfer (in Russian), Zhurn. Tekhn. Fiz., 1952, 22, No. 5.
30. D. A. Diomidovskii, Nonferrous Metallurgical Furnaces (in Russian), Metallurgizdat, 1956.
31. G. K. D'yakonov and G. A. Semenov, Steady-state heat transfer in a bed of spheres (in Russian), Izv. Akad. Nauk SSSR, Otd. Tekhn. Nauk, 1955, No. 7.
32. L. N. Erkova, A Study of Suspended Bed Phenomena (in Russian), Kand. Dissertat., Leningradskii Khimiko-Tekhnologicheskii Institut, Leningrad, 1954.
33. L. N. Erkova and N. I. Smirnov, The height of a suspended bed of spherical particles and its dependence on the process conditions (in Russian), Zhurn. Prikl. Khim., 1956, 29, No. 8, 1175-1182.
34. L. N. Erkova and L. I. Smirnov, The height of suspended beds of particles of various shapes and its dependence on process conditions (in Russian), Zhurn. Prikl. Khim., 1956, 29, No. 9, 1347-1353.
35. L. N. Erkova and L. I. Smirnov, The laws of suspended beds of solid particles (in Russian), Zhurn. Prikl. Khim., 1956, 29, No. 10.

36. N. M. Zhavoronkov, Fundamental Hydraulics of the Scrubbing Process and Heat Transfer in Scrubbers (in Russian), Izd. "Sov. Nauka," 1944.
37. N. M. Zhavoronkov, Gas pressure drop in flow through scrubber packings and size reduction of the solid material (in Russian), Khim. Prom., 1944, No. 1, 4-14.
38. N. M. Zhavoronkov, M. E. Aerov and N. N. Umnik, The pressure drop and packing density of granular beds (in Russian), Zhurn. Fiz. Khim., 1949, 23, 342.
39. N. M. Zhavoronkov and I. E. Furmer, The pressure drop in granular beds at low gas velocities (in Russian), Trudy Moskovskii Khimiko-Tekhnologicheskii Institut im. Mendeleeva, 1952, No. 12.
40. B. L. Zharkov and Kh. I. Kolodtsev, Calculation of gas formation in fluidized beds of coal (in Russian), Izv. V.T.I., 1951, No. 12.
41. L. K. Zhitkevich, Investigation of the Drying Process and Aerodynamics in Moving Beds of Peat (in Russian), Kand. Dissertat., Akad. Nauk BSSR, Minsk, 1953.
42. E. A. Zhikharev, Investigation of the Process of Pneumatic Transport of Granular Materials in Horizontal, Vertical and Inclined Pipelines (in Russian), Kand. Dissertat., Belorusskii Politekhnicheskii Institut, Minsk, 1955.
43. N. E. Zhukovskii, Fundamentals of Aerostatics (in Russian), GITTL, 1950.
44. S. S. Zabrodsky, Rapid Drying of Milled Peat in a Suspended State (in Russian), Kand. Dissertat., Akad. Nauk BSSR, Minsk, 1949.
45. S. S. Zabrodsky, Some problems in the complete drying of materials (in Russian), Izvestiya Akad. Nauk BSSR, 1950, No. 5, 113-123.
46. S. S. Zabrodsky, A method for roasting pulverized materials, Avtor. Svidetel'stvo SSSR No. 95113, 21.02.53.
47. S. S. Zabrodsky, An approximate comparison of fixed-bed and suspended-bed heat exchangers for utilizing heat from waste gases (in Russian), Tr. In-ta Energetiki, Akad. Nauk BSSR, No. 1, 1954, 126-149.
48. S. S. Zabrodsky, Comparison of heat exchangers of the fluidized bed and fixed bed types for utilization of heat from waste gases (in Russian), Tr. In-ta Energetiki, Akad. Nauk BSSR, No. II, 1955, 162-177.
49. S. S. Zabrodsky, The mechanism of heat transfer from walls to fluidized beds of granular materials (in Belorussian), Vestsi Akad. Navuk BSSR, Seryya Fizika Tekhnichnykh Navuk, 1956, No. 2, 65-80.
50. S. S. Zabrodsky, The resistance coefficient for solid particles in a gas stream (in Belorussian), Vestsi Akad. Navuk BSSR, Seryya Fizika Tekhnichnykh Navuk, 1956, No. 4, 39-50.
51. A. N. Zelikman, L. V. Belyaevskaya and O. E. Krein, A study of roasting molybdenite concentrates in a fluidized bed (in Russian), Tsvet. Met., 1956, No. 8.
52. R. L. Zenkov, The Mechanics of Poured Fillings (in Russian), Mashgiz, Moscow, 1952.

53. N. M. Indyukov, et al., Determination of the quantity of circulating catalyst in a cracking system with a moving bed of powdered catalyst (in Russian), Khim. i Tekhn. Topl., 1956, No. 4.
54. S. M. Kagan, Characteristics of gas flow through beds of moist materials (in Russian), Ts.K.T.I. Symposium "Za novoe sovetskoe energooborud.," 1939.
55. K. S. Kadymova, Effect of concentration on the laws of motion of sand in fluid streams (in Russian), Neft. Khoz., 1953, No. 3, 24-29.
56. I. M. Kampaneets, Experimental Studies of the Motion of Granular Materials in Tubes and Chambers of Various Types (in Russian), Kand. Dissertat., Energeticheskii Institut, Akad. Nauk SSSR, Moscow, 1953.
57. N. V. Karkhov, Gasification of finely powdered coal in fluidized beds (in Russian), Kislorod, 1944, No. 2.
58. Kachan, I. N., The rate of hindered fall of fine mineral grains in water (in Russian), in: Sb. nauchno-issl. rabot In-ta mekhanich. obrabot. polezn. iskopaemykh, 1953, No. 88.
59. V. G. Kashirskii, Methods of gasifying finely ground fuel in a fluidized bed (in Russian), Byulleten' Izobretenii, 1946, No. 10.
60. B. I. Kitaev, Heat Transfer in Shaft Furnaces (in Russian), Metallurgizdat, 1945.
61. B. K. Klimov, New Methods for the Thermal Treatment of Peat (in Russian), GONTI, 1939.
62. A. S. Klyachko, Convective heat transfer coefficients in gas-dispersed systems (in Russian), Zhurn. Tekhn. Fiz., 1945, No. 8.
63. A. V. Koganovskii and V. G. Petrenko, Adsorption of colloids in a suspended bed (in Russian), Kolloidn. Zhurn., 1953, 15, 99-107.
64. Kh. I. Kolodtsev and B. L. Zharkov, A study of the process of carbon combustion in high-voidage fluidized beds (in Russian), Izvestiya V.T.I., 1950, No. 10.
65. V. Ya. Kruglikov, A Study of the Problems in the Design of Reactors for Producing Synthesis Gas with Powdered Catalysts (in Russian), Kand. Dissertat., Mosneftint, Moscow, 1951.
66. L. I. Kudryashov, Improved calculations of heat transfer coefficients between gases and suspended particles using the thermal boundary layer method (in Russian), Izv. Akad. Nauk SSSR, Otd. Tekhn. Nauk, 1949, No. 11.
67. I. N. Kuz'minykh, Possibilities of Applying Bubbling and Fluidized Bed Methods in the Contact Process for Manufacturing Sulfuric Acid (in Russian), in: Tr. Soveshchaniya po metallurgii tsinka (1954), Metallurgizdat, Moscow, 1956.
68. P. Kh. Kurinov, A. P. Mikheev and S. A. Shashkin, Gasification of finely ground fuel and milled peat (in Russian), Tr. V.N.I.G.I., 1948, No. 1.
69. K. P. Lavrovskii and A. L. Rozental', The laws of chemical reactions in fluidized beds (in Russian), Khimiya i Khim. Tekhnol. Topl., 1956, No. 2.
70. K. P. Lavrovskii and A. L. Rozental', The mixing of gases in fluidized beds (in Russian), Khimiya i Khim. Tekhnol. Topl., 1956, No. 3.

71. K. P. Lavrovskii and A. L. Rozental', The laws of reactions in fluidized beds (in Russian), Trudy In-ta Nefti, Akad. Nauk SSSR, 1956, 8, 94.
72. K. P. Lavrovskii and A. L. Rozental', The mixing of gases in fluidized beds, (in Russian), Khimiya i Khim. Tekhnol. Topl., 1956, No. 3, 17-21.
73. I. P. Levsh, M. I. Niyazov and O. B. Erofeeva, Some problems in the hydrodynamics of suspended beds (in Russian), Trudy Sr. Az. Politekhn. In-ta, Tashkent, 1955.
74. G. Ya. Leizerovich, Fundamentals of the roasting process in fluidized beds (in Russian), Tsvet. Met., 1954, No. 6.
75. G. Ya. Leizerovich, Ovens and reactors based on the process of roasting in fluidized beds (in Russian), Tsvet. Met., 1955, 28, No. 2, 71-78.
76. G. Ya. Leizerovich, Roasting in Fluidized Beds (in Russian), Metallurgizdat, Moscow, 1955.
77. G, Ya. Leizerovich and I. V. Babina, Roasting of zinc concentrates in fluidized beds (in Russian), Collection of Papers, Metallurgizdat, 1953.
78. G. Ya. Leizerovich and E. Ya. Serebrennikova, Roasting of pyrites in a fluidized bed (in Russian), Bum. Prom., 1955, No. 12.
79. Z. S. Leont'eva, Combustion of carbon particles in a moving gas stream (in Russian), Izv. Akad. Nauk SSSR, Otd. Tekhn. Nauk, 1951, No. 12.
80. I. P. Linchevskii, Problems in the efflux of free-flowing solids (in Russian), Zhurn. Tekhn. Fiz., 1939, 9, No. 4, 343-347.
81. A. M. Lomov, Proposed furnaces and plants for roasting zinc concentrates in fluidized beds (in Russian), Tsvet. Met., 1956, No. 5.
82. M. Yu. Lur'e, Drying (in Russian), Gosenergoizdat, 1948.
83. A. V. Lykov (Luikov), Transfer Phenomena in Capillary-Porous Bodies (in Russian), GITTL, 1954.
84. A. V. Lykov (Luikov), Heat and Mass Transfer in the Drying Process (in Russian), Gosenergoizdat, 1956.
85. D. N. Lyakhovskii, Convective heat transfer between a gas and suspended particles (in Russian), Zhurn. Tekhn. Fiz., 1940, 10, No. 12.
86. D. N. Lyakhovskii, Convective heat transfer between spherical suspended particles and the surrounding medium (in Russian), Kotloturbostroenie, 1947, No. 5.
87. P. V. Lyashchenko, Gravitational Methods of Beneficiation (in Russian), Gostoptekhizdat, 1940.
88. I. G. Martyushin, Investigation of the Mechanism of the Motion of Gas and Solid Material (in Russian), Kand. Dissertat., Moskovskii Institut Khimicheskogo Mashinostroeniya, Moscow, 1952.
89. B. K. Marushkin, A Study of Fluidized Catalyst Bed Models (in Russian), Kand. Dissertat., Moscow, 1949.
90. D. I. Mendeleev, Fundamentals of Chemistry (in Russian), Vol. 2, 140 and 419, Goskhimizdat, 1947.
91. Metallurgical Treatment of Ores in a Suspended State (in Russian), Papers edited by A. N. Vol'skii, ONTI, 1936.

92. E. M. Minskii, Turbulent diffusion in porous media (in Russian), Doklady Akad. Nauk SSSR, 1951, 28, 3.
93. D. M. Mints, The suspension of a granular bed in an ascending fluid stream (in Russian), Doklady Akad. Nauk SSSR, 1952, 82, 17-20.
94. D. M. Mints, The hydrodynamic resistance of a granular bed suspended in a stream (in Russian), Doklady Akad. Nauk SSSR, 1952, 83, No. 4, 541.
95. D. M. Mints, The precipitation of concentrated suspensions (in Russian), Khim. Prom., 1954, No. 1, 33-36.
96. D. M. Mints and S. A. Shubert, Hydraulics of Granular Materials (in Russian), Izd. Min-va Kom. Khoz. RSFSR, 1955.
97. N. M. Mikhailov and I. M. Fedorov, New dryers for partly dried fuels (in Russian), Za Ekonom. Topl., 1951, No. 8.
98. M. A. Mikheev, Fundamentals of Heat Transfer (in Russian), Gosenergoizdat, 1956.
99. L. M. Moroz and Ya. I. Frenkel', The hydrodynamics of suspension droplets falling in a pure dispersion medium (in Russian), Kolloidn. Zhurn., 1949, 11, No. 3, 178-183.
100. I. P. Mukhlenov, D. G. Traber and E. S. Rumyantseva, The use of suspended catalyst beds for the oxidation of sulfur dioxide (in Russian), Khim. Prom., 1955, No. 8, 457-460.
101. R. N. Mullokandov, The hydraulic resistance of beds of spherical particles in isothermal and nonisothermal air streams (in Russian), Zhurn. Tekhn. Fiz., 1948, 18, No. 8, 1051-1062.
102. E. S. Nemet, Investigation of Adsorption Kinetics in Suspended Beds (in Russian), Kand. Dissertat., Leningradsk. Tekhn. Institut im. Lensoveta, 1955.
103. V. A. Nikolaeva, The Effect of Various Factors on the Process of Gasification of Finely Powdered Fuel in Fluidized Beds (in Russian), Kand. Dissertat., Institut Goryuchikh Iskopaemykh, Akad. Nauk SSSR, Moscow, 1954.
104. V. I. Oborin, The nature of the motion of unsymmetrical solid bodies in limited fluid streams (in Russian), Zhurn. Tekhn. Fiz., 1941, No. 9.
105. S. N. Obryadchikov and B. K. Marushkin, Screens for stabilizing fluidized beds of powdery catalysts, Avt. Svid. SSSR, No. 77693, 1950.
106. S. N. Obryadchikov and B. K. Marushkin, An investigation of processes with moving catalysts (in Russian), Neft. Khoz., 1946, 24, No. 11, 36-45.
107. D. I. Orochko, The Theoretical Basis for Carrying Out the Synthesis of Fluid Fuels (in Russian), Gostoptekhizdat, 1951.
108. N. V. Pavlovich, A Theoretical and Experimental Investigation of the Principles of Two-Stage Fluidized-Bed Combustion Chambers for Burning the Earthy Brown Coals of the Ukrainian SSR (in Russian), Kand. Dissertat., Moskovskii Energeticheskii Institut, Moscow, 1948.
109. N. V. Pavlovich, The combustion of earthy brown coals in fluidized bed furnaces (in Russian), Tr. In-ta Teploenergetiki Akad. Nauk USSR, 1950, Part 2.

110. N. V. Pavlovich, Results of an investigation of the combustion of brown coals in a two-stage fluidized-bed combustion chamber (in Russian), Trudy Akad. Nauk USSR, 1951, Vol. 1.
111. A. P. Palchevskii, Coefficients of hindered fall (in Russian), Gidrotekhn. Str-vo, 1953, No. 2.
112. I. V. Paramonov, Byulleten' Tsvet. Met., Alma-Ata, 1956, No. 3, 9-14.
113. M. K. Pis'men, Gasification in fluidized beds (in Russian), Khim. Tv. Topl., 1938, No. 2.
114. M. K. Pis'men, A method of calculating the process of gasification in a fluidized bed (in Russian), Tr. V.N.I.G.I., 1953, No. 5.
115. M. K. Pis'men, Gasification of brown coal in fluidized beds (in Russian), Tr. V.N.I.G.I., 1953, No. 5.
116. G. I. Pokrovskii and A. I. Aref'ev, The efflux of powdery solids (in Russian), Zhurn. Tekhn. Fiz., 1937, 7, No. 4, 424-427.
117. G. L. Polyak, Basic Calculations of Heat Exchange Equipment. Radiant Heat Transfer (in Russian), Report, Energeticheskii Institut, Akad. Nauk SSSR, 1946.
118. A. S. Predvoditelev, L. N. Khitrin, O. A. Tsukhanova, Kh. I. Kholodtsev and M. K. Grodzovskii, Combustion of Carbon (in Russian), Izd. Akad. Nauk SSSR, 1949.
119. V. L. Przhetslavskii, Inform. Byulleten', Karaganda, Giproshakht, 1956, No. 1.
120. S. Ya. Pshezhetskii and R. N. Rubinshtein, Carrying out heterogeneous catalystic reactions in flowing streams (in Russian), Zhurn. Fiz. Khim., 1946, 20, No. 12, 1421-1434.
121. I. M. Razumov, Hydraulic regimes in moving catalyst beds (in Russian), Neft. Khoz., 1953, No. 1.
122. I. M. Razumov, Ways to improve the operation of pneumatic conveyors in catalystic cracking plants (in Russian), Khim. i Tekhn. Topl., 1956, No. 12.
123. N. I. Reshetin and N. I. Syromyatnikov, Combustion of coke fines in suspension (in Russian), Za Ekonom. Topliva, 1949, No. 2.
124. A. L. Rozental', Investigations in the Field of Heterogeneous Processes with Powder Contact (in Russian), Kand. Dissertat., Institut Nefti, Akad. Nauk SSSR, 1954.
125. E. O. Rozental', Investigation of the Process of Drying Gelatin (in Russian), Kand. Dissertat., Moskovskii Tekhnologicheskii Institut Pishchevoi Promyshlennosti, Moscow, 1955.
126. V. P. Romadin, Problems of rapid suspended combustion of coal (in Russian), Teplosil. Khoz-vo, 1940, No. 8.
127. V. P. Romadin, Handling of Powdery Materials (in Russian), Gosenergoizdat, 1953.
128. P. G. Romankov, Hydraulic Processes in Chemical Technology (in Russian), Goskhimizdat, 1948.
129. P. G. Romankov, et al., Adsorption in fluidized beds (in Russian), Khim. Nauka i Prom., 1956, No. 3.
130. P. G. Romankov and V. N. Lepilin, Adsorption dynamics in a suspended adsorbent bed (in Russian), Zhurn. Prikl. Khim., 1956, 29, 599-604.

131. P. G. Ramonkov and P. A. Yablonskii, The laws of air classification of solid materials in straight-through separators (in Russian), Khim. Prom., 1956, No. 5.

132. G. I. Romashov, A theoretical analysis of the entrainment of powder particles from the cylindrical part of a vertical separator (in Russian), Zhurn. Tekhn. Fiz., 1940, No. 18.

133. Yu. E. Saar, An Investigation of the Gasification of Oil-Shale in a Suspended State (in Russian), Kand. Dissertat., Leningradskii Politekhnicheskii Institut, Leningrad, 1955.

134. S. A. Sashkin, The gasification of milled peat in a suspended bed (in Russian), Kislorod, 1944, 2, 29-37.

135. N. A. Semenénko and L. N. Sidel'kovskii, An investigation of semigas furnaces with suspended beds (in Russian), Za Ekonom. Topliva, 1949, No. 9.

136. N. A. Semenenko and L. N. Sidel'kovskii, Features and use tests of fluidized bed furnaces (in Russian), Teploenergetika, 1954, No. 3.

137. L. N. Sidel'kovskii, Experimental and Theoretical Bases for High Throughput Combustion of Solid Fuels in Fluidized Bed Furnaces (in Russian), Kand. Dissertat., Moskovskii Energeticheskii Institut, Moscow, 1948.

138. L. N. Sidel'kovskii, The conditions for fluidized bed furnaces to operate stably without slag (in Russian), Za Ekonom. Topliva, 1951, No. 6.

139. L. N. Sidel'kovskii, Increased reliability and efficiency of fluidized bed furnaces (in Russian), Trudy MEI (Moskovskii Energeticheskii Institut), 1956, No. 24.

140. L. N. Sidel'kovskii, High-throughput roasting of sulfur-containing raw-materials with utilization of the physical heat of the gases (in Russian), in: Transactions of Meeting on the Roasting of Materials in Fluidized Beds, Metallurgizdat, 1956.

141. L. N. Sidel'kovskii, N. A. Semenenko and M. Z. Shvartsman, High-throughput semigas fluidized furnaces (in Russian), Prom. Energetika, 1947, No. 5.

142. L. N. Sidel'kovskii and A. N. Shibaev, Investigation and Planning of Two-Stage Fluidized-Bed Furnaces (in Russian), Izd. Mosk. Energet. Inst., Moscow, 1950.

143. B. I. Skachkov, Roasting of zinc concentrates in fluidized beds in an air medium enriched with oxygen (in Russian), Tsvet. Met., 1958, No. 2.

144. N. I. Smirnov and Li De Ep, The motion of streams in beds of solid particles (in Russian), Zhurn. Prikl. Khim., 1951, 24, No. 1.

145. N. I. Smirnov and Li De Ep, Suspended beds of spherical particles (in Russian), Zhurn. Prikl. Khim., 1951, 24, No. 3.

146. N. I. Smirnov and Li De Ep, Suspended beds of particles of various shapes (in Russian), Zhurn. Prikl. Khim., 1951, 24, No. 4.

147. N. I. Syromyatnikov, The effect of reactive and buoyant boundary layer forces on the nature of the motion of burning solid fuel particles (in Russian), Izvestiya V.T.I., 1948, No. 10.

148. N. I. Syromyatnikov, A means of eliminating slag formation by cone-shaped combustion chambers completely screened in the combustion zone, Avt. Svid. SSSR, No. 80775, 1948.
149. N. I. Syromyatnikov, Furnaces with horizontally moving suspended beds of finely powdered fuels, Avt. Svid. SSSR, No. 85408, 1949.
150. N. I. Syromyatnikov, A theory of slag formation in furnaces and gas generators with suspended fuel beds (in Russian), Vestnik Inzh. i Tekhnikov, 1950, No. 4.
151. N. I. Syromyatnikov, Combustion of Fuels in the Suspended State (in Russian), in: Sb. Trudov U.P.I., Mashgiz, 1950.
152. N. I. Syromyatnikov, An investigation of a mechanism for ensuring nonslagging operation of fluidized bed furnaces (in Russian), Za Ekonom. Topliva, 1951, No. 12.
153. N. I. Syromyatnikov, Results of tests on furnaces of the fluidized, suspended and spouting types (in Russian), Za Ekonom. Topl., 1951, No. 2.
154. N. I. Syromyatnikov, Statistical theory of suspended bed formation (in Russian), Doklady Akad. Nauk SSSR, 1953, 93, No. 3, 421-424.
155. N. I. Syromyatnikov, Investigations and Some Practical Methods of Burning Pulverized Fuels (in Russian), Dokt. Dissertat., Energeticheskii Institut, Akad. Nauk SSSR, Moscow, 1954.
156. N. I. Syromyatnikov, Means of mechanizing furnace processes in metallurgical plants (in Russian), Trudy U.P.I., 1955, Sb. 53, 168-184.
157. N. I. Syromyatnikov, Distribution of solid fuel particles in a suspended bed (in Russian), Trudy U.P.I., 1955, Sb. 41.
158. N. I. Syromyatnikov, Increasing thermal stresses of furnace spaces (in Russian), Trudy U.P.I., 1955, Sb. 53, 157-167.
159. N. I. Syromyatnikov, The structure of suspended beds (in Russian), Trudy U.P.I., 1956, Sb. 61, 49-62.
160. T. F. Tagantseva, Heat transfer between falling peat particles and a fluid stream (in Russian), Torf. Prom., 1956, No. 5.
161. N. D. Taskaev, E. D. Mitusov and P. G. Lyubivyi, Semigas Grateless Furnaces for Burning Coal Fines (in Russian), Izd. Kirgiz. Fil. Akad. Nauk SSSR, Frunze, 1952.
162. B. I. Timin, Aerodynamic Conditions in the Process of Gasifying Pulverized Fuels in Fluidized Beds (in Russian), Kand. Dissertat., Ural'skii Politekhnicheskii Institut, Sverdlovsk, 1947.
163. O. M. Todes, Fundamental problems of adsorption and catalysts in moving and suspended beds (in Russian), in: Sb. Metody i Protsessy Khim. Tekhnologii, Izd. Akad. Nauk SSSR, 1955.
164. Transactions of the Technical Conference on Roasting Materials in Fluidized Beds (in Russian) (25-29 May, 1954), Metallurgizdat, 1956.
165. Kh. K. Truu, Practical methods of burning oil shales (in Russian), in: Doklad na Sessii po Goreniyu, Izd. Akad. Nauk SSSR, 1952.
166. Kh. K. Truu, Theory and Practice of Burning Oil Shales (in Russian), Kand. Dissertat., Energeticheskii Institut, Akad. Nauk SSSR, Moscow, 1952.

167. V. B. Fal'kovskii, Heat transfer in contact tubes under laminar conditions (in Russian), Zhurn. Tekhn. Fiz., 1953, 23, No. 12.
168. I. M. Fedorov, The problem of calculating the drying process in the suspended state (in Russian), Izvestiya V.T.I., 1949, No. 11.
169. I. M. Fedorov, Theory and Calculation of Drying Processes in the Suspended State (in Russian), Dokt. Dissertat., Moskovskii Institut Khimicheskogo Mashinostroeniya, Moscow, 1950.
170. I. M. Fedorov, Theory and Calculation of the Drying Process in the Suspended State (in Russian), Gosenergoizdat, 1955.
171. L. G. Khazin, Possibilities of using the properties of suspended beds for treating highly disperse powders (in Russian), Khim. Prom., 1951, No. 5.
172. L. N. Khitrin and Yu. P. Shelestin, The use of cyclone furnaces in solid fuel power stations (in Russian), Teploenergetika, 1955, No. 9.
173. E. I. Khodorov and V. M. Kosareva, Two-stage roasting of cement clinkers in suspended and overflowing beds (in Russian), Trudy Giprotsementa, 1955, No. 18, 3-33.
174. A. V. Chechetkin, Some problems in aerodynamics and heat transfer in reactors operating with suspended catalysts (in Russian), Trudy Moskovskii Khimiko-Tekhnologicheskii Institut im. Mendeleeva, 1955, No. 20.
175. A. F. Chudnovskii, Heat Transfer in Disperse Media (in Russian), GITTL, 1954.
176. V. V. Chukin, The combustion of solid fuel in a suspended state in cylindrical combustion chambers and in furnaces in the shape of an inverted cone (in Russian), in: Dokl. na Sessii po Goreniyu, Izd. Akad. Nauk SSSR, 1952.
177. Z. F. Chukhanov, Problems in the combustion of coal and coke and ways of developing techniques for burning and gasifying solid fuels (in Russian), Izvestiya Akad. Nauk SSSR, Otd. Tekhn. Nauk, 1953, No. 4.
178. N. A. Shakhova, An Investigation of Heat Transfer in Fluidized Systems (in Russian), Kand. Dissertat., Moskovskii Institut Khimicheskogo Mashinostroeniya, Moscow, 1954.
179. N. V. Shishakov, Gasification of pulverized fuels in a suspended state (in Russian), Voprosy Gazifikatsii, 1941, Sb. 4.
180. G. M. Shteingard, Study and Industrial Scale-Up of the Roasting of Zinc Concentrates in Fluidized Beds in the "Elektrotsink" Plant (in Russian), Tsentral'nyi Institut Informatsii Ministerstva Tsvetnoi Metallurgii, 1956.
181. A. P. Yufin, The motion of nonhomogeneous fluids through horizontal partly filled steel tubes (in Russian), Izvestiya Akad. Nauk SSSR, Otd. Tekhn. Nauk, 1949, No. 8, 1146-1159.
182. S. F. Yavorovskaya and A. L. Polubinskii, The use of fluidized methods in gas purification (in Russian), Khim. Prom., 1955, No. 2, 91-94.
183. I. A. Yavorskii, Pneumatic classification of unsorted coal during combustion on chain grates (in Russian), Za Ekonom. Topl., 1952, No. 2.

184. I. A. Yavorskii, Aerodynamics of streams of disperse media (in Russian), Report of the West Siberian Filial, Akad. Nauk SSSR, 1955.
185. I. A. Yavorskii and S. N. Shibanov, Preliminary results of an experimental investigation of pneumatic classification of unsorted coal on chain grates (in Russian), Izvestiya Tomsk. Politekhn. In-ta, 1952, vol. 69.
186. M. Abbott, A review of the hydraulic conveying of solids in pipe lines, J. Imp. Coll. Chem. Eng. Soc., 1953, 7, 101-110.
187. C. E. Adams, M. O. Germand and C. N. Kimberlin, Heating small reactors, high-temperature fluidized solids for continuous systems, Ind. Eng. Chem., 1954, 46, 2458-2460.
188. O. P. Agarwall, Humic acids by the fluidized oxidation of non-caking Tandas coal, J. Sci. Ing. Res., 1953, 12B, 443-446.
189. O. P. Agarwall and J. A. Storrow, Pressure drop in fluidized beds, Chem. and Ind., 1951, 278-286.
190. O. P. Agarwall and J. A. Storrow, Heat transfer in fluidized beds, Chem. and Ind., 1951, 321-324.
191. G. Agricola, De re metallica, Phillipum Bechium, Basel, 1556.
192. C. W. Albright, J. H. Holden, H. P. Simons and L. D. Schmidt, Pressure drop in flow of dense coal-air mixtures, Ind. Eng. Chem., 1951, 43, 1837.
193. H. H. Ambrose, Ph. D. Thesis, Univ. of Iowa, 1952.
194. N. R. Amundson, Solid fluid interaction in fixed and moving beds. Fixed beds with small particles, Ind. Eng. Chem., 1956, 48, No. 1, 26-35.
195. N. R. Amundson, Solid-fluid interactions in fixed and moving beds. Fixed beds with large particles, Ind. Eng. Chem., 1956, 48, No. 1, 35-43.
196. S. L. Andersen and R. H. Matthias, Prediction of catalyst activity in fluidized systems employing catalyst replacement rate, Ind. Eng. Chem., 1954, No. 6, 1296-1298.
197. N. K. Anderson and M. J. Sterba, Simplified fluid cracking unit meets requirements of smaller refineries, Petr. Ref., 1945, 24, 497.
198. T. Anderson and K. Balduc, Fluosolids roasting of zinc concentrates for contact acids, Chem. Eng. Progr., 1953, 49, No. 10, 527-532.
199. R. Andrieu, Compt. Rend., 27 congr. intern. chim. ind. (Brussels), 1954, I; Ind. chim., Belge, 1955, 20, Spec. No. 228-231.
200. J. R. Anton, Heat transfer in fluidized systems, Diss. Abstr., 1953, 13, 359-360.
201. J. W. Askins, G. P. Hinds, Jr. and F. Kunreuther, Fluid catalyst gas mixing in commercial equipment, Chem. Eng. Progr., 1951, 47, 401-404.
202. A. Baerg, J. Klassen and P. E. Gishler, Heat transfer in a fluidized solids bed, Can. Journ. Research., 1950, F28, 287-307.
203. K. R. Barker, J. J. Sebastian, L. D. Schmidt and H. D. Simons, Pressure feeder for powdered coal or other finely divided solids, Ind. Eng. Chem., 1951, 43, 1204-1209.
204. F. T. Barr and J. Jahning, Fluid coking and fluid coke, Chem. Eng. Progr., 1955, 51, 167-173.

205. R. Bart, Ph. D., Thesis, MIT, Cambridge, 1950.
206. W. Barth, Die strömungtechnische Probleme in der Verfahrentechnik, Chem. Ing. Techn., 1954, 26, No. 1, 29-34.
207. R. N. Bartholomew and D. L. Katz, Heat transfer from the wall of a tube to the fluidized bed, Chem. Eng. Progr. Symp. Ser., 1952, 48, No. 4, 3-10.
208. R. A. Beck, Evaluation of fluid catalyst (development of laboratory scale units), Ind. Eng. Chem., 1949, 41, 1242-1243.
209. G. Beilby, Aggregation and Flow of Solids, London, 1922.
210. D. H. Belden and L. S. Kassel, Pressure drops encountered in conveying particles of large diameter in vertical transfer lines, Ind. Eng. Chem., 1949, 4, 1174-1179.
211. I. Bena, Uvod do theorie fluidnnej vrstvy, Paliva, 1954, 34, 174.
212. I. Bena and V. Jesanek, Novodoby sposob prazenia pyritu (pece s prazenim vo vznose), Chem. Prumysl., 1953, 10, 345-347.
213. J. Beranek and I. Klumpar, Dynamicke vlastosti fluidni vrstvy. I. Kritika dosavadnich teorii fluidace, Chem. Listy, 1956, 50, 1540-1554.
214. C. Berg, New gas lift for solids, Petr. Refiner, 1952, 31, 131-136; Chem. Eng., 1953, 60, No. 5, 138-141.
215. R. A. Bernard and R. H. Wilhelm, Turbulent diffusion in fixed beds of packed solids, Chem. Eng. Progr., 1950, 46, No. 5, 233-244.
216. R. H. Blanding and B. E. Roethelii, Retorting oil shale by the fluid technique, Oil Gas J., 1947, 45, 84-88.
217. N. S. Blatch, Trans. Am. Soc. Civil Engrs, 1906, 57, 400-408.
218. K. Bochmann and A. Banerjel, Fluidized low-temperature carbonization of Indian coals, Coal Industry Sympos., Indian Mining J., 1955, 3, No. 9, 98-104.
219. P. Bottger, Zur Praxis des Wirbelschichtes, Chem. Ing. Techn., 1951, 23, 337-342.
220. D. F. Boucher, Fluid dynamics-fluidization. Ind. Eng. Chem., 1948, 40, 32-8; 1949, 41, 47-51.
221. W. Brötz, Grundlagen der Wirbelschichtverfahren, Chem. Ing. Techn., 1952, 24, No. 2, 57-81.
222. W. Brötz, Untersuchungen über Transportvorgänge in durchströmtem gekörntem Gut, Chem. Ing. Techn., 1956, 28, 165-174.
223. H. Bungartz, Getreidetrockner auf dem 27. Landmaschinensalon in Paris, Die Mühle, 1955, No. 15.
224. J. Galdas, Heat transfer to fluidized beds, Diss. Abstr., 1955, 15, 2139-2140.
225. J. R. Campbell and F. Rumford, The influence of solid properties on heat transfer from a fluidized solid medium, J., Soc. Chem. Ind., 1950, 69, 373-377.
226. L. E. Carlsmith and F. B. Johnson, Pilot plant developments of catalytic fluid cracking, Ind. Eng. Chem., 1945, 37, 451-455.
227. N. L. Carr and N. R. Amundson, Heat transfer and heat release in fluidized beds, Ind. Eng. Chem., 1951, 43, 1856-1862.
228. J. Cathala, Fluidization and research method of chemical engineering science, Chem. Eng. Sci., 1953, 2, 273-281.
229. J. Cathala and P. Jayles, Fluidization (the practical determination of the characteristics of a fluidized bed), Compt. Rend., 78 Congr. Soc. Sav., Paris, 1953, 481-492.

230. V. Charvat, Fluidni karbonisace a susene, Paliva, 1954, 34, 179-182.
231. V. Charvat and M. Houdek, Zaklady techniky vznasene vrstvy, Paliva, 1953, 33, 162-166, 253-258.
232. T. Chilton and A. Colburn, Pressure drop in packed tubes, Trans. Am. Inst. Chem. Engrs, 1931, 26, 178.
233. V. Chipalkatti, A new technique in textile dyeing, J. Sci. Ind. Research (India), 1955, 14B, 540-542.
234. J. Ch. Chu, Heat and mass transfer in fluidized beds, Collection Fluidization, Edit. by Othmer, 1956.
235. J. Ch. Chu, J. Kalil and W. A. Wetteroth, Mass transfer in a fluidized bed, Chem., Eng. Progr., 1953, 49, No. 3, 141-149.
236. J. Ciborowski, Podstawowe problemy procesu fluidyzacji, Przegl. Chem., 1948, 6, 164-172.
237. J. Ciborowski, O procesie fluidyzacji, Przem. Chem., 1949, 5, 28-33.
238. J. Ciborowski, Zagadnienia ciephlne procesu fluidyzacji, Przegl. Chem., 1949, 7, 1-7.
239. J. Ciborowski, O przenoszeniu ciepla w procesie fluidizacji, Przem. Chem., 1950, 6, 743-753.
240. J. Ciborowski, O analogiach meedzy ukladem fluidalnym a ukladem jednoskladnikowym welofazowym, Przem. Chem., 1956, 35, 9-13.
241. J. Ciborowski and T. Czarnota, O kalcynacji weglanow metoda fluidyzacji, Przem. Chem., 1954, 33, 81.
242. J. Ciborowski and L. Lesniewicz, O zgazowaniu paliva stalego w stanie fluidalnym, Przem. Chem., 1952, 31, 501-508, 544-551.
243. J. Ciborowski and B. Mlodzinski, O wypalaniu siarczkow metoda fluidyzacji, Przem. Chem., 1952, 31, No. 8, 49-58.
244. J. Ciborowski and B. Mlodzinski, O wypalaniu roznych siarczkow metoda fluidyzacji, Przem. Chem., 1953, 32, 443-449.
245. J. Ciborowski, B. Mlodzinski and A. Pfeffer, Spalanie siarczakow metoda fluidzacji w skali poltechniczney, Przem. Chem., 1954, 33, 28-29.
246. J. Ciborowski, B. Mlodzinski and J. Roszak, Fluidyzacja w skali przemyslowey, Przem. Chem., 1954, 33, 589.
247. J. Ciborowski and J. Roszak, O suszeniu gazow metoda fluidyzacji, Przem. Chem., 1952, 31, 244-253.
248. E. L. Clark, M. G. Pellipetz, H. H. Storch, S. Weller and S. Schreiber, Hydrogenation of coal in a fluidized bed, Ing. Eng. Chem., 1950, 42, 861-865.
249. J. S. Clarke, Slide valves. Their upkeep and adjustment, Oil Gas J., 1956, 73, No. 54, 112-116.
250. J. E. Cline, Vapor deposition of metals on ceramic particles, J. Electrochem. Soc., 1951, 98, 385-387.
251. G. Cohen de Lara, Houille Blanche, 1955, 2 (mars-avril), 167-175.
252. R. E. Conary, L. W. Devaney, L. E. Ruidisch, R. F. McClary and K. L. Kreutz, Thiophene from hydrocarbons and sulfur dioxide, Ind. Eng. Chem., 1950, 42, 467-471.
253. R. Cornforth and E. Lee, Firing kilns with fluid petroleum coke, Rock Products, 1955, 58, No. 11, 98-103, 120.

254. T. B. Counselman, How fluidization can serve the mineral industries, Engng Min. Journ., 1955, 156, 70-75, 96.
255. O. Craig and E. Smith, Burning fluid coke, Mech. Eng., 1956, 78, 921-925.
256. J. M. Culgan, Ph. D., Thesis, Georgia Inst. of Technology, 1952.
257. H. Cyr, C. Siller and T. Steele, Roasting metallic sulfides in a fluid column, Journ. Met., 1954, 200, 900-904.
258. T. Czarnota, J. Orlowski and J. Ciborowski, O kontaktowym utlenianiu naftalenu do bezwodnika ftalowego metoda fluidyzacji Przem. Chem., 1955, 34, 295-301.
259. L. S. Daniels, Fluid catalyst technique, Petr. Ref., 1946, 25, 65-72, 109, 435.
260. P. W. Dankwerts, J. W. Jenkins and G. Place, The distribution of residence time in an industrial fluidized reactor, Chem. Eng. Sci., 1954, 3, 26-35.
261. J. W. Delaplaine, Forces acting in flowing beds of solids, A. I. Ch. E. Journ., 1956, 2, No. 1, 127-138.
262. W. E. Deming and A. L. Mehring, Gravitational flow of fertilizers and other comminuted solids, Ind. Eng. Chem., 1929, 21, No. 7, 661.
263. F. Dent, Gasification of carbonaceous materials, Brit. Pat. 695,191, 1953.
264. R. Diekman and W. Forsyth, Laboratory prediction of flow properties of fluidized solids, Ind. Eng. Chem., 1953, 45, No. 6, 1174-1177.
265. M. S. Dimitri, R. P. Jongedyk and H. C. Lewwi, Distillation of fluidized hardwood, Chem. Eng., 1948, 55, No. 12, 124.
266. H. Dimmig, Fluid catalytic cracking, Petr. Times, 1945, 49, 1036-1038.
267. J. M. Dotson, New method measures the solid-gas concentration in high solid flow, Chem. Eng., 1949, 56, No. 10, 128-130.
268. W. M. Dow and M. Jakob, Heat transfer between a vertical tube and a fluidized air—solid flow, Chem. Eng. Progr., 1951, 47, 637-648.
269. J. M. Dunoyer, Theoretical and experimental study of the kinetics of gas-solid reactions, Ann. Chem., 1951, 6, No. 12, 165-219.
270. A. W. Dycke, Experimental fluid gasification of saw mill waste, Brit. Columb. Res. Couns. Tech. Bull., No. 18, 1950.
271. E. Egloff, Le cracking catalytique, Chemi et Ind., 1948, 59, 121-127.
272. J. Eichhorn and R. R. White, Particle to fluid heat transfer in mixed and fluidized beds, Sympos. Ser., 1952, 48, No. 4, 11-18.
273. J. M. Ellis, Some applications of fluidized technique, Chem. Eng. Min. Rev., 1949, 41, 295-299, 329-336.
274. S. Ergun, Determination of geometric surface area of crushed porous solids, Anal. Chem., 1952, 24, 388.
275. S. Ergun, Fluid flow through packed columns, Chem. Eng. Progr., 1952, 48, 89-94.
276. S. Ergun and A. A. Orning, Fluid flow through randomly packed columns and fluidized beds, Ind. Eng. Chem., 1949, 41, 1179-1184.

277. E. D. Etherington, R. J. Fritz, E. W. Nicholson and H. Scheeline, Fluid char adsorption process (a new solid fluid development for difficult gas separation), Chem. Eng. Progr., 1956, 52, 274.
278. G. C. Evans and L. F. Gerald, Mass transfer from benzoic acid granules to water in fixed and fluidized beds at low Reynolds numbers, Chem. Eng. Progr., 1953, 49, 135-140.
279. P. Evans and H. Kepper, Some practical applications of fluidization, J. Appl. Chem., 1952, 2, Supple. 1, 33-39.
280. L. Farbar, Flow characteristics of solid-gas mixtures in a horizontal and vertical circular conduit, Ind. Eng. Chem., 1949, 41, 1184-1191.
281. L. Farsky, Fluidacni zlab-novy spusob dopravy materialu v chemicken prumyslu, Chem. Prumysl., 1954, 4, 334-341.
282. R. Fehling, Der Strömungswiederstand ruhender Schüttungen, Forschungstechnik, 1939, 2, 33.
283. F. Fetting and E. Wicke, Strömungsformen und Wärmeübertragung in Gaswirbelschichten, Dechema Monogr., 1955, 24, 146-169.
284. F. Fetting and E. Wicke, Abbruch schnell verlaufender Reaktionen in Gaswirbelschichten, Chem. Ing. Techn., 1956, 28, 88-93.
285. W. Flesch, Gasification of finely divided solid fuel in a whirling bed, Gasif. Liq. Coal Sympos. Ann. Meet. AIME, 1952, 47-59.
286. W. Flesch, Vegasung von Staubkohlen, Glueckauf, 1954, 90, 537-543.
287. B. J. Flock, Improved fluid catalytic cracking unit reduces maintenance costs, Oil Gas J., 1946, 45, 72, 104.
288. A. L. Forster, New "fluid coking" process boosts domestic fuels yields from residual press, Petr. Engr., 1953, 25, No. 10.
289. W. Z. Forsythe and W. R. Hertwig, Attrition characteristics of fluid cracking catalyst, Ind. Eng. Chem., 1949, 41, 2100-2106.
290. L. J. Ford, Distillation tower for contact reactors of hydrocarbon gases with finely divided solids, U.S. Pat. 2,582,688, 1952.
291. R. Fort, Generalité sur la fluidisation, Bull. Soc. Chim. France, 1955, 448-452.
292. F. C. Franklin and L. N. Johanson, Flow of granular material through a circular orifice, Chem. Eng. Sci., 1955, 4, No. 3, 119-129.
293. M. Frey, Une technique nouvelle: la fluidification-application au nitrate d'ammonium et au nitrate de sodium, Chim. et Ind., 1955, 73, 1159-1166.
294. L. Friend, M. Leva, et al., Fluidization, Nomenclature proposed, Chem. Eng. News, 1949, 27, 686.
295. J. Fritz, Heat transfer between air and solid particles in 16-inch diameter fluid beds, Diss. Abstr., 1956, 16, 1115-1116.
296. J. Furukawa, The state of fluidized bed, Bull. Inst. Chem. Res., Kyoto Univ., 1952, 29, 71.
297. J. Furukawa and T. Omae, Characteristics of fluidized particles. II. Expansion of fluidized particles by a gas stream, J. Chem. Soc. Japan (Ind. Chem. Sec.), 1952, 55, 3-6.

298. J. Furukawa, Characteristics of fluidized particles. III. Miscibility. IV. Particles distribution, J. Chem. Soc. Japan (Ind. Chem. Sec.), 1953, 56, 557-560, 727-731.

299. J. Furukawa, T. Omae and J. Ueki, Synthesis of vinyl acetate in the gas phase. (Y). Application of fluid catalyzer, Chem. High Polymers (Tokyo), 1951, 8, No. 2, 111.

300. B. W. Gamson, Heat and mass transfer in a fluid-solid system, Chem. Eng. Progr., 1951, 47, No. 1, 19-28.

301. J. E. Garside and A. L. Roberts, A theory of fluidization and its application to gas purification, J. Appl. Chem., 1952, 2, Suppl. Iss. I, 40-47.

302. G. Genin, La nouvelle technique de fluidification, Industrie Chimique, 1951, 38, 157-163, 185-191.

303. G. F. Gerald, Fluid bed flowmeter, Ind. Eng. Chem., 1952, 44, 233-235.

304. B. Chosh, Pressure drop in the flow of suspensions, Part I, Trans. Ind. Inst. Chem. Engr, 1955-1956, 76-81.

305. A. E. Gilfilan, The fluo-solids reactor as a source of dioxide for sulphuric acid manufacture, J. Chem. Min. Soc. S. Africa, 1954, 54, 357-391.

306. E. R. Gilliland, Techniques of contacting fluids and solids, Can. Chem. Process Inds, 1951, 34, 632-639.

307. E. R. Gilliland and E. A. Mason, Gas and solid mixing in fluidized beds, Ind. Eng. Chem., 1949, 41, 1191-1196.

308. E. R. Gilliland and E. A. Mason, Gas mixing in beds of fluidized solids, Ind. Eng. Chem., 1952, 44, 218-224.

309. E. R. Gilliland, E. A. Mason and R. Oliver, Gas flow patterns in beds of fluidized solids, Ind. Eng. Chem., 1953, 45, 1177-1185.

310. P. E. Gishler, Fluidization technique applied to direct distillation of oil from bituminous sand, Can. J. Research, 1949, 27 EF, 104-111.

311. A. Godel, La fabrication du charbon actif par fluidification, Chim. et Industrie, 1948, 59, 437-439.

312. A. Godel, Fluidization used in making activated carbon, Chem. Eng., 1948, 55, No. 7, 110-111.

313. A. Godel, A new (fluidized) combustion technique, Eng. Boiler House Rev., 1956, 71, 145-153.

314. E. Graf, Ph.D. thesis, Eidgenössiche Technische Hochschule, Zurich, 1955.

315. M. Gregor, Vyznam fluidacnej techniky pre narodne hospodarstvo CSR, Paliva, 1954, 34, 171-173.

316. S. A. Gregory, Problems of plant design for fluidized processes, J. Appl. Chem., 1952, 2, Suppl., 1, 1-7.

317. W. B. Gregory, Pumping of clay slurry through 4-inch pipe, Mech. Engng, 1927, 49, 602-616.

318. E. Grohse, Analysis of gas-fluidized solid systems by X-ray adsorption, A. E. Ch. E. Journ., 1955, 1, 358-365.

319. H. W. Grote, J. Koekstra and G. T. Tobiasson, Midget fluid catalytic cracking units, Ind. Eng. Chem., 1951, 43, 545-550.

320. R. C. Gunnes, Fluidized solids technique in the petroleum industry, Chem. Eng. Progr., 1953, 49, 113-119.

321. A. D. Gupta, K. Y. Shrikhande and A. Lahiri, The fluidization of coal, J. Sci. Ind. Research (India), 1952, 11B, 582-533.
322. A. Guyer, E. Graf and A. Guyer, Jr., Untersuchungen an Schütt- und Wirbelschichten, Helvet. Chim., Acta, 1955, 38, 473-484.
323. C. C. Hall, Pilot plant for Fischer-Tropsch synthesis, Chem. Products, 1955, No. 4, 151.
324. C. C. Hall and J. Crumley, Some observations on fluidization as applied to the Fischer-Tropsch process, 1952, 2, Suppl. 1, 47-55.
325. C. C. Hall and L. Jolley, The fluidized solid technique and its industrial applications, Petroleum, 1950, 13, 217-223.
326. C. C. Hall and J. Rennie, A study of Fischer-Tropsch synthesis with fluidized iron catalysts, Proc. 3rd World Petr. Congr., Hague, 1951, sec. 4, 25-37.
327. C. C. Hall and A. H. Taylor, Design and operation of a fluid catalyst pilot plant for Fischer-Tropsch synthesis, J. Inst. Petrol., 1955, 41, 101-124.
328. J. Halperin and R. W. Stoughton, Batch versus continuous processing in fluidized nuclear reactors, Ind. Eng. Chem., 1956, 48, 2115-2122.
329. R. T. Hancock, The teeter condition, Mining Magaz., 1936, 55, 90-94; 1942, 67, 179.
330. R. T. Hancock, The physical basis of fluidization, Coke and Gas, 1949, 11, 16-17, 64-68, 386-388.
331. T. Hanratty, G. Latinen and R. Wilhelm, Turbulent diffusion in particulately fluidized beds of particles, A. I. Ch. E. Journ., 1952, 2, 372-380.
332. J. Happel, Pressure drop due to vapor flow through moving beds, Ind. Eng. Chem., 1949, 41, No. 6, 1161-1174.
333. J. Happel, Versatile chart aids Δp calculations, Chem. Eng., 1950, 57, No. 2, 137-138.
334. O. H. Hariu and M. C. Molstad, Pressure drop in vertical tubes in transport of solids by gases, Ind. Eng. Chem., 1949, 41, No. 6, 1148-1160.
335. L. P. Hatch, Trans. Am. Geophys. Union, 1943, 24, 537-547.
336. C. van Heerden, Some fundamental characteristics of the fluidized state, J. Appl. Chem., 1952, 2, Suppl. 1, 7-17.
337. C. van Heerden, A. P. Nobel and D. van Krevelen, Heat transfer in fluidized beds, Gen. Disc. Heat Transf., I. M. E. (London), ASME, 1951.
338. C. van Heerden, A. P. Nobel and D. van Krevelen, Studies on fluidization, I. The critical mass velocity, Chem. Eng. Sci., 1951, 1, 37-49.
339. C. van Heerden, A. P. Nobel and D. van Krevelen, Studies on fluidization. II. Heat transfer, Chem. Eng. Sci., 1951, 1, 51-66.
340. C. van Heerden, A. P. Nobel and D. van Krevelen, Mechanism of heat transfer in fluidized beds, Ind. Eng. Chem., 1953, 45, No. 6, 1237-1242.
341. P. M. Heertjes, H. G. Boer and A. H. Haas v. Dorsser, Temperature and humidity measurements in a drying fluidized bed, Chem. Eng. Sci., 1953, 2, No. 3, 97-107.

342. P. M. Heertjes and S. McKibbins, The partial coefficients of heat transfer in a drying fluidized bed, Chem. Eng. Sci., 1956, 5, 161-167.

343. J. Helman, F. Kunreuther, J. Marshall and Ch. Rehbein, More gasoline and less coke, Petr. Processing, 1956, 11, No. 6, 54-57.

344. A. Herrmann, Freiberger Forsch., Beihefte der Ztschr. Bergbau, 1953, No. 18, 75-91.

345. B. L. Hinkle, Ph.D. thesis, Georgia Inst. of Technology, 1953.

346. K. Hirai and K. Fukaya, Fluidization of solid particles, Chem. Engr. Japan, 1953, 17, 438-447.

347. G. W. Hodgson, B. Matchen, W. S. Peterson and P. E. Gishler, Oil from Alberta bitumen-simultaneous dehydratation and coking using fluidized solids, Ind. Eng. Chem., 1952, 44, 1492-1496.

348. H. Hoffmann, Beobachtungen bei der Trocknung durchströmter Getreideschüttungen, Die Mühle, 1954, No. 46.

349. N. Hornak and E. Whittenberger, Desulfurization of molten iron, J. Metals, 1956, 8, 425-429.

350. Houdry Process Corporation, Gas lifting of granular solids as a dense mass, Brit. Patent 714,745, 01.09.1954.

351. G. W. Howard, Transportation of sand and gravel in a four-inch pipe, Proc. Am. Soc. Civ. Engrs, 1938, 64, No. 7, 1377.

352. D. A. Howes, Catalytic cracking of petroleum oils, Ind. Chemist, 1947, 23, 217-225.

353. C. T. Hsu, Rate of mass transfer from a gas to a porous solid in fluidized beds, Diss. Abstr., 1953, 13, 207-208.

354. C. T. Hsu and M. C. Molstad, Process design data-rate of mass transfer from gas steam to porous solid in fluidized beds, Ind. Eng. Chem., 1955, 47, No. 8, 1550-1559.

355. D. Hyman, Ph.D. thesis, M.I.T., Cambridge, 1952.

356. M. Jakob, An attempt to explain discrepancy in observations of the heat transfer between a vertical surface and a fluidized bed of particles, J. Appl. Phys., 1952, 23, 1056.

357. F. E. James, Factors affecting the uniformity of fluidization in fluidized beds, Diss. Abstr., 1954, 14, 1654.

358. G. Jangg and W. Hausler, Die Beziehung zwischen Feststoffdichte und Strömungsgeschwindigkeit in Wirbelschichten, Österr. Chem. Ztg, 1956, 57, 213-216.

359. G. Jangg and W. Hausler, Ausdehnung and maximaler Festkörperdurchsatz in mehrstufigen Wirbelschichtsystemen, Chem. Ing. Techn., 1956, 28, 633-638.

360. J. Jilek, Nove spusoby zplynovani prochovych paliv, Paliva, 1953, 33, 97.

361. J. Jilek, Vyuziti fluidni techniky pro zplynovani prachu paliva ve Winklerove generatoru, Paliva, 1954, 34, 183-186.

362. P. Jiru, Fluidacni prazeni chudych domacich pyritu, Chem. Prumysl, 1954, 4, No. 29, 11.

363. P. Jiru, Zavislost tvaru fluidacniho reactoru na sitovem spektru naplne reaktoru, Chem. Prumysl., 1955, 5, No. 30, 174-175.

364. C. W. Jobes, Fluidized crystal dryer pays off, Chem. Eng., 1954, 61, No. 1, 166-168.

365. A. Johanssen, Wirbelschichtröstverfahren der Badischen Anilin- und Soda-Fabrik, Chem. Ing. Techn., 1952, 24, 104-109.
366. E. Johnson, A theory of fluidization and its application to sulphur recovery by solvent extraction, Inst. Gas Engrs, Publ. No. 378/179, 1951.
367. F. B. Johnson and R. G. Wood, New data on fluid coking, Petr. Refiner, 1954, 33, No. 11, 157-160.
368. M. F. Johnson and H. C. Mayland, Carbon burning rates of cracking catalyst in the fluidized state, Ind. Eng. Chem., 1955, 47, 127-132.
369. P. H. Johnson and C. P. Stark, Testing powdered cracking catalyst, Ind. Eng. Chem., 1953, 45, 849-855.
370. L. J. Jolley, Heat transfer in beds of fluidized solids, Fuel, 1949, 28, No. 5, 114-115.
371. L. J. Jolley and A. Poll, Fluidizing gasification of noncoking coals with steam in a small pilot plant, Gasif. Liquif. Coal Sympos. Ann. Meet. A.I.M.E., 1952, 60-72.
372. L. J. Jolley and J. E. Stanton, Fluidization of beds of coal and coke particles: some effects of size of particles and viscosity, density and velocity of gas, J. Appl. Chem. (London), 1952, 2, Suppl. 1, 62-68.
373. R. Jottrand, An experimental study of the mechanism of fluidization, J. Appl. Chem. (London), 1952, 2, Suppl. 1, 17-26.
374. R. Jottrand, Etude de quelques aspects de la fluidisation dans des liquides, Chem. Eng. Sci., 1954, 3, 12-16.
375. N. A. Jull and W. T. Martson, Sulphur dioxide from sulfides by fluosolids, Pulp Paper Mag. Can., 1951, 52, No. 8, 83-87.
376. A. Justat, Kataliza zawiesinowa i zapomniany patent polski, Przem. Chem., 1949, 28, 16-19.
377. J. C. Kalbach, Improving solid-gas contacting by fluidization, Chem. Met. Eng., 1944, 51, No. 6, 94-98.
378. J. C. Kalbach, Fluidization in chemical reactions, Chem. Eng., 1947, 54, No. 1, 105-108.
379. J. C. Kalbach, Handling of solid-gas reactions by fluidization, Chem. Eng., 1947, 54, No. 2, 136-138.
380. A. A. Kalinske, Settling rate of suspensions in solids contact units, Proc. Am. Soc. Civ. Eng., 1953, 79, No. 186, 1-7.
381. A. Kammiya, Air oxydation of coal fluidized bed, J. Fuel Soc. Japan, 1954, 33, 412-423.
382. J. Kapczynski, Fluidyzacyjne spalanie markazytu na skale poltechniczna, Przem. Chem., 1954, 10, 479-481.
383. N. Karr and N. Amundson, Heat transfer and heat release in a fluidised bed, Ind. Eng. Chem., 1951, 43, 1856-1862.
384. P. Kasten and N. Amundson, An elementary theory of adsorption in fluidised bed., Ind. Eng. Chem., 1950, 42, 1341-1346.
385. S. Katz and F. A. Zenz, Application of radioactive tracer, Petr. Ref., 1954, 33, No. 5, 203.
386. H. R. Keil, Operation of Fluo-solids roaster at Golden Cycle, Min. Eng., 1954, 6, No. 12, 1181-1185.
387. P. C. Keith, Gasoline from natural gas, Am. Gas. Assoc. Month., 1946, 28, 243.

388. A. F. Kelley, Measurements of solids in thermofor cracking process, Petr. Engr., 1945, 16, No. 13, 136-142.
389. K. N. Kettenring, E. L. Manderfield and J. Smith, Heat and mass transfer in fluidized systems, Chem. Eng. Progr., 1950, 46, 139-145.
390. G. Kidoo, Flow in fluidized reaction systems, Chem. Met. Eng., 1949, 56, No. 5, 112-114.
391. A. Kirnick, The reduction of nickel oxide in a fluidized bed, Chem. Eng. Progr., 1952, 48, 394-400.
392. R. P. Kite, Fluidization in noncatalytic operations, Chem. Eng., 1947, 54, No. 12, 112-115.
393. R. A. Koble, Heat transfer to fluidized solids, Chem. Eng., 1951, 58, No. 9, 174-175.
394. R. A. Koble, P. R. Johnes and W. A. Koehler, Blending and conveying of ceramic raw materials by fluidization, Am. Ceram. Soc. Bull., 1953, 32, 367-372.
395. C. Kojabashian, M.S. thesis, M.I.T., Cambridge, 1954.
396. H. Kramers, The viscosity of a bed of fluidized solids, Chem. Eng. Sci., 1951, 1, 35-37.
397. H. Krekeler, Hoechst continuous coking process. Petr. Ref., 1955, 34, No. 10, 139-140.
398. W. Kroll, Fliesserscheinungen an Haufwerken in schwingenden Gafässen, Chem. Ing. Techn., 1955, 27, 33-38.
399. H. Kunze, Das Wirbelschichtverfahren und seine technische Anwendung, Umschau, 1956, 56, No. 5, 147-149.
400. M. Kurokawa, New coal-chemical industry. A new aspect of coal utilisation, J. Fuel Soc. Japan, 1955, 34, 570-577.
401. G. Kuwai, Flow of solid particles through orifice-rate under low air pressure, Chem. Eng. (Tokyo), 1953, 17, 453-459.
402. A. Lange, Entwicklung der neuen Röstentechnik. Bergakademie Freiberg Forsch. B., Hüttenwessen Met., 1952, No. 3, 5-16.
403. J. A. Lee, Fluid catalysts make their debut in chemical production, Chem. Met. Eng., 1945, 52, No. 7, 100-101.
404. W. B. Lenhart and N. C. Rockwood, New and revolutionary method of lime manufacture, Rock Products, 1948, 51, No. 1.
405. M. Leva, Fluid dynamics-fluidization, Ind. Eng. Chem., 1950, 42, 55-61.
406. M. Leva, Elutriation of fines from fluidized system, Chem. Eng., Progr., 1951, 47, 39-45.
407. M. Leva, A correlation of heat transfer film coefficients in fluidized systems, Gen. disc. on heat transf. IME (London), ASME, 1951.
408. M. Leva and M. Grummer, Heat transfer to gases through packed tubes. Effect of particle characteristics, Ind. Eng. Chem., 1948, 40, No. 3, 415-419.
409. M. Leva and M. Grummer, Fluid flow-fluidization, Ind. Eng. Chem., 1952, 44, 69-70.
410. M. Leva and M. Grummer, A correlation of solids turnover in fluidized systems, its relation to heat transfer. Chem. Eng. Progr., 1952, 48, No. 6, 307-313.

411. M. Leva, M. Grummer and H. H. Storch, A study of fluidization of an iron Fischer-Tropsch catalyst, Chem. Eng. Progr., 1948, 44, No. 9, 707-716.

412. M. Leva, M. Grummer, M. Weintraub and M. Pollchik, Introduction to fluidization, Chem. Eng. Progr., 1948, 44, 511-520.

413. M. Leva, M. Grummer, M. Weintraub and M. Pollchik, Fluidization of solid nonvesicular particles, Chem. Eng. Progr., 1948, 44, 619-626.

414. M. Leva, T. Shirai and C. Wen, La prevision du debut de la fluidisation dans les lits solides granulaires, Genie Chim., 1956, 75, 33-42.

415. M. Leva and M. Weintraub, Fluid dynamics-fluidization, Ind. Eng. Chem., 1951, 43, 92-93.

416. M. Leva, M. Weintraub and M. Grummer, Heat transmission through fluidized beds of fine particles, Chem. Eng. Progr., 1949, 45, 563-572.

417. M. Leva, M. Weintraub, M. Grummer and M. Pollchik, Fluidization of an antracite coal, Ind. Eng. Chem., 1949, 41, 1206-1212.

418. M. Leva, M. Weintraub, M. Grummer, M. Pollchik and H. Storch, Fluid flow through packed and fluidized systems, U.S. Bur. Mines Bull., 1951, No. 504.

419. O. Levenspiel and J. S. Walton, Proc. Heat Transf. Fluid Mech. Inst. ASME, NY, 1949, 139-146.

420. O. Levenspiel and J. S. Walton, Bed wall heat transfer in fluidized systems, Chem. Eng. Progr. Symp. Ser., 1954, 50, No. 9, 1-13.

421. E. W. Lewis and E. W. Bowerman, Fluidization of solid particles in liquids, Chem. Eng. Prog., 1952, 48, 603-610.

422. W. K. Lewis, E. Gilliland and W. Bauer, Characteristics of fluidized particles, Ind. Eng. Chem., 1949, 41, 1104-1117.

423. W. K. Lewis, E. R. Gilliland and G. T. McBride, Gasification of carbon dioxyde in fluidized powder bed, Ind. Eng. Chem., 1949, 41, 1213-1226.

424. W. Lewis, E. R. Gilliland and R. R. Paxton, Low temperature oxydation of carbon, Ind. Eng. Chem., 1954, 46, 1327-1331.

425. W. K. Lewis, E. R. Gilliland and W. A. Reed, Reaction of methane with copper oxyde in a fluidized bed, Ind. Eng. Chem., 1949, 41, 1227-1237.

426. W. K. Lewis, E. R. Gilliland and M. R. Sweeney, Gasification of carbon metal oxydes in a fluidized powder bed, Chem. Eng. Progr., 1951, 47, 251-256.

427. L. Longchambron, Appareil de la fluidization en phase dense, Genie Chim., 1955, 73, No. 3, 80.

428. A. D. Longhouse, D. P. Brown, H. P. Simons and C. W. Albright, Application of fluidization to conveying grain, Agr. Engng, 1950, 31, No. 7, 349, 352.

429. J. Lowenstein, Fluidization velocity, Chem. Eng., 1955, 62, No. 4, 189-190.

430. P. Mars and D. v. Krevelen, Proc. Conf. Oxyd. Processes, 1954, 3, 4-59.

431. V. Martello and C. Cecotti, Dehydratation catalitique en phase vapeur, amélioration apporté par l'emploi de lits fluidisés, Chimie et Ind., 1954, 65, 528-533.
432. J. Martin, Etude experimentale des systèmes fluidisés solide gas, Chimie et Ind., 1953, 69, 57-71.
433. L. Massimilla and S. Bracale, Fluidizzazione libera e frenata di litti di particelle solide, Ric. Sci., 1956, 26, 487-504.
434. G. Matheson, W. Herbst and P. Holt, Characteristics of fluid solid systems, Ind. Eng. Chem., 1949, 41, 1099-1104.
435. J. F. Mathis and C. C. Watson, Effect of fluidization on catalytic cumene dealkylation, A. E. Ch. E. Journ., 1956, 2, 518-524.
436. K. Mathur and P. Gishler, A study of application of the spouted bed technique to wheat drying, Journ. Appl. Chem., 1955, 5, 624-636.
437. K. Mathur and P. Gishler, A technique for contacting gases with coarse solid particles, A. I. Ch. E. Journ., 1955, 1, 157-164.
438. T. Matsuyama, Heat transfer in fluidized catalyst beds, Chem. Eng. Japan, 1954, 18, 406-412.
439. W. J. Mattox, The fluid catalyst cracking process, Colloid Chem., 1950, 7, 477-495.
440. L. McCune and R. Wilhelm, Mass and momentum transfer in solid liquid systems-fixed and fluidized beds, Ind. Eng. Chem., 1949, 41, 1124-1134.
441. M. McDonald, Automatic control of a fluid catalytic cracking unit, Instrumentation, 1946, 2, No. 3, 11-13.
442. J. B. McKay, The Fluo-solid roaster at Campbeil Red Lake Mines Ltd, Trans. Can. Inst. Min. Met., 1951, 54, 457-460.
443. J. H. McLellan, Revivication of bone char in fluidized beds, Sugar J., 1950, 13, No. 4, 12-17, 30.
444. J. H. McLellan, An experimental fluidized bed for char revivication studies, Sugar J., 1951, 31, No. 9, 13-5, 18-19, 27.
445. W. E. McWhirter, J. R. Tusson and H. A. Parker, Destrehan model IV fluid catalyst cracker, Petr. Refiner, 1956, 35, No. 4, 201-205.
446. E. Meertens, La catalyse heterogène fluidisé par la méthode de courants, J. Chim. Phys. Biol., 1953, 47, No. 3-4, 286.
447. H. Meissner and H. Mickley, Removal of mists and dusts from air by beds of fluidized solids, Ind. Eng. Chem., 1949, 41, 1238-1242.
448. E. Mertens, Fluidized heterogeneous catalysis by the method of currents, J. Chim. Phys., 1950, 286-289.
449. T. Mertes and H. Rhodes, Liquid-particle behaviour, Chem. Eng. Progr., 1955, 51, 517-522.
450. H. S. Mickley and D. F. Fairbanks, Mechanism of heat transfer to fluidized beds, A. I. Ch. E. Journ., 1955, I, 374-384.
451. H. S. Mickley and Ch. Trilling, Heat transfer characteristics of fluidized beds, Ind. Eng. Chem., 1949, 41, 1135-1147.
452. C. O. Miller and A. K. Logwinuk, Fluidization studies of solid particles, Ind. Eng. Chem., 1951, 43, No. 5, 1220-1226.
453. R. Minet and J. Mirkus, Cost saving techniques for instrumenting a fluidized bed pilot plant, Chem. Eng. Progr., 1956, 52, 531-534.

454. B. W. Molstadt and J. F. Moser, Method of scouring equipment in a fluid coking, U.S. Pat. 2,735,806, 21.02.1956.
455. L. W. Morgan, G. M. Armstrong and H. C. Lewis, Distillation of hardwood in a fluidized bed, Chem. Eng. Progr., 1953, 49, 98-101.
456. I. B. Morris, C. M. Nicholls and F. W. Fenning, The application of fluidization technique to nuclear reactors, Chem. Age, 1956, 74, No. 1916, 739-742.
457. I. B. Morris, C. M. Nicholls and F. W. Fenning. The application of fluidization techniques to nuclear reactors—a preliminary assessment, Trans. Inst. Chem. Engrs, 1956, 34, No. 2, 168-184.
458. I. B. Morris, C. M. Nicholls and F. W. Fenning, Fluidization techniques and nuclear reactors, Atomics, 1956, 7, No. 6, 215-217, 223,
459. R. D. Morse, Fluidization of granular solids-fluid mechanics and quality, Ind. Eng. Chem., 1949, 41, 1117-1124.
460. R. D. Morse, Sonic energy in granular solid fluidization, Ind. Eng. Chem., 1955, 47, No. 6, 1170-1175.
461. R. D. Morse and C. Ballou. The uniformity of fluidization, its measurements and use, Chem. Eng. Progr., 1951, 47, 199-204.
462. J. Moss, Heat transfer-conduction, heat transfer between fluid and surface; fluidized beds; heat exchangers, Chem. Process, Eng., 1952, 33, 603-606.
463. J. Moss, Heat transfer in fluidized beds, Chem. Process Eng., 1954, 35, 10-13, 20.
464. E. V. Murphree, War developments in the petroleum industry, Ind. Eng. Chem., 1943, 35, 623-631.
465. E. V. Murphree, The fluid solid technique-applications in petroleum industry, J. Inst. Petrol., 1947, 33, 608-620.
466. E. V. Murphree, Fundamentals of fluid hydroforming, Petr. Refiner, 1951, 3, No. 12, 97-101.
467. E. V. Murphree, C. L. Brown, H. G. Fisher, E. J. Gohr, W. I. Sweeney, Fluid catalytic process (catalytic cracking of petroleum), Ind. Eng. Chem., 1943, 35, 768-773.
468. E. V. Murphree, H. G. Fisher, E. J. Gohr, W. Y. Sweeney and C. L. Brown, Fluid catalyst cracking for premium fuels, Petr. Refiner, 1943, 22, 357-364.
469. E. V. Murphree, E. Gohr, C. E. Jahning, H. Z. Martin and C. V. Tyson, Improved fluid process for catalytic cracking, Trans. Am. Inst. Chem. Engrs, 1945, 41, 19-33.
470. W. Murphy, Fluidization nomenclature and symbols, Ind. Eng. Chem., 1949, 41, 1249.
471. P. Murthy and A. Gupta, Gasification of solid fuels in a fluidized bed, J. Sci. Ind. Res. (India), 1955, 14B, 174-185.
472. S. Nagata and T. Matsuyama, Studies on the fluidized catalyst with mechanical agitation, Chem. Eng. Japan, 1952, 16, 301-306.
473. M. Nakajima, Mass transfer between particles and a fluid in fluidized beds, Reps, Himeji Inst. Technol., 1954, 4, 79-90.
474. K. Nakanishi, How Mitsubishi new Akita plant makes 99.997% electrolyte zinc, Min. World, 1955, 17, No. 11, 56-59, 70-71.

475. R. H. Newton, G. S. Dunham and T. D. Simpson, The TCC catalytic cracking process for motor gasoline production, Trans. A. I. Ch. E., 1945, 41, 215
476. E. W. Nicholson, J. E. Moise and R. L. Hardy, Fluidized solids pilot plants, Ind. Eng. Chem., 1948, 40, 2033-2039.
477. H. Nonnenmacher, Chemische Reaktionen in Wirbelschichten, Ztschr. Electrochem. Ber. Bunsenges., 1953, 57, No. 7, 512-518.
478. E. O. Novella, Production of pure sulphur dioxide by the fluidized technique, Annal. Real. Soc. Españ. Fis. y Quim., 1953, 49B, 733-750.
479. E. Novella, A. S. Ruiz and V. R. Romera, Production of pure sulphur dioxide by the fluidized solids technique, Annal. Real. Soc. Espan. Fis. y Quim., 1955, 51B, 351-362, 363-376.
480. H. Noyer and A. Haas von Dorsser, Temperature and humidity measurements in a drying fluidized bed, Chem. Eng. Sci., 1953, 2, 97-107.
481. W. W. Odell, Fluidized process. U.S. Pat. 2,631,921, 1953.
482. H. L. Olin, Fluidization, Petrol. Engr, 1951, 23C, No. 6, 5-12.
483. H. L. Olin and O. C. Dean, Heat transfer and fouling of fluidized beds, Petrol. Engr, 1955, 25, No. 3, 623-632.
484. T. Omae and J. Furukawa, Characteristics of fluidized particles. 4. Particle distribution in fluidized beds, Journ. Chem. Soc. Japan (Ind. Chem. Sec.), 1953, 56, No. 10, 727-731.
485. T. Omae and J. Furukawa, Characteristics of fluidized particles, J. Chem. Soc. Japan (Ind. Chem. Sec.), 1953, 56, 824-6, 909-912.
486. T. Omae and J. Furukawa, Rheological properties of fluidized particles, J. Chem. Soc. Japan (ind. Chem. Sec.), 1954, 57, 788-791.
487. T. Omae and J. Furukawa, Surface tension of fluidized particles, J. Chem. Soc. Japan (Ind. Chem. Sec.), 1954, 57, 785-787.
488. G. L. Osberg, Locating fluidized solids bed level in reactor—hot wire method, Ind. Eng. Chem., 1951, 43, 1871-1873.
489. G. L. Osberg and D. H. Charlesworth, Elutriation in a fluidized bed, Chem. Eng. Progr., 1951, 47, 566-570.
490. D. Othmer (editor), Fluidization, New York, 1956. Russian translation "Processes in Fluidized Beds" ("Protsessy v Kipyashchem Sloe"), Gostoptekhizdat, Moscow, 1958.
491. Y. Oyama and K. Nagano, Discharge rate of solid particles from a nozzle steeped in liquid, Repts. Sci. Research. Inst. (Tokyo), 1953, 29, 349-352.
492. I. D. Parent, N. Yagol and C. S. Steiner, Fluidization process—basic observations from laboratory equipment, Chem. Eng. Progr., 1947, 43, 429-436.
493. H. W. Parker, The effect of liquid on interparticle forces in gas-fluidized beds, Diss. Abstrs., 1956, 16, 2412.
494. V. F. Parry, J. B. Goodman and E. O. Wagner, Drying low rank coals in the entrained and fluidized state, Mining Eng., 1949, I, Sec. 3, No. 4, 89-98.
495. K. Peters, A. Orlichek and A. Schmidt, Wärmetransportfähigkeiten von Wirbelschichten, Chem. Ing. Techn., 1953, 25, No. 6, 3-3-316.

496. K. Peters and A. Schmidt, Untersuchungen über das Wirbelschichtverfahren, Oester. Chem. Ztg, 1953, 54, 233-238.
497. P. Gishler, Fluidized solids coking of Canadian heavy crude oils, Can. J. Technol., 1956, 34, 104-120.
498. E. S. Pettyjohn and E. B. Christiansen, Effect of particle shape on free-settling rates of isometric particles, Chem. Eng. Progr., 1948, 44, 157-172.
499. A. Pevtavy, Fluid carbonization. 4 Congr. Int. Chauf. Ind., Paris, 1952, No. 25.
500. A. Peytavy and J. Lahouste, La semicarbonisation de la houille à l'état fluidisé, Bull. Soc. Chim. France, 1955, 435.
501. P. H. Pinchbeck and F. Popper, Critical and terminal velocities in fluidization, Chem. Eng. Sci., 1956, 6, No. 2, 57-64.
502. M. Prange and A. Ziegler, Application de la fluidisation, oxydation et réduction en chimie minerale, Bull. Soc. Chim. France, 1955, 456-461.
503. J. W. Pringle, P. L. Barrick and H. F. Wington, Vacuum retorting of oil shale in a fluidized bed, Ind. Eng. Chem., 1952, 44, 1489-1491.
504. S. S. Pullar, Roasting by Fluo-solid process, Chem. Eng. Min. Rev., 1951, 43, 175-178.
505. J. Quinton and J. A. Storrow, Heat transfer to air flowing through packed tubes, Chem. Eng. Sci., 1956, 5, 245-257.
506. G. S. Rao, Fluidization of Hyderabad coals, Repts. Proc. Symp. Fuels India. 1950, 30-43.
507. I. M. Rausch, Gravity flow of solid beds in vertical towers, Ph. D. thesis, Princeton Univ., USA, 1948.
508. D. Read, Process for octane improvement, Petr. Engr, 1946, 18, 106-114.
509. D. Read, Midget fluid catalytic cracker makes process economical for the smaller refinery, Oil Gas J., 1946, 44, 144-149.
510. D. Read, Application of fluid catalytic cracking in present refinery operations, Oil Gas J., 1946, 44, 243-249.
511. P. Reboux, Phénomènes de fluidisation, Paris, 1954.
512. T. M. Reed and M. R. Fenske, Effect of agitation on gas fluidization of solids, Ind. Eng. Chem., 1955, 47, No. 2, 275-282.
513. J. M. Reid, Fluid gasification of oil, Inst. Gas Technol. Res. Bull., 1953, No. 16.
514. T. Reiss, Theoretical and experimental studies of fluidized catalyst systems, Bull. assoc. franc. tech. petrole, 1949, 76, 3-44.
515. G. H. Rema, Effect of gas and solid mixing in fluidized beds on chemical reactions, Chem. and Industry, 1955, 15, No. 3, 46-51.
516. W. Resnik and R. R. White, Mass transfer in system of gas and fluidized solids, Chem. Eng. Progr., 1949, 45, 377-390.
517. N. Rice and T. L. Johnson, Fluidized drying and briquetting of low rank western coal, Univ. Wyoming Natl. Resour. Research Inst. Bull., 1951, No. 5.
518. J. F. Richardson and W. N. Zaki, Sedimentation and fluidization, Trans. Inst. Chem. Engrs (London), 1954, 32, 35-53.
519. H. L. Riley, Manufacture of phtalic anhydride by the fluidized catalyst method, Chemistry and Industry, 1956, 49, 1464.

520. T. Roy, Production of pure SO_2 using metal oxides as oxygen carriers, Trans. Indian Inst. Chem. Engrs, 1954-55, 7, 21-32.
521. F. Sabel, Die Winklersche Wirbelschichttechnik, Chem.-Ing. Techn., 1952, No. 2, 93-97.
522. A. M. Sadler, Gas-solid fluidizing for transport, Chem. Eng., 1949-, 56, No. 5, 110.
523. F. Schaub, Anwendungen und Grenzen der Wirbelschichttechnik. Chem.-Ing. Techn., 1952, 24, No. 2, 98-103.
524. J. F. Schnacky, Avoiding overheating in a fluidized bed, Chem. Eng., 1947, 54, No. 10, 130.
525. J. F. Schnacky, New way to pressure seal solids flowing through continuous process, Chem. Eng., 1948, 55, No. 8, 124-125.
526. G. Schneebeli, Expériences sur la limite de validité de la loi de Darcy et l'apparation de la turbulence dans un écoulement de filtration, Houille Blanche, 1955, 10, No. 2, 141-149.
527. W. Schuster and P. Koslak, The measurements of fluidization quality, Chem. Eng. Progr., 1952, 48, 455-458.
528. G. M. Schwab, The present status of scientific research in industrial catalysis, Angew. Chem., 1953, 65, 502-507.
529. W. Schwartz, Das "Fluosolid" Verfahren, Ztschr. Erzbergbau Metallhüttenwes., 1948, 1, 116.
530. J. E. Seebold, J. W. Bertells, J. F. Snuggs and J. A. Rock, Fluid hydroforming: full scale unit, Oil Gas J., 1952, 51, No. 2, 135-136.
531. B. Sewerynski, Proces fluidyzacji, Hutnik, 1954, 21, No, 3, 79-82.
532. V. Shah, Fluidization and its applications, Bombay Technologist, 1956, 6, 65-69.
533. M. Shanahan, M.S. Thesis, M.I.T., Cambridge, 1954.
534. Shell Develop. Co., Process and apparatus for the regeneration of contact materials, Brit. Pat. 590,244, 1947.
535. R. M. Shepardson, C. W. Tyson and A. Vorhies, Fluid hydroforming pilot plant, fluidized catalyst technique, Oil Gas J., 1952, 51, No. 2, 110, 136-141.
536. P. W. Sherwood, La fabrication des nitroparaffines au stade industriel, Chimie et Industrie, 1956, 76, 85-86.
537. T. Shirai, Powder orifice nomograph, Chem. Eng. (Tokyo), 1952, 16, 86-89.
538. K. Y. Shrikhand, A. K. Gupta and A. Lahiri, Low temperature carbonisation of Indian coal under fluidized condition, J. Sci. Ind. Res. (India), 1954, 13B, 634-641.
539. W. Siemes, Das Wirbelschichtverfahren, Chemiker Ztg, 1955, 79, No. 24.
540. D. Singer, J. Benowsky and P. Jiru, Pracovny podminky fluidacni prazici pece, Chem. Prum, 1955, 5, No. 3, 60-63.
541. A. D. Singh, Partial devolatilization of coal by the fluidization process, Proc. Am. Gas Assoc., 1946, 400-409.
542. A. D. Singh, What coal fluidization may do for the power plant of tomorrow, Power Plant Eng., 1947, 51, No. 112-114.
543. A. D. Singh and L. J. Kane, Fluid devolatilization of coal-power plant practice, Trans. A.S.M.E., 1948, 70, 957-964.

544. M. Sittig, Fluidized solids, Chem. Eng., 1953, 60, No. 5, 219-231.
545. J. F. Snugs, Regeneration of spent catalyst in fluid catalytic cracking, Petr. Eng., 1947, 18, No. 6, 230-236.
546. K. E. Spells, Trans. Inst. Chem. Engrs (London), 1955, 33, No. 2, 79-84.
547. F. M. Stephens, Jr., The fluidized bed sulphate roasting of non-ferrous metals, Chem. Eng. Progr., 1953, 49, No. 9, 455-458.
548. R. Stocker, Erosion due to dust particles in a gas stream, Ind. Eng. Chem., 1949, 41, 1196-1199.
549. H. N. Stone, J. D. Batchelor and H. F. Johnstone, Low temperature carbonization rates in a fluidized bed, Ind. Eng. Chem., 1954, 46, 274-278.
550. L. Stott, Fluidized bed method of coating, Org. Finishing, 1956, 17, No. 6, 16-17.
551. F. Stracke and F. Schiffer, Fluid coke-steam plant fuel, Power, 1955, No. 12, 102-105, 198, 200, 202.
552. K. Takagi, Fluidized roasting of ore in a pilot plant, Chem. Eng. (Japan), 1952, 16, 214-225.
553. K. Takagi and K. Kuiyama, Fluidized roasting of ore in a pilot plant, Chem. Eng. (Japan), 1952, 16, 134-146.
554. K. Takahashi, Efflux of granular mass, Bull. Inst. Phys. Chem. Res. (Tokyo), 1933, 12, 984-994.
555. M. Tenenbaum and C. M. Squarry, Direct reduction of iron ore with fluidized solids technique, Iron Age, 1951, 167, No. 5, 72-73.
556. M. Tenenbaum and C. M. Squarry, Direct reduction of iron ore with fluidized solids technique, Met. Progr., 1952, 61, 147-148.
557. P. Terry and U. W. Urie, Control of level in fluid bed reactors, Austral. J. Appl. Sci., 1950, 1, 373-375.
558. A. Terui, Fluidized carbonization of coal, J. Fuel Soc. Japan, 1956, 35, 266-278.
559. I. Tesarik, Vznaseni lozi identickych kouli v turbulentni oblasti, Rozpr. Ceskosl. Ak. Ved, RTV, 1956, 66, No. 6, 76 s.
560. E. A. Thau, Grosswassergaserzeugung für chemische Synthesen, Oel and Kohle, 1942, 38, 589-601, 617-24, 685-90, 721-727.
561. C. L. Thomas, N. K. Anderson, H. A. Becker and J. McAffee, Cracking with catalyst, Petr. Ref., 1943, 22, 365-370.
562. C. L. Thomas and I. Hoekstra, Fluidized and fixed bed, method for contacting solids with gases and vapors, Ind. Eng. Chem., 1945, 37, 332-334.
563. B. H. Thompson and J. E. Garside, A study of fluidized gasification, Gas Journal, 1953, 782, Gas World, 1953, 138, 1386; 1954, 139, 140, 141.
564. B. H. Thompson and J. E. Garside, The fluidized gasification of coal, Inst. Gas Engrs. Copyright Publ. No. 474, 1955.
565. R. B. Thompson, A new application of fluidization, Chem. Eng. Progr., 1953, 49, 253-256.
566. R. Toomey and H. Johnstone, Gaseous fluidization of solid particle, Chem. Eng. Progr., 1952, 48, 220-226.
567. R. Toomey and H. Johnstone, Chem. Eng. Progr., 1952, 48, No. 5.

568. R. Toomey and H. Johnstone, Heat transfer between beds of fluidized solids and the walls of the container, Chem. Eng. Progr. Symp. Ser., 1953, 49, No. 5, 51-63.

569. R. P. Trainer, N. W. Alexander and F. Kunreuther, Fluid bed catalytic cracking plant, Ind. Eng. Chem., 1948, 40, 175-182.

570. H. Trawinski, Zur Hydromechanik aufgewirbelter Partikelschichten, Chem. Ing. Techn., 1951, 23, No. 17/18, 416-419.

571. H. Trawinski, Wärmeübergang zwischen Wirbelschichten und Gefässwandung, Chem. Ing. Techn., 1952, 24, 444-446.

572. H. Trawinski, Entmischung gasdurchströmter Partikelschichten und deren Zusammenhang mit der Wirbelschichtzähigkeit. Chem. Ing. Techn., 1953, 25, No. 4, 201-203.

573. H. Trawinski, Effektive Zähigkeit und Inhomogenität von Wirbelschichten. Chem. Ing. Techn., 1953, 25, No. 5, 229-237.

574. R. V. Trens, Heat transfer in gas-solid fluidized beds, Diss. Abstrs, 1955, 15, No. 10, 1814.

575. C. H. Tsu and M. C. Molstad, Process design data-rate of mass transfer from gas stream to porous solid in a fluidized bed, Ind. Eng. Chem., 1955, 47, 1550-1560.

576. R. E. Vener, Moving bed processes-present and potential applications. Engng Min. Journ., 1956, 157, No. 10, 89-99.

577. H. Verschoor, Appl. Sci. Research, 1950, A2, 155-161.

578. V. Vesely, O fluidizacii v priemysle ropy, Paliva, 1954, 34, 244-250.

579. H. A. Vreedenberg, Heat transfer by fluidized beds, Gener. discuss. on heat transf., IME and ASME, London, 1951.

580. H. A. Vreedenberg, Heat transfer between fluidized beds and vertically inserted tubes, Journ. Appl. Chem., 1952, 2, Suppl. Issue, 26-33.

581. A. H. Vroom, Bark pyrolisis by fluidization technique; new approach to bark utilization, Pulp. Pap. Mag. Can., 1952, 53, No. 5, 121-124.

582. C. J. Wall and W. J. Ash, Fluid-solid air sizer and dryer, Ind. Eng. Chem., 1949, 41, 1247-1249.

583. J. F. Walter, Development in fluid catalytic cracking, J. Inst. Petrol., 1946, 32, 295-303.

584. J. S. Walton, R. L. Olson and O. Levenspiel, Gas-solid film coefficients of heat transfer in fluidized coal beds, Ind. Eng. Chem., 1952, 44, 1474-1480.

585. W. W. Wamsley and L. N. Johanson, Fluidized bed heat transfer, Chem. Eng. Progr., 1954, 50, No. 7, 347-355.

586. K. T. Wang, Application of the fluidized powder technique in coal carbonization, J. Chem. Eng. China, 1948, 15, 33-40.

587. M. Watanabe and A. Kogoshi, Fluidized-bed roasting of pyrite, Bull. Res. Inst. Min. Dress. Met, Tohoku Univ., 1956, 12, 57-70.

588. M. Weber, Fluidacni zlab, Chemicky Prum., 1955, 5, No. 30, 35.

589. I. L. Weininger and W. G. Schneider, Thermal conductivity of granular beds filled with compressed gases, Ind. Eng. Chem., 1951, 43, No. 5, 1229-1233.

590. M. Weintraub and M. Leva, Fluid flow-fluidization, Ind. Eng. Chem., 1953, 45, 76-78.
591. M. Weintraub, Fluid dynamics-fluidization, Ind. Eng. Chem., 1954, 46, 112-117, 1955, 47, 561.
592. M. Weintraub and M. Leva, Pressure drop through fixed and fluidized beds, Chem. Met. Eng., 1950, 57, No. 1, 110-113.
593. J. Weisman, Effect of void volume and Prandtl modulus on heat transfer in tube banks and packed beds, A. I. Ch. E. Journal, 1955, 1, 342-348.
594. C. Y. Wen, Ph.D. Thesis, West Virginia Univ., USA, 1956.
595. C. Y. Wen and M. Leva, Fluidized bed heat transfer; A generalized dense phase correlation, A. I. Ch. E. Journal, 1956, 2, No. 4, 482-488.
596. J. Went and H. Bruym, Fluidized and liquid fuel reactors with uranium oxides, Nucleonics, 1954, 12, No. 9, 16.
597. C. W. Westphal, Metallurgical aspects of the roasting of ores by the fluosolid process, Mineria (Peru), 1952, 1, 6-18.
598. F. S. White and E. L. Kinsella, Solid fluidization applied to lime burning, Min. Engng, 1952, 4, No. 9, 903-906.
599. E. Wicke, Wärme und Stoffaustausch in aufgewirbelten Partikelschichten, Chem. Ing. Techn., 1951, 23, No. 17/18, 405-408.
600. E. Wicke and F. Fetting, Wärmeübertragung in Gaswirbelschichten, Chem. Ing. Techn., 1954, 26, No. 6, 301-309.
601. E. Wicke and K. Hedden, Strömungsformen und Wärmeübertragung in von Luft aufgewirbelten Schüttgutschichten, Chem. Ing. Techn., 1952, 24, No. 2, 82-91.
602. E. Wicke and H. Trawinski, Über die Vermischung des Strömungsmediums in Flüssigkeitswirbelschichten, Chem. Ing. Techn., 1953, 25, No. 3, 114-124.
603. R. H. Wilhelm, Rate processes in chemical reactors, Chem. Eng. Progr., 1953, 49, 150-154.
604. R. H. Wilhelm and M. Kwauk, Fluidization of solid nonvesicular particles, Chem. Eng. Progr., 1948, 44, 201-218.
605. R. H. Wilhelm and S. Valentine, The fluidized bed transition state in the vertical pneumatic transport of particles, Ind. Eng. Chem., 1951, 43, 1199-1203.
606. R. H. Williamson, Application of fluidized solids technique to coal gas purification, Gas World, 1948, 12, 809-811, 948-952.
607. R. H. Williamson and J. E. Garside, An application of the fluidized solids technique to coal gas purification, Gas World, 1949, 130, 1999-2001.
608. H. F. Wington, Oil shale retorting through application of fluidized solids technique, Chem. Engng. Progr., 1954, 50, 134-138.
609. W. F. Wolf and H. L. Gogenleiten, Experimental study of the flow of coal in chutes at Riverside Generating Station, Trans. ASME, 1945, 67, 585-599.
610. S. Yagi, Pyrite roasting by fluidization: I. Roasting tests of pyrrhotite and pyrite, II. Commercial scale fluidization, Chem. Eng. (Japan), 1951, 51, 218-232.
611. S. Yagi and D. Kunii. Fluidizing roasting of sulphide ore concentrate. I: theoretical considerations: II: roasting zinc blend concentrate in the fluidized bed, Chem. Eng. (Japan), 1952, 16, 283-293.

612. S. Yagi and D. Kunii, Proposed theory of fluidized roasting of sulphuric ore with uniform size, Journ. Chem. Soc. Japan (Ind. Chem. Sec.), 1953, 56, 131-135.
613. S. Yagi and D. Kunii, Studies on combustion of carbon particles in flames and fluidized beds, 5th Sympos. on Combust., Pittsburg, 1955, 231-244.
615. S. Yagi and D. Kuni,, Studies of rate coefficients of gasification for coke particles in fluidized bed, Chem. Eng. Japan, 1955, 19, No. 8, 401-413.
616. S. Yagi and D. Kunii, Studies of fluidized-solids reactors for particles with decreasing diameters, Chem. Eng. Japan, 1955, 19, No. 10, 500-506.
617. S. Yagi and D. Kunii, Studies on combustion of carbon particles in flame and fluidized beds, Reinhold Publ., N.Y., 1955.
618. S. Yagi, D. Kunii, Y. Ikeda and T. Kamata, Producer-gas reaction of coke particles in a fluidized bed, J. Fuel Soc. Japan, 1956, 35, 41-53.
619. S. Yagi and J. Muchi, Conditions of fluidization, Chem. Eng. Japan, 1952, 16, 307-312.
620. S. Yagi and J. Muchi, Fluidized bed type air preheater, Chem. Eng. Japan, 1954, 18, 532-537.
621. S. Yagi and T. Shirai, Fluid flow through fluidized solid beds; a nomograph, Chem. Eng. Japan, 1952, 16, 2-6.
622. S. Yagi and K. Takagi, Fluidization of powdered bed, Chem. Eng. Japan, 1951, 15, 212-217.
623. R. J. Young, Hydroforming relationship in particulate fluidization, Journ. Appl. Chem. (London), 1952, 2, Suppl. Iss. 1, 353.
624. W. N. Zaki, Sedimentation and fluidization, J. Imp. Coll. Chem. Eng. Sec., 1953, 7, 119-138.
625. F. A. Zenz, Two-phase fluid solid flow, Ind. Eng. Chem., 1949, 41, 2801-2806.
626. F. A. Zenz, Petrol. Refiner, 1953, 32, 123-128.

References published in 1957

627. S. I. Azos and R. L. Veller, The economic efficiency of roasting zinc concentrates in fluidized beds (in Russian), Tsvet. Met., 1957, No. 4.
628. V. S. Aliev, N. B. Al'tman and N. P. Kasimova, Some laws for the combustion of coke in continuously operating fluidized beds (in Russian), Az. N. Kh., 1957, No. 6.
629. V. S. Aliev, M. I. Rustamov and N. M. Indyukov, A study of the fluidization of powdered coke and aluminosilicate catalyst (in Russian), Az. N. Kh., 1957, No. 2.
630. V. S. Aliev, B. G. Ter-Sarkosov and L. P. Pilaeva, Development of a method for size separating dispersed coke particles during the coking of petroleum residues in fluidized beds (in Russian), in: Sb. "Vopr. issl. neftei i nefteproduktov, razrab. prots. pererabotki nefti i obsled. zav. ustr.," Part 1, Baku, Nauchno-Tekhnicheskii Otdel Neftprom. Az., 1957.

631. M. E. Aerov, L. I. Luk'yanov and G. A. Balueva, A laboratory reactor with a suspended catalyst bed (in Russian), Zav. Lab., 1957, 23, No. 3.
632. Ya. F. Batishchev, Heat transfer from walls to gas in tubes with packings of mixed characteristics (in Russian), Tr. Novocherk. Politekhn. In-ta, 1957, 70/84, 45-63.
633. Ya. F. Batishchev, A comparison of heat transfer in packed tubes and in smooth channels without packings (in Russian), Tr. Novocherk. Politekhn. In-ta, 1957, 70/84, 65-69.
634. M. E. Beker, The VAVAS Pneumatic Air-Fountain Dryer (in Russian), Pishchepromizdat, 1957.
635. A. K. Bondareva, Measurement of the thermal conductivity of suspended beds (in Russian), Doklady Akad. Nauk SSSR, 1957, 115, No. 4.
636. G. K. Boreskov and M. G. Slin'ko, The application of the fluidized bed method to heterogeneous catalytic processes (in Russian), Khim. Prom., 1957, No. 6, 321-330.
637. M. S. Verteshev and A. A. Komarovskii, A study of hydrodynamics and mass transfer in the process of dissolving salts in a suspended state (in Russian), Tr. Novocherk. Politekhn. In-ta, 1957, 44/58, 18-42.
638. M. S. Verteshev and A. A. Komarovskii, A study of mass transfer in the process of dissolving salts in a suspended bed (in Russian), Tr. Novocherk. Politekhn. In-ta, 1957, 73/87, 107-116.
639. A. F. Grigor'ev, et al., A laboratory column for studying reactions in fluidized catalyst beds (in Russian), Zav. Lab., 1957, 23, No. 3.
640. V. V. Gurvich, Tests on cyclone operation in foreign catalytic cracking plants (in Russian), Khimiya i Tekhnol. Topl. i Masel, 1957, Nos. 4 and 5.
641. S. S. Zabrodsky, The temperature driving force field and the efficiency of heat exchangers with moving dense beds of granular materials (in Russian or Belorussian), Vestsi Akad. Navuk BSSR, Seryya Fiz. Tekhn. Navuk, 1957, No. 1.
642. S. S. Zabrodsky, Some problems in the aerodynamics of fluidized (boiling) beds (in Russian or Belorussian), Tr. In-ta Energetiki Akad. Navuk BSSR, 1957, No. 3.
643. Yu. Ya. Kaganovich, Drying of salts in fluidized bed equipment (in Russian), Khim. Nauka i Prom., 1957, No. 6, 764-766.
644. V. V. Kashirskii, N. B. Lobachev and A. R. Yakorev, High-temperature thermal decomposition of pulverized peat in a stream of steam (in Russian), Zhurn. Prikl. Khim., 1957, 30, No. 1.
645. G. V. Labutin and A. G. Mal'tsev, Production of potassium sulfate from the chloride in a fluidized bed (in Russian), Tr. Vsesoyuz. Nauchno-Issl. Alyum. Magn. In-ta, 1957, No. 40.
646. K. P. Lavrovskii, A. L. Rozental' and A. Kh. Eglit, The interaction of iron ore and methane under fluidized bed conditions (in Russian), Doklad. Akad. Nauk SSSR, 1957, 112, No. 4.
647. P. I. Luk'yanov, I. V. Gusev and N. I. Nikitina, The motion of compact beds of granular material through equipment (in Russian), Khimiya i Tekhnol. Topl. i Masel, 1957, No. 12, 38-44.

648. V. E. Lunev and V. I. Chuprikov, Practical aspects of fluidized-bed roasting in the Ust'-Kamenogorsk complex (in Russian), Tsvet. Met., 1957, No. 2, 32-36.
649. I. G. Martyushin, Some problems in the hydromechanical design of fluidized bed equipment (in Russian), Tr. Moskovskii Institut Khimicheskogo Mashinostroeniya, 1957, No. 13, 145-158.
650. A. S. Mikulinskii, V. N. Yakunchikov, V. I. Val'shchikov and V. V. Efremkin, Refining amorphous boron by oxidative roasting in fluidized beds (in Russian), Tr. Ural'sk. Nauchno-Issl. Khim. In-ta, 1957 (1958), No. 5.
651. F. K. Mikhailov and N. I. Tsofin, Roasting of chalk fines in a fluidized bed (in Russian), Tr. Nauchno-Issl. In-ta Osn. Khim., 1957, No. 10.
652. D. I. Orochko, T. Kh. Melik-Akhnazarov and G. N. Poluboyarinov, A stagewise countercurrent method for contacting pulverized solid materials with gases and vapors in heterogeneous chemical processes (in Russian), Khim. i Tekhnol. Topl. i Masel, 1957, No. 12, 1-12.
653. M. K. Pis'men, Gasification of Brown Coal in Fluidized Beds (in Russian), in: Sb. Gazif. Tverd. Topl., Gostoptekhizdat, 1957.
654. M. K. Pis'men, V. G. Ermakov and Yu. I. Belyanin, Gasification of Baltic oil shales in fluidized beds (in Russian), Gaz. Prom., 1957, No. 3.
655. A. E. Protskii, Problems in drying milled peat in cyclones (in Russian), in: Sb. Nauchn. Rabot Beloruss. Politekhn. In-ta, 1957, No. 62.
656. R. B. Rozenbaum and O. M. Todes, The hindered fall of spheres in a cylindrical tube (in Russian), Doklad. Akad. Nauk SSSR, 1957, 115, No. 3, 504-507.
657. P. G. Romankov, V. N. Lepilin and E. S. Nemet, Some problems in the aerodynamics of suspended beds in narrow tubes (in Russian), Tr. Leningradsk. Tekhn. Instituta im. Lensoveta, 1957, No. 39, 28-33.
658. P. G. Romankov and E. S. Nemt, A study of adsorption processes in fluidized adsorbent beds (in Russian), Tr. Leningradsk. Tekhn. In-ta im. Lensoveta, 1957, No. 39, 34-43.
659. P. G. Romankov, N. B. Rashkovskaya and Z. A. Berezovskaya, Development of a practical process for drying basic dyes (in Russian), Tr. Leningradsk. Tekhn. In-ta im. Lensoveta, 1957, No. 39, 95-105.
660. I. I. Sidorchuk, The problem of the kinetics of fine particle communication in suspended (fluidized) beds (in Russian), Az. Neft. Khoz., 1957, No. 9.
661. S. P. Tevosov and I. F. Ibragimbekova, Adsorption of molecular iodine on activated carbon in a fluidized bed (in Russian), Tr. In-ta Khim. Akad. Nauk AzSSR, 1957, No. 16.
662. O. M. Todes and A. K. Bondareva, Special features of technological processes in fluidized beds (in Russian), Khim. Nauka i Prom., 1957, 2, No. 2, 223-232.
663. S. A. Feigin and S. P. Tyufyakin, Technico-economic guides to the production of heating gas from oil-shale fines (in Russian), Gazovaya Prom., 1957, No. 3.

664. E. I. Khodorov, Hydrodynamic conditions and heat transfer in roasting pulverized materials in suspended beds (in Russian), Tsement, 1957, No. 1.
665. E. I. Khodorov and V. M. Kosareva, Two-stage roasting in suspended beds and rotating furnaces (in Russian), in: Sb. "Khim. i Tekhnol. Silikatov," 1957, 163-169.
666. Ts. Tsvetkov and A. Peichev. The rate of reduction of copper oxide in a fluidized bed, Khimiya i Ind. (Sofia), 1957, 29, No. 7-8, 10-14.
667. A. V. Chechetkin, Heat transfer and aerodynamics of filtering streams (in Russian), Tr. Moskovskii Khimiko-Tekhnologicheskii Institut im. Mendeleeva, 1957, No. 24.
668. A. V. Chechetkin, High-Temperature and Heat Transfer Agents (in Russian), Gosenergoizdat, 1957.
669. M. A. Shakhova and A. I. Rychkov, An investigation of heat transfer in fluidized beds (in Russian), Tr. Moskovskii Institut Khimicheskogo Mashinostroeniya, 1957, 12, 119-132.
670. G. M. Shteingard, Study and Industrial Scale-Up of the Roasting of Zinc Concentrates in Fluidized Beds in the "Elektrotsink" Plant (in Russian), Izd. Tsentral'nyi Institut Informatsii Ministerstva Tsvetnoi Metallurgii, 1957.
671. F. D. Yasonov, Byulleten' Tsvetnoi Metallurgii, Izd. Tsentral'nyi Institut Informatsii Ministerstva Tsvetnoi Metallurgii, 1957, 7(84), 12-19.
672. R. Andrieu, Etude de quelques aspects des systèmes fluidises liquides—solides, Genie Chim., 1957, 77, No. 5, 134-139.
673. V. Anghel and M. Ionescu, Revista minelor., 1957, 9, No. 10, 467-470.
674. I. Bernaek and I. Klumpar, Fluidacni catalyticke reaktory a jejich vypocet, Chem. Prum., 1957, 2, 57-60.
675. S. Bracale and A. Cabella, Ricerca sci., 1957, 27, 1448-1456.
676. E. Buchet, Fluidisation, Rev. universelle mines, 1957, 13, 101-114.
677. J. Ciborowski, Fluidyzacja, PWT, Warszawa, 1957.
678. L. Farbar and M. Morley, Heat transfer to flowing gas-solids mixtures in a circular tube, Ind. Eng. Chem., 1957, 49, 1143-1150.
679. S. Friedlander and H. Johnstone, Depostion of suspended particles from turbulent gas streams, Ind. Eng. Chem., 1957, 49, 1151-1156.
680. M. Genbica, Aeracja materialow proszkowych, Cem. Vapno Gips, 1957, 13, 104-111.
681. T. Gopichand, K. Sarma and M. Rao, Quality studies in continuous fluidization, J. Sci. Ind. Research (India), 1957, 16-B, 97-98.
682. A. Handlos and T. Baron, Mass and heat transfer from drops in liquid—liquid extraction, A. I. Ch. E. Journal, 1957, 3, No. 1, 127-136.
683. A. Handlos, R. Kunstman and D. Schissler, Gas mixing characteristics of a fluid bed regenerator, Ind. Eng. Chem., 1957, 49, 25-30.
684. T. I. Hanratty and A. Bandukwala, Fluidization and sedimentation of spherical particles, A. I. Ch. E. Journ., 1957, 3, 293-296.

685. J. Happel and H. Brenner, Viscous flow in multiparticle systems—motion of spheres and a fluid in a cylindrical tube, A. I. Ch. E. Journ., 1957, 3, No. 4, 506-513.
686. A. Jakob and L. Osberg, Effect of gas thermal conductivity on local heat transfer in a fluidized bed, Can. Journ. Chem. Engng, 1957, 35, No. 1, 5-9.
687. A. Johnssen and W. Danz, Über das Wirbelschicht-Röstverfahren d. BASF, Chem. Ing. Techn., 1957, 29, 563-572.
688. B. Johnson, U. S. Atom. Ener. Comm. Rep., 1957, H. W.-51834.
689. A. A. Jonke, Fluidized bed technique for treatment of aqueous nuclear wastes, Nucl. Sci. Engng, 1957, 2, 303.
690. P. A. Kolodzie, Jr. and M. v. Winkle, Discharge coefficients through perforated plates, A. I. Ch. E. Journ., 1957, 3, No. 3, 305-312.
691. H. Kramers, Application de la technique de fluidisation à la construction d'un bain thermostatique pour hautes temperatures, Genie Chim., 1957, 77, 25-33.
692. L. Lapidus and I. C. Elgin, Mechanics of vertical-moving fluidized systems, A. I. Ch. E. Journ., 1957, 3, No. 1, 63-68.
693. M. Leva, Correlations of the dense-plate fluidized state and their applications, Can. Journ. Chem. Eng., 1957, 35, 71-76.
694. M. Leva, Flow through packings and beds, Chem. Engng, February, 1957, 263-268.
695. R. W. Ludt, Drying of sand on a hot surface, A. I. Ch. E. Journ., 1957, 3, No. 3, 343-347.
696. L. Massimilla and S. Bracale, Il mescolamento della fase solida nei sistemi, solido-gas fluidizzati, liberi e frenati, Ricerca sci., 1957, 27, No. 5, 1509-1526.
697. L. Massimilla, S. Bracale and A. Cabella, Sul trasporto del calore nei sistemi; solido-gas fluidizzati, liberi e frenati, Ricerca sci., 1957, 27, No. 6, 1853-1864.
698. J. Mullin, Effect of maldistribution on the performance of packed columns, Ind. Chemist, 1957, 33, 408-417.
699. W. W. Odell, Contact of fluids with solids, U.S.A. Pat. No. 2,783,187; 26.02.1957.
700. A. Peptavy and P. Foch, A study of semicarbonization in a fluidized bed, Chem. Eng. in Coal Ind., Pergamon Press, London, 1957.
701. S. Raseev (1); S. Raseev, S. Feyer-Hoffman, P. Valter and L. Verman (2), Hidraulica stratului fluidizat. 1. Deducerea teoretica a relatiel dintre viteza fluidului si densitatea volumetrica a stratului fluidizat. 2. Verifiacarea experimentala, pentru fluide neocompresibile, particule omogene si lipsa aportului de solid in strat, Studii cercet, chim., 1957, 5, No. 4, 569-579, 581-609.
702. R. Rhodes and E. Goden, Characterization of agitation effects in shaken flasks, Ind. Eng. Chem., 1957, 49, 1233-1236.
703. A. Rutherford and N. R. Amundson, Some remarks on longitudinal mixing or diffusion in fixed beds, A. I. Ch. E. Journ., 1957, 3, No. 2, 280-282.
704. M. Sekiya and K. Setoyama, Fluidization of particles of gypsum and related substances, Yogyo Kyokai Shi., 1957, 65, 69-75.

705. E. Singer, D. B. Todd and V. P. Guinn, Use of tracers in refinery studies catalyst mixing patterns in commercial catalytic cracking units, Ind. Eng. Chem., 1957, 49, 11-19.
706. I. D. Stauffer, Physical characteristics of supported fluidized beds, Diss. Abstrs, 1957, 17, 326.
707. D. Struve, Mechanics of two-phase flow; concurrent counter-gravity flow of ideal vertical fluidized systems, Diss. Abstrs, 1957, 1521-1522.
708. K. Taceda, Studies on the behavior of the mass of mutually hindering particles in liquid by means of liquid-solid fluidized bed, Bull. Yamagata Univ. (Engng), 1957, 4, No. 2, 273-280.
709. K. Takeda, Fluidization quality of fluidized bed, Chem. Engng (Japan), 1957, 21, No. 3, 124-129.
710. I. Tesarik, Turbulent flow through fluidized beds of identical spheres, Acta technica, 1957, No. 5, 410-445.
711. A. P. Ting and R. H. Luebbers, Viscosity of suspensions of spherical and other isodimensional particles in liquid, A. I. Ch. E. Journal, 1957, 3, No. 1, 111-116.
712. C. Y. Wen and L. T. Fan, Calculate fluidized heat transfer by nomographs, Chem. Eng., 1957, 64, No. 7, 254-257.
713. U. Wesser and G. Mardus, Zum Wärmeübergang in Wirbelschichten, Chem. Ing. Techn., 1957, 29, No. 5, 332-335.
714. E. Wicke, A new process principle with the fluidized bed, Chem. Eng. Sci., 1957, 6, 160-169.
715. S. Yagi, D. Kunii, H. Masumoto and A. Kometoni, Ann. Rept. Enging Res. Inst. Fac. Enging Univ. Tokyo, 1957, 16, No. 1, 40-49.
716. G. Yasui, Characteristics of air pockets in fluidizedbeds, Diss. Abstrs, 1957, 17, 1290.
717. F. A. Zenz, Fluid catalyst design data, Petr. Refiner, 1957, 36, No. 4-11. Series of papers. Russian translation "Catalysts for Cracking in Fluidized Beds" ("Katalizatory Krekinga v Kipyashchem Sloe"), Izd. GOSINTI, Moscow, 1958.
718. F. A. Zenz, Chapter "Fluidization," 365-400, "Encyclop. Chem. Technol." edit. Kirk. Othmer. 1st. Suppl. vol., NY., 1957.

References published in 1958

719. G. A. Adamov, Measurements of the densities and specific weights of suspensions, fluidized beds, liquids and gases in ascending streams (in Russian), Izmerit. Tekhnika, 1958, No. 4.
720. L. A. Akopyan, A. N. Planovskii and A. G. Kasatkin, Problems in the design of packed columns operating in the optimum regime (in Russian), Khim. Nauka i Prom., 1958, 3, No. 6, 745-747.
721. V. S. Aliev, N. M. Indyukov, V. A. Safonov and I. S. Shevtsov, Production of highly concentrated sulfur dioxide by oxidizing roasting pyrites in a fluidized bed (in Russian), in: Sb. Tr. Az. Nauchno-Issl. In-ta po Perebotke Nefti, No. 3, Aznefteizdat, 1958.
722. A. R. Babenko and V. I. Smirnov, A study of the process of sulfatization of copper and zinc during roasting in fluidized beds (in Russian), Trudy U.P.I., 1958, Sb. 73.

723. A. R. Babenko and V. I. Smirnov, The theory of sulfating roasting of collective concentrates in fluidized beds (in Russian), Byull. Tsvet. Met., 1958, No. 22.

724. I. V. Babina and A. D. Besser, A method of roasting zinc concentrates in fluidized beds with simultaneous removal of lead and cadmium (in Russian), Tsvet. Met., 1958, 31, No. 12, 20-24.

725. E. A. Banist and P. N. Platonov, Efflux of granular materials (in Russian), Izvestiya VUZov, Pishchevaya Tekhnologiya, 1958, No. 5.

726. Ya. F. Batishchev, An Investigation of Heat Transfer from Walls to Gas in Packed Tubes (in Russian), Kand. Dissertat., Moskovskii Institut Khimicheskogo Mashinostroeniya, Moscow, 1958.

727. R. S. Bernshtein, V. V. Pomerantsev and S. L. Shagalova, A generalized method for calculating the aerodynamic resistance of packed beds (in Russian), in: Sb. "Voprosy aerodin. i teploper. v kot.-top. prots.," Gosenergoizdat, 1958.

728. A. K. Bondareva, Particle motion and heat transfer in suspended beds (in Russian), Kand. Dissertat., Leningrad, 1958.

729. V. M. Borishanskii, Resistance and motion of air through beds of spheres (in Russian), in: Sb. "Vorposy aerodin. i teploper. v kot.-top. prots.," Gosenergoizdat, 1958, 290-298.

730. I. A. Burovoi, Automation of roasting processes in fluidized beds (in Russian), in: Sb. Materialov po Avtomatisatsii Proizvod. Protsessov, 1958, No. 3.

731. M. S. Verteshev, A Study of the Hydrodynamics and Mass Transfer in a Batchwise Process for Dissolving Salts in a Monodisperse Suspended Bed (in Russian), Kand. Dissertat., Novocherkassk. Politekhn. In-t, Novocherkassk, 1958.

732. Z. N. Galachalova, P. N. Fedoseev and V. V. Ferrein, The effects of variable drying regimes for grains on the physiological processes in seeds and the planting of wheat (in Russian), Vestn. Sel'khoznauki, 1958, 3, No. 10, 137-142.

733. A. M. Gasparyan and A. A. Zaminyan, The mechanism of falling particles in viscous media, Dokl. Akad. Nauk ArmSSR, 1958, 26, No. 1, 39-46.

734. N. I. Gel'perin, V. Ya. Kruglikov and V. G. Ainshtein, Heat transfer between a fluidized bed and the surface of a single tube in longitudinal and transverse gas flow (in Russian), Khim. Prom., 1958, No. 6, 358-363.

735. V. D. Goroshko, R. B. Rozenbaum and O. M. Todes, Approximate hydraulic relationships for suspended beds and hindered fall (in Russian), Izvestiya VUZov, Neft' i Gaz, 1958, No. 1, 125-131.

736. Kh. A. Grigoryan, Z. E. Aliev, A. M. Kuliev and N. A. Agakishiev, An investigation of the transport of a granular adsorbent in a dense bed (in Russian), in: Sb. Tr. Azerb. Nauchno-Issl. In-ta po Pererabot. Nefti, 1958, No. 3, 399-429.

737. E. M. Gutman, An investigation of diffusional transfer in capillary-porous media (in Russian), Inzh.-Fiz. Zhurn., 1958, 1, No. 10, 62-68.

738. M. A. Kodusov, I. D. Kopylov, N. D. Nikol'skaya, et al., Fluidized bed method in the mercury, antimony, tungsten,

molybdenum and cobalt industries (in Russian), Tr. In-ta "Gipronikel'," 1958, No. 3.

739. S. V. Donskov, Heat transfer between tube bundles and transverse streams of powdery material (sand) (in Russian), Teploenergetika, 1958, No. 11, 76-81.

739a. F. G. Egorov and G. M. Rakoch', A graphico-analytical method for controlling the roasting of sulfide concentrates in fluidized beds (in Russian), Byull. Tsentr. In-ta Informatsii Tsvetmet., 1958, No. 18, 119.

740. L. N. Erkova and N. I. Smirnov, Suspended beds of solid particles and their laws (in Russian), Tr. Leningradsk. Tekhn. In-ta im. Lensoveta, 1958, No. 45.

741. S. S. Zabrodsky, The fundamental laws of heat transfer in fluidized beds (in Russian), Inzh.-Fiz. Zhurn., 1958, 1, No. 3, 40-51.

742. S. S. Zabrodsky, Analysis of experimental data on heat transfer in fluidized beds (in Russian), Inzh.-Fiz. Zhurn., 1958, 1, No. 4, 22-30.

743. S. S. Zabrodsky, Heat Transfer to Fluidized Beds of Granular Materials (in Russian), Dokt. Dissertat., Akad. Nauk BSSR, Minsk, 1958.

744. S. S. Zabrodsky, Heat transfer to fluidized beds of granular materials (in Russian), Tr. In-ta Energetiki Akad. Nauk BSSR, No. 8, 1958.

745. A. A. Zaminyan, Hindered Fall and Vertical Movement of Suspensions (in Russian), Kand. Dissertat., Leningradsk. Tekhn. Institut im. Lensoveta, Leningrad, 1958.

746. A. N. Zelikman, V. I. Bibikova, et al., An investigation of the behavior and reduction of rhenium in the roasting of molybedenite concentrates in fluidized bed furnaces (in Russian), Tsvet. Met., 1958, 31, No. 11, 47-52.

747. G. S. Zelinskii and P. N. Platonov, Aerodynamics of powder beds (in Ukrainian), Dokl. Akad. Nauk USSR, 1958, No. 2.

748. A. Zelnik, An Investigation of Convective Heat Transfer in a Gas Stream Containing Suspended Particles of a Solid Material (in Russian), Kand. Dissertat., Moskovskii Institut Khimicheskogo Mashinostroeniya, Moscow, 1958.

749. I. M. Ismailov and G. N. Abaev, "Hyperflow" systems for controlling transport (in Russian), in: Sb. Tr. Azerb. Nauchno-Issl. In-ta Neft. Prom., 1958, No. II, 318-324.

750. B. V. Kantorovich, Fundamentals of the Theory of Combustion and Gasification of Solid Fuels (in Russian), Izd. Akad. Nauk SSSR, 1958.

751. N. M. Karavaev, I. Ya. Pil'skii and I. A. Lytkin, The coking of coal tar residues in coke beds (in Russian), Koks i Khimiya, 1958, 10.

752. Kim Gen Chon and Kim Khon Sen, The motion of solid particle-air systems in vertical tubes, Khvakhak ka khvakhak konop, 1958, No. 2, 81-95.

753. D. K. Kollerov and V. A. Zhitenskaya, The pressure drop of gas streams in beds of pulverized oil shale (in Russian), Khim. i Tekhnol. Topl. i Masel, 1958, No. 7, 15-21.

754. V. D. Korchagin, The effect of bed structures on its hydraulic properties (in Russian), Izvestiya VUZov, Pishchevaya Tekhnologiya, 1958, No. 3, 124-127.
755. V. P. Kramskoi, Elements of the Theory of the Processes of Cracking and Reforming over Circulating Fine-Grained Contact-Catalytic Materials (in Russian), Aznefteizdat, Baku, 1958.
756. S. A. Kruglov, Heaters with granular heat transfer media (problems in their design and use) (in Russian), Tr. Mosk. Neft. In-ta, 1958, No. 23, 101-115.
757. M. G. Kryukova, Some problems in heat transfer between a gas and particles (in Russian), Inzh.-Fiz. Zhurn., 1958, 1, No. 4, 10-16.
758. I. N. Kuz'minykh, E. L. Yakhontova, A. I. Rodionov and E. I. Ermakova, Drying superphosphates in fluidized beds (in Russian), Izvestiya VUZov, Khimiya i Khim. Tekhnol., 1958, No. 3.
759. L. A. Kul'skii, A. M. Koganovskii and M. I. Rybchinskii, Countercurrent adsorbers with double-decked suspended beds (in Russian), Khim. Prom., 1958, No. 8.
760. Yu. P. Kurochkin, A Study of Heat Transfer Between a Heated Tube and a Stream of Granular Material (in Russian), Kand. Dissertat., Vsesoyuznyi Teplotekhnicheskii Institut, Moscow, 1958.
761. E. S. Lev, Gas permeation through beds of free-flowing materials (state of the problem) (in Russian), in: Sb. "Vopr. Aerodin. i Teploper. v Kot.-top. Protsessakh," Gosenergoizdat, 1958, 241-251.
762. R. S. Livshits and E. V. Smidovich, The effect of the regime of fluidization on the behavior of the catalytic cracking process (in Russian), Tr. Mosk. Neft. In-ta, 1958, No. 3.
763. I. L. Lyuboshits, Artifical drying of milled peat (in Russian), Tr. In-ta Energetiki Akad. Nauk BSSR, No. 5, 1958.
764. D. Ya. Mazurov, The use of fluidized beds to rapidly cool cement clinkers (in Russian), Nauchn. Soobshch. Vsesoyuzn. Nauchno-Issl. In-ta Tsement. Prom., 1958, No. 3, 7-14.
765. A. M. Malets, A. N. Ternovskaya, L. N. Chudov, et al., Reconstruction of the mechanical furnaces at the Shchelkovskii chemical plant for roasting pyrites in fluidized beds (in Russian), Khim. Prom., 1958, No. 3, 18-22.
766. Yu. Mamedaliev, V. S. Aliev and S. E. Sultanov, Alkylation of benzene in an experimental apparatus with a fluidized bed of fine circulating catalyst under pressure (in Russian), Doklad. Akad. Nauk AzSSR, 1958, 14, No. 9.
767. I. G. Martyushin and N. N. Varygin, The temperature dependence of the critical velocity in the fluidized drying of materials (in Russian), Trudy Moskovskii Instituta Khimicheskogo Mashinostroeniya, 1958, 15, No. 1, 47-53.
768. I. P. Mulhlenov and E. Ya. Tarat, The pressure drop through grids (in Russian), Zhurn. Prikl. Khim., 1958, 31, No. 4, 542-549.
769. D. I. Orochko, T. Kh. Melik-Akhnazarov and A. P. Zinov'ev, Reactors for chemical processes in fluidized beds (in Russian), Khim. Nauka i Prom., 1958, 3, No. 6, 694-703.

770. D. I. Orochko, T. Kh. Melik-Akhnazarov and G. N. Poluboyarinov, Stagewise countercurrent contact apparatus with fluidized beds of pulverized materials (in Russian), Khim. i Tekhnol. Topl. i Masel, 1958, No. 2, 22.
771. I. I. Pavlov, An investigation of the flow structure of two-phase mixtures in tubes (in Russian), Nauch, Tr. Mosk. Lesotekhn. In-ta, 1958, No. 9, 189-202.
772. A. N. Planovskii and L. A. Vlasenkov, Kinetics of continuous adsorption processes in fluidized beds (in Russian), Khim. i Tekhnol. Topl. i Masel, 1958, No. 9, 7-13.
773. V. V. Pechkovskii, S. A. Amirova and V. V. Parkacheva, An investigation of the sulfating roasting of cobalt concentrates in fluidized beds (in Russian), Nauch. Dokl. Vyssh. Shkol., Khimiya i Khim. Tekhnol., 1958, No. 3.
774. L. S. Pioro, An Investigation of Heat Exchangers with Mobile Packings (in Russian), Kand. Dissertat., Kievsk. Politekhn. In-t, Kiev, 1958.
775. A. N. Planovskii and L. A. Vlasenkov, Kinetics of continuous adsorption processes in fluidized beds (in Russian), Khim. i Tekhnol. Topl. i Masel, 1958, No. 9.
776. L. K. Pliskin, Some principles in the automatic control of processes in fluidized beds (in Russian), Dokl. Akad. Nauk AzSSR, 1958, 14, No. 4.
777. V. L. Przhetslavskii, The drying of coal and other friable materials in air chutes (in Russian), in: Sb. Materialov Vsesoyuzn. Nauchno-Tekhn. Soveshchaniya po Sushke (1956), Energeticheskaya Sektsiya, Profizdat, 1958, 241-244.
778. R. B. Rozenbaum and O. M. Todes, Theoretical analysis of the hindered fall of spheres in viscous media (in Russian), Zapiski Len. Gorn. In-ta, 1958, 36, No. 3, 16-27.
779. P. G. Romankov and N. B. Rashkovskaya, Intensification of the process of drying paste-like materials (in Russian), in: Sb. "Sushka v Khimich. i Legk. Prom.," Profizdat, 1958, 23-43.
780. G. P. Sechenov and V. S. Al'tshuler, The use of high pressures in fluidized bed gas generators (in Russian), Gazovaya Prom., 1958, No. 11.
781. B. I. Skachkov, The roasting of zinc concentrates in fluidized beds in an air medium enriched with oxygen (in Russian), Tsvet. Met., 1958, No. 2.
782. M. G. Slin'ko, The use of fluidized beds for laboratory heating and cooling in investigations (in Russian), Zhurn. Fiz. Khim., 1958, 32, No. 12, 2841-2842.
783. S. A. Sultanov, Yu. G. Mamedaliev and A. B. Terteryan, Application of fluidized bed technology to alkylation reactions (in Russian), Tr. Az. Nauchno-Issl. In-ta po Pererab. Nefti, Aznefteizdat, 1958, No. 3.
784. Yu. I. Turskii, Semicoking of coal fines in a moving bed (in Russian), Khim. i Tekhnol. Topl. i Masel, 1958, No. 12.
785. V. V. Fefilov and A. I. Sokolova, Drying wood chips in a fluidized bed (in Russian), Giprolizn. i Lesokhim. Prom., 1958, No. 7.
786. G. M. Fialko, Automatic stabilization of the sulfur dioxide concentration in furnaces for the pulverized roasting of pyrites (in Russian), Khim. Prom., 1958, No. 4, 48-50.

787. G. K. Filonenko, Relationship between the temperature and moisture content of a material during its drying (in Russian), Tr. Odessk. Tekhnol. In-ta Pishch. i Kholod. Pr-sti. Tekhnologiya Konservir., 1958, 9, No. 1, 3-11.
788. Ya. Tsiborovskii (J. Ciborowski), The Processes of Chemical Technology (in Russian), Goskhimizdat, 1958.
789. Ya. Tsiborovskii and Ya. Roshak (J. Ciborowski and J. Roszak), An investigation of heat transfer between a heated surface and a fluidized bed (in Russian), Inzh.-Fiz. Zhurn., 1958, 1, No. 12, 8-17.
790. Ya. Tsiborovskii and A. Seletskii (J. Ciborowski and A. Selecki), Some data on liquid desorption from particle surfaces in fluidized beds (in Russian), Zhurn. Prikl. Khim., 1958, 31, No. 4, 518-525.
791. S. A. Tsukerman, Powder Metallurgy (in Russian), Izd. Akad. Nauk SSSR, 1958.
792. M. S. Sharlovskaya, An investigation of heat transfer in fluidized beds by the unsteady-state-regime method (in Russian), Izv. Sib. Otd. Akad. Nauk SSSR, 1958, No. 7, 62-74.
793. M. S. Sharlovskaya, Heat transfer and hydrodynamics in the transitional zone of a fluidized bed of granular material (in Russian), Izv. Sib. Otd. Akad. SSSR, 1958, No. 10, 88-95.
794. N. A. Shakhova, Multichamber drying of free-flowing materials in fluidized beds (in Russian), Tr. Moskovskii Instituta Khimicheskogo Mashinostroenie, 1958, 15, No. 1, 65-68.
795. V. V. Shibanovas, The pressure drop in granular fluidized beds (in Russian), Inzh.-Fiz. Zhurn., 1958, 1, No. 2, 99-102.
796. L. A. Shilov, An Investigation of the Combustion of Shale Coke in Fluidized Beds (in Russian), Kand. Dissertat., Leningradskii Tekhnologicheskii In-t im. Lensoveta, Leningrad, 1958.
797. R. I. Shkadov, The motion of suspended particles in a gas stream (in Russian), Zapiski Len. Gorn. In-ta, 1958, 36, No. 3.
798. A. D. Shtokarev, Equipment for roasting concentrates and schists in fluidized beds (in Russian), Prom. Ekon. Byull. (Sverdlovsk. Sovnarkhoz.), 1958, No. 6.
799. I. A. Yavorskii, Aerodynamics of fluidized beds with solid phases (in Russian), Izv. Sib. Otd. Akad. Nauk SSSR, 1958, No. 6, 104-115.
800. I. A. Yavorskii, The aerodynamics of streams in disperse media (in Russian), Izv. Vost. Fil. Akad. Nauk SSSR, 1958, No. 4.
801. I. A. Yantovskii, The drying of industrial coals (in Russian), in: Sb. Mater. Vsesoyuzn. Nauchno-Tekhn. Soveshchaniya po Sushke, Energetich. Sektsiya, Profizdat, 1958, 169-182.
802. K. Andersson, Extension of Ergun pressure-drop equation to ideally fluidized beds, T. V. F. (Strockholm), 1958, 29, 255-256.
803. P. J. Bakker and P. M. Heertjes, Porosity measurements in a fluidized bed, Brit. Chem. Eng., 1958, 3, 240-245.
804. O. Beber, K. Watson and B. Woertz, Apparatus for low velocity fluidization, U.S.A. Pat. 2,856,273; 14.10.58.
805. J. Bena, Rovnomerna fluidizacia gul'ovych castic, Chem. Prumysl., 1958, 8, No. 10, 516-523.
806. J. Beranek, Fluidizing plant and its design, Brit. Chem. Engng, 1958, 3, No. 7, 358-363.

807. T. Blickle, B. Cegledi and E. Kiss, Szemcses anyagok szaritasa fluidizalt retegben, Magyar Kemikusok Lapja, 1958, 13, 413-417.
808. S. Bracale, A. Cabella and L. Massimilla, Sul mescolamento della fase gassosa nei sistemi solido-gas fluidizzati, Chim. Ind. (Milan), 1958, 40, 621-627.
809. S. Bretsznajder, L. Lesniewicz and M. Jaszczak, Studia nad procesami wibracyjnymi w warstwie sypkiej, Chem. Stosow., 1958, 2, No. 3, 259-273.
810. W. Brötz, Grundriss der chemischen Reaktionstechnik, Weinheim, 1958, 86-117.
811. J. J. Carberry and R. H. Bretton, Axial dispersion of mass in flow through fixed beds, A. I. Ch. E. Journ., 1958, 4, No. 3, 367-375.
812. J. Ciborowski and J. Roszak, Poglad na problematike cieplna techniki fluidizacyjnej, Przem. chem., 1958, 37, No. 2, 103-108.
813. J. Ciborowski and J. Roszak, O przenikaniu ciepla wukladzie fluidalnym. I. Dynamika ladunku gruboziarnistego w podwyzszonych temperaturach i wysokich warstwach, Chem. stosow., 1958, 2, No. 3, 305-317.
814. J. Ciborowski and A. Selecki, Nektore dane dotyczace odparowania cieczy na ladunku fluidalnym, Chem. stosow., 1958, 2, No. 2, 135-145.
815. J. Ciborowski and S. Wronski, Redukcja siarczanu sodu wodorem w ladunku fluidalnym, Przem. Chem., 1958, 37, No. 3, 520-522.
816. J. D. Corney, Fluidized handling of Fullers earth, Ind. Chemist, 1958, 34, No. 406, 656-658.
817. M. Cox, A fluidized bed adsorbent airdrying plant, Trans. Inst. Chem. Eng., 1958, 36, No. 1, 29.
818. T. Czarnota, A. Wolny and J. Ciborowski, O lacznym mieszaniu i suszeniu soli metoda fluidyzacji, Przem. Chem., 1958, 37, No. 9, 583-589.
819. R. G. Deissler and J. S. Boegli, An investigation of effective thermal conductivities of powders in various gases, Trans. ASME, 1958, 80, No. 7, 1417-1423, Discuss. 1423-1425.
820. D. D. Dunlop, L. I. Griffin and J. F. Moser, Particle size control in fluid coking, Chem. Eng. Progr., 1958, 54, No. 8, 39.
821. E. Ebach and R. White, Mixing of fluids flowing through beds of packed solids, A. I. Ch. E. Journ., 1958, 4, No. 2, 161-169.
822. R. Emerick, Fluidization, Carnegie Tech., 1958, 22, No. 6, 34-36, 49.
823. C. G. Frye, W. C. Lake and H. C. Eckstrom, Gas-solid contacting with ozone decomposition reaction, A. I. Ch. E. Journ., 1958, 4, No. 4, 403-408.
824. J. Furukawa and T. Omae, Liquid-like properties of fluidized systems, Ind. Eng. Chem., 1958, 50, No. 5, 821-828.
825. M. B. Glaser and G. Thodos, Heat and momentum transfer in the flow of gases through packed beds, A. I. Ch. E. Journ., 1958, 4, No. 1, 63-68.
826. K. Gorling, S. Wallden and J. Svenske, Swed. Pat. 163,430, 10.07.1958.

827. J. Happel, Viscous flow in multiparticle systems: slow motion of fluids relative to beds of spherical particles, A. I. Ch. E. Journ., 1958, 4, No. 2, 197-201.
828. R. Jottrand, Calculation of terminal falling velocity of particles in fluids, Brit. Chem. Engng, 1958, 3, 143-145.
829. Y. Kato and T. Takemura, Elutriation of solid particles from fluid-liquid-solid systems in a continuous reactor, Repts. Fac. Engng, Yamanaushi Univ., 1958, No. 9, 39-42.
830. J. Klassen and P. E. Gishler, Heat transfer from column wall to bed in spouted, fluidized and packed systems, Can. Journ. Chem. Engng, 1958, 36, 12-19.
831. R. Lemlich and J. Caldas, Jr., Heat transfer to a liquid fluidized bed, A. I. Ch. E. Journ., 1958, 4, No. 3, 376-380.
832. H. L. List, Heat transfer to flowing gas-solid mixtures, Diss. Abstrs, 1958, 18, 1750.
833. L. Madonna and R. Lama, The derivation of an equation for predicting minimum spouting velocity, A. I. Ch. E. Journ., 1958, 4, No. 4, 497.
834. G. Matz, Fliessbett-Sublimation-ein neues Trenn-Verfahren, Chem.-Ing. Techn., 1958, 30, No. 3, 319-329.
835. A. H. Maude and R. L. Whitmore, Generalized theory of sedimentation, Brit. J. Appl. Phys., 1958, 9, 477-480.
836. A. B. Metzner and W. L. Friend, Theoretical analogies between heat, mass and momentum transfer, and modifications for fluids of high Prandtl or Schmidt numbers, Can. Journ. Chem. Engr, 1958, 36, 235-240.
837. S. D. Raseev, S. Feyer, P. Valter and L. Verman, Hidraulica stratului fluidizat. II. Verificarea experimentala pentru fluide neocompresible particule omogene si lipsa aportului des solid in strat, Lucr. Inst. petrol. gas, Bucuresti, 1958, 4, 121-151.
838. S. D. Raseev, Hidraulica stratului fluidizat. III. Verificare experimentala pentru fluide compresible si lipsa aportului de solid in strat, Lucr. Inst. petrol. gas, Bucuresti, 1958, 4, 153-174.
839. W. J. Rice and R. H. Wilhelm, Surface dynamics of fluidized beds and quality of fluidization, A. I. Ch. E. Journ., 1958, 4, No. 4, 423-429.
840. J. Richardson and A. Bakhtiar, Mass transfer between fluidized particles and gas, Trans. Instn. Chem. Engrs, 1958, 36, No. 4, 283-295.
841. J. F. Richardson and A. E. Mitson, Heat transfer from a liquid-fluidized system to a tube wall, Fluid Handling, 1958, No. 97, 37-40.
842. J. F. Richardson and A. E. Mitson, Sedimentation and fluidization. Part II. Heat transfer from a tube wall to a liquid-fluidized system, Trans. Inst. Chem. Engrs, 1958, 36, No. 4, 270-282.
843. E. Ruckenstein and V. Schorr, Despre modelul propus de Levenspiel si Walton pentru transferul de caldura dintre un strat fluidizat si perete, Acad. Resp. Pop. Romine, Inst. energ. Stud. cercet., 1958, No. 1, 7-13.
844. K. Sano and K. Okajima, The fluidization of pyrite cinder particles, Nippon. Kogyo Kaishi, 1958, 74, 485-491.
845. C. G. Sinclair and D. B. Robinson, Minimum fluidizing velocities for coal in air, Can. Journ. Chem. Engng, April 1958, 51-58.

846. P. Smith, Jr. and M. Winkle, Discharge coefficients through perforated plates at Reynolds numbers of 400 to 3000. A. I. Ch. E. Journ., 1958, 4, 266-268.
847. D. Struve, L. Lapidus and J. Elgin, The mechanics of moving vertical fluidized systems, III. Application to cocurrent countergravity glow, Can. Journ. Chem. Engng, 1958, 36, No. 4, 141-152.
848. Gh. Suciu and V. Schorr, Asupra unor aspecte ale fluidizarii stationare, Revista Chimie, 1958, 9, No. 9, 482-490.
849. P. Szolcsanyi, Lazitott agyak mechanikaja buborekmodell alapjan, Magyar kem. folyoirat, 1958, 64, No. 4, 131-136.
850. H. Vreedenberg, Heat transfer between a fluidized bed and a horizontal tube, Chem. Eng. Sci., 1958, 9, No. 1, 52-60.
851. N. Wakao, T. Oshima and S. Yagi, Mass transfer from packed beds of particles to a fluid, Chem. Engng (Japan), 1958, 22, No. 12, 780-785.
852. N. Wakao, T. Oshima and S. Yagi, Axial turbulent diffusion coefficients in water-fluidized beds, Chem. Engng (Japan), 1958, 22, No. 12, 786-789.
853. R. R. Wells, M. M. Johnson and F. T. Sterling, Recovering mercury from cinnabar-stibnite ore by flotation and fluidized-bed roasting, U. S. Bur. Min. Rept. Invest., 5433, 19 pp. 1958.
854. L. Wender and G. T. Cooper, Heat transfer between fluidized-solids beds and boundary surfaces-correlation of data, A. I. Ch. E. Journ., 1958, 4, No. 1, 15-23.
855. E. Wicke, Neue Anwendung von Wirbelschichten, Umschau, 1958, 58, No. 2, 51-53.
856. G. Yasui and L. N. Johanson, Characteristics of gas pockets in fluidized beds, A. I. Ch. E. Journ., 11958, 4, No. 4, 445-452.
857. F. A. Zenz and N. A. Weil, A theoretical-empirical approach to the mechanism of particle entrainment from fluidized beds, A. I. Ch. E. Journ., 1958, 4, No. 4, 472479.

References published in 1959

858. A. Ya. Averbukh, The use of fluidized bed processes in the technology of organic materials (in Russian), Tr. Leningradsk. Tekhn. In-ta, im. Lensoveta, 1959, No. 54, 63-70.
859. V. A. Aglitskii, et al., Some features of the technology of treating combined copper-zinc concentrates in the Mid-Urals copper-smelting plants (in Russian), Tsvet. Met., 1959, No. 9.
860. V. G. Ainshtein, An Investigation of Heat Transfer Processes Between Fluidized Beds and Single Tubes Submerged in the Beds (in Russian), Kand. Dissertat., M.I.T. Kh. T. im. Lomonosova, Moscow, 1959.
861. V. S. Aliev, A. F. Azizov, A. P. Kasimova and Sh. K. Kyazimov, An investigation of the contact-catalytic process for converting alcohol into divinyl in fluidized beds of pulverized catalyst (in Russian), Az. Khim. Zhurnal, 1959, No. 3.
862. V. S. Aliev, A. F. Azizov, A. P. Kasimova and Sh. K. Kyazimov, Contact catalytic conversion of ethyl alcohol into divinyl in

fluidized beds of pulverized catalysts in continuously operating equipment (in Russian), Az. Khim. Zhurnal, 1959, No. 4.

863. V. S. Aliev, N. M. Indyukov and M. I. Rustamov, Determination of the critical velocity for forming a fluidized bed of heat transfer agent (in Russian), Khim. i Tekhnol. Topl. i Masel, 1959, No. 8, 65-67.

864. V. S. Aliev, A. P. Kasimova and Sh. K. Kyazimov, An investigation and development of a process for dehydrogenation of n-butane to butylene in a fluidized bed of pulverized chrome-aluminum catalyst (in Russian), Az. Neft. Khoz., 1959, No. 7.

865. V. S. Al'tshuler and G. P. Sechenov, Some conditions for the normal operation of fluidized bed gas generators (in Russian), Tr. In-ta Goryuch. Iskopaemykh Akad. Nauk SSSR, 1959, 11, 139-147.

866. K. A. Andrianov, S. A. Golubtsov, N. N. Tishina and I. V. Trofimova, Direct synthesis of phenyltrichlorsilane in a fluidized bed (in Russian), Zhurn. Prikl. Khim., 1959, 32, No. 1.

867. K. A. Andrianov, S. A. Golubtsov and I. V. Trofimova, The reaction of silicon with hydrogen chloride in a fluidized bed (in Russian), Zhurn. Prikl. Khim., 1959, 32, No. 7.

868. K. A. Andrianov, S. A. Golubtsov, I. V. Trofimova and N. P. Lobusevich, Direct synthesis of methylchlorsilane in a fluidized bed (in Russian), Zhurn. Prikl. Khim., 1959, 32, No. 10.

869. V. N. Anokhin, D. G. Traber, I. P. Mukhlenov and E. S. Rumyantseva, Conversion of carbon monoxide in a suspended bed of catalyst (in Russian), Tr. Leningradsk. Tekhn. In-ta im. Lensoveta, 1959, No. 54, 37-46.

870. I. V. Babina, Metallurgy of the Nonferrous Metals (in Russian), Metallurgizdat, 1959.

871. I. V. Babina, A. D. Besser, E. I. Alyushin, et.al., Roasting of zinc concentrates in fluidized beds with removal of the lead and cadmium and consolidation of the calcined material (in Russian), Tsvet. Met., 1959, No. 6.

872. A. P. Baskakov, Z. Sporysh and Chü-Tse-Hsiang, Determination of heat transfer coefficients from solid heat transfer media to charges (in Russian), in: "Trudy Mezhvuz. Konfer. po Energotekhn. Ispol'zov. i Rats. Met. Szhig. Melkozern. Topl.," U.P.I., 1959, 108-114.

873. A. P. Baskakov and N. I. Syromyatnikov, A simplified method for calculating the heating time of materials in fluidized beds (in Russian), Izvestiya VUZov, Energetika, 1959, No. 8, 75-81.

874. Ya. Beranek and D. Sokol, The theory of fluidized beds (in Russian), Khim. Prom., 1959, No. 2, 61-68.

875. N. N. Bogdanov, et al., The Gasification of Milled Peat (in Russian), Gosenergoizdat, 1959.

876. B. I. Bondarenko, Catalytic Cracking Plants (in Russian), Gostoptekhizdat, 1959.

877. N. N. Varygin and I. G. Martyushin, A calculation of heat transfer surface area in fluidized bed equipment (in Russian), Khim. Mashinostroenie, 1959, No. 5, 6-9.

878. P. P. Budnikov, A. I. Polinkovskaya, A. A. Benuni and G. A. Petrikhina, The expansion of clays and volcanic materials in fluidized beds (in Russian), Stroit. Materialy, 1959, No. 9.

879. I. A. Burovoi, The automatic control of the roasting regime of pyrites in fluidized beds (in Russian), in: Metallurgiya Tsvet. Met., Metallurgizdat, 1959.

880. I. A. Vakhrushev, An Investigation of Heat Transfer Between Gaseous and Solid Phases When Both Phases are in Motion (in Russian), Kand. Dissertat., Vses. Nauchno-Issl. In-t po Pererab. Nefti i Gaza i Poluch. Iskusstv. Zhidk. Topl., Moscow, 1959.

881. L. A. Vlasenkov, An Investigation of the Kinetics and Process of Continuous Adsorption in Fluidized Beds of Fine-Grained Adsorbents (in Russian), Kand. Dissertat., Moskovskii In-t Khimicheskogo Mashinostroeniya, Moscow, 1959.

882. L. A. Vlasenkov and A. N. Planovskii, An investigation of the process of continuous adsorption in a fluidized bed of fine-grained adsorbent (in Russian), in: Khim. i Tekhn. Pererabotki Nefti, Gostoptekhizdat, 1959.

883. V. F. Volkov, Utilization of the residues from gas-generating plants—tars formed by their gasification in fluidized beds (in Russian), in: "Trudy Mezhvuz. Konfer.," U.P.I., 1959, 234-236.

884. I. M. Vostryakov, Roasting carbonaceous pyrites in a fluidized bed (in Russian), Khim. Prom., 1959, No. 7.

885. D. Yu. Gamburg, I. I. Ryabtsev, A. P. Kotkovskii and S. A. Shchepakov, Gasification of milled peat in fluidized bed gas generators (in Russian), Tr. In-ta Torfa, Akad. Nauk BSSR, 1959, No. 7.

886. E. N. Glazkov and A. S. Sinakevich, Fluidized-bed roasting of arsenical gold-bearing concentrates (in Russian), Tsvet. Met., 1959, No. 8.

887. M. A. Glinkov, Fundamentals of the General Theory of the Thermal Operation of Furnaces (in Russian), Metallurgizdat, 1959.

888. Z. R. Gorbis, V. I. Zhidko and G. S. Zelinskii, An investigation of the aerodynamics of fluidized beds of grain (in Russian), Izvestiya VUZov, Pishchevaya Tekhnologiya, 1959, No. 2, 110-115.

889. Kh. A. Grigoryan, A. M. Kuliev, Z. E. Aliev and B. A. Pinsker, An investigation of the process of gas separation in fluidized adsorbent beds (in Russian), in: Sb. Tr. In-t Neftekhim. Prots., Akad. Nauk AzSSR, 1959, No. 4.

890. V. M. Dement'ev, Problems in the aerodynamics of fluidized beds (in Russian), Teploenergetika, 1959, 6, No. 1, 50-56.

891. V. M. Dement'ev, The design of multilayer furnaces for roasting limestone in fluidized beds (in Russian), Inzh.-Fiz. Zhurn., 1959, 2, No. 12, 31-37.

892. S. S. Zabrodsky, The effectiveness of heat conduction in a fully developed fluidized bed (in Russian), Inzh.-Fiz. Zhurn., 1959, 2, No. 1, 32-38.

893. I. I. Ioffe and A. F. Grigorov, The theory of reaction rates in fluidized beds of catalyst (in Russian), Khim. Prom., 1959, No. 3, 243-250.

894. I. N. Kitler and Yu. A. Lainer, The use of fluidized beds in the aluminum industry (in Russian), Byull. Tsvet. Met., 1959, No. 20.

895. M. A. Klevleev and A. I. Skoblo, Drying suspensions in fluidized beds and in suspension (in Russian), in: Khim. i Tekhn. Pererabotki Nefti, Gostoptekhizdat, 1959.

896. T. A. Kozeiko and I. I. Sidorchuk, Some characteristics of distributors for contacting in fluidized beds (in Russian), Az. Neft. Khoz., 1959, No. 8, 40-41.

897. A. A. Komarovskii and M. S. Verteshev, The rate of mass transfer during dissolution in stationary and suspended beds (in Russian), Khim. Prom., 1959, No. 6, 530-533.

898. N. B. Kondukov, D. P. Dubrovskaya, et al., Vapor-phase separation of benzene from sulfur compounds in a stream of coke oven gas with a fluidized adsorbent (in Russian), Koks i Khimiya, 1959, No. 10.

899. V. I. Korshunov and A. Z. Yurovskii, The physical properties of air-suspensions as heavy dry media for gravitational beneficiation (preliminary communication) (in Russian), Izv. Akad. Nauk SSSR, Otd. Tekhn. Nauk, Metallurg. i Topl., 1959, No. 1, 115.

900. S. P. Kosarev, Horizontal recuperative furnaces for heating and subsequent cooling of free-flowing materials in fluidized beds, Avt. Svid. SSSR, No. 116245, 19.01.59.

901. V. Ya. Kruglikov, M. A. D'yakonov, V. D. Ainshtein and A. V. Volynskii, Fluidized sand bed heaters, Byull. Gosudarstvenn. In-ta Prikladn. Khimii, Leningrad, 1959.

902. M. A. Kuvakin, The calcination of coke in fluidized beds (in Russian), Byull. Tsvet. Met., 1959, No. 15.

903. E. F. Kurgaev, Diffusion in fluidized beds (in Russian), Inzh.-Fiz. Zhurn., 1959, 2, No. 4, 3-7.

904. G. Ya. Leizerovich, I. V. Babina and E. Ya. Serebrenikova, Roasting of Zinc Concentrates in Fluidized Beds (in Russian), Metallurgizdat, 1959.

905. I. G. Lesokhin, The use of pyrites cinders for producing pulverulent iron (in Russian), Trudy Leningradsk. Tekhn. In-ta im. Lensoveta, 1959, No. 54, 82-93.

906. I. G. Lesokhin, Mass transfer during the combusion of iron sulfides in suspended beds (in Russian), Trudy Leningradsk. Tekhn. In-ta im. Lensoveta, 1959, No. 54, 71-81.

907. I. G. Lesokhin, Problems in producing pulverulent iron from pyrites cinders in suspended beds (in Russian), Zhurn. Prikl. Khim., 1959, 32, No. 1, 60-65.

908. D. I. Lisovskii and N. S. Krysenko, Reducing-sulfating roasting of oxidized nickel ores in fluidized beds (in Russian), Izvestiya VUZov, Tsvet. Met., 1959, No. 6.

909. P. I. Luk'yanov, I. V. Gusev and N. I. Nikitina, The contact pressure of moving beds of granular materials against the walls of shaft-type equipment (in Russian), Khim. i Tekhnol. Topl. i Masel, 1959, No. 1, 63-68.

910. A. V. Lykov (Luikov) and Yu. A. Mikhailov, The Theory of Energy and Mass Transfer (in Russian), Izd. Akad. Nauk BSSR, 1959.

911. M. V. Lykov (Luikov), Methods for the approximate thermal design of fluidized bed dryers (in Russian), Inzh.-Fiz. Zhurn., 1959, 2, No. 3, 10-18.

912. A. M. Malets, An investigation of roasting pyrites in a fluidized bed (in Russian), Khim. Prom., 1959, No. 1.

913. A. M. Malets, A. N. Ternovskaya and L. N. Churov, Reconstruction of the mechanical furnaces in the Shchelovsk. chemical complex for the roasting of pyrites in fluidized beds (in Russian), Khim. Prom., 1959, No. 3.
914. M. F. Maslovskii and N. A. Shakhova, The drying of suspensions and solutions in fluidized beds of inert granular materials (in Russian), Khim. Mashinostr., 1959, No. 6, 27-29.
915. V. K. Mikhailov and G. Ya. Leizerovich, Metallurgy of the Nonferrous Metals (in Russian), Metallurgizdat, 1959.
916. I. P. Mukhlenov, Phase interactions and classifications of two-phase disperse systems used in the chemical industry (in Russian), Trudy Leningradsk. Tekhn. In-ta im. Lensoveta, 1959, No. 54, 5-13.
917. I. P. Mukhlenov, A. Ya. Averbukh and E. S. Tumarkina, Applications of the foam method of contacting liquids and gases in organic technology (in Russian), Trudy Leningradsk. Tekhn. In-ta im. Lensoveta, 1959, No. 54, 125-128.
918. I. P. Mukhlenov, T. N. Rozova and L. S. Lazarev, Removal of dust from gases in foam gas washers (in Russian), Trudy Leningradsk. Tekhn. In-ta im. Lensoveta, 1959, No. 54, 94-102.
919. I. P. Mukhlenov, D. G. Traber and E. S. Rumyantseva, Kinetics of the oxidation of sulfur dioxide in a suspended catalyst bed (in Russian), Trudy Leningradsk. Tekhn. In-ta im. Lensoveta, 1959, No. 54, 53-62.
920. I. P. Mukhlenov, D. G. Traber and V. S. Sarkits, The effects of hydrodynamic factors on heat transfer from a suspended bed to a heat transfer surface (in Russian), Trudy Leningradsk. Tekhn. In-ta im. Lensoveta, 1959, No. 54, 24-37.
921. I. P. Mukhlenov, D. G. Traber, V. B. Sarkits and T. P. Bondarchuk, Heat transfer from fluidized catalyst beds to heat transfer surfaces (in Russian), Zhurn. Prikl. Khim., 1959, No. 6, 1291-1295.
922. I. P. Mukhlenov and E. S. Tumarkina, The effects of surface tension on the hydrodynamics of suspended beds of liquids (foam beds) (in Russian), Trudy Leningradsk. Tekhn. In-ta im. Lensoveta, 1959, No. 54, 117-124.
923. I. P. Mukhlenov, E. S. Tumarkina, K. K. Kil'shtedt, V. M. Khalepa and L. F. Nikitina, The recovery of sulfuric acid mist (in Russian), Trudy Leningradsk. Tekhn. In-ta im. Lensoveta, 1959, No. 54, 103-116.
924. E. N. Osipova, Ya. Lukesh and B. M. Bel'man. The estimation of the efficiency of fluidized bed furnaces (in Russian), Trudy Leningradsk. Tekhn. In-ta im. Lensoveta, 1959, No. 54, 129-134.
925. V. V. Pechkovskii and S. A. Amirova, A study of batchwise oxidation-sulfation roasting of sulfide ores in fluidized beds in model furnaces (in Russian), Zhurn. Prikl. Khim., 1959, 32, No. 7.
926. A. N. Planovskii, D. I. Orochko and P. I. Luk'yanov, Processes in Fluidized Beds (in Russian), Kratkaya Informatsiya Gosudarstv. Nauchno-tekhn. Komiteta Sovet Ministrov ob Itogakh Vsesoyuznoi Konferentsii po Geterogennym protsessov c Tverdoi Fazoi v Psevdoozhizhennom (kipyashchem) Sloe, Vsesoyuzn. In-t Nauchnoi i Tekhnicheskoi Informatsii, Moscow, 1959.

927. M. V. Polyakov, V. V. Shalya and Z. Z. Vysotskii, An investigation of the catalytic conversion of methanol into formaldehyde in a fluidized catalyst bed (in Russian), Zhurn. Prikl. Khim., 1959, 32, No. 10.

928. I. M. Razumov and L. I. Larionova, Calculation of the entrainment of fine-grained material from reactors with fluidized catalyst beds (in Russian), Khim. Mashinostr., 1959, No. 4, 19.

929. K. A. Reznikovich and O. M. Todes, The energy characteristics of vertical pneumatic conveyors (in Russian), Izvestiya VUZov, Neft' i Gas, 1959, No. 8.

930. A. I. Rychkov and N. A. Shakhova, Calculation of fluidizing velocities for materials of close and wide particle size distributions (in Russian), Inzh.-Fiz. Zhurn., 1959, 2, No. 9, 92-96.

931. V. B. Sarkits, Heat Transfer from Suspended Beds of Granular Materials to Heat Transfer Surfaces (in Russian), Kand. Dissertat., Leningrad, 1959.

932. V. B. Sarkits, D. G. Traber and I. P. Mukhlenov, Heat transfer from suspended catalyst beds to heat transfer surface (in Russian), Zhurn. Prikl. Khim., 1959, No. 10, 2218-2225.

933. V. B. Sarkits, D. G. Traber and E. A. Protopopova, The mechanism of heat transfer between suspended beds of granular materials and walls (in Russian), Trudy Leningradsk. Tekhn. In-ta im. Lensoveta, 1959, No. 54, 14-23.

934. E. Ya. Serebrenikova, Roasting granulated copper concentrates and schists in fluidized beds on a semiindustrial scale (in Russian), Tsvet. Met., 1959, No. 12.

935. G. P. Sechenov and V. S. Al'tshuler, Gasification of pulverized fuel in fluidized beds at pressures up to 15 atm. gauge (in Russian), Gaz. Prom., 1959, No. 11.

936. G. P. Sechenov and V. S. Al'tshuler, The effect of pressure on the behavior of fluidized beds (in Russian), Tr. In-ta Goryuchikh Iskopaemykh, Akad. Nauk SSSR, 1959, 11, 188-197.

937. G. P. Stel'makh, An approximate calculation of the velocities of particles in a suspended state (in Russian), Inzh.-Fiz. Zhurn., 1959, 2, No. 10.

938. B. F. Stepochkin, Some generalizations in the hydraulics of fluidized beds (in Russian), Khimiya i Khim. Tekhnol., 1959, 2, No. 3.

939. B. F. Stepochkin, Generalized relationships for the flow of fluids through granular beds (in Russian), Nauchn. Dokl. Vyssh. Shkoly, Khimiya i Khim. Tekhnol., 1959, No. 1, 210-214.

940. N. I. Syromyatnikov, Practical methods for the combustion of pulverized fuels (in Russian), in: Trudy Mezhvuz. Konfer., U.P.I., 1959, 115-127.

941. N. I. Syromyatnikov, The combustion of fuel in fluidized beds (in Russian), in: Trudy Mezhvuz. Konfer., U.P.I., 1959, 136-143.

942. N. I. Syromyatnikov and V. F. Volkov, Fluidized Bed Processes (in Russian), Metallurgizdat, 1959.

943. I. Tesarzhik, The theory of suspended beds of granular and flaky materials in fluid streams (in Russian), Izv. Akad. Nauk SSSR, Otd. Tekhn. Nauk, Mekhan. i Mashinostr., 1959, No. 6, 27-35.

944. K. V. Topchieva, I. P. Planovskii and V. V. Lushnikov, A study of the reaction kinetics for cracking cumene on a fluidized aluminosilicate catalyst (in Russian), Vestn. Moskovsk. Gosudarstv. Universiteta, Ser. Mat., Mekh., Astr., Fiz., Khim., 1959, No. 3.

945. D. G. Traber, I. P. Mukhlenov and E. S. Rumyantseva, Kinetics of oxidizing sulfur dioxide gas in a suspended catalyst bed (in Russian), Trudy Leningradsk. Tekhn. In-ta im. Lensoveta, 1959, No. 54, 53-62.

946. D. G. Traber, E. S. Rumyantseva and I. P. Mukhlenov, The effect of particle size of a vanadium catalyst on its activity in the oxidation of sulfur dioxide gas in a suspended bed (in Russian), Trudy Leningradsk. Tekhn. In-ta im. Lensoveta, 1959, No. 54, 47-52.

947. A. I. Fefelov and E. M. Lyapustina, The rate of roasting flotation pyrites in a suspended state (in Russian), Zhurn. Prikl. Khim., 1959, 32, No. 3.

948. V. K. Khokhlov and A. E. Myagkov, Complete roasting of clinkers in fluidized beds (in Russian), Nauchn. Soobshch. Vsesoyuzn. Nauchno-Issledovatel'sk. In-ta Tsementa, 1959, No. 6.

949. Ya. Tsiborovskii and S. Vronskii (J. Ciborowski and S. Wronski), Reduction of sodium sulfate by hydrogen in a fluidized bed (in Russian), Zhurn. Prikl. Khim., 1959, 32, No. 3, 473.

950. Ya. Tsiborovski and Ya. Roshak (J. Ciborowski and J. Roszak), An investigation of heat transfer between a heated surface and a fluidized bed (in Russian), Inzh.-Fiz. Zhurn., 1959, 2, No. 1, 15-24.

951. Ya. Tsiborovskii and Ya. Roshak (J. Ciborowski and J. Roszak), An investigation of heat transfer between a gas and solid particles in a fluidized bed (in Russian), Inzh.-Fiz. Zhurn., 1959, 2, No. 2, 3-9.

952. M. S. Sharlovskaya, Convective Heat Transfer in a Fluidized Bed (in Russian), Kand. Dissertat., Tomsk. Politekhn. In-t, Tomsk, 1959.

953. M. S. Sharlovskaya, Heat transfer in the transition zone of a fluidized bed (in Russian), in: Tr. Mezhvuz. Konfer., U.P.I., 1959, pp. 166-176.

954. K. E. B. Andersson, Pressure drop in fluidized and packed beds, Trans. Roy. Inst. Techn., Stockholm, No. 131, 25 pp. 1959.

955. Anon., Application of fluid bed reactors: a symposium, Mining World, 1959, 21, No. 4, 25-26.

956. P. J. Bakker and P. M. Heerjes, Porosity determination of fluidized beds, Brit. Chem. Engng, 1959, 4, No. 10, 524-529.

957. J. Bena, J. Ilavsky, E. Kossaczky and O. Zakutny, K niektorym problemom pri rovnomerne fluidnej vrstve gul'ovych castic, Chem. zvesti, 1959, 13, 170-185.

958. K. B. Bengtson and L. N. Johanson, High temperature clorination in a fluidized bed, Chem. Engng Progr., 1959, 56, No. 3, 49-58.

959. J. Beranek and D. Sokol, Prah fluidace nepravidelnych castic, Chem. prumysl., 1959, 9, No. 1, 5-10.

960. T. Blickle, Festkörperreaktionen in Wirbelschichten aus verpillten Festkörpern, Chem. Techn., 1959, 11, No. 3, 131-134.

961. T. Blickle and P. Kaldi, Mischung und Verweilzeit in Wirbelschichtapparaturen, Chem. Techn., 1959, 11, No. 4, 181-184.

962. H. R. Burch, Applications of fluid bed reactors-fluosolids roasting at Weed Heights, Mining World, 1959, 21, No. 4, 29.

963. E. J. Cairns and J. M. Prausnitz, Velocity profiles in packed and fluidized beds, Ind. Engng Chem., 1959, 51, No. 12, 1441.

964. R. P. Charbonier, Study of a simplified theory of gravity concentration, Trans. Can. Inst. Min. Met., 1959, 62, 199-202.

965. J. Ciborowski and J. Roszak, O przenikaniu ciepla w ukladzie fluidalnym, Chem. Stosowana, 1959, No. 3.

966. J. F. Davidson, R. C. Paul, M. J. S. Smith and H. A. Duxbury, The rise of bubbles in a fluidized bed, Trans. Instn Chem. Engrs, 1959, 37, No. 6, 323-328.

967. D. Davis, Heat transfer to externally heated beds of glass beads fluidized with air, Nomogram, Chem. Process Engng, 1959, 40, No. 2, 63.

968. D. Davis, Heat transfer to internally heated beds of glass beads fluidized with air, Nomogram, Chem. Process Engng, 1959, 50, No. 3, 105.

969. M. Ditze and W. Landau, Theoretische und konstruktive Erwägungen zum Bau eines Wirbelkokers, Chem. Techn., 1959, 11, No. 3, 125-131.

970. J. M. Dotson, Factors affecting density transients in a fluidized bed, A. I. Ch. E. Journ., 1959, 5, No. 2, 169-174.

971. E. Eckert and R. Drake, Introduction to the Transfer of Heat and Mass, 2nd edition, N. Y., McGraw-Hill, 1959, 530 pp.

972. R. Ernst, Der Mechanismus des Wärmeüberganges an Wärmeaustauscher in Fliessbetten (Wirbelschichten), Chem. Ing. Techn., 1959, 31, No. 3, 166-173.

973. L. T. Fan and C. J. Swartz, Fluidization in beds of rice and wheat, Can. Journ. Chem. Engng, 1959, 37, No. 5, 204.

974. L. T. Fan, Y. Yang and Ch. Wen, Semifluidization: mass transfer in semifluidized beds, A. I. Ch. E. Journ., 1959; 5, No. 3, 407-409.

975. J. Ferron, Gas—solid heat transfer and gas-mixing in fluidized beds, Diss. Abstrs, 1959, 19, 2556-2557.

976. H. Franz, Application of fluid bed reactors—pyrite treated in Kennecott reactor, Mining World, 1959, 21, No. 4, 30-31.

977. W. H. Gauvin, J. S. Pasternak, L. B. Torobin and L. Yaffe, Radioactive tracer technique for particle-velocity measurement in solid-gas systems, Can. Journ. Chem. Engng, 1959, 37, 95-98.

978. B. Ghosh and G. L. Osberg, Can. Journ. Chem. Engng, 1959, 37, No. 6, 205-207.

979. A. Gomezplata, Effect of fluidization on the catalytic cracking of cumene, Diss. Abstrs, 1959, 19, 2040.

980. M. Gregor and J. Bena, Fluidizacna technika, Nasa veda, 1959, 6, No. 3, 101-106.

981. A. Hamburger and W. H. Groot, Continue adsorptie in tegenstroom in een kolom met gefluidiseerde bedden, Ingenieur (Utrecht), 1959, 71, Ch. 11-Ch. 16.

982. C. C. Harris, Sedimentation rates, fluidization and flow through porous media, Nature (Engl.), 1959, 184, No. 4687, Suppl. 10, 716.

983. C. C. Harris, Correlation of sedimentation rates by dimensionless groups, Nature (Engl.), 1959, 183, 530-531.
984. P. Heertjes, La fluidizzazione, Chimica a industria, 1959, 41, No. 2, 101-108.
985. F. B. Hill and R. H. Wilhelm, Radiative and conductive heat transfer in a quiescent gas-solid bed of particles: theory and experiment, A. I. Ch. E. Journ., 1959, 5, No. 4, 486-496.
986. E. S. Howell, Sulphatising copper sulfides, Mining World, 1959, 21, No. 4, 31-32.
987. P. Johnson, J. Parsons and J. Roberts, Try fluid bed reactors for making chloromethanes, Ind. Engng Chem., 1959, 51, 499-506.
988. M. Leva, Fluidization, N. Y., 1959.
989. M. Leva and C. Y. Wen, Fluidized bed heat transfer correlation, A. I. Ch. E. Journ., 1959, 5, No. 1, 7M.
990. A. L. Loeffler and B. F. Ruth, Particulate fluidization and sedimentation of spheres, A. I. Ch. E. Journ., 1959, 5, No. 3, 310-314.
991. J. G. Lowenstein, Horizontal flow of slurries—design so solids can not settle out, Chem. Engng, 1959, 66, No. 1, 133-135.
992. L. Massimilla, V. Betta and C. Della Rocca, Sull'efflusso di sistemi fluidizzati, solido-gas attraverso fori di piccolo diametro, Ricerca Sci., 1959, 29, 1030-1039.
993. W. May, Fluidized-bed reactor studies, Chem. Engng Progr., 1959, 55, No. 12, 49-56.
994. D. McAskill, Fluosolids processing for selective roasting, Mining World, 1959, 21, No. 4, 32-33.
995. W. Muhlrad, La theorie et la practique des suspensions fluo-liquides, Génie Chimique, 1959, 81, 165-177.
996. J. O. Osburn, Predict rate of mass transfer, Chem. Engng, 1959, 66, No. 1, 136-138.
997. R. Overcashier and D. Todd, Some effects of baffles on a fluidized system, A. I. Ch. E. Journ., 1959, 5, No. 1, 54-60.
998. J. Oxley and I. Campbell, Fluidized beds for metal production, Journ. Metals, 1959, 11, 135-137.
999. H. Parker and W. Stevens, Effect of liquid on interparticle forces in gas-fluidized beds, A. I. Ch. E. Journ., 1959, 5, No. 3, 314-318.
1000. B. G. Price, L. Lapidus and I. Elgin, Mechanics of vertical moving fluidized systems, II. Application to countercurrent operation, A. I. Ch. E. Journ., 1959, 5, No. 1, 93-97.
1001. J. A. Quinn, Mechanics of two-phase flow: cocurrent cogravity flow of ideal vertical fluidized systems, Diss. Abstrs, 1959, 20, No. 6, 2192.
1002. J. F. Richardson and P. Ayers, Heat transfer between particles and a gas in a fluidized bed, Trans. Instn. Chem. Engrs, 1959, 37, No. 6, 314-322.
1003. R. C. Romano, Frequency response analysis of a fluidized bed, Diss. Abstrs, 1959, 19, 2559.
1004. J. Roszak and J. Ciborowski, Odwodnianie zolcieni zelazowej w stanie fluidalnym, Przem. chem., 1959, 38, 363-370.
1005. D. S. Scott and D. A. Guthrie, Removal of a mist in a fluidized bed, Can. Journ. Chem. Engng, 1959, 37, No. 5, 200-203.

1006. W. Siemes, Förderung von körnigem Gut in geneigten Fliessbetten, Chem. Ing. Techn., 1959, 31, 212-213.

1007. J. F. Skelly, To select a reactor use systems engineering, Mining World, 1959, 21, No. 4, 27-28.

1008. P. L. Slis, Th. W. Willemse and H. Kramer, Response of the level of a liquid fluidized bed to a sudden change in the fluidizing velocity, Appl. Sci. Research, 1959, 8A, 209-218.

1009. S. L. Smith and F. L. Stubbs, For smelting—three ways to use a fluid bed reactor, Mining World, 1959, 21, No. 6, 42-43.

1010. I. Tesarik, Vznaseni lozi zrnitych a vlockovitych castic ve vode, Rozpr. CSAV, 1959, TV, 69, No. 3.

1011. B. Thorley, I. B. Saunby, K. B. Mathur and G. L. Osberg, An analysis of air and solid flow in a spouted wheat bed, Can. Journ. Chem. Engng, 1959, 37, No. 5, 184-192.

1012. D. B. Todd, Reducing entrainment from fluidized beds, U.S.A. Pat. 2,874,480, 24.02.1959.

1013. L. B. Torobin and W. H. Gauvin, Fundamental aspects of solid-gas flow, Can. Journ. Chem. Eng., 1959, 37, 129-141.

1014. R. E. C. Weaver, L. Lapidus and I. C. Elgin, The mechanics of vertical moving liquid—liquid fluidized systems: I. Interphase contacting of droplets passing through a second quiescent fluid, A. I. Ch. E. Journal, 1959, 5, No. 4, 533-539.

References published in 1960

1015. Yu. A. Avgustov, Plastic coating of metallic objects by submersion in fluidized powders (in Russian), Khim. Mashinostr., 1960, No. 2.

1016. I. A. Aladzahalov, Decisions made by the "Giprotsvetmet" Institute in planning the technological parts of new metallurgical plants with fluidized bed roasting furnaces (in Russian), in: Sb. "Primenenie v SSSR Protsessov Obzhiga v Kipyashchem Sloe," Tsentral'n. In-t Informatsii Tsvetnoi Metallurgii, Moscow, 1960.

1017. R. A. Apriamov, Roasting of molybdenite concentrates in fluidized beds in the "Elektrotsink" plant (in Russian), Ref. as No. 1016.

1018. I. V. Babina and V. B. Kononov, Methods of treating antimony ores in fluidized beds (in Russian), Ref. as No. 1016.

1019. I. V. Babina and V. G. Marchenko, Roasting of zinc concentrates in fluidized beds with removal of lead and cadmium and the particle growth of the roasted material in the Belovsk. zinc plant (in Russian), Ref. as No. 1016.

1020. A. P. Baskakov, The effect of solids mixing in a fluidized bed on heat transfer between gas and particles (in Russian), Inzh.-Fiz. Zhurn., 1960, 3, No. 9, 10-16.

1021. I. M. Bernshtein, New schemes for deveopling automatic control systems for fluidized bed roasting processes (in Russian), Ref. as No. 1016.

1022. O. F. Bints, The roasting of nickel converter mattes in the KS experimental furnace at the "Yuzhuralnikel" combine (in Russian), Ref. as No. 1016.

1023. N. M. Boguslavskii and T. Kh. Melik-Akhnazarov, Fluidization in the Chemical Industry (in Russian), GOSINTI, Moscow, 1960.
1024. A. K. Bondareva and O. M. Todes, Thermal conduction in fluid bed heat transfer (in Russian), Inzh.-Fiz. Zhurn., 1960, 3, No. 2, 105-110.
1025. V. A. Borisevich, Experimental determination of the weight flow rate of a free-flowing material (sand) in gravity efflux from a vertical tube through throttling devices (in Russian), Inzh.-Fiz. Zhurn., 1960, 3, No. 11, 89-92.
1026. M. P. Bryantseva, Research on the granulation, drying and roasting of granulated materials Mid-Urals copper smelting complex (in Russian), Ref. as No. 1016.
1027. I. A. Burovoi, Automatic control of fluidized bed processes (in Russian), Byull. Tsvet. Met., 1960, No. 9.
1028. I. A. Burovoi, Automatic control of fluidized bed processes (in Russian), Ref. as No. 1016.
1029. I. A. Burovoi and I. A. Denisova, A simplified method for calculating the excess heat in roasting zinc concentrates in fluidized beds (in Russian), Ref. as No. 1016.
1030. I. A. Burovoi and G. Ya. Krichevskii, Development and investigation of a system for removing excess heat from fluidized beds at high thermal loads (in Russian), Ref. as No. 1016.
1031. V. T. Vdovichenko and N. P. Gadenko, The production of chloroderivatives of methane by oxychlorination of natural gas (in Russian), Gaz. Prom., 1960, No. 4.
1032. V. F. Volkov, Some problems in the hydrodynamics of fluidized beds (in Russian), in: Sb. "Voprosy Energotekhnol. Ispol'z. Topliv Sibiri," 1960, 129-136.
1033. N. I. Gel'perin, V. G. Ainshtein, E. N. Gel'perin and S. D. L'vova, The hydrodynamics of granular materials in conical and conico-cylindrical equipment (in Russian), Khim. i Tekhnol. Topliv i Masel, 1960, No. 8, 51-57.
1034. N. I. Gel'perin, V. G. Ainshtein and A. V. Zaikovskii, Apparatus with fluidized beds of free-flowing materials in centrifugal fields (in Russian), Khim. Mashinostroenie, 1960, No. 3, 1-4.
1035. N. I. Gel'perin, V. G. Ainshtein and A. I. Rad'ko, Equipment for contacting gases or liquids with granular materials in fluidized beds, Avt. Svid. SSSR No. 131683, 10.09.60.
1036. L. S. Getskii, A. G. Batyuk, V. P. Savraev, V. G. Fel'man, A. V. Bratchik and V. A. Polulyakh, The use of fluidized beds for sulfating lead powders (in Russian), Ref. as No. 1016.
1037. A. S. Ginzburg, Drying of Food Products (in Russian), Pishchepromizdat, 1960.
1038. F. Z. Grek, Methods for the continuous mixing of materials in fluidized beds, and equipment for carrying out these methods, Avt. Svid. SSSR No. 128001, 28.04.60.
1039. M. A. Grishin, An experimental investigation of the vortex drying of vegetables (in Russian), Inzh.-Fiz. Zhurn., 1960, 3, No. 5, 40-45.
1040. M. A. Grishin, The hydrodynamics of coarse granular materials in a vortex stream of air (in Russian, Inzh.-Fiz. Zhurn., 1960, 3, No. 6, 82-85.

1041. I. N. Gutsaev, Waste-heat boilers in the "Elektrotsink" plant (in Russian), Ref. as No. 1016.
1042. Kh. D. Dudiev, Exploitation of fluidized bed furnaces in the Chelyabinsk electrolytic zinc plant (in Russian), Ref. as No. 1016.
1043. V. I. Eremin, An improvement of furnace construction for roasting zinc concentrates in fluidized beds (in Russian), Ref. as No. 1016.
1044. S. S. Zabrodsky, The problem of heat transfer in fluidized beds (in Russian), Inzh.-Fiz. Zhurn., 1960, 3, No. 2, 111-114.
1045. S. S. Zabrodsky, D. I. Solomonov and V. P. Begun, The drying of milled peat in fluidized beds using low-temperature flue gases (in Russian), Inzh.-Fiz. Zhurn., 1960, 3, No. 10, 19-26.
1046. A. N. Zelikman, L. V. Belyaevskaya, O. E. Krein and V. M. Petrov, The roasting of molybdenum concentrates in fluidized beds (in Russian), Ref. as No. 1016.
1047. I. F. Zemskov, Tubes for the transfer of adsorbents in multi-stage continuously operating equipment for adsorption in fluidized beds of solid adsorbents (in Russian), Khim. Mashinostr., 1960, No. 2.
1048. I. M. Zikeev, Design of furnaces and fluidized bed equipment (in Russian), Ref. as No. 1016.
1049. V. I. Ignat'ev and N. I. Zverev, A laboratory air separator with a fluidized bed (in Russian), Teploenergetika, 1960, No. 2, 55-58.
1050. I. I. Ioffe and L. M. Pis'men, Statistical methods for studying the macrokinetics of processes in fluidized catalyst beds (in Russian), Khim. Prom., 1960, No. 4, 287-293.
1051. F. E. Keneman, N. G. Zalogin, V. N. Vorob'ev and O. S. Antoshina, The mechanism of gravity efflux of free-flowing materials (in Russian), Inzh.-Fiz. Zhurn., 1960, 3, No. 3.
1052. I. M. Kirko and M. V. Filippov, Properties of suspended beds of ferromagnetic particles in a magnetic field (in Russian), Zhurn. Tekhn. Fiz., 1960, 30, No. 9, 1081-1084.
1053. M. G. Kryukova, The rate of heat transfer between a gas and solid particles (in Russian), in: Sborn. "Energotekhnol. Ispol'z. Topliva," 1960, Part 1, 215.
1054. P. P. Lanin, The roasting of nickel converter mattes in fluidized beds at the Noril'sk complex (in Russian), Ref. as No. 1016.
1055. B. P. Levin and N. D. Nikol'skaya, The roasting of nickel and cobalt materials in fluidized beds (in Russian), Ref. as No. 1016.
1056. B. P. Levin and N. D. Nikol'skaya, Fluidized bed methods in the plans of the "Gipronikel'" Institute (in Russian), Ref. as No. 1016.
1057. F. M. Loskutov, V. S. Lovchikov and V. D. Kiseleva-Koroleva, The reduction roasting of zinc cinders in fluidized beds (in Russian), Ref. as No. 1016.
1058. A. V. Lykov (Luikov), Molar-molecular heat and mass transfer in capillary-porous bodies (in Russian), Inzh.-Fiz. Zhurn., 1960, 3, No. 10, 88-92.

1059. A. M. Malets and A. N. Ternovskaya, Development of methods for roasting pyrites in fluidized beds in the chemical industry (in Russian), Ref. as No. 1016.

1060. Yu. P. Mamedaliev, M. M. Guseinov and S. M. Mamedov, The production of carbon tetrachloride by chlorination of pentane in a fluidized catalyst bed (in Russian), Doklad. Akad. Nauk SSSR, 1960, 139, No. 2.

1061. V. K. Mikhailov, Fluidized bed roasting of mercury ores from the Nikotovskii deposits (in Russian), Byull. Tsvet, Met., 1960, No. 4.

1062. V. K. Mikhailov and G. Ya. Leizerovich, Volatilization of mercury from ores in fluidized beds (in Russian), Ref. as No. 1016.

1063. I. N. Munits, G. A. Avstreikh and V. I. Berkh, Automation of experimental fluidized bed equipment (in Russian), Ref. as No. 1016.

1064. I. P. Mukhlenov, D. G. Traber and V. B. Sarkits, Heat transfer between suspended beds of granular materials and heat transfer surfaces (in Russian), Zhurn. Prikl. Khim., 1960, 33, No. 10, 2206-2212.

1065. V. N. Petrov, Features of the hydrodynamic regimes of fluidized beds with horizontal perforated partitions (in Russian), Tr. Mosk. In-ta Neftekhim. i Gazovoi Promyshlennosti, 1960, No. 28, 102-115.

1066. O. Ya. Polotnyuk, Kinetic methods for the quantitative determination of break-through in reaction systems with various rates of mixing of the reagents (in Russian), Inzh.-Fiz. Zhurn., 1960, 3, No. 7.

1067. G. D. Rabinovich, Heat transfer in beds of granular materials (in Russian), Inzh.-Fiz. Zhurn., 1960, 3, No. 11, 18-25.

1068. R. I. Rapoport and A. A. Gurevich, Automatic level control in fluidized bed equipment (in Russian), Ref. as No. 1016.

1069. P. G. Romankov, N. B. Rashkovskaya and Z. A. Berezovskaya, New methods for dying paste-like pigments (in Russian), Lakokrasochn. Mat. i ikh Primenenie, 1960, No. 3.

1070. E. Ya. Serebrenikova, Improved roasting of zinc concentrates in fluidized beds (in Russian), Ref. as No. 1016.

1071. E. Ya. Serebrenikova, Roasting of copper concentrates and schists in fluidized beds (in Russian), Ref. as No. 1016.

1072. I. I. Sidorchuk, The problems of elevated temperatures in the fluidized beds of cracking catalyst regenerators (in Russian), Az. Neft. Khoz., 1960, No. 1.

1073. A. S. Sinakevich, A. D. Vishnevskii and L. I. Chumakov, The roasting of arsenical-pyrites gold concentrates in fluidized beds (in Russian), Ref. as No. 1016.

1074. M. G. Slin'ko, The part played by mass transfer processes in the catalytic oxidation of naphthalene to phthalic anhydride (in Russian), Khim. Prom., 1960, No. 1.

1075. A. A. Smirnova, Work in the "Giprotsvetmet" Institute on the further improvement of fluidized bed furnaces (in Russian), Ref. as No. 1016.

1076. Yu. I. Sobchuk and A. T. Ul'yanov, Tests on heat utilization in the roasting of pyrites in fluidized beds (in Russian), Ref. as No. 1016.

1077. M. I. Stul', The application of fluidized beds for roasting flotation pyrites in the Shchelkovsk. chemical complex (in Russian), Ref. as No. 1016.

1078. D. A. Sushchinskii, Work on processes in fluidized bed furnaces carried out at the All-Union Scientific Research Institute for Nonferrous Metals (VNIITsvetmet) (in Russian), Ref. as No. 1016.

1079. A. P. Sychev and I. M. Cherednik, Roasting of lead sulfide concentrates in fluidized beds (in Russian), Ref. as No. 1016.

1080. N. I. Syromyatnikov, High-frequency methods of investigating heat transfer in fluidized beds (in Russian), Tr. U.P.I., 1961, Part 96, 70-73.

1081. O. M. Todes, Papers on investigations of fluidized beds in the Peoples' Republic of Poland (in Russian), Inzh.-Fiz. Zhurn., 1960, 3, No. 12, 123-129.

1082. D. G. Traber, V. B. Sarkits and I. P. Mukhlenov, Heat transfer from suspended beds of granular materials to heat transfer surfaces (in Russian), Zhurn. Prikl. Khim., 1960, 33, No. 10, 2197-2205.

1083. R. N. Urubkov, Exploitation of fluidized bed furnaces in the Ust'-Kamenogorsk lead-zinc plant (in Russian), Ref. as No. 1016.

1084. P. N. Fedoseev, High-Temperature Microperiodic Drying of Seed-Grade Wheat with Small Moisture Differences in the Grains (in Russian), Kand. Dissertat., Novosibirsk, 1960.

1085. M. V. Filippov, Effect of magnetic fields on suspended beds of ferromagnetic particles (in Russian), in: Sb. "Prikladnaya Magnitogidrodinamika," Riga, 1960, 215-236.

1086. M. I. Fridland and A. I. Skoblo, Scale-up of particle entrainment from fluidized beds (in Russian), Khim. Mashinostr., 1960, No. 5, 18-21.

1087. M. I. Fridland and A. I. Skoblo, An investigation of particle entrainment from fluidized beds (in Russian), Izvestiya VUZov, Neft' i Gaz, 1960, No. 1, 71-78.

1088. M. I. Fridland and A. I. Skoblo, Mechanical entrainment of particles by gases in fluidized bed equipment (in Russian), Trudy Moskovsk. In-ta Neftekhimicheskoi i Gazovoi Promyshlennosti, No. 28, 1960, 93.

1089. M. I. Fridland and A. I. Skoblo, The fall of particles through the openings of distributing grids (in Russian), Izvestiya VUZov, Neft' i Gaz, 1960, No. 9, 97-101.

1090. E. I. Khodorov, V. M. Kasareva, L. S. Fraiman, B. I. Viktorenko and N. N. Chernobaeva, The roasting of raw cement mixtures in rotating furnaces and in shaft reactors with fluidized beds (in Russian), Ref. as No. 1016.

1091. E. I. Khodorov and V. Ya. Shapirov, The roasting of nepheline limestone charges in fluidized beds in the experimental equipment of the S. M. Korov aluminum plant at Volkhovsk (in Russian), Ref. as No. 1016.

1092. D. M. Chizhikov, I. N. Kitler, Yu. A. Lainer, V. E. Kuvinov and I. N. Kiryazina, The application of fluidized beds to various alumina manufacturing processes (in Russian), Ref. as No. 1016.

1093. Ya. Kh. Shvartsman, The combustion of elementary sulfur in fluidized bed furnaces (in Russian), Byull. Tsvet. Met., 1960, No. 4.
1094. G. M. Shteingart, The roasting of zinc concentrates in fluidized beds in the "Elektrotsink" plant (in Russian), Ref. as No. 1016.
1095. I. A. Yuzhaninov, I. N. Munits and G. B. Labutin, Pilot-scale fluidized bed equipment for producing alumina (in Russian), Ref. as No. 1016.
1096. F. M. Yasonov, The exploitation of fluidized bed furnaces in the "Ukrtsink" plant (in Russian), Ref. as No. 1016.
1097. P. I. Bakker and P. M. Heertjes, Porosity distributions in a fluidized bed, Chem. Eng. Sci., 1960, 12, No. 4, 26-271.
1098. H. Bannister, Heat transfer in beds of fluidized particles, Industrial Chemist, July 1960, 331-337.
1099. P. K. Baumgarten and R. L. Pigford, Density fluctuations in fluidized beds, A. I. Ch. E. Journ., 1960, 6, No. 1, 115-123.
1100. L. Belik, Das Wirbelstossverfahren zur Trocknung von Schüttgut, Chem. Ing. Techn., 1960, 32, No. 4, 253-257.
1101. J. Beranek, Zur Theorie der Wirbelschicht, Chem. Techn., 1960, 12, No. 9, 515-520.
1102. J. Beranek, Dynamic properties of fluidized beds, III. Influence of particle shape on the dynamic properties of the bed, Coll. Czech. Chem. Comm., 1960, 25, No. 9, 2319-2333.
1103. S. H. Bransom, Factors in the design of continuous crystallizers, Brit. Chem. Engng, 1960, 5, No. 12, 838-844.
1104. E. I. Cairns and I. M. Prausnitz, Longitudinal mixing in fluidization, A. I. Ch. E. Journ., 1960, 6, No. 3, 400-405.
1105. E. I. Cairns and I. M. Prausnitz, Macroscopic mixing in fluidization, A. I. Ch. E. Journ., 1960, 6, No. 4, 554-560.
1106. I. G. Collier, New thinking on heat transfer, Industrial Chemist, 1960, 36, No. 422, 181-184.
1107. G. Davies and D. Robinson, A study of aggregative fluidization, Can. Journ. Chem. Engng, 1960, 38, No. 6, 175-183.
1108. C. B. Decken, H. I. Hantke, I. Binckebanck and K. P. Bachus, Bestimmung des Wärmeübergangs von Kugelschüttungen an durchströmendes Gas mit Hilfe der Stoffübergangs-Analogie, Chem. Ing. Techn., 1960, 32, No. 9, 591-594.
1109. I. Dlouhy and W. H. Gauvin, Evaporation rates in spray drying, Can. Journ. Chem. Engng, 1960, 38, No. 4, 113-120.
1110. G. A. Donnadieu, Etude des échanges thermiques entre fluide et particules solides dans les milieux fluidisés, Jour. recherch. du CNRS, 1960, No. 51, 159-169.
1111. G. A. Donnadieu, Etude de la circulation des particules dans un lit fluidisé, Jour. rech. centre nat. rech. scient., 1960, No. 53, 295-300.
1112. L. T. Fan, Y. C. Yang and C. Y. Wen, Mass transfer in semi-fluidized beds for solid-liquid system, A. I. Ch. E. Journ., 1960, 6, No. 3, 482-487.
1113. R. M. Criffith, Mass transfer from drops and bubbles, Chem. Engng Sci., 1960, 12, No. 3, 198-213.
1114. R. F. Hoffman, L. Lapidus and I. C. Elgin. The mechanics of vertical moving fluidized systems: IV. Application to batch

fluidized systems with mixed particle sizes, A. I. Ch. E. Journ., 1960, 6, No. 2, 321-324.

1115. R. B. Lancashire, E. A. Lezberg and I. F. Morris, Heat transfer coefficients for a full-scale pebble-bed heater, Ind. Eng. Chem., 1960, 52, No. 5, 433-434.

1116. W. Landau and E. Franz, Ergebnisse von Wirbelschichtversuchen mit technischen Schüttgütern, Chem. Techn., 1960, 12, No. 4, 200-206.

1117. M. Leva, Pressure drop and power requirements in a stirred fluidized bed, A. I. Ch. E. Journ., 1960, 6, No. 4, 688-692.

1118. M. Leva, Some considerations of dense-phase fluidized-bed correlations. Part I. Brit. Chem. Engng, 1960, 5, No. 10, 703-707.

1119. F. F. Liu and C. Orr, Jr., Apparent viscosity of gas-solid fluidized systems, J. Chem. Engng Data, 1960, 5, No. 4, 430-432.

1120. L. A. Madonna and R. F. Lama, How to calculate pressure drop in spouted beds, Ind. Eng. Chem., 1960, 52, No. 2, 169-172.

1121. L. Massimilla and I. W. Westwater, Photographic study of solid gas fluidization, A. I. Ch. E. Journ., 1960, 6, No. 1, 134-138.

1122. H. S. Mickley, D. F. Fairbanks and R. D. Hawthorn, The relation between the transfer coefficient and thermal fluctuations in fluidized bed heat transfer, Preprints 4th Nat. Heat. Transf. Conf., A. I. Ch. E. Journ., 1960, No. 8.

1123. F. Nakashio and W. Sakai, Residence time of particles in fluidized bed, Chem. Engng (Japan), 1960, 24, No. 6, 452-455.

1124. L. Neuzil, The behavior of a mixture of particles of granular material of varying size in a fluidized bed (preliminary communication), Collect. Czech. Chem. Communs, 1960, 25, No. 8, 2207-2209.

1125. T. Otake and S. Tone, A new method of evaluating effective thermal conductivities and wall heat transfer coefficients in a packed bed, Chem. Engng (Japan), 1960, 24, No. 3, 156-160.

1126. P. N. Rowe and P. F. Wace, Gas-flow patterns in fluidized beds, Nature (Engl.), 1960, 188, No. 4752, 737-738.

1127. E. Ruckenstein and I. Teoreanu, Cu privire la transportul de caldura sau de materie dintre agentul de fluidizare si particulele solide ale stratului fluidizat, Studii ce cercet. fiz. Acad. RPR, 1960, 11, No. 1, 117-128.

1128. E. Ruckenstein and I. Teoreanu, Cu privire la transferul caldura sau de materie dintre particule si fluid intr un strat fluidizat, Rev. chim. (RPR), 1960, 11, No. 3, 137-143.

1129. W. Schotte, Thermal conductivity of packed beds, A. I. Ch. E. Journ., 1960, 6, No. 1, 63-67.

1130. W. W. Shuster and F. C. Haas, Point viscosity measurements in a fluidized bed, J. Chem. and Engng. Data, 1960, 5, No. 4, 525-530.

1131. N. R. Sunkoori and R. Kaparthi, Heat transfer studies between particles and liquid medium in a fluidizedbed, Chem. Eng. Sci., 1960, 12, 166-174.

1132. V. Vanecek, M. Markvart and E. Lederer, Fluidni suseni, Chem. Prumysl., 1960, 10, No. 10, 521-525.

1133. K. Vasudeva, Some aspects of fluidization, Bombay Technologest, 1959-1960, 10, Silver Jubilee Numb., 179-186.
1134. H. A. Vreedenberg, Heat transfer between a fluidized bed and a vertical tube, Chem. Eng. Sci., 1960, 11, 274-285.
1135. C. Y. Wen and R. F. Hashinger, Elutriation of solid particles from a dense-phase fluidized bed, A. I. Ch. E. Journ., 1960, 6, No. 2, 220-226.
1136. C. Y. Wen and E. N. Miller, Heat transfer in solids—gas transport lines, Ind. Engng Chem., 1961, 53, No. 1, 51-53.
1137. F. Westfried and C. Cazacu, Asupra resistentei hidroaerodinamice a straturilor de particule fluidizate intr—un curent ascendent de fluid, Stud. si cercet. energet., 1960, 10, No. 1.
1138. F. A. Zenz and D. F. Othmer, Fluidization and fluid-particle systems, Reinhold Publish. Corp., N. Y., 1960.

References published in 1961

1139. R. A. Bakhtiozan and Z. R. Gorbis, Experimental investigation of convective heat transfer in dusty streams (in Russian), in: Materialy Soveshchaniya po Teplo- i Massobmenu (Doklad), Minsk, 1961.
1140. M. I. Beilin, Some problems in the hydrodynamics of fluidized beds (in Russian), Khimiya i Tekhnol. Topl. i Masel, 1961, No. 4, 19-23.
1141. A. K. Bondareva and O. M. Todes, The thermal conductivity of fluidized beds (in Russian), Ref. as No. 1139.
1142. V. A. Borisevich, An investigation of heat transfer during the motion of dispersed materials in tubes (in Russian), Ref. as No. 1139.
1143. V. F. Borodkin, Chlorination of the phthalocyanine of copper in a fluidized bed (in Russian), Izvestiya VUZov, Khimiya i Khim. Tekhnol., 1961, No. 4, 4, 665-667.
1144. N. I. Gel'perin, V. G. Ainshtein and L. P. Timokhova, The hydrodynamics of fluidized granular materials in conical equipment (in Russian), Khim. Mashinostr., 1961, No. 4, 12-15.
1145. M. Ya. Ginzburg, Computing devices for the automatic determination of complex process parameters in fluidized beds (in Russian), Khim. i Tekhnol. Topl. i Masel, 1961, 6, No. 10, 33-37.
1146. Z. R. Gorbis, The mechanics of heat transfer in two-phase streams of gas and solid particles (in Russian), Ref. as No. 1139.
1147. V. D. Gorochko and Yu. P. Gupalo, A radiometric method of determining viscosity (in Russian), Zavodskaya Laboratoriya, 1961, No. 1, 38-40.
1148. V. F. Denisov and M. I. Penzina, The problem of cooling gases from zinc concentrate roasting furnaces in fluidized beds (in Russian), Tsvet. Met., 1961, 34, No. 9, 42-48.
1149. V. I. Zhidko, An investigation of the drying of grain in fluidized beds (in Russian), Ref. as No. 1139.
1150. S. S. Zabrodsky, Fluidized beds—a means for intensifying the cooling of furnace gases (in Russian), Ref. as No. 1139.

1151. S. S. Zabrodsky, The pressure drop in fluidized beds in conical equipment (in Russian), Inzh.-Fiz. Zhurn., 1961, 4, No. 2, 92-94.

1152. S. S. Zabrodsky, Fluid bed heat Transfer, International Developments in Heat Transfer, A.S.M.E., Part IV, 1961, 737-741.

1153. I. F. Zemskov, A. S. Stepanov and V. I. Gnezdov, The adsorption of organic mercury compounds in multitray adsorbers with fluidized beds of activated carbon (in Russian), Zhurn. Prikl. Khim., 1961, 34, No. 6, 1354-1356.

1154. I. I. Ioffe, A. F. Grigorov, R. P. Rumyantsev and E. T. Ivanova, The effects of macrokinetic factors on the rate of naphthalene oxidation in fluidized catalyst beds (in Russian), Khim. Prom., 1961, No. 4, 258-261.

1155. V. A. Kalender'yan and Z. R. Gorbis, An experimental investigation of heat transfer in longitudinally moving beds of powder (in Russian), Ref. as No. 1139.

1156. I. I. Kershanskii and G. S. Gubaidulin, The roasting of gold-bearing sulfide concentrates in fluidized beds (in Russian), Tsvet. Met., 1961, 34, No. 6, 45-50.

1157. N. B. Kondukov, The hydrodynamics and conditions of similarity in the transitional regime of fluidization (in Russian), Inzh.-Fiz. Zhurn., 1961, 4, No. 3, 31-38.

1158. N. B. Kondukov, Bases for the design of equipment for removing sulfur from organic products in beds of fluidized adsorbents (in Russian), Khim. Mashinostr., 1961, No. 1, 10-12.

1159. V. D. Koroleva and F. M. Loskutov, Reduction roasting of zinc residues in fluidized beds (in Russian), Tsvet. Met., 1961, 34, No. 7, 33-40.

1160. G. Ya. Krichevskii and A. G. D'yachko, A mathematical description of the dynamic characteristics of furnaces for the fluidized bed roasting of granulated copper-zinc charges (in Russian), Tsvet. Met., 1961, 34, No. 9, 34-42.

1161. M. Kh. Levinter and Zh. F. Galimov, The transport of spherical catalysts in flowing streams (in Russian), Khim. i Tekhnol. Topl. i Masel, 1961, 6, No. 9, 20-26.

1162. M. S. Lodyseva and I. A. Burovoi, A mathematical description for purposes of automation of one class of fluidized bed furnace (in Russian), Zhurn. Vses. Khim. Ob-va im. Mendeleeva, 1961, 6, No. 5, 553-556.

1163. A. V. Lunev and G. Kh. Vermishev, The preliminary enrichment of polymetallic ores in fluidized beds (in Russian), Tsvet. Met., 1961, 34, No. 4, 66-67.

1164. A. V. Lykov (Luikov), The Theoretical Basis of Constructional Thermophysics (in Russian), Izd. Akad. Nauk BSSR, 1961.

1165. M. I. Mantsevich, Automation of processes for subliming antimony from ores by the method of roasting in fluidized beds (in Russian), Tsvet. Met., 1961, 34, No. 4, 12-17.

1166. I. G. Martyushin, Tests of hydrodynamic calculations of fluidized bed roasting furnaces (in Russian), Tsvet. Met., 1961, 34, No. 6, 32-39.

1167. I. P. Mukhlenov, D. G. Traber, M. F. Mikhalev and Ya. M. Shmekker, The oxidation of sulfur dioxide in equipment with a fluidized bed of catalyst (in Russian), Khim. Prom., 1961, No. 1, 42-46.

1168. I. P. Mukhlenov, D. G. Traber, E. S. Rumyantseva and V. M. Pomerantsev, The hydrodynamics of suspended catalyst beds under pressure (in Russian), Zhurn. Prikl. Khim., 1961, 34, No. 6, 1181-1185.

1169. Zh. R. Nadzhar'yan and V. G. Bryndin, Roasting of nickel concentrates obtained by separation of flotation converter mattes in fluidized bed furnaces (in Russian), Tsvet. Met., 1961, 34, No. 3, 52-55.

1170. O. S. Nichiporenko, Aerodynamics of fluid bed reactors for the direct reduction of iron, Dopovidi Akad. Nauk URSR, 1961, No. 4, 495-500.

1171. I. N. Piskunov, The shape of the vertical cross section of fluidized bed furnaces for roasting sulfides (in Russian), Tsvet. Met., 1961, No. 4, 17-20.

1172. G. D. Rabinovich, Heat and mass transfer in dense granular beds (in Russian), Ref. as No. 1139.

1173. I. M. Razumov and L. I. Larionova, Calculation of the quantity of fine-grained material entrained from a fluidized bed by a gas stream (in Russian), Khim. Mashinostr., 1961, No. 1, 13-14.

1174. I. M. Razumov, O. A. Chefranov, A. N. Bushin and P. S. Baranova, Calculation and tests of continuous stream catalyst transporters (in Russian), Khim. i Tekhnol. Topl. i Masel, 1961, 6, No. 5, 39-43.

1175. P. G. Romankov, V. N. Lepilin and N. B. Rashkovskaya, Suspended bed problems (in Russian), Izvestiya VUZov, Khimiya i Khim. Tekhnol., 1961, 4, No. 2, 298-302.

1176. B. S. Sazhin, An investigation of the process of drying free-flowing materials in spouting dryers and development of practical drying equipment (in Russian), Ref. as No. 1139.

1177. P. Solchani, A new method for describing the expansion of fluidized beds (in Russian), Khim. Prom., 1961, No. 4, 267-269.

1178. Ch. Ts. Hsiao and V. I. Smirnov, An investigation of sulfating roasting of cobalt-containing mattes from nickel and copper plants in fluidized beds (in Russian), Tsvet. Met., 1961, 34, No. 1, 35-39.

1179. I. Ya. Tyuryaev and A. V. Erofeeva, The kinetics of butane dehydrogenation in a fluidized catalyst bed (in Russian), Zhurn. Prikl. Khim., 1961, No. 2, 370-375.

1180. I. Ya. Tyuryaev, A. L. Tsailingold' and A. V. Builov, Nonhomogeneity of suspended catalyst beds (in Russian), Khim. Prom., 1961, No. 5, 356-359.

1181. I. Ya. Tyuryaev, A. L. Tsailingold' and A. B. Builov, The effectiveness of gas mixing in reactors with suspended beds of fine-grained catalysts (in Russian), Zhurn. Prikl. Khim., 1961, 34, No. 3, 558-564.

1182. I. G. Fadeev, I. M. Razumov, A. I. Skoblo, O. A. Chefranov and K. L. Reznikovich, Calculation of the pressure drop in the transportion of granular materials in flowing streams (in Russian), Khim. Mashinostr., 1961, No. 2, 26-28.

1183. M. I. Fridland, I. M. Razumov and A. I. Skoblo, Calculation of particle entrainment in fluidized bed equipment (in Russian), Khim. i Tekhnol. Topl. i Masel, 1961, No. 2, 36-38.

1184. O. S. Chekhov, V. N. Anokhin, B. N. Shekun and R. Z. Khiterer, An investigation of the hydrodynamics in fluidized beds of solid particles under high pressure (in Russian), Khim. Prom., 1961, No. 7, 492-494.

1185. W. I. Anderson, Shallow fluid bed waste heat boiler, Brit. Chem. Engng, 1961, 6, No. 6, 403-404.

1186. K. E. B. Andersson, Sedimentation and ideal fluidization, Svensk. Kem. Tidskrift, 1961, 73, 334-345.

1187. K. E. B. Andersson, Pressure drop in ideal fluidization, Chem. Engng Sci., 1961, 15, No. 3-4, 276-297.

1188. R. C. Bailie, L. T. Fan and J. J. Stewart, Uniformity and stability of fluidized beds, Ind. Engng, Chem., 1961, 53, No. 7, 567-569.

1189. R. C. Bailie, L. T. Fan and J. J. Stewart, Instability of fluidized beds, J. Chem. Engng Data, 1961, 6, No. 3, 469-473.

1190. H. A. Becker, An investigation of laws governing the spouting of coarse particles, Chem. Engng Sci., 1961, 13, No. 4, 245-262.

1191. B. O. Beyaert, L. Lapidus and I. C. Elgin, The mechanics of vertical moving liquid—liquid fluidized systems: II. Countercurrent flow, A. I. Ch. E. Journ., 1961, 7, No. 1, 46-48.

1192. P. D. Bloore and J. S. M. Botterill, Similarity in behavior between gas bubbles in liquid and fluidized solid systems, Nature (Eng.), 1961, 190, No. 4772, 250-251.

1193. J. S. M. Botterill, Fluidization, Brit. Chem. Engng, 1961, 6, No. 5, 327-331.

1194. K. McG. Bowling and P. L. Waters, Some problems affecting the carbonisation of coal in a fluidized bed, Brit. Chem. Engng, 1961, 6, No. 4, 265.

1195. S. H. Bransom and S. Pendse, Dynamics of liquid—solid fluidization, Ind. Engng, Chem., 1961, 53, No. 7, 575-576.

1196. I. I. van Deemter, Mixing and contacting in gas—solid fluidized beds, Chem. Engng, Sci., 1961, 13, No. 3, 143-154.

1197. E. D. Ermeng, Designing a fluidized adsorber, Chem. Engng, 1961, 68, No. 11, 87-94.

1198. I. F. Frantz, Fluid-to-particle heat transfer in fluidized beds, Chem. Engng, Progr., 1961, 57, No. 7, 35-42.

1199. E. Franz and G. Thurm, Einfache Auslegung von Wirbelschichtanlagen mit Hilfe von Nomogrammen, Chem. Techn., 1961, 13, No. 1, 4-8.

1200. S. K. Friedlander, A note on transport to spheres in Stokes flow, A. I. Ch. E. Journ., 1961, 7, No. 2, 347-348.

1201. A. R. Gordon, R. W. Pickering and O. G. Woodward, Fluid roasting of zinc plant residues at Risdon, Tasmania. Brit. Chem. Engng, 1961, 6, No. 4, 261-263.

1202. D. Harrison, J. F. Davidson and J. W. Kock, On the nature of aggregative and particulate fluidization, Trans. Instn Chem. Engrs, 1961, 6, No. 3, 202-211.

1203. D. Harrison and L. S. Leung, Bubble formation at an orifice in a fluidized bed, Nature, 1961, 190, No. 4774, 433-434.

1204. N. I. Hassett, Flow patterns in particles beds, Nature, 1961, 189, No. 4769, 997-998.
1205. H. Kahnt, Das Wirbelschicht-Adsorptionverfahren und seine kleintechnishe Anwendung bei der Braunkohlenschwelgas-Entbenzinierung, Chem. Techn., 1961, 13, No. 3, 143-149.
1206. S. Kiuchi and I. Okahara, J. Mining Met. Inst. Japan, 1961, 77, No. 872, 115-120.
1207. J. Lakomy and S. Erben, Fluidni pec na stanoveni bodu samovzniceni, Ropa a uhlie, 1961, 3, No. 1, 19-21.
1208. I. B. Lewis, Fluidization of solids, Times Sci. Rev., 1961, No. 40, 15-18.
1209. I. B. Lewis, Fluidization research in the chemical engineering division A.E.R.E. Harwell. Trans. Instn Chem. Engrs, 1961, 39, No. 3, 166-167.
1210. L. A. Madonna, R. F. Lama and W. L. Brisson, Solids-air jets, Brit. Chem. Engng, 1961, 6, 524-528.
1211. L. Massimilla, Sull'eff lusso di sistemi fluidizzati solido-gas da fori di piccolo e grande diametro, Chim. e industria, 1961, 43, No. 5, 483-492.
1212. L. Massimilla, V. Betta and C. Della Rocca, A study of streams of solids flowing from solid-gas fluidized beds, A. I. Ch. E. Journ., 1961, 7, No. 3, 502-508.
1213. L. Massimilla, A. Solimando and E. Squillace, Gas dispersion in solid-liquid fluidized beds, Brit. Chem. Engng, 1961, 6, No. 4, 232-239.
1214. R. M. G. Meek, The measurement of heat transfer coefficients in packed beds by cyclic method, Intern. Develop. in Heat Transf., ASME, 1961, Part IV, 770-780.
1215. H. Oetiker, Fluidlift—Förderung: Anwendungen und Erfahrungen in der Praxis, Mühle, 1961, 98, No. 32, 417-419.
1216. C. Otake and S. Tone, Mechanism of heat transfer in solid particle beds, Chem. Engng (Japan), 1961, 25, No. 1, 56-64.
1217. I. A. Quinn, L. Lapidus and I. C. Elgin, The mechanics of moving vertical fluidized systems, V. Concurrent cogravity flow, A. I. Ch. E. Journ., 1961, 7, No. 2, 260-263.
1218. A. M. Rao and R. Kaparthi, Studies on heat transfer coefficients of liquid fluidized beds, J. Scient. and Industr. Res., 1961, D20, No. 7, 261-265.
1219. R. E. Riccetti and G. Thodos, Mass transfer in the flow of gases through fluidized beds, A. I. Ch. E. Journ., 1961, 7, No. 3, 442-444.
1220. J. F. Richardson and J. Szekely, Mass transfer in a fluidized bed, Trans. Instn Chem. Engrs, 1961, 39, No. 3, 212-222.
1221. R. A. Rizzoli, Practical fluidized beds, Proc. Roy. Austral. Chem. Inst., 1961, 28, No. 4, 156-165.
1222. P. N. Rowe, The anatomy of fluidizedbeds, Industrial Chemist, 1961, 37, No. 434, 167-169.
1223. P. N. Rowe, Drag forces in a hydraulic model of a fluidized bed. Part II. Trans. Instn, Chem. Engrs, 1961, 39, No. 3, 175-180.
1224. P. N. Rowe and G. A. Henwood, Drag forces in a hydraulic model of a fluidized bed. Part I. Trans. Instn Chem. Engrs, 1961, 39, 43-54.

1225. P. N. Rowe and W. M. Stapleton, The behavior of 12-inch diameter gas fluidized beds, Trans. Instn Chem. Engrs, 1961, 39, No. 3, 181-187.

1226. L. Russel, New developments in the bulk transport of chemical reagents, Trans. Instn Chem. Engrs, 1961, 39, No. 3, A51, 53-54.

1227. I. H. Stockel, Some aspects of the one-dimensional fluidized flow of divided solids, Rheol. acta, 1961, 1, No. 4-6, 415-422.

1228. K. S. Sutherland, Solid mixing studies in gas fluidized beds. Part I. A preliminary comparison of tapered and non-tapered beds, Trans. Instn Chem. Engrs, 1961, 39, No. 3, 188-194.

1229. P. Szolcsanyi, Folyadekkal lazitott szemeses agyak kiterjedese a lyukelmelet alapjan, Magyar kem. fol., 1961, 67, No. 4, 171-173.

1230. P. Szolcsanyi, Lazitott szemcses agyak hidromechanikai elmeletenek fejlodese, Magyar. kem. lapja, 1961, 16, No. 2, 84-90.

1231. S. R. Tailby and M. A. T. Cocquerel, Some studies of solids mixing in fluidized beds, Trans. Instn Chem. Engrs, 1961, 39, No. 3, 195-201.

1232. W. I. Thomas, P. I. Grey and S. B. Watkins, Effect of particle size distribution in fluidization. Part I. The preparation and analysis of particles, Brit. Chem. Engng, 1961, 6, No. 2, 80-86.

1233. W. I. Thomas, P. I. Grey and S. B. Watkins, Effect of particle size distribution in fluidization. Part II. Classification and elutriation, Brit. Chem. Engng, 1961, 6, No. 3, 176-181.

1234. P. F. Wace and S. I. Burnett, Flow patterns in gas-fluidized beds, Trans. Instn Chem. Engrs, 1961, 39, No. 3, 168-174.

1235. J. Wadsworth, An experimental investigation of the local packing and heat transfer processes in packed beds of homogeneous spheres, Int. Develop. in Heat Transf., ASME, 1961, Part IV, 760-769.

1236. G. Winterstein and G. Bïesz, Die Berechnung von Wirbelschichtanlagen nach der Ähnlichkeitstheorie, Chem. Techn., 1961, 13, No. 2, 64-68.

1237. S. Yagi, D. Kunii, K. Nagahara and H. Naito, Studies on axial mixing of particles in moving beds, Chem. Engng (Japan), 1961, 25, No. 6, 469-476.

1238. S. Yagi, D. Kunii and H. Naito, Studies on axial mixing of particles in moving beds, Annual Rept, Engng Res. Inst. Fac. Engng Univ. Tokyo, 1961, 19, No. 2, 39-46.

1239. S. Yagi and D. Kuni, Studies on heat transfer in packed beds, Int. Develop. in Heat Transf., ASME, 1961, Part IV, 750-757.

1240. S. Yagi, D. Kunii and N. Wakao, Radially effective thermal conductivities in packed beds, Int. Develop. in Heat Transf., ASME, 1961, Part IV, 742-749.

1241. J. S. M. Botterill, K. A. Redish, D. K. Ross, J. R. Williams, The mechanism of heat transfer to fluidized beds, Preprint 3rd Congr. Europ. Fed. Chem. Engn., 1962, B-49—B-54.

1242. L. Fan, Ch. Lee and R. Bailie, Axial solid distribution in gas-solid fluidized beds, A. I. Ch. E. Journ., 1962, 8, No. 2, 239-244.

Additional references

1243. T. B. Anderson and R. Jackson, The nature of aggregative and particulate fluidization, Chem. Eng. Sci., 1964, 19, 509-511.
1244. R. Jackson, The mechanics of fluidized beds. Part I. The stability of the state of uniform fluidization. Trans. Instn. Chem. Engrs., 1963, 41, 13-21.
1245. P. N. Rowe, Gas-solid reaction in a fluidized bed, Chem. Engng Progr., 1964, 60, N 3, 75-80.
1246. P. N. Rowe, The effect of bubbles on gas-solid contacting in fluidized beds. Chem. Engng. Progr. Sympos. Ser., 1962, 58, N 38, 42-56.
1247. P. N. Rowe and K. S. Sutherland, Solid mixing studies in gas-fluidized beds. Part II. Behavior of deep beds of dense materials, Trans. Instn. Chem. Engrs., 1964, 42, N 2, 55-63.
1248. P. N. Rowe and B. A. Partridge, Gas flow through bubbles in a fluidized bed. I. Flow through an ideal bubble. Chem. Engng. Sic. 1963, 18, 511-524.
1249. J. F. Davidson and D. Harrison, Fluidized particles, Cambridge Univ. Press, 1963.
1250. P. N. Rowe and B. A. Partridge, Particle movement caused by bubbles in a fluidized bed, Preprint 3rd Congr. Europ. Fed. Chem. Engng., B 22, 1962.
1251. S. S. Zabrodsky, N. V. Antonishin and V. A. Nemkovich, A device for carrying out heat transfer between two gas streams, Avt. Svid. No. 851607, 19.10.1963.
1252. P. N. Rowe, Comments on heat transfer between solid particles and a gas in a nonuniform (aggregated) fluidized bed, Int. J. Heat Mass Transfer, 1963, 6, 989-991.
1253. A. I. Tamarin, Selection of the number of stages in a multi-effect fluidized bed heat exchanger (in Russian), Inzh.-Fiz. Zhurn., 1963, 6, No. 4, 88-91.
1254. N. V. Kharchenko and K. E. Makhorin, The problem of intensifying heat transfer between fluidized beds and submerged bodies at high temperatures (in Russian), Inzh.-Fiz. Zhurn., 1964, 7, No. 5, 11-17.

Index